The Physics of the Two-Dimensional Electron Gas

NATO ASI Series

Advanced Science Institutes Series

A series presenting the results of activities sponsored by the NATO Science Committee, which aims at the dissemination of advanced scientific and technological knowledge, with a view to strengthening links between scientific communities

The series is published by an international board of publishers in conjunction with the NATO Scientific Affairs Division

A	**Life Sciences**	Plenum Publishing Corporation
B	**Physics**	New York and London
C	**Mathematical and Physical Sciences**	D. Reidel Publishing Company Dordrecht, Boston, and Lancaster
D	**Behavioral and Social Sciences**	Martinus Nijhoff Publishers
E	**Engineering and Materials Sciences**	The Hague, Boston, Dordrecht, and Lancaster
F	**Computer and Systems Sciences**	Springer-Verlag
G	**Ecological Sciences**	Berlin, Heidelberg, New York. London,
H	**Cell Biology**	Paris, and Tokyo

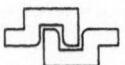

Series B: Physics

The Physics of the Two-Dimensional Electron Gas

Edited by

J. T. Devreese

and

F. M. Peeters

University of Antwerp
Antwerp, Belgium

Plenum Press
New York and London
Published in cooperation with NATO Scientific Affairs Division

Proceedings of a NATO Advanced Study Institute on
The Physics of the Two-Dimensional Electron Gas,
held June 2–14, 1986,
in Oostduinkerke, Belgium

Library of Congress Cataloging in Publication Data

NATO Advanced Institute on the Physics of the Two-Dimensional Electron Gas
 (1986: Oostduinkerke, Belgium)
 The physics of two-dimensional electron gas.

 (NATO ASI series. Series B, vol. 157)
 "Proceedings of a NATO Advanced Study Institute on the Physics of the
Two-Dimensional Electron Gas, held June 2–14, 1986, in Oostduinkerke, Bel-
gium"— T.p. verso.
 "Published in cooperation with NATO Scientific Affairs Division."
 Bibliography: p.
 Includes index.
 1. Electron gas—Congresses. 2. Energy-band theory of solids—Congresses.
3. Semiconductors—Congresses. 4. Hall effect—Congresses. 5. Electron-
phonon interactions—Congresses. I. Devreese, J. T. (Jozef T.) II. Peeters, F. III.
North Atlantic Treaty Organization. Scientific Affairs Division. IV. Title. V. Ti-
tle: Two-dimensional electron gas. VI. Series: NATO advanced science in-
stitutes series. Series B, Physics; v. 157.
QC176.8.E4N38   1986   530.4′1   87-13983
ISBN-13: 978-1-4612-9061-2   e-ISBN-13: 978-1-4613-1907-8
DOI: 10.1007/ 978-1-4613-1907-8

© 1987 Plenum Press, New York
Softcover reprint of the hardcover 1st edition 1987
A Division of Plenum Publishing Corporation
233 Spring Street, New York, N.Y. 10013

PREFACE

The 1986 Advanced Study Institute on "The Physics of the Two-Dimensional Electron Gas" took place at the Conference Centre "Ter Helme", close to Oostende (Belgium), from June 2 till 16, 1986.

We were motivated to organize this Advanced Study Institute in view of the recent experimental and theoretical progress in the study of the two-dimensional electron gas. An additional motivation was our own theoretical interest in cyclotron resonance in two-dimensional electron systems at our institute.

It is my pleasure to thank several instances and people who made this Advanced Study Institute possible. First of all, the sponsor of the Advanced Study Institute, the NATO Scientific Committee. Furthermore, the co-sponsors: Agfa Gevaert, Bell Telephone Mfg. Co. N.V., Burroughs Belgium, Control Data, Digital Equipment Corporation, Esso Belgium, European Research Office (USA), Kredietbank, National Science Foundation (USA).

Special thanks are due to the members of the Program Committee and the members of the Organizing Committee.

I would also like to thank Mrs. H. Evans for typing assistance.

J.T. Devreese
Professor of Theoretical Physics

February 26, 1987

CONTENTS

THE INTEGER AND THE FRACTIONAL QUANTUM HALL EFFECT

SPECIAL TOPICS

THE QUANTUM HALL EFFECT *

Klaus von Klitzing

Max-Planck Institut für Festkörperforschung, Heisenberg-
strasse 1, Postfach 80 06 65, D-7000 Stuttgart 80, F.R.G.

Ladies and Gentlemen:

First of all thank you very much for the invitation and for the possi-
bility to give this opening lecture at this meeting of the Advanced Study
Institute with the title "The Physics of the Two-
Dimensional Electron Gas".

Now this subject "The Two-Dimensional Electron Gas" has been a hot
topic in semiconductor physics for about twenty years and since 1975 we
have every two years an International Conference just with this title "The
Physics of Two-Dimensional Systems", but during the last three or four
years there has really been an explosion in this field. A large number of
publications, a large number of groups are now working in this field and
the reason is that this physics of two-dimensional systems is not only a
very interesting subject for basic research, there are also a lot of applications and therefore a large number of industrial laboratories are interested in this field. Today I will give you more or less an introduction.
It is very difficult to give a lecture for specialists and simultaneously
for non-specialists, so I will speak at the beginning a little bit about
the two-dimensional systems in general
and about some applications. After this I will give an introduction to the Quantum Hall Effect, including
some applications and at the end I will point out the problems
in connection with the question how to understand the
Quantum Hall Effect in detail and this will
be the subject of a large number of talks during this Conference. I am
quite sure that in the near future a large number of groups will work on
these two-dimensional systems and the quantized Hall effect is one reason
for this.

In Fig. 1 I list some typical structures which are related to two-
dimensional systems. If you hear words like quantum wells, NIPI which
means differently doped semiconductors or HEMT; all these structures are

---

* This text is based on a tape recording of the opening lecture by Prof.
  K. von Klitzing.

2D-SYSTEMS:

- Quantum Wells

- NIPI

- HEMT, TEGFET, MODFET, SDHT

- **Quantum Hall Device**

1D-SYSTEMS:

- Quantum Well Wires

- Transport perpendicular to

  quantum wells (tunneling

  structures, superlattices, nipi, ...)

Fig. 1. Examples of two (2D)- and one-dimensional (1D) systems.

related to two-dimensional systems and I have already the title of the next Conference for Professor Devreese: **"One-dimensional systems"**. If you are going to smaller and smaller structures you reach this situation where you have the one-dimensional systems, not quantum wells but quantum wires or if you are discussing electronic transport for very thin layers: tunneling structures, superlattices and so on, you have problems connected to one-dimensional systems and this will be the direction of future research.

Now, if you are speaking about two-dimensional systems, I think we should have the following picture in mind (see Fig. 2). You have electrons confined within very narrow potential layers. If this potential layer is

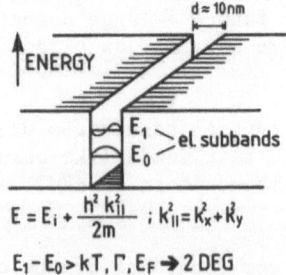

Fig. 2. Schematic view of a two-dimensional system (2DEG).

of the order of 10 nanometres = 100 Å you observe new quantum phenomena because the De Broglie wavelength of the electrons has to fit into the quantum well. This leads to quantum levels E_0, E_1, ... which are called electric subbands. If the energy separation between these quantum levels (i.e. E_1-E_0) for the electrons which are moving in this very thin layer is larger than all other energies, like the thermal energy (kT) or some broadening (Γ) or some other energies; if this separation is large enough, then we are speaking about a **two-dimensional system**. The energy in the z-direction is fixed and the electrons are only free to move in the xy-plane, called parallel plane. So this is a typical example of a two-dimensional system and I mentioned at the beginning already that a large number of applications are connected with such a two-dimensional system.

I will just mention here the **quantum well laser diode** (see Fig. 3). You already have such a system today in some equipment; in a compact disc for example. Such a laser structure consists of a narrow potential well for the electrons and for the positive charge of the holes. In this potential well the lowest electric subband is E_0 and this energy for the holes is H_0. If the channel becomes narrower then the energy E_0 increases and H_0 decreases. The energy difference E_0-H_0 corresponds to the energy of the light which is emitted by such a structure. And $h\nu = E_0-H_0$ can be changed just by changing the width of the quantum well. I should mention that the first experiments on this laser diode were done in 1979. And thus it took only five years from the first experiment to the product! You see that there is a strong interaction between basic research and applied research.

As a possible example of such a quantum well I mentioned the radiation of such a system as a laser diode but you can also have the **absorption**. The optical transition from one state to a higher state if you illuminate the system (see Fig. 4). There is an interesting effect if you apply an electric field across such a structure because then the difference between the energy levels is changed. If you apply an electric field perpendicular to such a layer then you are shifting these discrete energy

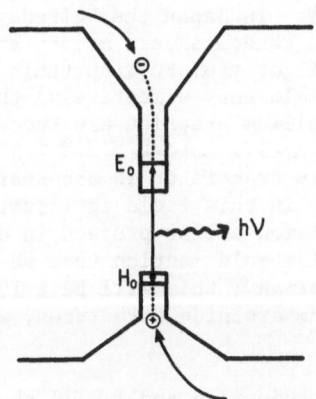

Energy difference E_0-H_0 tunable
with the width of the quantum well

Fig. 3. The quantum well laser diode.

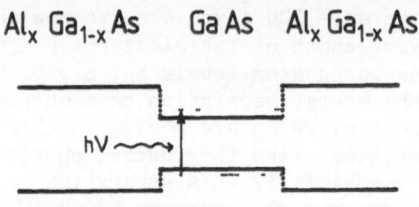

$Al_x Ga_{1-x} As \quad Ga As \quad Al_x Ga_{1-x} As$

with electric field perpendicular to the quantum well

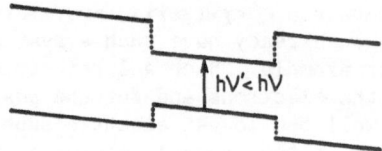

Fig. 4. Absorption of light by a two-dimensional structure.

levels to a smaller energy difference and consequently you are changing
the absorption. Thus, you can change the intensity of light just by electrical signals. For example, an electrical puls of length 200 picosecond
simultaneously gives an optical puls due to the change of absorption by the
electric field. These are only some possible applications of two-dimensional systems.

Today I have to speak about the **quantum Hall effect**.
If we are discussing the quantum Hall effect, I should start with a discussion of the HEMT (high electron mobility transistor) structure - a device
which is very important for applications. And just last year we could find
in a publication that the first commercial HEMT is on the market. HEMT
technology is very important for microwave generation, for example. The
HEMT is exactly the structure which is used for the quantum Hall effect.
Once more the first experiments on these devices were done in 1980 and now,
within five years, you can buy such a device on the market. This structure
seems to be better than other devices in high frequency applications like
the gallium arsenide MESFET. In Japan they already started to use this
HEMT structure to produce a random excess memory so you already have an integration. Very recently I got from Fujitsu (this is a company in Japan
which is working in this field very effectively) the information that integrated circuits based on gallium arsenide are successfully tested. HEMT
means **high electron mobility transistor**
and it seems to be that this transistor is necessary for the next generation
of computers. The activity in this field is illustrated by a Japanese
research program, a 200 million dollar project in order to produce computers
for the next generation. I should mention that we are also starting a
project in this field in Germany; this will be a 100 million dollar project
just for research on gallium arsenide structures, which is important for
new devices.

So this is a short introduction and I will show you the basic idea of
such a **HEMT structure** which is identical with the known
structure of a Silicon field effect transistor. The geometry is very
similar and I will discuss this picture (see Fig. 5) which is a cross-
section of a field effect transistor. You can use this picture for both

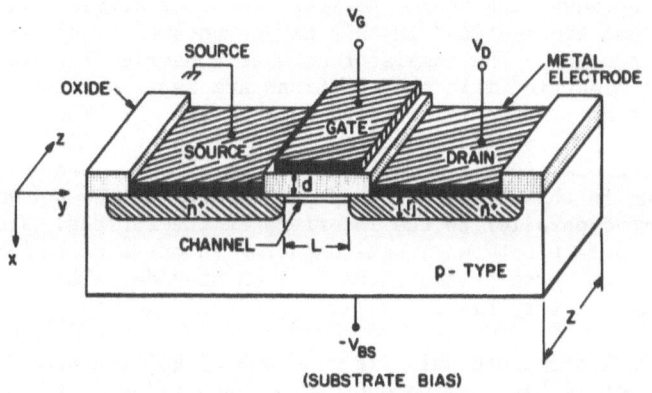

Fig. 5. Field effect transistor.

Silicon MOSFET or for the high electron mobility transistor: the HEMT transistor. The most important part, the area indicated by the word "channel", is the two-dimensional electron gas and this is formed at the surface of the semiconductor. The semiconductor is Silicon here, but maybe also gallium arsenide, and on top of the semiconductor I have some kind of insulator. This may be Silicon dioxide or another semiconductor like aluminum arsenide which acts as an insulator. And on top of this insulator we have a metal plate so that we have the situation of a capacitator. A metal plate on top, then the insulator and then a semiconductor, and if one applies a positive voltage on top of this structure, then we have a positive charge at the gate and the corresponding number of negative charges at the other side of this capacitator. So we are producing electrons at the surface and we can change the concentration of these electrons at this surface which corresponds to the number of charges on the top layer. This means that we can change the conductivity by changing the concentration of the electrons at this interface and these electrons are confined within a very narrow channel at the surface of the semiconductor. The channel is so narrow that we have a two-dimensional electron gas at the surface of such a structure. If you look in detail, you will find something like what is shown in Fig. 6. Due to the positive charges at the gate, the conduction

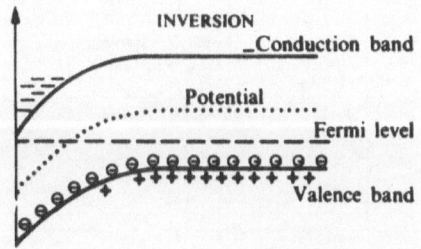

Fig 6. Energy bands in a MOSFET.

band is bent downwards where the free electrons attain lower energy values. The electrons are confined in this narrow potential well at the surface of the semiconductor. The insulator acts as a barrier for the electrons and due to the electric field the electrons are kept as close as possible to the interface. I have illustrated this in Fig. 7. The electron wave function or the distribution of electrons is also indicated in Fig. 7. You see that the electrons are confined within 40 Å = 4 nanometres at the surface and in this situation they cannot move in the potential well, they can only move parallel to the interface at the surface. For Silicon we have an interface between a semiconductor on one side and an insulator on the other side which is amorphous Silicon dioxide. The quality of this material is not very high.

The HEMT structure which is made out of gallium arsenide has a similar structure but we have semiconductors on both sides of the interface. But one of the semiconductors acts like an insulator for electrons; so we have for example on one side aluminum arsenide and on the other side gallium arsenide. The quality of this interface between the two materials is much better in the gallium arsenide system than in the Silicon-Silicon dioxide system. The reason is that gallium arsenide and aluminum arsenide have nearly the same lattice constant (see Fig. 8). So you can grow different single crystals of different materials and this is the reason that most of the applications now connected with these structures, which consist of different materials, are based on gallium arsenide - aluminum arsenide. Because they have the same lattice constant, there is no mismatch between those two materials at the interface. Such structures can be made with so-called **molecular beam epitaxy** (MBE) machines. Such a machine consists of a high vacuum chamber. The vacuum of it is a thousand times better than in space outside the earth atmosphere. A substrate material is placed in the chamber and different components of the material are evaporated on this substrate. In this way you can grow the crystals and you can change the composition of this material within **one atomic layer** ! You can produce sandwiches of materials, for example gallium arsenide - aluminum arsenide and you can change this within some angstrom thickness.

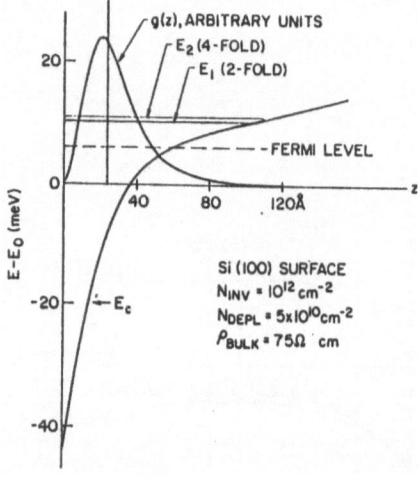

Fig. 7. Interface potential and electron wavefunction for a Si-MOSFET.

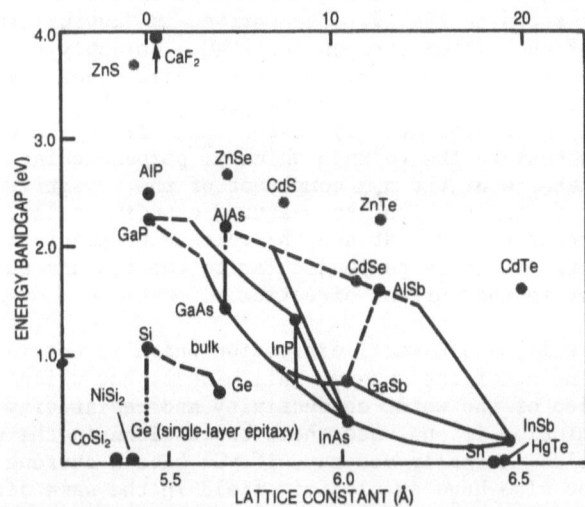

Fig. 8. Energy band gap and lattice constant for different semi-conductors.

We are producing structures of two-dimensional systems which look like the one shown in Fig. 9, which is a typical gallium arsenide - aluminum gallium arsenide heterostructure. One starts with a substrate material on which we are growing, with the MBE machine, undoped gallium arsenide, this will be the active layer. Then we have another material aluminum arsenide or aluminum gallium arsenide which acts like an insulator. And on top of this you can put the gate metal. Like in the Silicon field effect transistor we have at the interface between the two materials the two-dimensional system.

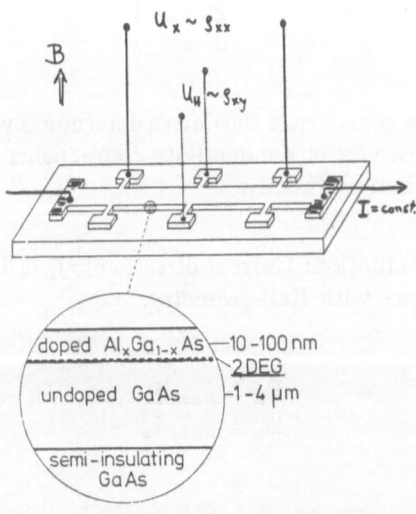

Fig. 9. GaAs-Al$_x$Ga$_{1-x}$As heterostructure.

The geometry used for the quantum Hall effect experiments is typically like the one shown in Fig. 9. The current is flowing from one side to the other side of the system through the 2DEG. We apply a magnetic field perpendicular to this plane, i.e. to the two-dimensional layer, and we are measuring different signals, for example the voltage drop U_x, which is proportional to the resistivity tensor ρ_{xx}. If you have a magnetic field you can also measure the voltage which is perpendicular to the current flow which is a measure of the ρ_{xy} component of the resistivity tensor. Thus you have two voltages which are measured experimentally, the voltage in the direction of the current and the voltage perpendicular to the current if the magnetic field is perpendicular to the two-dimensional system and perpendicular to the current direction.

Now, I will make a small digression which is not very important for the understanding of the quantum Hall effect, but which is necessary for an explanation of the words conductivity and resistivity which leads very often to confusion if one uses these expressions in the presence of a magnetic field. Normally you say, if you have a current flowing in one direction you also have an electric field in the same direction and the proportionality constant of this is the conductivity. But in a magnetic field you have a **conductivity tensor** and you have a different equation. You can write for example (see Fig. 10): the current vector is equal to the conductivity tensor times the electric field, or you can write this also in the other direction: the electric field is equal to the resistivity tensor times the current. The most important point is that the geometry of the device determines whether one can measure the components of the resistivity - or of the conductivity tensor. If we have a device which is used for the Hall effect, a current flows through the device (see Fig. 9) and thus the current cannot flow across the boundaries

Transport coefficients (homogeneous system) in a magnetic field:

$$\mathbf{j} = \sigma \cdot \mathbf{E}$$
$$\mathbf{E} = \rho \cdot \mathbf{j}$$
$$\sigma \cdot \rho = 1$$

The *geometry* of the device determines whether resistivity or conductivity components can be measured directly.

Assumption: Current direction $\mathbf{j} = \mathbf{j}_x$ is fixed for device with Hall-geometry:

$$E_x = \rho_{xx} \cdot j_x$$
$$E_y = \rho_{yx} \cdot j_x$$

Hall devices: $\longrightarrow \rho_{xx}, \rho_{xy}$

Fig. 10. Transport coefficients for a 2DEG in a magnetic field.

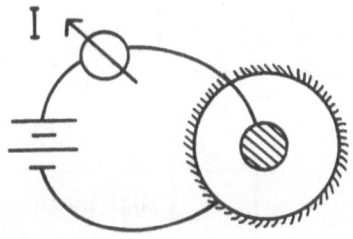

Electric field direction is fixed
⟶ current in direction of the electric field is
connected with σ_{xx}.

$$\sigma_{xx} = \frac{\rho_{xx}}{\rho_{xx}^2 + \rho_{xy}^2}$$

$$\sigma_{xy} = -\frac{\rho_{xy}}{\rho_{xx}^2 + \rho_{xy}^2}$$

Fig. 11. The Corbino-device.

of the sample. So there is only one direction for the current and it is
convenient to use the equation where one multiplies the resistivity tensor
with the current density (see Fig. 10). In this case the measured voltage
in the current direction has to do something with the resistivity component
ρ_{xx} and the Hall voltage U_H is proportional to the resistivity component
ρ_{xy} . So, for Hall devices we are normally measuring these ρ_{xx} and ρ_{xy}
components.

There is another geometry, the so-called Corbino-geome-
try. The Corbino-device is a circular device, here you have one electrical contact in the center and another electrical contact on the outside of
the circle. So, in this geometry, the electric field direction is fixed.
In this case it is natural to use the equation in Fig. 10 which includes
the conductivity tensor. From such a Corbino-device we get direct information on the conductivity tensor σ_{xx}. The relation between the conductivity tensor and the resistivity tensor is given in Fig. 11.

I will now come to the quantum Hall effect and
I will start with an introduction for non-specialists: what
happens if you measure the quantum
Hall effect? I will not discuss what is inside this black box
(see Fig. 12), I will just explain the experimental situation. So you have
such a black box which contains a two-dimensional electron gas with a high
magnetic field, which is produced by a superconducting coil (see Fig. 12).
At low temperatures, we are working around 4 Kelvin, a current is flowing
through the device and one is measuring two voltages: the Hall voltage (U_H)
and the voltage in the direction of the current (U_x). If you are doing
this in a strong magnetic field and at low temperature, you measure a Hall
voltage which increases with the current (see Fig. 13). If you increase
the current (I) then the Hall voltage (U_H) also increases and you have a

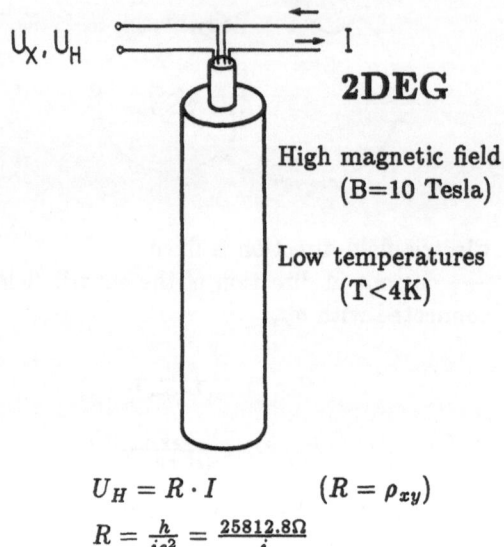

$$U_H = R \cdot I \qquad (R = \rho_{xy})$$

$$R = \frac{h}{ie^2} = \frac{25812.8\Omega}{i}$$

$i = 1, 2, 3, \dots$ fundamental QHE

$i = \frac{1}{3}, \frac{2}{3}, \frac{2}{5}, \frac{3}{5}, \frac{3}{7}, \dots$ fractional QHE

i=1 $\longrightarrow$ R = 1 Klitzing

Fig. 12. The Hall resistance.

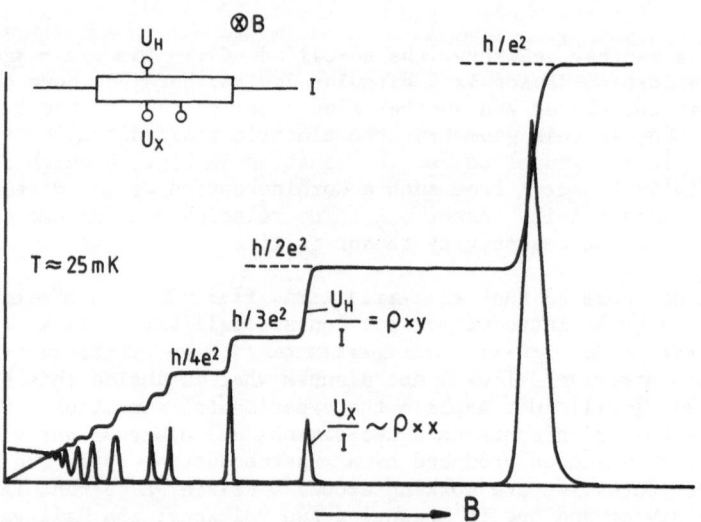

Fig. 13. The resistivities ρ_{xx} and ρ_{xy} in the quantum Hall regime for the case of a GaAs-Al$_x$Ga$_{1-x}$As heterostructure.

proportionality constant which has the dimension of a resistance (R). This resistance is identical with the ρ_{xy} component of the resistivity tensor. In the experimental situation this resistance does not change from sample to sample but it has a fixed value in the magnetic field ranges where plateaus in ρ_{xy} (B) are observed and this fixed value depends only on the fundamental constants: the Planck constant h and the elementary charge e.

The proportionality constant h/e^2 is about 25 kilo-Ohm's (Ω) and is independent of the geometry of the device, the material of the device and all microscopic details. The interesting point is that we have a **new type of resistor**: a resistor which has only a fixed value. We have the so-called **fundamental quantum Hall effect** if the integer i in Fig. 12 is 1, 2, 3, ... (in experiments normally up to 16). It is possible to observe another effect, which will be discussed at this Conference but I will not discuss this today, the so-called **fractional quantum Hall effect**. In that case one can see some special resistors $R = \dfrac{h}{ie^2}$ with i-values of $i = 1/3, 2/3, 2/5, \ldots$. At a meeting in Munich somebody recently mentioned that perhaps one should call the fundamental resistance $R_0 = \dfrac{h}{e^2} =$ 25812,8 Ohm just one von Klitzing. I was surprised recently to get from a company an announcement that they have von Klitzing in stock now. You can buy von Klitzings and if you look at the price-list you can see that a quarter von Klitzing costs 20 DM and that you get a half von Klitzing for 21 DM and so on. The fact that you can buy for example a quarter von Klitzing with a value of 6453,2 Ohm is just a joke since these resistors are wire resistors and have nothing to do with the quantum Hall resistor. A wire resistor can have any value, depending mainly on the material and the geometry at the wire. There are some problems with such resistors which are not present for the quantum Hall resistor - I will discuss this later.

Even if the quantum Hall effect is not working, all the activities in this field are very good for producing new magnets. Superconducting magnets specially designed for quantum Hall experiments have led to an increase of the quality of the superconducting magnet and the price has been reduced by a factor of 3 during the last four years, mainly (I think) due to the activities in this field.

Back to the main part of the experiment: the sample. Different materials (the size is typically of the order of millimetres) can be used for the quantum Hall experiment: Silicon field effect transistor, gallium arsenide heterostructure, indium gallium arsenide, ... All of these samples show the **same result** if you are looking at the quantum Hall effect.

Fig. 13 shows the experimental result obtained for such a two-dimensional system in the case of the gallium arsenide heterostructure. Two quantities are measured with increasing magnetic field. First, the voltage in the direction of the current, which shows some oscillations. In certain magnetic field ranges, the voltage drop is zero, so we have a current flowing through the system but we have no voltage drop in the direction of the current. If you measure the other voltage, the so-called **Hall effect**, this voltage is zero without magnetic field but increases and you see the step-like increase. If you measure the resistance value at these plateaus, then you have plateaus at $\dfrac{h}{e^2} \approx 25$ KΩ, $\dfrac{h}{2e}$, $\dfrac{h}{3e^2}$, $\dfrac{h}{4e^2}$, ... If you measure this resistance to very high accuracy you will find that for different samples you will always get the same result. There is an uncertainty of less than $10^{-6} - 10^{-7}$.

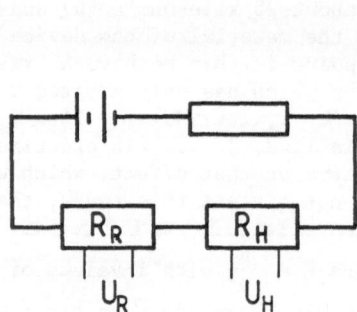

Fig. 14. Experimental arrangement
with a wire resistor
(R_R) and a quantum Hall
resitor (R_R).

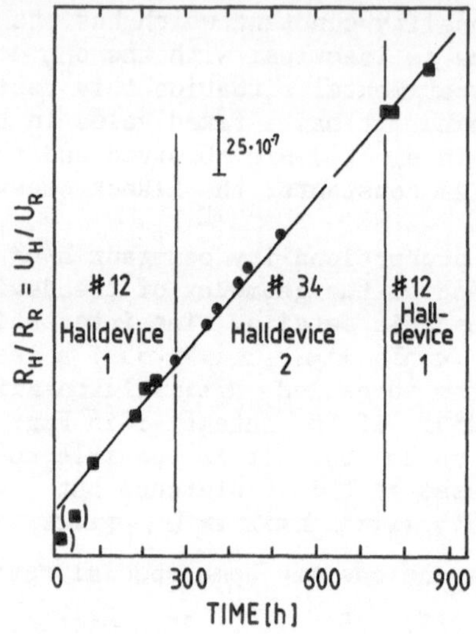

Fig. 15. Time dependence of a wire
resistor as compared to a
quantum Hall resistor.

I will show you one interesting result which demonstrates the stability of the quantized Hall resistance: if you are comparing a quantum Hall device with a wire resistor you have the experimental arrangement as shown in Fig. 14 which is an electrical circuit with two resistors: a wire resistor (R_R) and a quantum Hall resistor (R_H). If the resistors have the same values, the same currents will flow through them and the voltage drops U_R and U_H should be the same. Now you can measure these voltages as a function of time and you get the result as shown in Fig. 15. The ratio of the two voltages or the ratio of these two resistors is not constant, it changes with time! Those experiments were done at the PTB in Germany and two different Hall devices were used, Hall device nr. 1 and Hall device nr. 2. You see that it has a monotonic behaviour as function of time. What is the explanation of this change? The explanation is that the reference resistor R_R has changed in time. So the quantum Hall resistor is more stable, more reproducible than such a wire resistor. This is the problem with wire resistors: they are changing with time! There is some recrystallization, some changes in the material, the stress is changing and therefore all the wire resistors are changing with time. This is demonstrated in the picture shown in Fig. 16 which gives the change of the standard resistors of the different national laboratories with time. There are negative changes and also positive changes. The stability of the quantum Hall resistors is given by the horizontal line. If one can agree that a quantum Hall resistor is a stable and reproducible resistor, then everyone in the world can use this resistor as a resistance reference. This is one of the applications of the quantum Hall effect. The **quantized Hall resistance is more stable** and **more reproducible** than any wire resistor and there is a good chance that, within the next two years, this will be accepted as an **international reference resistor.**

This application is very similar to the application of **the Josephson effect**. The Josephson effect has something to do with superconductivity (see Fig. 17). If we radiate two coupled super-

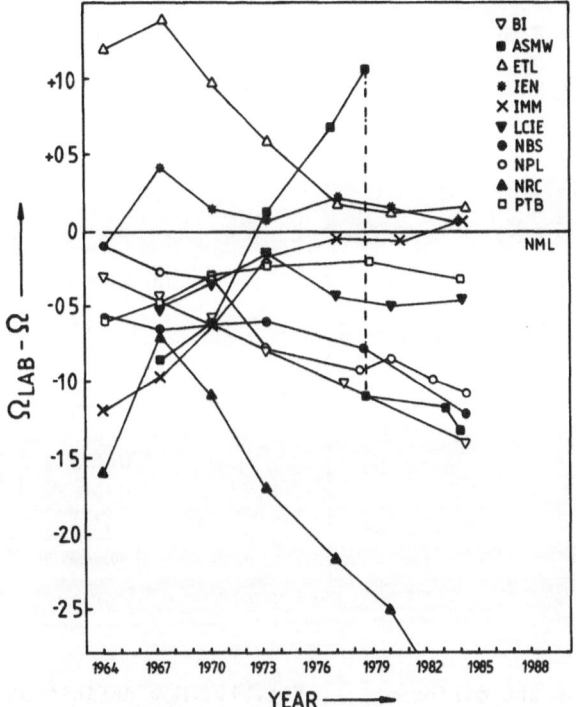

Fig. 16. Change of different standard resistors as function of time.

Quantized Hall Resistance $R_H = \frac{h}{ie^2}$ is more *stable* and more *reproducible* than any wire resistor.

$\Longrightarrow$ **International reference resistor** R_H

Application of the QHE is similar to the application of the *Josephson Effect.*

Josephson-effect

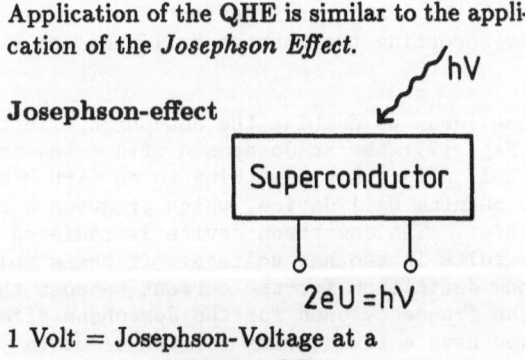

1 Volt = Josephson-Voltage at a
$$\nu = 483594.0\,GH_z$$

Fig. 17. Application of the quantum Hall effect and the Josephson effect.

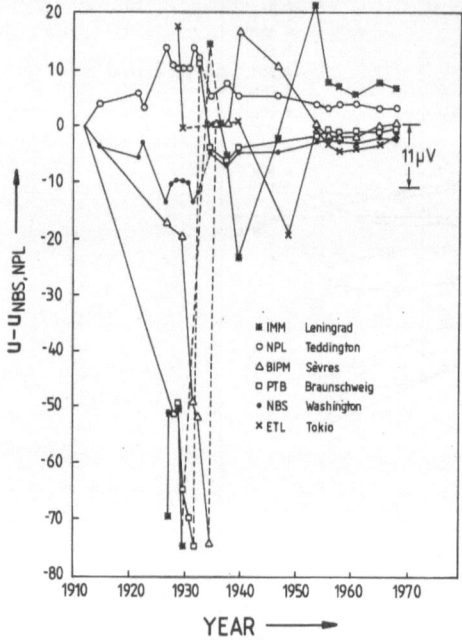

Fig. 18. Change of the different
voltage standards as
function of time.

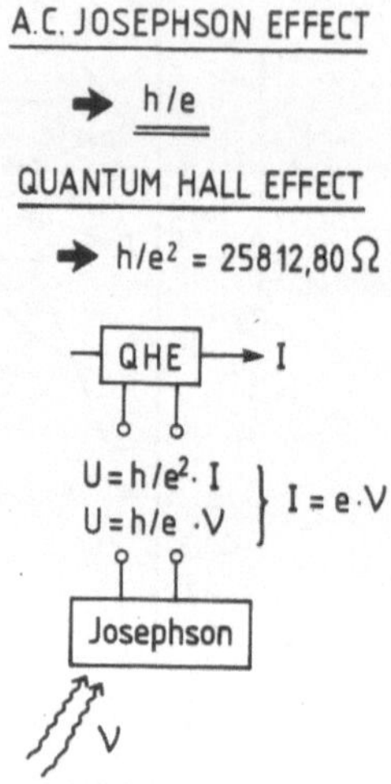

Fig. 19. New definition for the
current by using the
Josephson effect and the
quantum Hall effect.

conductors with some microwaves then one can measure a voltage and this
voltage depends only: i) on the frequency, which is well-known and ii) on
the fundamental constants: the elementary charge and Planck constant. One
can now define a **Volt** as a **Josephson voltage at
a certain frequency**. This has been accepted as the new
reference for voltages. In Fig. 18 I show what is the result for such e
quantum phenomenon for the accuracy of the voltage. This figure shows the
different voltages of the different chemical cells which were used as
standards at the different national laboratories. One clearly sees the
drift, which is very similar to the drift of the 1 Ohm resistor. In 1978
one accepted that one should change to the Josephson voltage and now we
have the same voltage all over the world. The same evolution will take
place, I think, with the quantum Hall effect that we will remove the drift
of the Ohm just by accepting the quantum Hall resistor as a reference
resistor.

There are some ideas to combine the Josephson effect with the quantum
Hall effect (see Fig. 19): the ac Josephson effect has something to do with
h/e, the quantum Hall effect has something to do with h/e^2. A current (I)
flows through the quantum Hall device, which produces a certain voltage
which depends on h/e^2. The Josephson device is radiated with some microwave (ν), which results in another voltage. If these voltages are identical,
you will have a new definition for the current because the current here is
proportional to the frequency used for the Josephson effect and the elementary charge. If you have a fixed value for the elementary charge, the
frequency is normally well-known, then you have a **new definition for the unit of current: the 1
amp**. At present, the current is defined as the force between two wires.

Absolute value for the:

quantized Hall resistance $R_H = \frac{h}{ie^2}$

$R_H(i = 2) = 12906.4030(35)\Omega \qquad$ PTB
$R_H(i = 4) = 6453.2004(11)\Omega \qquad$ NBS
$R_H(i = 8) = 3226.6001(7)\Omega \qquad$ ETL

$$\frac{h}{2e^2} = \begin{array}{ll} 12906.4030 \pm 0.0035\Omega & \text{(PTB)} \\ 12906.4008 \pm 0.0022\Omega & \text{(NBS)} \\ 12906.4004 \pm 0.0028\Omega & \text{(ETL)} \end{array}$$

Fig. 20. Absolute vaue of the quantized Hall resistance.

If the current of 1 amp is flowing through the two wires which are a distance of 1 metre apart then the force should be a given value. But this is a very difficult experiment. The uncertainty of the realization of 1 amp in this way is very large. Today it is not possible to produce a current based on this definition with an uncertainty of less than 10^{-6}.

Different laboratories have tried to measure the absolute value of the Hall resistance. As shown in Fig. 20, laboratories in different countries got identical results. I just used the data by PTB, National Bureau of Standards (NBS) and the Electro-Technical Laboratory (ETL) in Japan. The results are for different samples and for different values of the integer, i = 2, 4 and 8. I have calculated the corresponding value for $h/2e^2$ and you see that the different laboratories agree to about an uncertainty of 2×10^{-7}.

I will just mention another application of the quantized Hall resistance, because the measured quantity h/e^2 is identical with another very fundamental constant in physics, the **fine structure con-**

The measured quantity h/e^2 is identical with the inverse **finestructure constant:** α^{-1}

$$\alpha^{-1} = \frac{h}{e^2} \cdot \frac{2}{\mu_o c} \simeq 137$$

Classical interpretation of α

Hydrogen atom

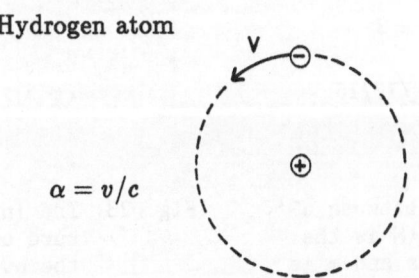

$\alpha = v/c$

Fig. 21. The fine structure constant α.

stant. The inverse fine structure constant is given by $\alpha^{-1} = \dfrac{h}{e^2} \times \dfrac{2}{\mu_0 c}$, with $\mu_0 = 4\pi \times 10^{-7}$ H/m the constant permeability of vacuum and $c = 2.99792458 \cdot 10^8$ m/s. Given the two fixed numbers μ_0 and c, then you see that h/e^2 is identical to the fine structure constant. The fine structure constant is just a number; it is about 1/137. A classical interpretation can be given to this fine structure constant in terms of the hydrogen atom (see Fig. 21). In the center there is a positive charge and an electron moving around this positive charge. If you calculate, in a classical way, the velocity of this electron around this positive charge then this velcocity relative to the velocity of light is exactly the fine structure constant.

There have been a lot of discussions about the reason why the inverse fine structure constant has a value of 137, why not 148 or something else? The same question arises, what is the reason why the mass m of an electron relative to a proton mass M is about 1/1836. Let me just mention to you that already in 1928, there was some discussion whether we can relate the ratio M/m to the fine structure constant (see Fig. 22). Different physicists tried to find some relation between these dimensionless units. I also got some recommended combinations to calculate the fine structure constant on the basis of other units (see Fig. 22), of π for example and you see that you can combine them in such a way that you will get nearly the correct value. I think a lot of physicists are still fascinated by this number and I recently found a publication discussing Pauli. He died in 1958 in a hospital in Zurich in room number 137 and there he said: "Very soon I will solve the problem of the number 137 as the inverse fine structure constant". Shortly after this statement he died. At the time when I prepared my Nobel lecture for Stockholm I looked at the publication of the Nobel lecture by Rubia (he got the Nobel prize in 1984) and then I found in the acknowledgement a large number of names. I counted them and once more there were 137 co-workers in this paper. You see that it is a fascinating number, but α^{-1} is not exactly 137. The recommended value for the inverse fine structure constant is $1/\alpha \approx 137.036$. I think a new value for the fine structure constant will be published this year and I also think that the quantum Hall effect already contributed to this least square adjustment of the fundamental constants (see Fig. 23).

$$\alpha^{-1} = 4\pi^3 + \pi^2 + \pi = 137.0363$$
$$\alpha^{-1} = 2\left[\sinh\left(\tfrac{\pi}{2}\right)^2 - 1\right] = 137.0384$$
$$\alpha^{-1} = \sqrt{5} \cdot 2^{1/6} \cdot e^4 = 137.03597$$

Perles (1928)
$$M/m = 2\pi(\pi - 1) \cdot \alpha^{-1}$$
Fürth (1929)
$$M/m = 64\pi/15\alpha - 2$$
Haas-Jordan (1937)
$$M/m = 4\pi\alpha^{-1} \cdot \sqrt{17/15}$$
Born (1940)
$$M/m = \tfrac{\pi}{18} \cdot \alpha^{-2}$$

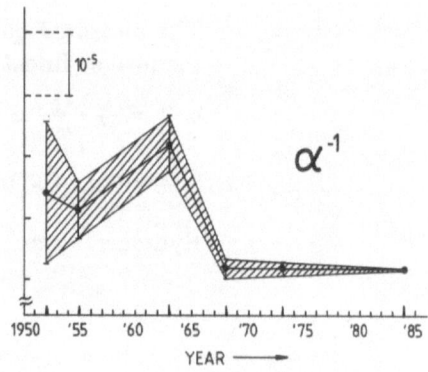

Fig. 22. Relations between α^{-1}, π and M/m (M is the proton mass and m is the electron mass).

Fig. 23. The inverse fine structure constant α^{-1} and the evolution of the uncertainty on it as function of time.

Why is it interesting to have a value for the fine structure constant? It is very important for different areas in physics: you can determine the fine structure constant by different experiments. I will just mention here some experiments from optics: there is a fine structure (fs) splitting in atomic physics, a hyperfine splitting (hfs), the anomalous magnetic moments of the electron (a_e) and all these quantities depend on the fine structure constant. However, for the interpretation of the experiments one needs a complicated theory - the quantum electrodynamic theory (QED). Other experiments can be used to calculate the fine structure constant more or less directly like measurements of the Faraday constant (F) or the gyromagnetic ratio of the protons (λ_p) or h/e^2 (see Fig. 24). Thus you have different experiments which can give you information about the fine structure constant and all these experiments should give the same value. If you do not get the same value, then you have to think about the experiment or about the theory of the experiment. One can learn something if different expements are trying to get the same number for the fine structure constant. If you have the new value then you are also influencing the values of other fundamental constants like the electron mass (m_e), the Planck-constant (h), the electron charge (e), the Compton wavelength (λ_c), ... You can learn a lot in different fields of physics if you have very accurate values for this fundamental constant.

But now back to the experiment of the quantum Hall effect. This is a show curve (see Fig. 25); normally you are not doing this at very low temperature. Here the temperature is 8 mK, normally you do the experiment at 1 or 2 K. At very low temperature you have nicely developed steps. At high temperatures the Hall effect is just a straight line without steps. The flatness of the plateaus is only developed if you reduce the temperature or if you increase the magnetic field. There are a lot of high precision measurements on these plateaus and empirically one found that the deviation from the quantized value ρ_{xy} is connected to the value of the resistivity ρ_{xx}^{min} in the following way: $\Delta\rho_{xy} < 0.5 \, \rho_{xx}^{min}$. ρ_{xx}^{min} is the re-

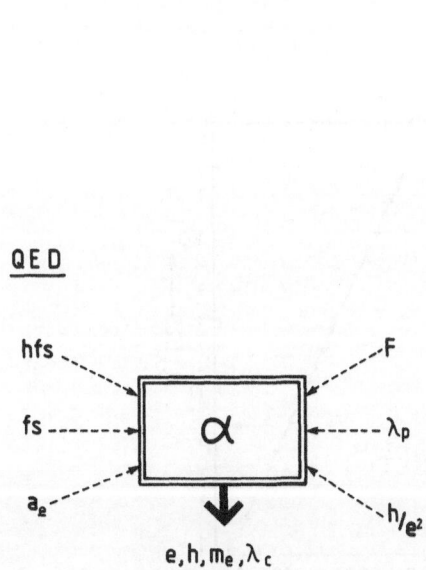

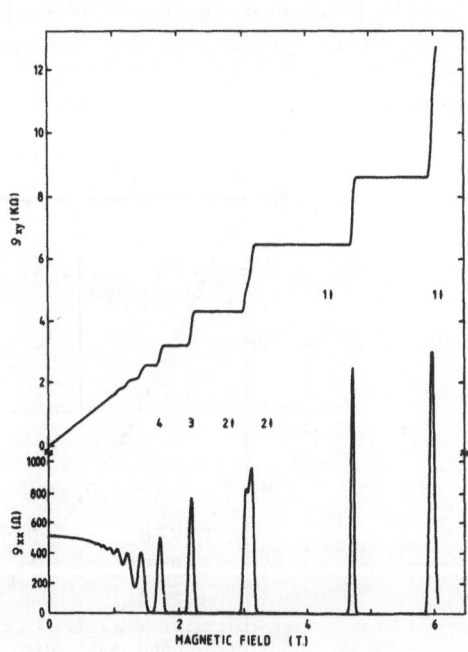

Fig. 24. Impact of theory and experiment in determining the value of the fine structure constant.

Fig. 25. The quantum Hall effect at very low temperature (8 mK).

17

sistivity at the centre of the plateau which in reality is never exactly zero! If you are going to higher temperatures then ρ_{xx}^{min} increases and you can measure the temperature dependence of ρ_{xx}^{min} which looks like a zero in Fig. 25. If ρ_{xx}^{min} is exactly zero we also believe that the plateau in ρ_{xy} is exactly quantized, it has the value of the quantized Hall resistance. If you measure for example ρ_{xx} between 30 K and 8 K you are observing a strong variation of this value (see Fig. 26): a 3-orders of magnitude change in ρ_{xx}^{min} over the temperature range 30 K - 8 K. If you are further reducing the temperature you are going to get near zero, but ρ_{xx}^{min} is still a finite value. At high temperature, between 30 K and about 8 K, you see really an **activated behaviour**. The thermally activated conductivity can be described as an exponential function of some activation energy ΔE over kT. But if you analyze the data at very low temperatures, you see some deviation from a straight line. There the behaviour can be explained by a function

$$\rho_{xx} \sim \frac{1}{T} \exp\left(-\frac{T_0}{T}\right)^{\frac{1}{2}}$$

In an amorphous system such a deviation from an exponential curve has usually been explained by a so-called **variable range hopping**. I will just mention that deviations from activated resistivity are observed at low temperatures, normally below 2 K and you can explain this by variable range hopping. I will not discuss this here in detail. There are some publications, for example by Ono in 1982, which explains the above temperature behaviour and he exactly gets the behaviour observed experimentally. Simultaneously, you can look at the plateaus and the flatness of the plateaus. How flat are these plateaus? You also see an activated behaviour in the slope of the Hall plateaus but this activated behaviour remains activated even in the temperature range where we see a resistivity which shows this so-called variable range hopping. After this discovery, some theoreticians started to calculate the influence of this variable range hopping on the Hall effect and really they found that this

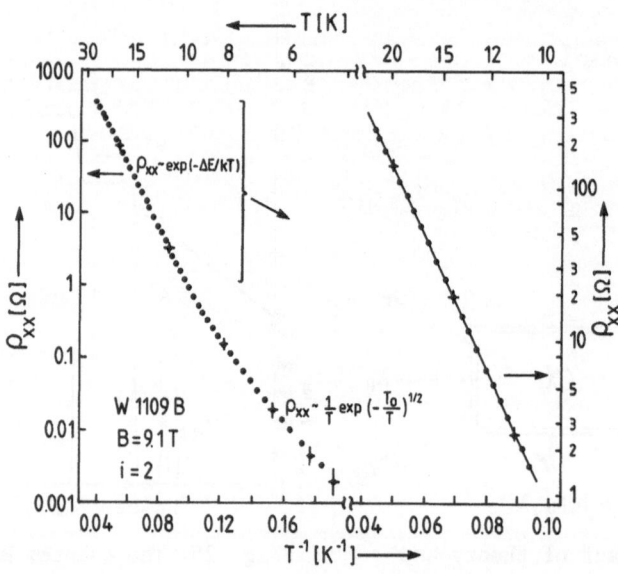

Fig. 26. Temperature dependence of the resistivity ρ_{xx} at the Hall plateau.

influence is very small. It is an experimental result that we have an activated behaviour for the slope and we can reduce the slope just by decreasing the temperature or increasing the magnetic field (because the activation energy increases with increasing magnetic field). It is only a question of temperature and therefore we are going to temperatures normally of 2 K. At 2 K the plateau is already so flat that we have no chance to see the final slope so that within 10^{-8} we have in a 10% range of the magnetic field identical values for this quantized Hall resistance.

I have said nothing up to now about **the physics of the quantum Hall effect**. This is a difficult problem; we have about ten lectures during the next two weeks which are related to the quantum Hall effect: the fractional and the normal quantum Hall effect. I will just give you an introduction to the quantum Hall effect and I will show you that this explanation cannot be correct. The result is correct but there are a lot of assumptions which are not correct. It seems to be very difficult to have a microscopic theory of the quantum Hall effect. We can learn a lot about the physics of the two-dimensional system just by explaining the quantum Hall effect and by finding the correct result. You can get the correct result in a very simple picture and I will give you this explanation but I will show you after this that different points are not exactly correct, but fortunately these simplifications are not very important for the final result. If you are discussing the Hall effect you will always have the following rough picture: there is a current flowing, the charges have a certain velocity. If we apply a magnetic field, then there will be a Lorentz force acting on these charges. The electrons will move perpendicular to the current direction and perpendicular to the magnetic field. Due to the boundary of the sample we are building up an electric field which leads to an opposite force so that finally the electrons are going from one end of the sample to the other. We have two forces which are in equilibrium, the Lorentz force and the force of the

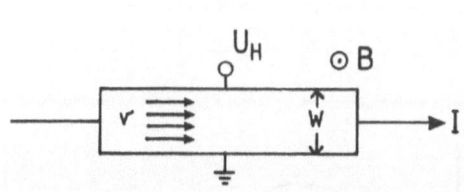

Lorentz force = force in Hall field

$$e \cdot v \cdot B = e \cdot E_H$$

current density: $j = n \cdot e \cdot v$

$$j \cdot \frac{B}{n \cdot e} = E_H$$

$j = I/W$; $E_H = U_H/W$

Hall resistance: $R_H = \frac{U_H}{I} = \frac{B}{n \cdot e}$

$n=$ two-dimensional carrier density

Fig. 27. Simple calculation for
the Hall resistance.

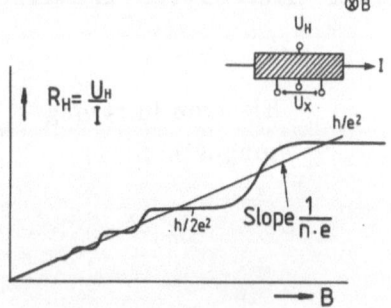

STRONG MAGNETIC FIELD

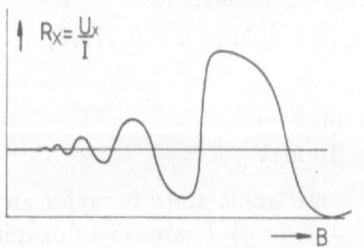

Fig. 28. A rough picture of the
quantum Hall effect.

Hall field on the electrons. One can perform a simple calculation of the Hall voltage for two-dimensional systems and you will get the result shown in Fig. 27. You can define a Hall resistance and this was the quantity which was measured. This Hall resistance is proportional to the magnetic field and inversely proportional to the carrier density. If we measure the Hall effect now as a function of the magnetic field, we expect on the basis of this equation some linear increase of the Hall voltage as a function of the magnetic field (see Fig. 28). This is the normal Hall effect and from the slope we will get information on the carrier density. This is very important, mainly in semiconductor physics, where you can change the number of electrons in a semiconductor by doping. This is a standard way to determine the carrier density.

But, experimentally you observe the following: we have not such a monotonic increase of the Hall voltage, we have a step-like increase and we have these oscillations in R_x. These are Shubnikov-de Haas oscillations and in order to explain these, we have to introduce some quantum mechanics. But I will not replace the lecture for one week in quantum mechanics now, because this will be done during the Conference. I will just explain what is the background, what is the origin of these plateaus and the origin of these oscillations. You have to solve the problem of electrons in a strong magnetic field. So, what happens if an electron is moving in a strong magnetic field? These electrons are moving around a circle, and the quality of the device is so high that they can move without scattering, without change in the momentum. Like in a hydrogen atom, you have some **quantization**. In hydrogen atoms you also have some electrons moving in circles and you have some fixed energies for this motion. If you calculate the electrons in the magnetic field you basically have the problem of the harmonic oscillator and you also find that the energy has only fixed values. So an electron in a strong magnetic field can only have fixed values of energy, which is given by the equation:

$$E_n = (n + \tfrac{1}{2})\, \hbar\omega_c;$$

n is the **Landau quantum number** and $\hbar\omega_c$ is the so-

Electron in strong magnetic fields

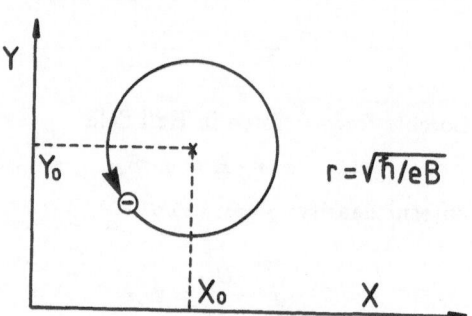

Landau quantization:
$E = (n + \tfrac{1}{2})\hbar\omega_c; \; n = 0, 1, 2, 3, \dots$

$B = 10\,Tesla$
$r \simeq 8nm$
$E \simeq 10\ \text{meV}$

Each electronic state occupies an area $\Delta X \cdot$
$\Delta Y = F_0 = \frac{h}{eB}$ (=area of a flux quantum h/e)
$[X_0, Y_0] = i \cdot \frac{\hbar}{eB}$

Fig. 29. Quantum mechanics of an electron in a strong magnetic field.

called **cyclotron energy.** For a magnetic field of 10 Tesla,
we find that the radius of this classical orbit is of the order of 8 nanometres and the energy difference between these energy levels is of the
order of 10 millivolts. But the most important result of such a calculation is also that each electronic state, each circle of an electron,
occupies a certain area. The Pauli principle says that it is forbidden to
have two electrons in the same place or in the same set of quantum numbers.
This is a very important result. Each electron state occupies an area
h/eB. If we increase the magnetic field, the orbit becomes smaller. Therefore the area occupied by an electron becomes smaller. Normally you are
not using this picture of an electron moving on a circle; there are different pictures and during the Conference I think mainly another picture will
be used of an electron with a plane wave in one direction and a harmonic
oscillator function in the other direction. So this picture only serves
to explain, in a very simple way, that you need a certain area for each
circle.

In this simple picture you have certain energies which are allowed:
E_0, E_1, E_2, ... and only these energies are allowed (see Fig. 30). Each
of these energies can be occupied by a large number of electrons. The degeneracy per unit area of each of these states is the number of electrons
which can occupy one energy level. So, if you have a high magnetic field
the area necessary for one state is small. Therefore a large number of
electrons can occupy the same energy. The quantum Hall effect is observed
if the magnetic field and the carrier density are chosen in such a way
that such energy levels are fully occupied. If an integer number of these
levels are fully occupied, then we can observe the quantum Hall effect.

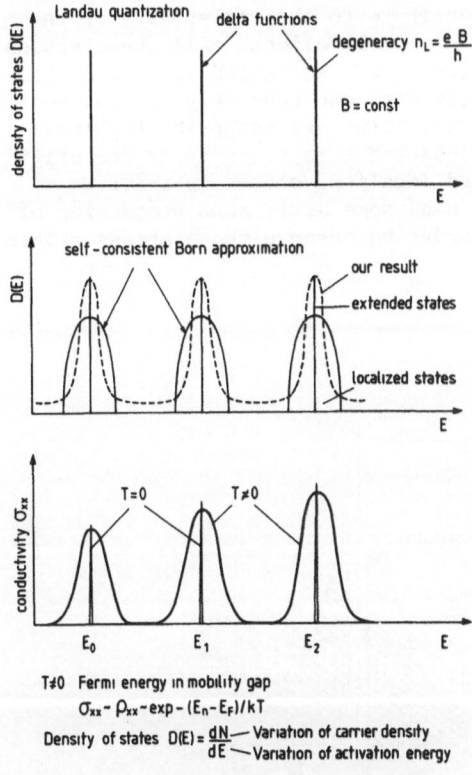

Fig. 30. Density of states of
electrons in two dimen-
sions and the corres-
ponding conductivity ρ_{xx}.

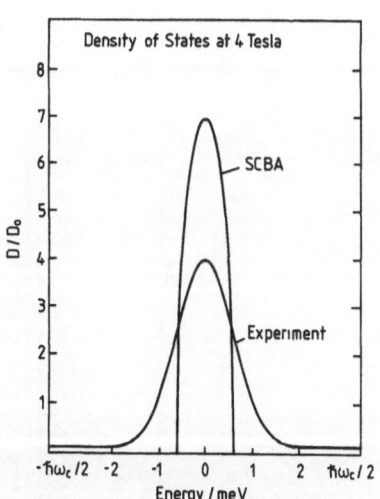

Fig. 31. Broadening of the
Landau level.

We can change the occupation just by changing the magnetic field, or by
changing the carrier density of the system. This can be calculated just
by using a simple equation. I showed you that the Hall resistance is
equal to the magnetic field over the carrier density and the electronic
charge. You just have to put the numbers into this equation and you will
get the quantized Hall resistance. You will get h/ie^2. The number "i" now
has a physical meaning; this corresponds to the number of occupied Landau
levels.

In reality, everything is much more complicated: you not only have
the discrete energy levels; you have a broadening of these levels (see Fig.
30) because we never have an ideal system; we have some impurities; we have
some charges in our materials; we have some dislocations and this leads to
a broadening of these discrete levels. Already ten years ago, Ando calculated the broadening within a certain approximation (the self-consistent
Born approximation (SCBA)) and he got for the density of states the shape
depicted in Fig. 31. Today we believe that the density of states looks
more like the other curve in Fig. 31. Today, in the afternoon, Professor
Gornik will discuss in more detail the problems related to the density of
states. Up to now, it is not clear what is the origin of the density of
states in the region between the energy gaps.

I mentioned at the beginning that these energy gaps, these discrete
energy levels are very important, because the quantum Hall effect is observed
if one of these levels is exactly occupied. Now, what is the interpretation
today? The interpretation is that we have to distinguish between so-called
localized states and **extended states** (see
Fig. 32). Only in the center of the Landau levels we have so-called extended states and only these states contribute to the resistivity or the
conductivity, thus the value of the quantum Hall effect. All these states,
which are in the tails of the Landau levels, are so-called localized states
and these do not contribute to the resistivity and thus they are not contributing to the quantum Hall effect. So, if we are occupying the localized states, the Hall effect remains fixed and this seems to be the origin
of the plateaus; we need these so-called localized states in order to
produce the quantum Hall plateaus. We need some dirt, some broadening of
these levels and some localization in order to observe the plateaus. This

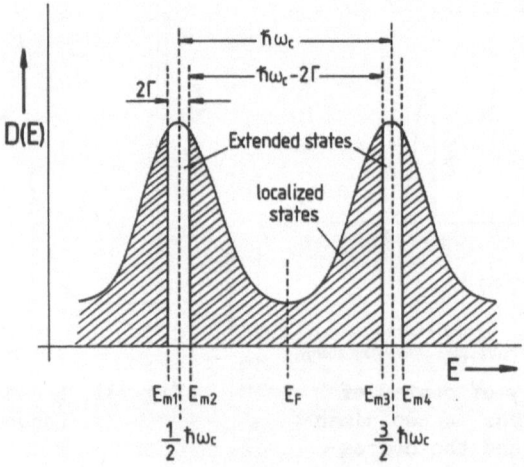

Fig. 32. Localized and extended states in a Landau level.

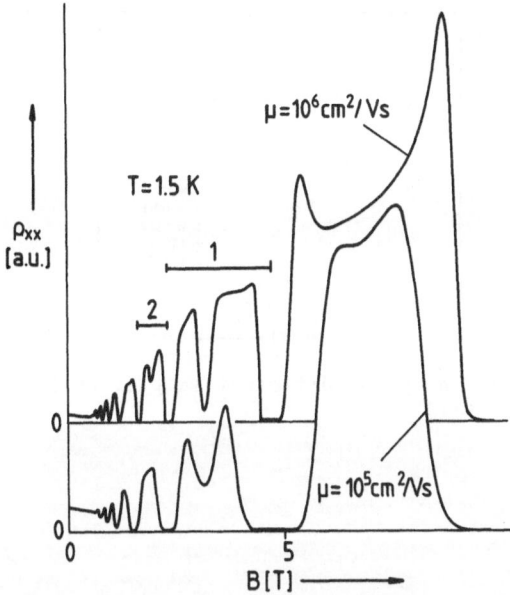

Fig. 33. The resistivity ρ_{xx} for two different samples with an order
of magnitude difference in the mobility

is demonstrated by the experiment shown in Fig. 33. We have two samples:
a high quality sample and a low quality sample. The mobility is ten times
smaller in the low quality system. But you see we have much wider plateaus
for a sample which has lower mobility (note that ρ_{xy} has a plateau in the
region where $\rho_{xx} \approx 0$). Thus we need some dirt in our samples in order to
produce these quantum Hall plateaus.

Thus our present picture for the density of states is as follows:
there are localized states and there are extended states in the center.
For the value of the Hall plateau and the value of the resistivity, only
the occupation of these extended states are important. Now we can understand the activation of the resistivity. If the Fermi level is in the
middle between two Landau levels (see Fig. 32) we can excite electrons into
the states of the next Landau level and the occupation of those extended
states will change the resistivity and will lead to a change in the plateau.
This mechanism is responsible for the finite slope of the plateaus and for
a non-zero value of the resistivity ρ_{xx}. I will just show you that we have
measured this activation energy (see Fig. 34). You see that for different
samples with different magnetic fields, we really got an activation energy
which is related to this energy difference between these Landau levels,
which is nearly identical to the cyclotron energy. I will not discuss in
detail how we now measured the density of states. I think Professor Gornik
will do this. We analyzed the activation energy, we analyzed how the Fermi
energy moves if you change the filling of these levels and from this you
can calculate the density of states of these levels.

Now, finally I will just mention **some problems**. I tried
to explain the quantum Hall effect in a very simple way and this simple
explanation must be incorrect. I explained that we have a compensation of
the Lorentz force by the Hall field. In a simple picture you assume that
there will be a homogeneous current flowing in the sample. But if you
calculate the current distribution you will find that this is not possible.
There must be a high current at the edges. If you try to solve the problem

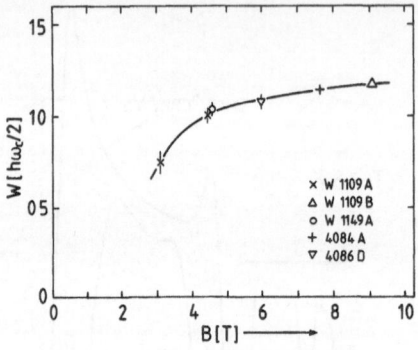

Fig. 34. Activation energy of different samples as function of the
magnetic field.

self-consistently, you find that we cannot have a homogeneous electric
field for example. Locally, we still have the equilibrium between the
Lorentz force and the Hall field. Thus, even if one has an inhomogeneous
current distribution the quantum Hall effect will not be influenced by it.

We were interested for example to measure this current distribution:
we have a higher current at the edges of the sample, a lower current in the
centre and we just measured this potential by using different probes at
different points on the sample. Look how the voltage difference develops
in the quantum Hall regime and I will just show you a strange result which
has been confirmed by three or four different groups now. Fig. 35 shows
a typical quantum Hall experiment. We measure the Hall voltage as a function of the magnetic field and there is a step-like increase. Now, if you
measure across half of the sample, between the points 4 and 6, normally you

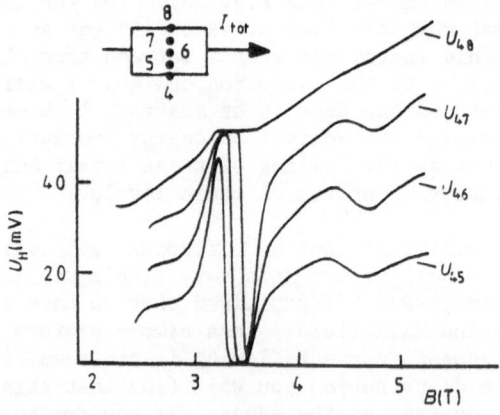

Fig. 35. Distribution of the Hall voltage over the sample.

expect that you measure just half of the voltage because you expect only half of the current is flowing through the system. If you are doing this, you see the strange behaviour as shown in Fig. 35. It seems to be that at the plateau you measure the same voltage between the points 6 and 8 and between the points 4 and 6. The interpretation is that all the current is flowing on one side. At this magnetic field all the current is flowing in the upper part of the sample. This is a very strange result and we have demonstrated that this has something to do with some inhomogeneities in the samples. The samples are never so inhomogeneous that we have the same carrier density on both sides of the sample. We have changed the carrier density and demonstrated that you can have an inverse structure. So everything is much more complicated in a real system because we do not have an ideal system. The final result that the quantum Hall effect is correct is not influenced because the total current is flowing between both edges and this is the important point. It will be very difficult to develop a microscopic theory of the quantum Hall effect, which includes all these microscopic details about some inhomogeneities, some impurities and so on. A large number of theoretical work is now done to start such a microscopic picture and I think Professor Hajdu will speak about some calculations in this direction.

There are still a large number of other problems. I will not discuss them here; I will just summarize the results. We believe that the quantized Hall resistance is really quantized on the condition that we have no energy dissipation. Thus we have a current flowing in the sample without voltage drop in the direction of the current, so if ρ_{xx} is really exactly zero. But, there are a lot of problems which are not solved. We have energy dissipation in real experiments and this energy dissipation is never zero. We have complicated boundary conditions; we always have a finite temperature. We sometimes have electrical by-pass; the current is flowing not only through the two-dimensional system but also at other paths through the sample. We have to use a finite current in our system and this finite current also disturbs the situation. But the energy dissipation can be so strongly reduced that the corrections for the application of the quantum Hall effect become unimportant. The reason is that one can reduce the temperature or increase the magnetic field so that ρ_{xx} can be reduced to a value which is so small that we are not able to measure it. The conclusion is that the discussion of this quantum Hall effect must include an analysis of the resistivity as a function of temperature of the current, of the shape of the sample. There is still a lot of work to be done to understand this in detail. Fortunately the quantum Hall effect at the level of the accuracy necessary for applications is working but there are still a lot of problems that have to be solved in the future.

Thank you for your attention.

THEORY OF THE INTEGER QUANTUM HALL EFFECT -

AN INTRODUCTORY SURVEY

J. Hajdu

Institut für Theoretische Physik
Universität zu Köln
D-5000 Köln 41, Federal Republic of Germany

INTRODUCTION

The aim of these lectures is to give an introduction to the present state of the theory of the integer quantum Hall effect (IQHE). The elementary theory of the Hall effect and the quantum mechanics of free electrons in a strong magnetic field are recalled. The qualitative explanation of the IQHE due to the localization of electrons in a random potential is presented. The high field limit in which classical percolation thecry can be applied is described in detail. The so-called gauge argument is described and critically commented upon. Further topics are: the topological, the scattering theoretical and the field theoretical approaches to the IQHE. The essential conclusion is that localization is a possible explanation of the IQHE but the theory is still far from being complete.

KLAUS VON KLITZING'S DISCOVERY

The Hall resistivity R_H of a silicon Metal Oxide Semiconductor Field Effect Transistor (MOSFET) measured as a function of the gate voltate U_G shows at low temperatures($\approx 1K$) and high magnetic fields ($\approx 10T$) characteristic plateaux. This observation contradicts the established theory which predicts $R_H \sim 1/U_G$. In 1980 Klaus von Klitzing discovered that the plateau values of R_H are entirely independent of the properties of the sample and are given by

$$R_H(\text{plateau}) = \frac{1}{i} \frac{h}{e^2}, i = 1,2,3,\ldots \tag{1.1}$$

/1/. Present day experimental accuracy of this quantization of the Hall resistivity is about 10^{-7}. Thus (1.1) provides a high precision measurement of the Sommerfeld fine structure constant e^2/hc. Furthermore, in the plateau regimes of R_H the (longitudinal) resistivity R practically vanishes (the corresponding voltage U is less than 10^{-14} V). To appreciate these results, as shown in Fig. 1, and to understand the physical phenomenon behind it, let me recall a few well-known facts.

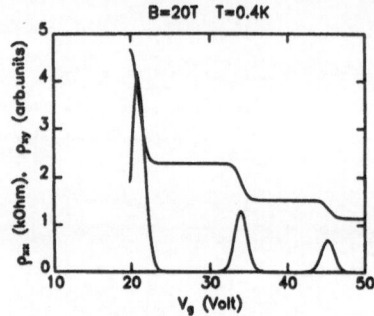

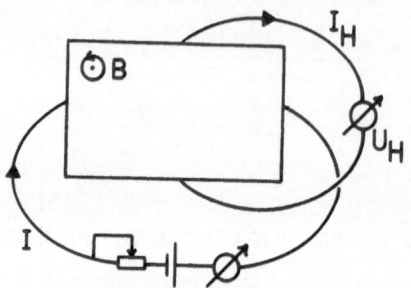

Fig. 1. Resistivity and Hall re-
     sistivity of a MOSFET-
     integer quantum Hall
     effect.

Fig. 2. Standard Hall measurement
     with two current loops.

THE HALL EFFECT

In 1879 the American physicist Edwin Herbert Hall discovered that, in
the presence of a magnetic field B, a perpendicular electric current I im-
posed on an isotropic conductor gives rise to a voltage drop

$$U_H = R_H I \tag{2.1}$$

which is proportional to the current and perpendicular both to the current
and to the magnetic field. The voltage drop parallel to the current is
given by Ohm's law,

$$U = RI \tag{2.2}$$

The semiconducting interface of an n type MOSFET is practically a two-
dimensional electron system. The MOSFET can be viewed as a plate condensor;
the electron concentration in the interface is proportional to the gate
voltage. Assuming the interface to be rectangular and denoting the lengths
of the edges by L_x and L_y ,

$$R = \rho \frac{L_x}{L_y} \ , \quad R_H = \rho_H \tag{2.3}$$

Thus, in two dimensions, the Hall resistivity – being independent of the
extensions of the system – is a specific quantity. Shorting U_H by an external resistor leads to a (Hall) current. In an isotropic system the most
general linear relation between fields ($E_x = L_x U$, $E_y = L_y U_H$) und current
densities ($J_x = I/L_y$, $J_y = I_H/L_x$) is

$$\underline{E} = \underline{\underline{\rho}} \ \underline{J} \tag{2.4}$$

where $\underline{E} = (E_x, E_y)$, $\underline{J} = (J_x, J_y)$ and

$$\underline{\underline{\rho}} = \begin{pmatrix} \rho & \rho_H \\ -\rho_H & \rho \end{pmatrix} \tag{2.5}$$

Introducing the specific conductivities by inverting the resistivity tensor

$$\underline{J} = \underline{\underline{\sigma}} \ \underline{E} \tag{2.4'}$$

$$\underline{\underline{\sigma}} = (\underline{\underline{\rho}})^{-1}$$

$$\sigma = \sigma_{xx} = \frac{\rho}{\rho^2 + \rho_H^2}, \quad \sigma_H = \sigma_{yx} = \frac{\rho_H}{\rho^2 + \rho_H^2} \tag{2.6}$$

If $\rho = 0$ (plateau regimes) and $\rho_H \neq 0$ then also $\sigma = 0$ and $\sigma_H = 1/\rho_H$. Thus,

$$\sigma_H(\text{plateau}) = i\,\frac{e^2}{h} \tag{2.7}$$

where now also $i = 0$ corresponds to a measured value.

The Hall effect is a consequence of the Lorentz force which appears in the equation of motion of an electron

$$m\underline{\dot{v}} = -e\underline{v} \times \underline{B} - e\underline{E} \tag{2.8}$$

The homogeneous equation ($\underline{E} = 0$) describes a circular motion in a plane perpendicular to $\underline{B}$ with frequency of revolution $\omega_c = eB/m$ (cyclotron frequency). The center coordinates are conserved quantities; the radius of the cyclotron orbit is a function of the kinetic energy. For $\underline{E} = (E,0)$ a particular solution of the inhomogeneous equation is $\underline{v} = (0, v_D)$, $v_D = -\frac{E}{B}$ (= const). Thus the general motion is a superposition of cyclotron motion and an inertial motion perpendicular to $\underline{E}$ and $\underline{B}$ (cf. Fig. 3). The x coordinate of the cyclotron center is still conserved. The time average of the velocity is

$$\bar{v}_x = 0, \quad \bar{v}_y = v_D \tag{2.9}$$

This yields for the average current density $\underline{j} = (-e/Ar)N\underline{v}$ of N independent electrons

$$J_x = 0, \quad J_y = \frac{en}{B}\,E \tag{2.10}$$

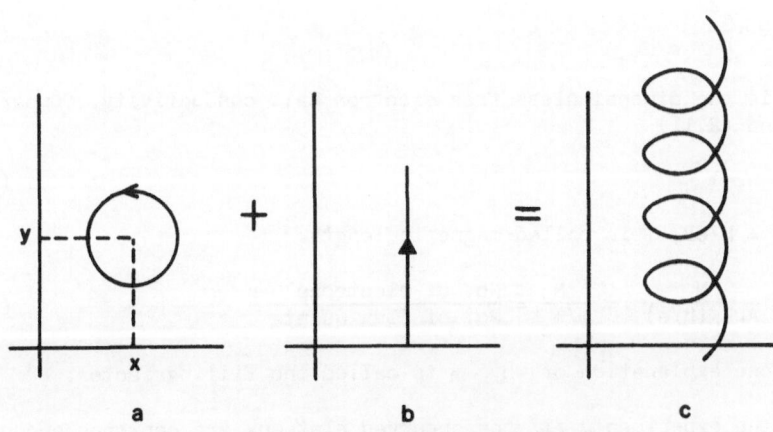

Fig. 3. Motion in crossed electric and magnetic fields.
a.: $\underline{E} = 0$,   b: stationary motion (drift),
c.: general case (superposition of a. and b.)

where $n = N/Ar$ is the electron concentration. Thus, the longitudinal conductivity of a free electron gas σ^o vanishes and the Hall conductivity is given by

$$\sigma_H^o = \frac{en}{B} \qquad (2.11)$$

This is, as a function of n, a straight line; there is no indication of any quantization.

To get a finite conductivity in the direction of the electric field, we have to take into account a friction force. Indeed, adding to the rhs of (2.8) - $(m/\tau)\underline{v}$, the stationary solution ($\underline{v} = 0$) yields

$$\sigma = \frac{\sigma_0}{1 + (\omega_c \tau)^2} \quad , \quad \sigma_H = \omega_c \tau \sigma \qquad (2.12)$$

where $\sigma_0 = e^2 n \tau / m$. The situations B = 0 and B large, $\omega_c \tau \gg 1$ are complementary:

B = 0: $\quad \sigma = \sigma_0 \sim \tau \quad , \quad \sigma \to \infty \qquad$ for $1/\tau \to 0$

i.e. friction is needed to get a non-diverging conductivity.

$B \neq 0; \quad \omega_c \tau \gg 1: \quad \sigma \approx \sigma_0/(\omega_c \tau)^2 \sim 1/\tau , \sigma \to 0 \qquad$ for $1/\tau \to 0$

i.e. friction is needed to get a non-vanishing conductivity.
At low temperatures the friction is due to the scattering by a random potential U. In the weak potential limit $1/\tau \sim U^2$. Consequently, for B = 0 $\sigma \sim U^{-2}$ and for $\omega_c \tau \gg 1$ $\sigma \sim U^2$. In the latter case the most simple perturbation theory can be applied. In high magnetic fields the center of the cyclotron orbit <u>migrates</u> in the direction of the electric field - giving rise to the (longitudinal) conductivity. In the limit $\omega_c \tau \to \infty$ the cyclotron center <u>drifts</u> in the direction perpendicular to the fields.

What are the relevant units? In a two-dimensional system the physical dimension of the conductivity is (charge)^2/action. Therefore, the atomic unit of the conductivity is e^2/h and we may write

$$\sigma_H^o = \eta \frac{e^2}{h} \qquad (2.13)$$

where η is the dimensionless free electron Hall conductivity. Comparing (2.13) and (2.11)

$$\eta = n 2\pi l^2 \qquad (2.14)$$

where $l^2 = h/eB$, l is called magnetic length.

$$\eta = \frac{N}{ArB/(h/e)} = \frac{N}{\phi/\phi_0} = \frac{\text{No. of electrons}}{\text{No. of flux quanta}} \qquad (2.15)$$

This is one explanation of why η is called the filling factor.

In the experiments /1/ the observed plateaux are centered around $\eta = i$ (integer filling). Surprisingly, at any integer filling, the observed Hall conductivity $\sigma_H = i(e^2/h)$ of a real system (MOSFET, heterojunction) exactly coincides with the calculated Hall conductivity of the highly fictitious free electron system - see the schematic Fig. 4.

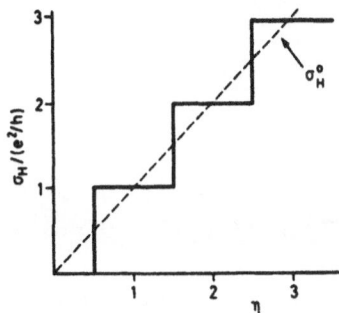

Fig. 4. Idealized quantum Hall effect and Hall conductivity of free
electrons (-----) as a function of the electron concentration.

QUANTUM MECHANICS OF AN ELECTRON IN A MAGNETIC FIELD

The Hamiltonian of a free electron in a constant magnetic field $\underline{B}$ and electric field $\underline{E}$ is

$$H = \frac{p^2}{2m} (\underline{p} + e\underline{A})^2 - e\phi \qquad (3.1)$$

$$\underline{B} = \text{curl } \underline{A} \ , \ \underline{E} = - \text{grad } \phi$$

$(e > 0)$. The task is to solve the 2d eigenvalue problem

$$H\Psi_\alpha (x,y) = \varepsilon_\alpha \Psi_\alpha (x,y) \qquad (3.2)$$

for specified boundary conditions.
Let $A = (0,Bx,0)$, $\underline{B} (0,0,B)$ and
$\phi = -Ex$, $\underline{E} = (E,0,0)$.
Boundary conditions: In the y direction periodic

$$\Psi_\alpha (x,y + L_y) = \Psi_\alpha (x,y) \qquad (3.3)$$

and in the x direction
<u>either</u>

$$\Psi(x,y) \to 0 \text{ for } x \to \infty$$
$$\text{corresponding to } L_x \to \infty \quad \text{(Landau model)} \qquad (3.4)$$

<u>or</u>

$$\Psi(x,y) = 0 \text{ for } x = \frac{L_x}{2} \quad \text{(Dirichlet)}$$
$$\text{corresponding to finite } L_x \text{ (Teller model).} \qquad (3.5)$$

In both cases Ψ should be normalized to unity.

Since H is independent of y, the momentum in this direction is conserved, $[H,p_y] = 0$. Consequently the 2d eigenvalue problem (3.2) factorizes

$$\Psi_\alpha (x,y) = \frac{1}{\sqrt{L_y}} e^{iky} u_\alpha (x) \qquad (3.6)$$

$$p_y \Psi_\alpha = \hbar k \Psi_\alpha \qquad (3.7)$$

where - due to (3.3) -

$$k = \frac{2\pi}{L_y}\mu \ , \ \mu = 0, \pm 1, \ldots \tag{3.8}$$

$$Hu_\alpha = \varepsilon_\alpha u_\alpha \tag{3.9}$$

$$H = \frac{p_x^2}{2m} + \frac{m\omega_c^2}{2}(x - X)^2 + eEx \tag{3.10}$$

Here $X = -l^2 k$ is the x coordinate of the cyclotron center (cf. sec. 6). α denotes two quantum numbers, $\alpha = (\nu, k)$.

Let us first consider the Landau model (3.4). The solution of (3.9) then is

$$u_\alpha(x) = u_\nu(x - X - v_D/\omega_c) \tag{3.11}$$

where $u_\nu(x)$ with $\nu = 0, 1, 2, \ldots$ is the normalized oscillator wave function, and

$$\varepsilon_\alpha = h\omega_c(\nu + \tfrac{1}{2}) + eEX - \frac{m}{2}v_D^2 \tag{3.12}$$

The first term on the rhs of (3.12) is the quantized kinetic energy of the cyclotron motion, the second term is the potential energy of the cyclotron center in the constant electric field and the third term is a constant. For E = 0 the spectrum (3.12) is equidistant and degenerate; the degeneracy of each Landau level

$$\varepsilon_\nu = h\omega_c(\nu + \tfrac{1}{2}) \tag{3.13}$$

per unit area can be obtained as follows. First let L_x be finite and $|X| \leq L_x/2$. The number of possible μ values is $L_x L_y (2\pi l^2)$. Thus the degeneracy per area is $1/2\pi l^2 \sim B$; for each ν there are $1/2\pi l^2$ states per unit area. This is the reason why $\eta = n2\pi l^2$ is called the filling factor.

The velocity in the state Ψ_α is

$$\langle\alpha|v_x|\alpha\rangle = 0 \tag{3.14}$$

$$\langle\alpha|v_y|\alpha\rangle = \langle\alpha|\frac{\partial H}{\partial p_y}|\alpha\rangle = \frac{1}{\hbar}\frac{\partial\varepsilon_\alpha}{\partial k}$$

$$= -\frac{l^2}{\hbar}\frac{\partial\varepsilon_\alpha}{\partial X} = v_D \tag{3.15}$$

Let us turn now to the Teller model. The b.c. (3.5) correspond to infinite potential walls at $x = \pm L_x/2$ (confinement). The spectrum is

$$\varepsilon_\alpha = \varepsilon_\nu(X) + eEX \tag{3.16}$$

where $\varepsilon_\nu(X)$ is shown schematically in Fig. 5 (It cannot be given in an analytic form). As compared to the Landau spectrum (3.13) the degeneracy is lifted. In the bulk ($|X| < L_x/2-1$), however, the spectrum is, up to exponential corrections in $1/L_x$, the same as in the Landau case. Properties: $\varepsilon_\nu(-X) = \varepsilon_\nu(X)$, $\varepsilon_\nu(X) \to \infty$ for $|X| \to \infty$. The wave functions

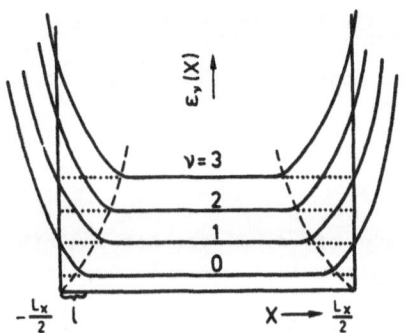

Fig. 5 The energy spectrum of free electrons in the Teller model.

$u_\alpha(x) = u_\nu(x,X)$ are the so-called Weber functions. The velocity in the state Ψ_α is now

$$\langle \alpha | v_x | \alpha \rangle = 0$$

$$\langle \alpha | v_y | \alpha \rangle = - \frac{l^2}{\hbar} \frac{\partial \epsilon_\nu(X)}{\partial X} + v_D \qquad (3.17)$$

The first term on the rhs of (3.1)) is non-vanishing at the edges ($x = \pm L_x/2$) and plays an important role in the free electron diamagnetism /2.3/.

DIFFERENT WAYS OF CALCULATING THE HALL CONDUCTIVITY

It is very instructive to compare different ways of calculating the Hall conductivity. To simplify the presentation we continue considering free electrons.

(a.) Using the Kubo formula for the Landau model

For non-interacting electrons the Kubo formula for the conductivity tensor reads

$$\sigma_{kl} = \frac{e^2}{Ar} \int_o^\infty dt \; e^{-\eta t} Tr\{\frac{1}{i\hbar}[f,r_l]v_k(t)\} \qquad (4.1)$$

where $k,l = x,y$, $f = f(H_o)$,

$$f(\epsilon) = \{\exp[(\epsilon - \xi)/k_B T] + 1\}^{-1} \qquad (4.2)$$

is the Fermi distribution with normalization

$$Tr \; f = N \qquad (4.3)$$

H_o is the free electron Hamiltonian (3.10) for zero electric field, $v_k(t) = \exp(iHt/\hbar)v_k\exp(-iHt/\hbar)$ and $\eta \to +0$. Obviously $v_x(t) + iv_y(t) = (v_x + iv_y)\exp(i\omega_c t)$. Using the cyclic invariance of the trace, $Tr(ABC) = Tr(BCA)$, (4.1) and (4.3) we get

$$\sigma_{xx} = \sigma^{\dot 0} = 0, \; \sigma_{yx} = \sigma_H^{\dot 0} = \frac{en}{B} \qquad (4.4)$$

This is identical with the classical result (cf. sec. 2).

33

(b.) Using the Kubo Formula for the Teller Model

Since in the Teller model the coordinate x is bounded we can, in (4.1), execute the time integral

$$\int_o^\infty dt \; e^{-\eta t} v_x(t) = \eta \int_o^\infty dt \; e^{-\eta t} x(t) - x \tag{4.5}$$

The second term on the rhs of (4.5) does not contribute to (4.1). The first term has only non-vanishing matrix elements on the energy shell,

$$\lim_{\eta \to +0} \eta \int_o^\infty dt \; e^{-\eta t} <\alpha|X(t)|\alpha'> = \lim \eta \int_o^\infty dt \; e^{-\eta t} e^{i(\varepsilon_\alpha - \varepsilon_{\alpha'})t/\hbar} X_{\alpha\alpha'} \tag{4.6}$$

$$= \begin{cases} 0 & \text{if } \varepsilon_\alpha \neq \varepsilon_{\alpha'} \\ X_{\alpha\alpha'} & \varepsilon_\alpha = \varepsilon_{\alpha'} \end{cases}$$

Furthermore, using $-i\hbar v_y = [H_o, y]$ we get

$$\frac{1}{i\hbar} <\alpha|[f,y]|\alpha> = -\frac{f_\alpha - f_{\alpha'}}{\varepsilon_\alpha - \varepsilon_{\alpha'}} <\alpha|v_y|\alpha'> \tag{4.7}$$

and, with (3.17) for $v_D = 0$,

$$\sigma_{yx}(= -\sigma_{xy}) = \frac{e^2}{Ar} \sum_\alpha (-\frac{\partial f_\alpha}{\partial \varepsilon_\alpha}) \frac{1}{\hbar}^2 \frac{\partial \varepsilon_\alpha}{\partial X} X \tag{4.8}$$

(Recall that $\varepsilon_\nu(X)$ has only the degeneracy $\varepsilon_\nu(X) = \varepsilon_\nu(-X)$ and $\partial \varepsilon_\alpha / \partial X$ is essentially non-vanishing only for $|X| \gtrsim L_x/2$). Using

$$\frac{1}{Ar} \sum_\alpha = \frac{1}{Ar} \sum_\nu \sum_k + \frac{1}{L_x} \sum_\nu \int \frac{dk}{2\pi} = \frac{1}{2\pi l^2 \nu} \sum_\nu \frac{1}{L_x} \int_{-\infty}^\infty dX \tag{4.9}$$

(for $L_y \to \infty$) and $\varepsilon_\nu(X) \to \infty$ for $|X| \to \infty$, and integrating by parts with respect to X we get again $\sigma_{yx} = \sigma_H^o = \frac{en}{B}$.

In the limit of zero temperatures

$$-\frac{\partial f}{\partial \varepsilon} = \delta(\varepsilon - \varepsilon_F) \tag{4.10}$$

Thus, in the Teller model in this limit, only the states at the Fermi energy ε_F contribute to the Hall conductivity. Using the formula

$$\delta(\varepsilon_\alpha - \varepsilon_F) = \sum_1 \frac{\partial(X - X_{\nu 1})}{|\partial \varepsilon_\alpha / \partial X_{\nu 1}|} \tag{4.11}$$

where $X_{\nu 1}$ are the roots of the equation $\varepsilon_\nu(X) = \varepsilon_F$ (4.8) can be rewritten as

$$\sigma_{yx} = \frac{e^2}{h} \sum_\nu \sum_1 \text{sign}(\partial \varepsilon_\alpha / \partial X_{\nu 1}) X_{\nu 1}/L_x \tag{4.12}$$

If, in the bulk, ε_F lies in the gap between the Landau levels $\varepsilon_{\nu+1}$ and $\varepsilon_{\nu+2}$ then - since $X_{\nu 1} \simeq \pm L_x/2$ -

$$\sigma_{yx} = i \frac{e^2}{h} \tag{4.13}$$

where $i = \nu + 1$. This is, of course, in agreement with $\sigma_H^0 = \frac{en}{B} = \frac{e^2}{h}$ for $\eta = i$.

(c.) Hall Current in Thermal Equilibrium

In crossed electric and magnetic fields, $f(\epsilon_\alpha)$ describes an equilibrium state. Taking the average of the current density $j_y = \frac{-e}{Ar} v_y$ in the Landau model we get, using (3.15)

$$J_y = -\frac{e}{Ar} \text{Tr}(fv_y) = -\frac{e}{Ar} \sum_\alpha \langle \alpha | v_y | \alpha \rangle$$

$$= -\frac{e}{Ar} N v_D = \frac{en}{B} E \tag{4.14}$$

which corresponds to the previous result.

In the Teller model, however, the equilibrium Hall current vanishes:

$$J_y = \frac{-e}{Ar} \sum_\alpha f_\alpha \{ v_D - \frac{l^2}{\hbar} \frac{\partial \epsilon_\nu(X)}{\partial X} \} \tag{4.15}$$

((3.17) has been used). The contribution of the first term in the curly bracket is $\frac{en}{B}$. To calculate the contribution of the second we expand f_α in powers of E and retain only the linear term,

$$f_\alpha = f(\epsilon_\nu(X)+eEX) \simeq f(\epsilon_\nu(X)) + \frac{\partial f}{\partial \epsilon_\nu(X)} eEX \tag{4.16}$$

Integrating by parts we get for the contribution in question $-\frac{en}{B}$. Thus, for the Teller model in thermal equilibrium $\sigma_{yx} = 0$. The bulk and the edge contributions to σ_{yx} compensate each other exactly.

(d.) Hall Current in a Spatially Homogeneous Sationary State

The equilibrium distribution function $f(\epsilon_\alpha)$ considered above describes an inhomogeneous density distribution (in a similar way as the barometric formula). In reality, however, the density distribution in the bulk may very well be homogeneous. Such a distribution is established if the chemical potential increases with X in the same way as the potential energy,

$$\xi(X) = \xi + eEX \tag{4.17}$$

Inserting this into $f(\epsilon_\alpha)$ the electric field drops out. Since $\epsilon_\nu(X)$ is even, the average of $\partial \epsilon_\nu(X)/\partial X$ with respect to this distribution vanishes and we get the same result ($\sigma_{yx} = \sigma_H^0 = en/B$) as for the Landau model in thermal equilibrium. Furthermore, as can easily be confirmed, the linearization of

$$J_y = \frac{e}{Ar} \sum_\alpha f(\epsilon_\nu(X)) \frac{l^2}{\hbar} \frac{\partial \epsilon_\alpha}{\partial X} \tag{4.18}$$

with respect to E leads to the Kubo formula.

As the above examples indicate, the Hall conductivity depends both on the boundary conditions and on the thermodynamic state of the system. In any case, for free electrons we obtained $\sigma_{yx} = en/B$ or $\sigma_{yx} = 0$; free electrons do not show the quantum Hall effect.

It is worth while considering the free electron Hall conductivity as

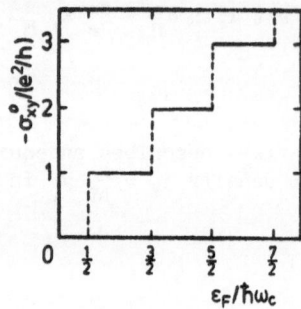

Fig. 6. Free electron Hall conductivity as a function of the
Fermi energy.

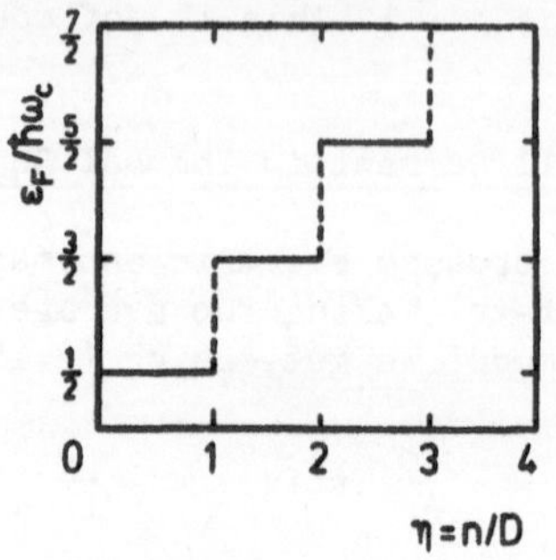

Fig. 7. Free electron Fermi energy as
a function of the electron
concentration.

a function of the Fermi energy. This is obviously a step function (cf.
(4.12) and Fig. 6.). The straight line $\sigma_H^o \sim n$ comes about because $\varepsilon_F(n)$
is also a step function (Fig. 7). The chain role gives

$$\frac{d\sigma_{yx}}{dn} = \frac{d\sigma_{yx}}{d\varepsilon_F} \frac{\partial \varepsilon_F}{\partial n} = "0 \cdot \infty" \tag{4.19}$$

In the (idealized) quantum Hall effect (Fig. 4) σ_{yx} is a step function as
a function of n (and not of ε_F!) with $d\sigma_{yx}/dn = 0$ in the plateau regimes.
As we shall see this is due to the fact that in more realistic systems at
$T = 0$ $\sigma_{yx}(\varepsilon_F)$ is a step function but $\varepsilon_F(n)$ is smooth and, therefore, the
product of the derivatives which occurs in (4.19) vanishes (in the plateau
regimes).

DISORDERED SYSTEMS

A random potential $U(x,y)$ gives rise to broadening of the Landau levels
(3.13). The average denstiy of states

$$n(\varepsilon) = \frac{1}{Ar} \langle Tr \delta(\varepsilon - \mathcal{H}) \rangle_{RP} \tag{5.1}$$

where

$$\mathcal{H} = H_o + U \tag{5.2}$$

and $\langle...\rangle_{RP}$ denotes averaging with respect to the random parameters in U,
can be calculated by using perturbation theory. A typical result is a sum
of Gaussian spectral functions

$$n(\varepsilon) = \frac{1}{2\pi l^2} \sum_\alpha \frac{1}{\sqrt{2\pi} \Gamma_\nu} \exp(-\varepsilon^2/2\Gamma_\nu^2) \tag{5.3}$$

where the band width Γ_ν depends on $B/4$. For $\Gamma_\nu \ll \hbar\omega_c$ there are quasi-gaps
between the Landau bands. (Fig. 8) In the Teller model the broadening van-
ishes near the edges and there are $2(\nu+1)$ edge states in the gap following
the νth band of bulk states (Fig. 9). When using perturbation theory to
evaluate the averaged Kubo formula (4.1) one obtains, for $T = 0$, the results
schematically given in Figs. 10 and 11 /5,6,4/. For integer filling, $\eta = i$,

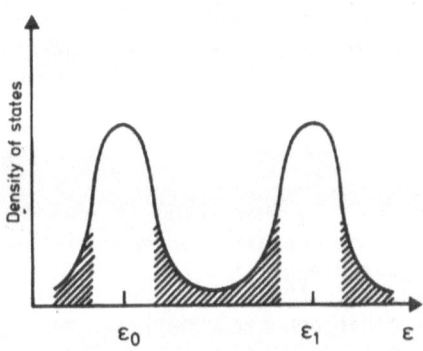

Fig. 8. Landau bands with localized
       ($\mathit{/////////}$) and delocalized
       states.

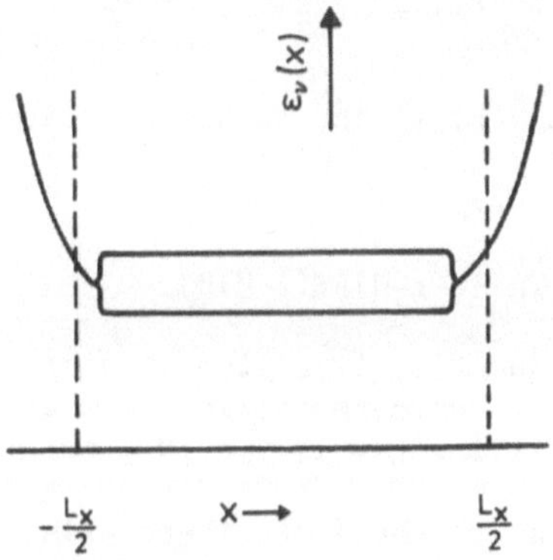

Fig. 9. Broadening in the Teller
       model.

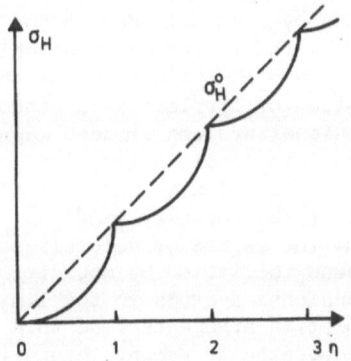

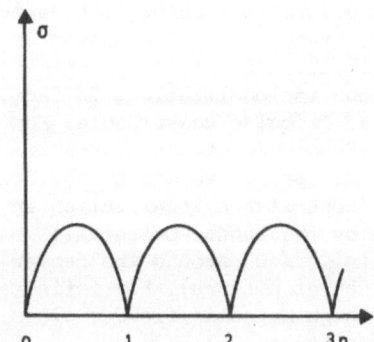

Fig. 10. Perturbative calculation
        of the Hall conductivity
        (schematic).

Fig. 11. Perutubative calculation
        of the longitudinal con-
        ductivity (schematic).

the conductivities take the (observed) values $\sigma_{xx} = 0$, $\sigma_{yx} = i\ e^2/h$ –
but no plateaux show up.

There is now a general proof which shows that, under certain conditions, these values are actually exact. The proof is based on the Středa
formulae which are exact rearrangements of the Kubo formula (4.1) for an
arbitrary additive system (independent electrons). For such systems the
Greenwood formula

$$\sigma_{kl} = \int d\epsilon\, (-\frac{\partial f}{\partial \epsilon})\sigma_{kl}(\epsilon)d\epsilon \tag{5.4}$$

holds where $\sigma_{kl}(\epsilon)$ is the conductivity at zero temperature and Fermi energy ϵ. According to Streda /7/

$$\sigma_{yx}(\varepsilon) = e(\frac{\partial N}{\partial B})_\varepsilon + \sigma_{yx}^I(\varepsilon) \qquad\qquad (5.5)$$

where $N(\varepsilon)$ is the number of states per area with energies below (integrated density of states) and both $\sigma_{yx}^I(\varepsilon)$ and $\sigma_{xx}(\varepsilon)$ are given by expressions of the type

$$Tr\ \{\delta(\varepsilon-H)A\delta(\varepsilon-H)B\} \qquad\qquad (5.6)$$

Consequently, if ε lies in a spectral gap ($\varepsilon \neq \varepsilon_g$) then both σ_{xx} and σ_{yx}^I vanish. Furthermore, if there are such gaps at energies corresponding to the filling $\eta = i$, i.e. $N = eB/h$, (5.5) yields $\sigma_{yx} = i\ e^2/h$.

As we shall see there exist several other non-perturbational ways leading to the result $\sigma_{yx} = i\ e^2/h$ for $\eta = i$: scattering theory (Section 9), the gauge argument (Sect. 7), the topological approach (Sect. 8) (these all assume $\sigma_{xx} = 0$), the $B \to \infty$ limit combined with classical percolation arguments (Sect. 6), numerical analysis /8,9/, and last but not least the field theoretical approach (Sect. 9). The last two accomplish even more: they explain the observed plateaux. Let me mention in passing that for an arbitrary density of randomly distributed point scatterers (δ-potentials) the equilibrium Hall current of an infinite system (Landau model) can be shown to be linear in the electric field and equal to the free electron Hall current if, at $T = 0$, the Fermi energy lies in a spectral gap /10/.

As pointed out above, the perturbation theory does not lead to the observed plateaux. Why? The answer is indicated by the vanishing longitudinal conductivity in the plateau regimes: The most obvious explanation for this phenomenon is localization. In fact, localization of carriers in a random potential is quite unavoidable. For two-dimensional systems at zero magnetic field and zero temperatures the theory predicts localization, i.e. $\sigma(\varepsilon_F) = 0$, for all values of the Fermi energy. At any finite but sufficiently low temperatures localization in the tails of the impurity bands (caused by the random potential) is predicted. The regime of delocalized states ($\sigma(\varepsilon_F) \neq 0$) around the center of each band is limited by mobility edges (Bányai picture). For infinite two-dimensional systems in the limit of very high magnetic fields, classical percolation arguments lead to a similar picture /11/, Fig. 9 /cf. Sect. 6). According to recent theoretical achievements /12,13/, at finite magnetic fields a bunch of states at the center of the Landau bands ($\varepsilon \simeq \varepsilon_\nu$) is always delocalized - even at zero temperature.

Although the theory of localization in strong magnetic fields is still in progress we can take for granted that a random potential gives rise to both level broadening and localization. Whereas the first effect can be treated by perturbation theory, the second - being similar to a phase transition - is beyond the range of this method. Once, however, the existence of localization is assumed, the plateaux in the Hall conductivity can easily be explained /14/. Localized states are similar to bound states which do not carry electric current. Since, in the case of an additive system, the contributions of the states to the current are independent, we can take them one by one. As long as the Fermi energy varies in a localization regime, σ_{xx} vanishes and σ_{yx} remains constant (J_y is equal to the current of all delocalized states below the Fermi energy). If there is a gap at energies corresponding to $\eta = i$ we can conclude that the plateau values of σ_{yx} are given by $i\ e^2/h$. This explains the observed plateaux since - due to the broadening of the Landau levels - ε_F is a smooth function of n and, therefore, plateaux in $\sigma_{yx}(\varepsilon_F)$ give rise to plateaux in $\sigma_{yx}(n)$. If all states were localized, $\sigma_{yx} = 0$ would follow. Since this is not the case, we may a posteriory conclude that delocalized states must exist in the transition regimes between adjacent plateaux (these regimes are, at sufficiently low temperatures, very narrow). This conclusion is consistent with the observation $\sigma_{xx} \neq 0$ in the transition regimes.

The qualitative explanation of the quantum Hall effect given above is based on the assumption of the existence of localized states in an infinite system (Landau model). In the case of the Teller model (L_x finite) delocalized edge states exist at all energies (above the ground state energy). In order to explain the quantum Hall effect in this model, it is sufficient to assume that the edge states behave in the same way as in the case of free electrons (perhaps up to corrections of the order $1^2/L_x^2$). Since the contribution of these states to the Hall current is $i\, e^2/h$ (for $\eta = i$) we may assume that all bulk states are localized (except for states near ε_ν). In the next section this picture will be confirmed in the limit of very high magnetic fields.

HIGH FIELD LIMIT

We continue to consider a single electron in a random potential,

$$\mathcal{H} = \frac{m}{2}v^2 + U(x,y) \tag{6.1}$$

For high magnetic fields it is advantageous to introduce "center and relative" coordinates defined by

$$x = X + \xi, \qquad y = Y + \eta \tag{6.2}$$

$$\xi = v_y/\omega_c, \qquad \eta = -v_x/\omega_c \tag{6.3}$$

/15/. (X,Y) and (η,ξ) are canonically conjugate variables,

$$[X,Y] = i1^2, \quad [\eta,\xi] = i1^2 \tag{6.4a,b}$$

For U = 0, classically, $|\xi(t)|$, $|\eta(t)| \leq 1$ (cf. Sect. 1). Assuming that U(x,y) is slowly varying, $|\text{grad } U\,|1 << |\,U|$, in first approximation

$$\mathcal{H} = \frac{m\omega_c^2}{2}(\xi^2 + \eta^2) + U(X,Y) \tag{6.5}$$

The eigenvalues of the part quadratic in ξ and η are determined by (6.4b) to be ε_ν(3.13). The equations of motion corresponding to the Hamiltonian

$$\mathcal{H} = \varepsilon_\nu + U(X,Y) \tag{6.6}$$

are

$$X = \frac{1^2}{\hbar}\frac{\partial U}{\partial Y}, \qquad Y = -\frac{1^2}{\hbar}\frac{\partial U}{\partial X} \tag{6.7}$$

For closed equipotential lines, U(X,Y) = const. (6.4a) can approximately be satisfied by the semiclassical quantization role

$$\oint Y dX = 2\pi 1^2 (\mu+\gamma) \tag{6.8}$$

where $\mu = 0,1,2,\ldots$, $0 \leq \gamma < 1/2$. Thus, the area between two adjacent "allowed" equipotential lines is $2\pi 1^2$. Since the density of states is $1/2\pi 1^2$ this area contains one single quantum state; an equipotential line corresponds to a single wave function.

In the limit $B \rightarrow \infty$ $[X,Y] = ih/eB \rightarrow 0$ and, therefore, X and Y can be treated as classical variables. This is a reliable first order approximation as long as $|\partial^2 U/\partial X^2|$, $|\partial^2 U/\partial Y^2| < |U|/l^2$. In the classical approximation equations (6.7) describe a one-dimensional flow along an equipotential line. In the presence of an electric field $E = (E,0,0)$ U has to be replaced by $U + e\,EX$. The distribution function f obeys the continuity equation

$$\frac{\partial \bar{f}}{\partial t} + \frac{\partial \bar{f}}{\partial X}\dot{X} + \frac{\partial \bar{f}}{\partial Y}\dot{Y} + v_D\frac{\partial \bar{f}}{\partial Y} = 0 \tag{6.9}$$

and the normalization condition

$$\sum_\nu \iint\frac{dXdY}{2\pi l^2}\,\bar{f} = N \tag{6.10}$$

(cf. (6.8)). Putting $E(t) = Ee^{nt}$, the linear response solution $f = f(H) + g$, $g \sim E$ of (6.9) is at $t = 0$

$$\bar{f} = f + eE\frac{\partial f}{\partial U}(X-\overline{X}) \tag{6.11}$$

where X is the infinite time average of $X(-t)$,

$$\overline{X} = \lim_{n\rightarrow+0} \int_0^\infty dt\; e^{-nt}X(-t) \tag{6.12}$$

$X(t) \equiv X(t,X,Y)$, $Y(t) \equiv Y(t,X,Y)$ are the solutions of (6.7) with initial conditions $X(0) = X$, $Y(0) = Y$. Calculating the average of the current density

$$j_y = \frac{-e}{Ar}(\dot{Y} + v_D) \tag{6.13}$$

$$J_y = \iint\frac{dXdY}{2\pi l^2}\,j_y \tag{6.14}$$

we assume that the system is confined in the X direction by sufficiently high walls at $|X| = L_x/2$. This confinement corresponds to vanishing wave functions at $|X| = L_x/2$, i.e. to the Teller model. Integrating by parts we obtain

$$\sigma_{yx} = -\frac{e^2}{h}\sum_\nu\frac{1}{Ar}\iint dXdY\frac{\partial f}{\partial X}\,\overline{X} \tag{6.15}$$

This is nothing else but the high field limit of the Kubo formula for σ_{yx} (cf. Sect. 4).

In the following we restrict ourselves to the zero temperature limit, $\partial f/\partial X = (\partial f/\partial U)(\partial U/\partial X)$,

$$\frac{\partial f}{\partial U} = -\partial(\epsilon_\nu + U(X,Y) - \epsilon_F) \tag{6.16}$$

in which the integrand of (6.14) is non-vanishing only at the equipotential lines defined by

$$1: \quad \epsilon_F = \epsilon_\nu + U(X,Y) \tag{6.17}$$

As long as these lines are well separated we can transform the integrals with respect to X and Y into integrals tangential and normal to the equipotential lines,

$$\int dX \int dY = \sum_1 \int ds \int \frac{dU}{|grad\ U|} \tag{6.18}$$

and get

$$\sigma_{yx} = \frac{e^2}{h} \sum_\nu \sum_1 \frac{1}{Ar} \int ds n_x(s)\bar{X}(s,\nu) \tag{6.19}$$

/16/ (cf. also /17/). Similarly

$$n = \frac{1}{2\pi l^2} \sum_\nu \sum_1 \frac{1}{Ar} \int ds n_x(s)X(x,\nu) \tag{6.20}$$

Here ds is the equipotential line element and $\underline{n} = grad\ U/|grad\ U|$ the unit vector normal to the line. Obviously, closed equipotential lines (which correspond to localized states) do not contribute to σ_{yx} since for such lines X is independent of X and

$$\oint ds \underline{n}(s) = 0 \tag{6.21}$$

But such lines do contribute to n (i.e. they can be occupied by electrons). This is exactly what we need. According to classical percolation theory open lines in the bulk (which correspond to delocalized states) can only exist at $U(X,Y) = 0$ (if large area and $\langle U(X,Y)\rangle_{rp} = 0$ is assumed). By (6.17) this means that percolation in the bulk occurs only at the center of the energy bands (the width of which is $U_{max} + |U_{min}|$) (Fig. 12). Due to the confinement, open equipotential lines exist at the edges $|X| = L_x/2$ at any value $\epsilon_F \geq \epsilon_0$ of the Fermi energy. They correspond to skipping orbits (edge currents). Since for the open equipotential lines at the edges $\bar{X} = X = \pm L_x/2$, $n_x = \pm 1$ and $\int ds = L_y$, (6.19) yields

$$\sigma_{yx} = i\frac{e^2}{h} \quad , \qquad i = \sum_\nu \theta(\epsilon_F - \epsilon_\nu) \tag{6.22}$$

i.e. the quantized plateau values. The open equipotential lines at the edges are topologically different from the closed ones in the bulk. The most impressive way to see this is to recall that a system with periodic boundary conditions on the wave function in the y direction is – with respect to connectivity – equivalent to a cylinder around the x axis (Fig. 13). The percolating lines (a) at the edges of the cylinder cannot be con-

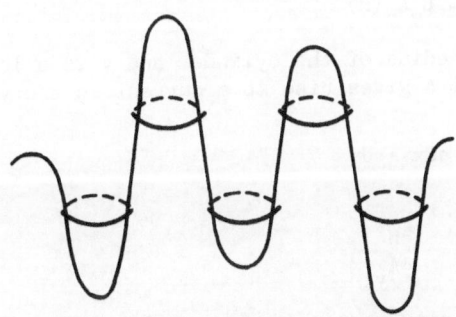

Fig. 12. Localization by a random potential
in a strong magnetic field.

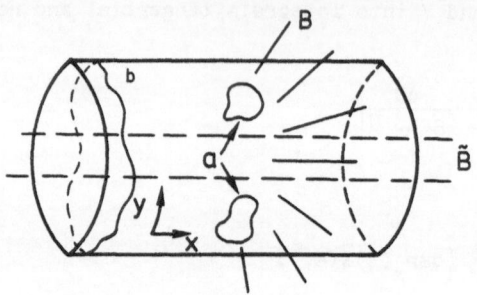

Fig. 13. Cylinder geometry with topologically
different orbits a und b.

verted with closed lines (b) by any local perturbation (topological stability of the quantization (6.22), /16/).

In the case of free electrons X is a conserved quantity and therefore,
$\bar{X} = X$. Combining (6.19) and (6.20) we get $\sigma_{xx} = \sigma_H^0 = en/B$. The formula
(6.19) is the proper generalization of the free electron expression (4.12)
for disordered systems in the limnit $B \to \infty$. Notice that in the present model in this limit $\sigma_{xx} = 0$.

THE GAUGE ARGUMENT

The universality of the quantum Hall effect suggests explaining it by
a fundamental principle rather than by a detailed transport theory. This
is the motivation behind the gauge argument formulated by Laughlin /18/.
In this argument the cylinder geometry with a radial magnetic field B
(Fig. 13) is essential from the physical point of view and not only a com-
fortable mathematical way of visualizing the periodic boundary condition on
the wave function in the y direction.

Let us introduce a homogeneous magnetic field $\tilde{\underline{B}}$ along the cylinder axis
which vanishes at the cylinder sturface (i.e. in the two-dimenensional
system under consideration). The vector potential $\tilde{\underline{A}}$ corresponding to $\tilde{\underline{B}}$ is,
on the cylinder, a pure tangential U(1) gauge field and is related to the
flux through the cylinder by

$$\tilde{\phi} = \int d\underline{f}\tilde{\underline{B}} = \oint ds\tilde{\underline{A}} = L_y \tilde{A}_y(R) \tag{7.1}$$

Here $R = L_y/2\pi$ is the radius of the cylinder and y is a local cartesian
coordinate. A change of A gives rise to a current in the y direction,

$$H = \frac{1}{2m}(p+e\underline{A}+e\tilde{\underline{A}})^2 + eEx + U \tag{7.2}$$

$$ev_y = \frac{e}{m}(p_y+eB+e\tilde{A}_y) = \frac{\partial H}{\partial \tilde{A}_y} \tag{7.3}$$

(For the sake of simplicity we consider a non-interacting system). The
equilibrium average of the current density $j_y = -(e/Ar)v_y$

$$J_y = - \frac{1}{Ar} \, Tr\{f \, \frac{\partial H}{\partial A_y}\} \tag{7.4}$$

– where f denotes the Fermi function – can be expressed in terms of the thermodynamic potential

$$\Omega = - \frac{1}{\beta} \, Tr \, \log \, \{1 + \exp \, [\beta \, (\zeta - H)]\} \tag{7.5}$$

$(\beta = 1/k_B T)$

$$J_y = - \frac{1}{Ar} \, \frac{\partial \Omega}{\partial A_y} = - \frac{1}{L_x} \, \frac{\partial \Omega}{\partial \tilde{\Phi}} \tag{7.6}$$

On the cylinder

$$\tilde{A}_y \rightarrow \tilde{A}'_y = \tilde{A}_y + \delta \tilde{A}_y = \tilde{A}_y + \delta \tilde{\Phi}/L_y \tag{7.7}$$

is a gauge transformation. The invariance of the Schrödinger equation with the Hamiltonian (7.2) requires

$$\psi \rightarrow \psi' = \exp \, (-i\delta \tilde{\Phi} y/L_x \hbar)\psi \tag{7.8}$$

Since both ψ and ψ' have to satisfy the periodic boundary condition

$$\psi(x, y + L_y) = \psi(x, y) \tag{7.9}$$

a change of the flux $\tilde{\Phi}$ by

$$\delta \tilde{\Phi} = \mu \Phi_o, \qquad \mu = 0, \pm 1, \, \ldots \tag{7.10}$$

is an exact symmetry transformation of the system. In other words: increasing $\delta \tilde{\Phi}$ from 0 to $\Phi_o = h/e$ maps back the system onto its initial state. Consequently, as Laughlin argues, the corresponding change $\Delta \Omega$ defined by averaging (7.6) over a unit flux quantum,

$$J_y = \frac{1}{\Phi_o} \int_{\tilde{\Phi}}^{\tilde{\Phi} + \Phi_o} d\tilde{\Phi}' \, J_y = - \frac{1}{L_x} \, \frac{\Delta \Omega}{\Phi_o} \tag{7.11}$$

must also be due to a symmetry transformation of the system. Furthermore, since to T = 0 ($\Omega = \epsilon$) and vanishing dissipation ($\sigma_{xx} = 0$), the only possible slow excitation process leading to a finite change of ϵ is that of shifting a number (i) of electrons along the cylinder axis by which $\Delta \epsilon = -ieEL_x$. In order to make this process periodic, Laughlin assumes L_x to be finite and requires that the number of electrons pushed out at one edge of the system reenter at the other ("pumping"). Thus, for T = 0 and vanishing dissipation

$$J_y = - \frac{(-ieEL_x)}{L_x \Phi_o} = i\frac{e^2}{h} \, E \tag{7.12}$$

For free electrons (U = 0) with $L_x \rightarrow \infty$ a change $\delta \tilde{A}_y = Ba$ of the gauge field can be compensated by the translation $x \rightarrow x - a$ which yields $\delta H \sim -eEa$, $\partial H/\partial \tilde{A}_y = -eE/B$ and, therefore

$$J_y = \frac{en}{B}E = \eta \frac{e^2}{h} \, E \tag{7.13}$$

Comparing (7.12) and (7.13) we see that in the case of free electrons the gauge argument holds only for integer filling, $\eta = i$. The reason for this is the degeneracy of the many electron ground state for any other value of the filling factor. Due to the degeneracy the transformation $\Phi \rightarrow \Phi + \Phi_0$ does not necessarily lead to the initial state. Furthermore, due to the degeneracy, an arbitrarily small amount of disorder leads to finite dissipation ($\sigma_{xx} \neq 0$) unless localization takes place. Thus, the correct requirement for the gauge argument to be applicable is that the ground state of the system be non-degenerate.

There are some problems with the gauge argument:
1. The systems investigated in experiments do not have cylinder geometry.
2. If L_x is infinite (as needed for the pumping process), the equilibrium Hall conductivity vanishes (since bulk and edge currents compensate each other exactly).
3. The pumping process is not explicitly included in the physical description of the conduction problem.
4. The adiabatic change of the flux $\overset{\lower.5ex\hbox{$\smallsmile$}}{\Phi}$ is required but not formulated in terms of a dynamic process.
5. The averaging of the current over a unit flux is not sufficiently motivated.
6. The shift and pumping back of the electrons is assumed – and not proved – to be the only possible periodic non-dissipative excitation process at $T = 0$.
7. The value of the integer i occuring in (7.12) remains indetermined (it is not related to the actual value of the filling factor).

In my opinion, the most important benefit of the gauge argument is that it has challenged many physicists to understand, and to prove – and improve upon – it.

TOPOLOGICAL APPROACH

The desire to put the gauge argument on solid ground led to a general topological approach to the integer quantum Hall effect. The formulation of this approach presented briefly in this section /19/ is essentially equivalent to the original works of Niu, Thouless and Wu /20/ and Avron and Seiler /21/.

The basic idea of the topological approach is to take into account the two current loop structure of the standard Hall measurement (cf. Fig. 2). One way to do this is to consider electrons on a torus which is a doubly-connected manifold. Another topologically equivalent possibility is to repeat a rectangular system in both directions infinitely many times. The large system has the discrete translational symmetries of a (two-dimensional) rectangular lattice with lattice constants L_x and L_y. In the presence of a perpendicular magnetic field the translations have to be assisted by appropriate gauge transformations. For rational values of the filling factor these symmetry operations commute /22/ and the well-known Bloch theorem of solid state physics applies. The wave functions are labeled by three quantum numbers: a band index n and a two-dimensional wave number vector $\underline{k} = (\theta_1/L_x, \theta_2/L_y)$, $0 \leq \theta_1 \leq 2\pi$, $1 = 1,2$

$$H\psi_\alpha = \varepsilon_\alpha \psi_\alpha \tag{8.1}$$

$$\psi_\alpha(\underline{r}) = e^{i(\theta_1 x/L_x + \theta_2 y/L_y)} u_\alpha(\underline{r}) = U(\theta) u_\alpha(\underline{r}) \tag{8.2}$$

$\alpha = (n, \underline{\theta})$, $\underline{\theta} = (\theta_1, \theta_2)$, $\underline{r} = (x,y)$. Using the notation

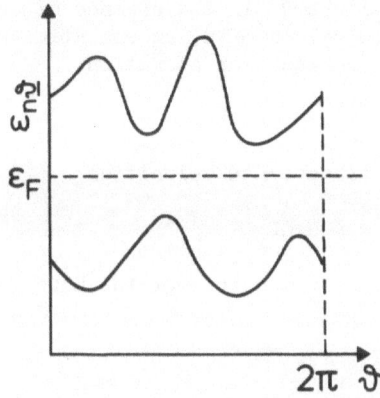

Fig. 14. Bloch bands with Fermi energy in a gap.

$$q(\underline{\theta}) = U^{-1}(\underline{\theta}) \, qU(\underline{\theta}) \qquad (8.3)$$

u_α is the solution of the eigenvalue problem

$$H(\theta) \, u_\alpha(\underline{r}) = \varepsilon_\alpha u_\alpha(\underline{r}) \qquad (8.4)$$

with periodic boundary conditions in x and y for any fixed value of θ .
According to the Bloch theorem

$$\varepsilon_\alpha = \varepsilon_{n\underline{\theta}} = \varepsilon_{n\underline{\theta}+\underline{g}} \qquad (8.5)$$

and

$$u_\alpha = u_{n\underline{\theta}} = u_{n\underline{\theta}+\underline{g}} e^{-i(g_1 x/L_x + g_2/yL_y)} \qquad (8.6)$$

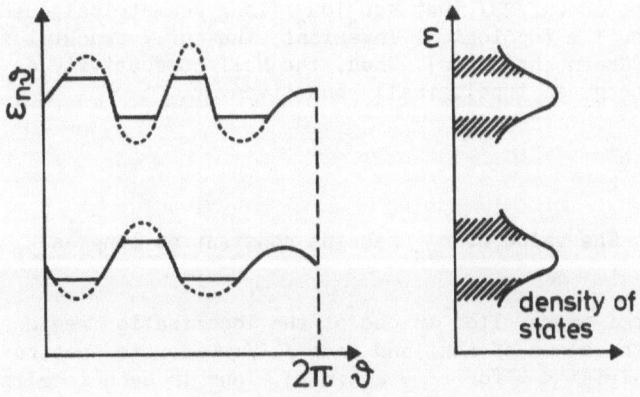

Fig. 15. Bloch bands and localized states with density
of states in the thermodynamic limit.

where $g_l = 2\pi$ integer, $l = 1,2$. In the absence of degeneracy ("generic" case) $\varepsilon_{n\underline{\theta}}$ is, for fixed values of n, an analytic function of $\underline{\theta}$ (energy bands, Fig. 14). Using the equation of motion

$$-i\hbar\underline{v} = [H; \underline{r}] \tag{8.7}$$

one easily obtains the transformed velocities

$$v_x = \partial_1 H/\hbar L_x, \quad v_y = \partial_2 H/\hbar L_y \tag{8.8}$$

($\partial_l = \partial/\partial\theta_l$) and the transformed form of the Kubo formula (4.1),

$$\sigma_{kl} = \int_0^2 \frac{d\theta_1}{2\pi} \int_0^2 \frac{d\theta_2}{2\pi} \sigma_{kl}(\underline{\theta}) \tag{8.9}$$

$$\sigma_{kl}(\underline{\theta}) = \frac{e^2}{h}2\pi i \sum_{n\neq m} \frac{(\partial_1\hat{H})_{nm}(\partial_k\hat{H})_{mn} - (\partial_1\hat{H})_{nm}(\partial_k\hat{H})_{mn}}{(\varepsilon_{n\underline{\theta}} - \varepsilon_{m\underline{\theta}})^2} f(\varepsilon_{n\underline{\theta}}) \tag{8.10}$$

$q_{mn} = \langle m\underline{\theta}|q| n\underline{\theta}\rangle$. In the present model (equivalent to a finite torus) $\sigma_{xx} = 0$.

In the limit of zero temperature

$$f(\varepsilon) = \theta(\varepsilon_F - \varepsilon) \tag{8.11}$$

and two possibilities have to be distinguished.
<u>1st</u>: The Fermi energy ε_F lies in a band gap (Fig. 15).

In this case (8.9, 10) can be rewritten as

$$\sigma_{yx} = \frac{e^2}{h} \sum_{\substack{n \\ \varepsilon_\alpha \leq \varepsilon_F}} \{\frac{i}{2\pi} \int_T \langle du_\alpha|du_\alpha\rangle\} \tag{8.12}$$

where $T = (0 \leq \theta_1 \leq 2\pi) \times (0 \leq \theta_2 \leq 2\pi)$ corresponds to the first Brillouin zone and

$$|du_\alpha| = |\partial_1 u_\alpha\rangle d\theta_1 + |\partial_2 u_\alpha\rangle d\theta_2 \tag{8.13}$$

Now it can be shown /23/ that $\langle du_\alpha|du_\alpha\rangle$ is a geometrical quantity and its integral over T a topological invariant; the curly bracktet in (8.12) is an integer (Chern character). Thus, the Hall conductivity as a function of the Fermi energy is topologically quantized,

$$\sigma_{yx} = k\frac{e^2}{h} \tag{8.14}$$

(k integer). The value of σ_{yx} remains constant as long as ε_F varies in the same gap.

<u>2nd</u>: The Fermi energy lies in one of the localization regimes Δ_j. These are defined in the sense of Aoki and Ando /14/: Discrete spectrum and $\langle\alpha |x|\beta\rangle$, $\langle\alpha|y|\beta\rangle < \infty$ for $\varepsilon_\alpha = \varepsilon_\beta$ in Δj. Thus in each localization regime

$$\langle \alpha | \underline{v} | \beta \rangle = \frac{i}{\hbar}(\varepsilon_\alpha - \varepsilon_\beta) \langle \alpha | \underline{r} | \beta \rangle = 0 \qquad (8.15)$$

for $\varepsilon_\alpha = \varepsilon_\beta$. Since

$$\langle \alpha | v_k | \alpha \rangle = \langle u_\alpha | \hat{v}_k(\theta) | u_\alpha \rangle = \frac{L_k}{\hbar} \langle u_\alpha | \partial_k \hat{H} | u_\alpha \rangle$$

$$= \frac{L_k}{\hbar} \partial_k \varepsilon_\alpha \qquad (8.16)$$

(k = x,y; 1,2), in each localization regime

$$\partial_k \varepsilon_\alpha = 0 \qquad (8.17)$$

Therefore, localized states do not contribute to (8.10). The Hall conductivity retains a quantized value (8.14) as long as ε_F varies in a localization regime. In the limit L_x, $L_y \to \infty$ the Brillouin zone shrinks to a point and the structure of the spectrum becomes as shown in Fig. 15.

The relation of the plateau values of σ_{yx} to ε_F, or respectively to the controlled parameters n or B, is - so far - entirely indetermined. If, however, the spectrum consists of Landau bands separated by (quasi-) gaps we can determine the (approximate) value of $\sigma_{yx}(\varepsilon_F)$ for the case that ε_F is in one of the gaps (and, therefore, $\sigma_{xx}(\varepsilon_F) = 0$ as needed for consistency) by using perturbation theory. The actual value of $\sigma_{yx}(\varepsilon_F)$ is then that particular exact value ke^2/h which is closest to the approximate one. Finally, to relate the results obtained to the observed plateaux, we have to assume that, in the thermodynamic limit, the Fermi energy is a continuous function of n and B.

Notice that the assumption of localization regimes which exist is absolutely indispensible in order to explain the observed quantum Hall effect by the topological approach. In fact, this effect seems to be a common product of topological quantization and localization.

A few additional remarks: Neglecting the Coulomb interaction between the electrons is not substantial. The topological quantization at T = 0 requires only the ground state to be non-degenerate and separate from the rest of the spectrum /21/. The theory can also be applied to the dynamic Hall conductivity /19/. The quantization breaks down if the energy $\hbar\omega$ corresponding to the frequency of the external field becomes comparable with the widths of the localization regimes. The role of the edge states within the topological approach is - to the best of my knowledge - not yet understood. Formally, of course, there are no edge states on a torus. They may, however, enter the picture when the torus is cut off to show that the curly bracket in (8.12) is a Chern character /21, 19/.

SCATTERING THEORETICAL APPROACH

As we saw, the Hall conductivity of real (disordered) systems at integer filling (η = i) is identical with that of free electrons. Shortly after the discovery of the quantum Hall effect Prange /24/ pointed out that this phenomenon is due to the exact compensation of the missing contribution of the localized (bound) states to the Hall current by an increased contribution of the delocalized (scattering) states in each fully occupied Landau band. The advantage of the scattering theory is to make manifest how this compensation comes about.

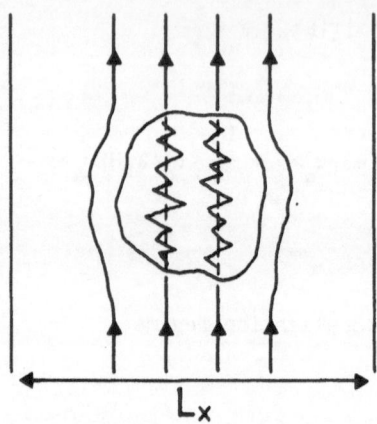

Fig. 16. Scattering by a localized object in crossed
(weak) electric and (strong) magnetic fields.

Let us consider a system with a finite scattering range in the bulk
and periodic boundary conditions in the y direction. As long as $eEL_x \ll \hbar\omega_c$
and the scattering is elastic, asymptotic energy conservation ($L_y \to \infty$,
$t \to \infty$)

$$\hbar\omega_c(\nu'+\tfrac{1}{2}) + eEX' = \hbar\omega_c(\nu+\tfrac{1}{2}) + eEX \tag{9.1}$$

with $\nu' = \nu$ requires $X' = X$. The only possible process is forward scattering
with some phase shift $\delta(k)$ of the wave function /25/ (cf. Fig. 16). Back-
ward scattering is excluded because $k \to -k$ corresponds to an energy change
$2eEL^2k$. For finite L_y the scattering gives rise to an energy shift proportional to $1/L_y$. The periodic boundary condition (7.9) requires the gauge invariant phase of the scattered wave beyond the scattering range to satisfy

$$kL_y + \delta(k) - 2\pi(\delta\Phi/\Phi_0) = 2\pi\mu \tag{9.2}$$

Changing the gauge flux $\tilde{\Phi}$ by Φ_0 is equivalent to changing μ by unity. With
the corresponding change of the energy

$$\Delta\varepsilon = \varepsilon(k_{\mu+1}) - \varepsilon(k_\mu) = v_D\hbar(k_{\mu+1}-k_\mu)$$

$$= \frac{hv_D}{L_y}\{1 - \frac{1}{2\pi}[\delta(k_{\mu+1})+\delta(k_\mu)]\} \tag{9.3}$$

the contribution of a single scattering state to the averaged current (7.11)
is $\bar{J}_y(k_\mu) = -\Delta\varepsilon/L_x\Phi_0$. Summing over the $N-N_B$ delocalized states within a
single fully occupied Landau band

$$J_y = -\frac{ev_D}{Ar}\{(N-N_B) - \frac{1}{2\pi}[\delta(k_{max}) - \delta(k_{min})]\} \tag{9.4}$$

On the other hand, according to Levinson's theorem, the total phase shift
of the scattering states is related to the number of bound states by

$$\delta(k_{myx}) - \delta(k_{min}) = -2\pi N_B \tag{9.5}$$

Therefore,

$$J_y = -\frac{ev_D N}{Ar} = -2\pi N_B \tag{9.6}$$

/26,27/. This is the desired result. The weakness of the scattering approach is that in a real system the scatterers are not concentrated in a finite area as assumed in Fig. 16. Moreover, the localization transition to $\sigma_{xx} = 0$ cannot be described by this approach.

FIELD THEORETICAL APPROACH

The most ambitious approach to the quantum Hall effect is the field theory by Levine et al. /13/: the delocalization at (or near) the Landau levels and the localization elsewhere is not assumed, nor simulated by bound states, but proved to be a (statistical) property of two-dimensional random systems in strong magnetic fields. Hall conductivity and longitudinal conductivity are investigated simultaneously. - Unfortunately, however, the theory is both conceptually and technically rather elaborate; I do not know if all assumptions are justified and all calculations correct, but the idea appears to me convincing.

A system of independent electrons in a short range random potential $U(\underline{r})$,

$$\langle U(\underline{r})U(\underline{r}')\rangle = (aU_0)^2 \, \delta(\underline{r}-\underline{r}') \tag{10.1}$$

is equivalent to a field theory (with a broken symmetry). The effective Lagrangian which describes fluctuations around the mean field behaviour is at B = 0

$$L = \sigma^{(0)} L_1 \tag{10.2}$$

where $\sigma^{(0)}$ is the mean field conductivity (obtained by perturbation theory). In 2 dimensions $\sigma^{(0)}$ is shown to be renormalized to the (exact) value $\sigma = 0$. This means localization for all energies (at T = 0). The situation is substantially different in the presence of a magnetic field. The breaking of isotropy to axial symmetry around the fields gives rise to a second term in L,

$$L = \sigma^{(0)} L_1 + \sigma_H^{(0)} L_2 \tag{10.3}$$

($\sigma^{(0)}$ and $\sigma_H^{(0)}$ are given by expression of the type (2.12)). If the boundary conditions for the fields permit edge currents, L_2 can be proven to be a topological invariant. As a consequence $\sigma_H^{(0)}$ is renormalized to $\sigma_H = 0$ modulo e^2/h, i.e. $\sigma_H = ie^2/h$. The discontinuities happen at $\varepsilon_F = \varepsilon_v$; at these "phase boundaries" σ is finite, and vanishes elsewhere. This furnishes the idealized quantum Hall effect.

CONCLUSIONS

- The integral quantum Hall effect is a localization phenomenon.
- The high accuracy is due to topological stability.
- A predictive theory of localization in strong magnetic fields is urgently needed.

Open questions:

- Mechanism in the transition regimes at T = 0.
- Role of the Coulomb interaction.
- Mechanism of transition to fractional quantum Hall effect.

Most of the work has still to be done!

REFERENCES

/1/ K. von Klitzing, G. Dorda, and M. Pepper, Phys. Rev. Lett. 45:494 (1980).
/2/ E. Teller, Z. Phys. 67:311 (1931).
/3/ M. Heuser and J. Hajdu, Z. Phys. 270:289 (1974).
/4/ R. Gerhardts, Z. Phys. B21:275, 285 (1975).
/5/ T. Ando and Y. Uemura, J. Phys. Soc. Jpn. 36:959 (1974);
    T. Ando, J. Phys. Soc. Jpn. 37:622 (1974).
/6/ T. Ando, Y. Matsumoto, and Y. Uemura, J. Phys. Soc. Jpn. 39:279 (1975).
/7/ P. Streda, J. Phys. C15:L717 (1982).
/8/ T. Ando, J. Phys. Soc. Jp. 52:1740 (1983); 53:3101, 3216 (1984);
    H. Aoki and T. Ando, Phys. Rev. Lett. 54:831 (1985).
/9/ A. MacKinnon, L. Schweitzer, and B. Kramer, Surf. Sci. 142:189 (1984);
    L. Schweitzer, B. Kramer and A. MacKinnon, J. Phys. C17:4111 (1984);
    L. Schweitzer, B. Kramer and A. MacKinnon, Z. Phys. B59:379 (1985).
/10/ J. Kosch, U. Gummich and J. Hajdu, Z. Phys. B62:295 (1986).
/11/ M. Tsukada, J. Phys. Soc. Jpn. 41:1466 (1976).
/12/ Y. Ono, J. Phys. Soc. Jpn. 51:2055, 3544 (1982); 52:2492, Suppl. 247 (1983); 53:2342 (1984).
/13/ H. Levine, S.B. Libby, and A.M.M. Pruisken, Nuclear Phys. B240:30,49, 71 (1984).
/14/ H. Aoki and T. Ando, Solid State Comm. 38:1079 (1981).
/15/ R. Kubo, H. Hasegawa, and N. Hashitsume, J. Phys. Soc. Jpn. 14:56 (1959).
/16/ S.M. Apenko and Yu. E. Lozovik, J. Phys. C18:1197 (1985); Sov. Phys. JETP 62:328 (1985).
/17/ Y. Ono, in: Anderson Localization, Y. Nagaoka and H. Fukuyama, ed., Springer, Berlin, Heidelberg, New York (1982).
/18/ R.B. Laughlin, Phys. Rev. B23:5632 (1981).
/19/ W. Pook, Diploma thesis, Köln (1986); W. Pook and J. Hajdu, to be publ.
/20/ Qian Niu, D.J. Thouless, and Yong-Shi Wu, Phys. Rev. B31:3372 (1985).
/21/ J. Avron and R. Seiler, Phys. Rev. Lett. 54:259 (1985).
/22/ J. Zak, Phys. Rev. 134:A1607 (1964).
/23/ D.J. Simms and N.M.J. Woodhouse, Lecture Notes in Physics 53 (1976).
/24/ R.E. Prange, Phys. Rev. B23:4802 (1981).
/25/ J. Chalker, J. Phys. C16:4297 (1983); Surf. Sci. 142:182 (1984).
/26/ W. Brenig, Z. Phys. B50:305 (1983); W. Brenig and K. Wysokiński, preprint (1986).
/27/ R. Joynt and R.E. Prange, Phys. Rev. B29:3303 (1984).

EXPERIMENTAL ASPECTS OF THE FRACTIONAL QUANTUM HALL EFFECT

Gregory S. Boebinger

Dept. of Physics and Francis Bitter National Magnet Lab.
Massachusetts Institute of Technology
Cambridge, Massachusetts 02139   USA

CHAPTER 1: INTRODUCTION

1.1 The Fractional Quantum Hall Effect

The fractional quantum Hall effect, FQHE, is one of the most exciting phenomena to be discovered recently in solid state physics. The FQHE is observed in high-mobility ($\mu > 10^5 cm^2/Vs$) two-dimensional systems at low temperatures (T < 2 K) and high magnetic fields (B > 5 T). The FQHE is phenomenologically similar to the integral quantum Hall effect, IQHE: Plateaus are observed in the Hall resistivity, ρ_{xy}, concomitant with minima in the diagonal resistivity, ρ_{xx} (see Figure 1.1). While the IQHE exists at magnetic fields corresponding to integral Landau-level filling, ν, the FQHE is observed at fractional Landau-level filling $\nu = p/q$ where q is always odd ($\nu = nh/eB$, n = area density, and eB/h = Landau-level degeneracy). In the T → 0 limit, for both the IQHE and the FQHE, the plateaus in ρ_{xy} are accurately quantized to $h/\nu e^2$ and the minima in ρ_{xx} reveal the presence of zero resistance states in the two-dimensional system. This low temperature behavior indicates the existence of energy gaps above the ground state of the system. While the IQHE can be phenomenologically understood through consideration of the single particle density of states, the FQHE requires a many-body approach to account for the phenomena.

Several theories have been developed assessing the nature of the electronic state underlying the FQHE. In particular, a theory due to Laughlin interprets the FQHE as the signature of an incompressible quantum fluid at fractional filling factors $\nu = p/q$ with q odd. At finite temperatures it predicts thermal excitation of fractionally charged quasi-electrons and quasiholes across and energy gap above the ground state. Although many interesting properties of the electron liquid can be deduced from the existing model, presently the only experimentally accessible quantity is the size of the energy gap associated with a given fractional state. Also, the difficult temperature and magnetic field constraints have limited research on the FQHE to magneto-transport experiments.

These lecture notes contain a review of the experimental aspects of the FQHE. Discussions of theoretical efforts will be limited to providing a context for understanding experimental results.

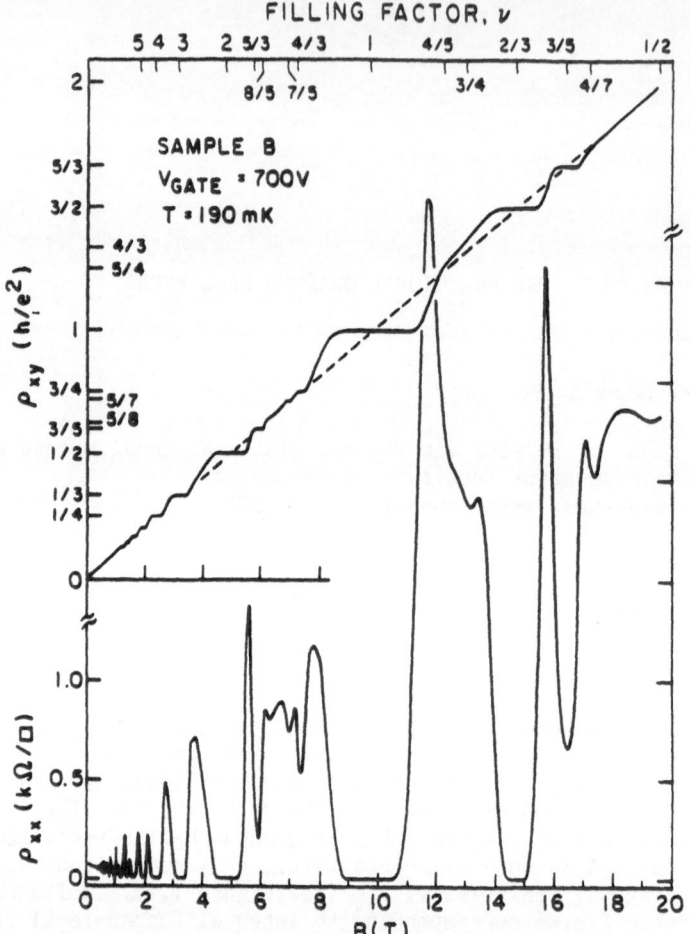

Fig. 1.1. The fractional quantum Hall effect:
(a) Hall resistivity vs magnetic field, (b) diagonal.
resistivity vs magnetic field. The Landau level filling
factor is given at the top. The FQHE consists of the
plateaus and minima observed at fractional Landau level
filling factor.

1.2 Historical Perspective

The Classical Limit. In 1879 E.H. Hall reported his famous experiment whereby the sign of the charge and the density of the charge carriers
in a conductor could be determined. A simple analysis based on the free-
electron model indicates that, in the $\omega_c \tau \ll 1$ classical limit, the
transverse (Hall) resistivity, ρ_{xy}, will depend linearly on magnetic field
and the longitudinal resistivity, ρ_{xx}, will be magnetic field independent.
[Kittel, 1976] In two-dimensional systems, these results are unchanged,
as the z-component of the current is unaffected by the magnetic field in
the z-direction.

The Quantum Limit. In the quantum limit, defined by $\omega_c \tau \gtrsim 1$, the
free-electron model breaks down as the Landau level structure of the
single electron density of states becomes resolved. This structure is
much more dramatic in two-dimensional systems, in which energy gaps exist

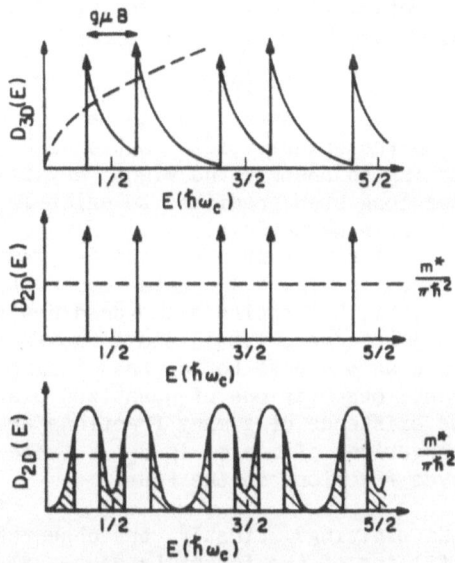

Fig. 1.2. The single electron density of states
in a magnetic field for (a) an ideal 3D system,
(b) an ideal 2D system, (c) a 2D system with
disorder. The dashed lines give the density of
states for B = 0.

in the density of states for ideal 2D systems (see Figure 1.2). The
number of Landau levels filled, ν, is proportional to n/B. For this
reason, many physical phenomena which depend upon the position of the
Fermi level, such as specific heat, magnetization, and magneto-transport,
exhibit oscillations periodic in n or B^{-1}. While studying these structures in the transport coefficients at high magnetic fields on a high-
mobility Si-MOSFET, von Klitzing discovered the IQHE. [von Klitzing, 1980]
The IQHE consists of the plateaus and minima observed at integral Landau
level filling factor (see Fig. 1.1).

The existence of broad zero resistance minima in ρ_{xx} suggests that
the Fermi energy is pinned in between Landau levels for a finite range of
magnetic field. For this to be the case, there must exist electron states
in the gaps of the density of states. In order for these mid-gap states to not
affect transport phenomena, these mid-gap states must be localized. This
might qualitatively account for the independence of ρ_{xy} on magnetic field
over a finite range. The truly startling feature of the IQHE is the precision of the quantization in units of $h/\nu e^2$. This precision allows a
high-accuracy determination of the fine-structure constant,

$$\alpha = \frac{\mu_0 c e^2}{2h} \sim \frac{1}{137}$$

where $\mu_0 = 4\pi \times 10^{-7}$ H/m is the permeability of the vacuum, c is the speed
of light in vacuum, e is the electric charge, and h is Planck's constant.
The most recent results give $\alpha^{-1} = 137.035968(23)$ which agrees well with
earlier results from the gyromagnetic ratio of the proton. [Tsui, 1982a]

According to an elegant gedanken experiment by Laughlin, the precision of the IQHE results from the exact quantization of electric charge and the existence of a mobility gap in the density of states. [Laughlin, 1981; Laughlin, 1984a]

The Extreme Quantum Limit. In 1982, Stormer and Tsui brought a new high-mobility modulation-doped GaAs/Al$_3$Ga$_7$As heterojunction which had been grown by Gossard to the Francis Bitter National Magnet Laboratory at MIT. They were attempting to observe the Wigner crystal, an electron crystal state which has long been predicted to exist in an ideal 2D electron system in very high magnetic fields. Instead, they discovered the fractional quantum Hall effect at $\nu = \frac{1}{3}$ and $\frac{2}{3}$. [Tsui, 1982b] Their original data is contained in Fig. 1.3. Later observations at lower temperatures on higher mobility samples discovered new structures at $\nu = \frac{2}{3}, \frac{2}{5}, \frac{3}{5}, \frac{4}{3}, \frac{5}{3}$, and $\frac{4}{9}$ [Stormer, 1983a] and established the precise quantization of the FQHE at $\nu = \frac{1}{3}$ to better than 1 part in 10^4. [Tsui, 1983] At present, observations of quantized plateaus in ρ_{xy} have firmly established the existence of a many fractions and many more have been suggested by observations of minima in ρ_{xx}. Table 1.1 contains the current list of observed fractions in the FQHE.

Why is the FQHE so exciting? Firstly, the observation of a quantized plateau at one-third filling of the lowest Landau level represents a new experimental observation of a fractional quantum number. Clearly, the single particle density-of-states model cannot account for the existence of structures phenomenologically similar to the IQHE at fractional filling factors, where D(E) is structure-less. It follows that the electronic state underlying the FQHE must be of many electron origin. Furthermore, a generalization of the Laughlin gedanken experiment suggests that the new quantized plateaus in the FQHE result from the transport of fractionally charged quasiparticles in the many-electron system.

1.3 Outline of the Manuscript

With an introduction to the FQHE completed, we turn attention in the next chapter to the body of theory that has been developed in the past few years. This theory is largely centered around a many-body ground state wave function first proposed by Laughlin. [Laughlin, 1983] The theoretical work will be outlined, with particular attention given to the excitations above the ground state that would be probed by activation energy experiments.

Chapter 3 is devoted to the experimental configuration required to observe the FQHE: the simultaneous realization of low temperatures, high magnetic fields, and high mobility samples.

The experimental results are presented in Chapter 4, which is divided into four sections. Section 4.1 presents the latest discoveries of new fractions in the FQHE. Section 4.2 discusses an observed competition between neighboring minima in ρ_{xx}. The data is the same as that in Boebinger, 1985a; however, the discussion has been enlarged to include the implications of this data on the qualitative scaling theory of the FQHE. Section 4.3 presents the complete activation energy data on the highest mobility samples. The discussion includes recent theoretical attempts to model the data. Section 4.4 concerns recent data and discussion on the mobility dependence of the activation energies. Theoretical attempts to quantitatively determine the effects of disorder on the FQHE are also discussed. The data from these two sections is from Boebinger, 1985b, Boebinger, 1986 and Boebinger, unpublished. Section 5.1 compares the activation energy data from other research groups to the data of sections 4.3 and 4.4. Section 5.2 concludes the lecture notes.

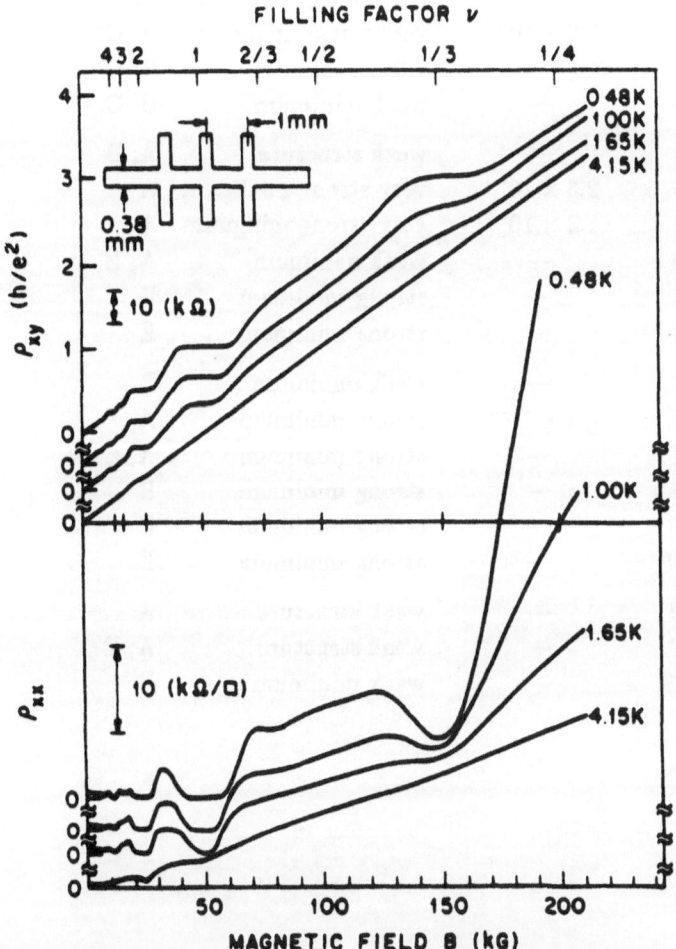

Fig. 1.3. [Original fractional quantum Hall effect measurements of Tsui, Stormer and Gossard.] ρ_{xy} and ρ_{xx} vs B, taken from a GaAs-A$\ell_{.3}$Ga$_{.7}$As sample with N = 1.23 x 10^{11}cm^{-2}, μ = 90,000 cm^2/Vs, using I = 1 μA.

Table 1.1. Observed fractions in the FQHE. The references are A. [Chang, 1984a]; B. [Mendez, 1984]; C. [Ebert, 1984]; D. [Boebinger, 1985a]; E. [Clark, 1986]; F. [Stormer, 1983a]; G. [Boebinger, unpublished].

Filling factor	Accuracy of ρ_{xy} quantization	Structure in ρ_{xx}	Refs.
1/3	3.0×10^{-5}	very strong minimum	A
2/3	3.0×10^{-5}	very strong minimum	A
4/3	9.0×10^{-4}	very strong minimum	C, D
5/3	1.0×10^{-3}	very strong minimum	A, C, D
7/3	—	weak minimum	B, C
8/3	—	weak minimum	B, C, D
1/5	—	weak structure	A, B
2/5	2.3×10^{-4}	very strong minimum	A
3/5	1.3×10^{-3}	very strong minimum	A
4/5	—	weak minimum	A, E
7/5	—	strong minimum	E
8/5	—	strong minimum	E
2/7	—	weak minimum	F
3/7	3.3×10^{-3}	strong minimum	A
4/7	—	strong minimum	A, E
9/7	—	strong minimum	E
10/7	—	strong minimum	E
11/7	—	strong minimum	E
4/9	—	weak structure	A
5/9	—	weak structure	A, E
13/9	—	weak minimum	E

CHAPTER 2: THEORETICAL BACKGROUND

2.1 The Laughlin Theory

The Laughlin Wavefunction. Shortly after the discovery of the FQHE, a study of two-dimensional systems of up to six electrons by numerical diagonalization of the Hamiltonian found that the ground state of the system is not crystalline, but liquid-like. [Yoshioka, 1983] Almost simultaneously, Laughlin proposed a many-body ground state wave function to describe the FQHE, which provides great insight into the underlying nature of the ground state. [Laughlin, 1983] Setting the cyclotron energy $\hbar\omega_c = \hbar(eB/m^*) = 1$ and the magnetic length $l_0 = (\hbar/eB)^{\frac{1}{2}} = 1$, the Laughlin ground state wave function is

$$\Psi_m = \left\{ \prod_{j<k}(z_j - z_k)^m \right\} \exp\left(-\tfrac{1}{4}\sum_l |z_l|^2 \right) ,$$

where $z_l = x_l - iy_l$ is the location of the lth electron, m is odd to preserve the antisymmetry of Ψ, and the functional form restricts the electrons to the lowest Landau level. By mapping the problem onto a classical one-component plasma, Laughlin demonstrates that the wavefunction describes the ground state at $\nu = 1/m$. He further calculates that the total energy per particle for the wave function is lower than that of a charge density wave ground state.

To generate a quasihole or quasielectron located at z_0, Laughlin employs raising and lowering operators representing the addition and removal of a single flux quantum from the system. The resulting wave functions are

$$\Psi_m^{+z_0} = \exp\left(-\frac{1}{4}\sum_l |z_l|^2\right)\left\{\pi(z_i - z_0)\right\}\left\{\prod_{j<k}(z_j - z_k)^m\right\},$$

and

$$\Psi_m^{-z_0} = \exp\left(-\frac{1}{4}\sum_l |z_l|^2\right)\left\{\pi\left(\frac{z_0}{\partial z_i} - \frac{z_0}{l_0^2}\right)\right\}\left\{\prod_{j<k}(z_j - z_k)^m\right\},$$

for the quasihole and quasielectron, respectively. The quasiparticle excitations are found to have charge $\pm e/m$. There is a finite energy gap above the ground state which is proportional to the only relevant energy scale, the inter-electron Coulombic energy, $e^2/\varepsilon l_0$. (The cyclotron and Zeeman energies are considered to be too large to be relevant.)

Laughlin's model accounts for the FQHE at $\nu = 1/q$ for all q odd (and $\nu = 1 - 1/q$ by electron-hole symmetry). He notes that the quantum fluid ground state will eventually yield to the Wigner crystal at very low filling factors. More recent calculations indicate that the FQHE will terminate at $\nu \sim \frac{1}{7}$, beyond which the Wigner crystal is calculated to have a lower ground state energy. [Levesque, 1984; Lam, 1984]

Fractional Quasiparticle Statistics. The quasiparticles in the Laughlin model, in addition to possessing fractional charge, also obey fractional statistics. The wave function at $\nu = p/q$ changes by a complex phase factor, $\exp(i\nu\pi)$, upon the interchange of two quasiparticles. The quasiparticles can also be described by wave functions obeying Bose or Fermi statistics. [Halperin, 1984; Arovas, 1984]

The Hierarchical Model. The observations of the FQHE at $\nu = p/q$ are currently understood in terms of a hierarchy of condensed ground states. [Haldane, 1983; Laughlin, 1984b; Halperin, 1984] At each level of the hierarchy, the new ground states result from the condensation of a quasiparticle gas from the preceding level. For example, at $\nu = \frac{1}{3}$ the ground state results from the condensation of a gas of electrons. As the magnetic field is decreased from $\nu = \frac{1}{3}$, quasielectrons of charge $-\frac{1}{3}e$ are formed. At $\nu = \frac{2}{5}$ this gas of quasielectrons condenses to form the new $\frac{2}{5}$ ground state. As the magnetic field continues to decrease, $-\frac{2}{5}e$ quasielectrons form, until they condense to form a new ground state at $\nu = \frac{3}{7}$. The hierarchy continues in this manner. If the magnetic field is increased from $\nu = \frac{1}{3}$, quasiholes of charge $+\frac{1}{3}e$ are formed. At $\nu = \frac{2}{7}$ this gas of quasiholes condenses to form the new ground state and so on. The first levels of the hierarchy are depicted in Figure. 2.1. This hierarchy theoretically results in the existence of a condensed state at every rational filling, $\nu = p/q$. However, the hierarchy will be terminated by the critical quasiparticle density below which the dilute gas of quasiparticles at each level of the hierarchy would be expected to form a Wigner crystal, thus precluding a Laughlin ground state at those fractional fillings. [Halperin, 1984] For this reason, for example, at $\nu = \frac{100}{299}$ one expects a Wigner crystal of $-\frac{1}{3}e$ quasiparticles, rather than a

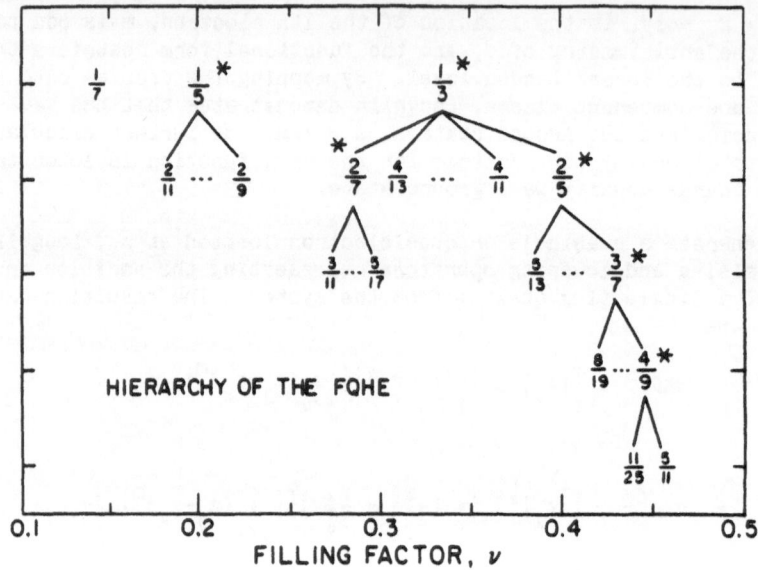

Fig. 2.1. The hierarchy of states in the FQHE. The stars indicate
the observed fractions. The '...' indicate the existence of
additional intermediate fractions.

new high order Laughlin ground state with $\pm\frac{1}{243}e$ excitations. In experiments, the actual termination of the hierarchy may well result from the
effects of disorder, as will be discussed in chapter 4.

The hierarchial picture implies that no "daughter" state can exist
unless its "parent" state exists, in accord with experimental observations. Also, if the quasiparticles in the FQHE can be approximated as
point charges, the energy gaps would be expected to scale as

$$\frac{(e^*)^2}{\varepsilon l_0^*} = \frac{(e/q)^2}{\varepsilon} \left\{\frac{(e/q)B}{h}\right\}^{\frac{1}{2}} \propto q^{-\frac{5}{2}}B^{\frac{1}{2}},$$

where q is given by $\nu = p/q$. Thus, a determination of the energy gap for
quasiparticle pair production, Δ, at $\nu = \frac{1}{3}$ would establish the magnitude
of the entire spectrum of energy gaps in the FQHE.

Quasiparticle Pair Production Energies. Within the framework of the
Laughlin theory, many calculations of the energy gap at $\nu = \frac{1}{3}$ have been
published. The energy gap, as previously stated, is given by $\Delta = Ce^2/\varepsilon l_0$,
where C is the constant of proportionality to be determined. From hypernetted-chain calculatins, Laughlin determines C = 0.056 and Chakraborty
determines C = 0.053. [Laughlin, 1984b; Chakraborty, 1985] Calculations
by Haldane and Rezayi on systems of finite numbers of electrons yield C =
0.105, when the data is extrapolated to the N → ∞ limit. [Haldane, 1985]
A single-mode approximation, in analogy with Feynman's theory of superfluid ⁴He by Girvin, MacDonald, and Platzman yields C = 0.106. [Girvin,
1985] The best estimate to date of the quasiparticle pair production
energy results from Monte Carlo calculations by Morf and Halperin, which
give C = 0.099 ± 0.009. [Morf, 1986] These results will be compared to
experimental data in chapter 4.

The Dispersion Relation of the Excitations. The quasiparticle pair creation energies discussed thus far correspond to the creation of an infinitely-separated quasiparticle pair. Laughlin considers the existence of excitons formed from the quasiparticles, finding that the interparticle distance is proportional to the wave vector of the center of mass motion. [Laughlin, 1984c] The resulting dispersion relation reveals a decrease in the energy gap above the ground state at smaller wave vectors, corresponding to the exciton states of the quasiparticle pair. Laughlin finds a minimum energy gap equivalent to $C = 0.014$ at $kl_0 \sim 0.4$; however, the approximation used is inaccurate as $kl_0 \to 0$.

More recently, two independent calculations yield a dispersion relation in which the minimum energy excitation occurs at finite wave vector, $kl_0 \sim 1.4$. Girvin et al. employ a single mode approximation, analogous to Feynman's theory for the excitation spectrum of superfluid ^{4}He. In this approximation, the dynamic structure factor is assumed to consist of a single frequency mode. This approximation is accurate at long wavelengths when any higher energy continuum modes can be neglected. [Girvin, 1985; Girvin, 1986] Haldane and Rezayi study systems of finite numbers of electrons on a sphere and find a consistency among the results for four to eight electrons. [Haldane, 1985] Their results are corroborated by recent calculations for up to nine electrons. [Fano, 1986] Figure 2.2 contains the dispersion relation for small wave vectors, showing the agreement between the two calculations. The energy gap at infinite quasiparticle separation as estimated from the Monte Carlo calculations is indicated in the figure. The "magneto-roton minimum" appears to be a precursor to the closing of the gap associated with the onset of the Wigner crystal at $\nu \sim \frac{1}{7}$. The magneto-roton minimum will be discussed further in conjunction with data in chapter 4.

The Laughlin model is the widely accepted theory of the FQHE. The ground state and excited state wavefunctions account for the various phenomena of the FQHE in a very orderly manner. The several energy gap calculations are well-behaved and have converged upon results that are believed to be very good estimates of the exact values in the Laughlin model.

2.2 The Cooperative Ring Exchange Theory

A promising alternative theoretical approach to the FQHE has very recently been developed. [Kivelson, 1986] This theory starts from a ground state underlying the FQHE which is the Wigner crystal. It is found that there are large contributions to the electron correlation energy due to electrons shifting coherently along a closed path in the crystal lattice. The contributions from large ring exchanges can be orders of magnitude larger than contributions from electron pair exchanges, due to the reduced tunneling barrier for ring exchanges. These cooperative ring exchanges are found to add in phase for certain rational fillings of the lowest Landau level, leading to energetically favorable configurations at these rational fillings. The effect is largest at $\nu = \frac{1}{3}$ and is also thought to be present at $\nu = \frac{1}{5}, \frac{2}{5}, \frac{2}{7}, \frac{1}{7}, \frac{3}{7},$ and $\frac{3}{5}$. There are finite energy gaps above the ground state corresponding to the creation of fractionally charged quasiparticle excitations. These quasiparticle excitations correspond to local dilations of the Wigner crystal and the creation energies are estimated to be $\sim 0.5 \ \nu^2 e^2/\varepsilon l_0$. At $\nu = \frac{1}{3}$, this agrees very closely with half the quasiparticle pair creation energy from the Laughlin model. The cooperative ring exchange theory is too new to fairly judge, although it holds promise in providing an alternate accounting for the FQHE. Given the presence of the large cooperative ring exchanges in the Wigner crystal, it is possible that the crystal may have melted to form a liquid-like ground state. [Halperin, 1986] Perhaps,

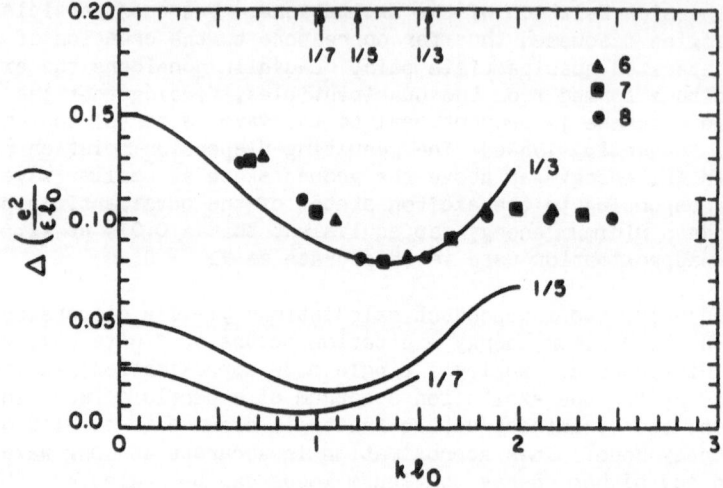

Fig. 2.2. The dispersion relation for excitations in the FQHE. The solid lines for $\nu = \frac{1}{3}$, $\frac{1}{5}$, and $\frac{1}{7}$ are from [Girvin, 1986]; the data points for 6-, 7-, and 8-electron systems at $\nu = \frac{1}{3}$ are from [Haldane, 1985]. The estimated particle pair creation energy at infinity from [Morf, 1986] is indicated by the vertical error bar at the right. The arrows at the top indicate the magnitude of the reciprocal lattice vector of the corresponding Wigner crystal.

therefore, the Laughlin model and the cooperative ring exchange theory represent two theoretical approaches to describe the same ground state.

This concludes the bulk of the discussion of the theory of the FQHE. The experimental chapters follow, in which additional aspects of the theory and experiment are compared and contrasted. The scaling theory of the FQHE is introduced in chapter 4, as are corrections of the Laughlin theory to account for higher Landau level mixing, finite thickness of two-dimensional layer, and finite electron mobility of the system.

CHAPTER 3: EXPERIMENTAL CONFIGURATIONS

As we stated in the introduction, successful experiments on the FQHE require low temperatures, high magnetic fields, and high mobility samples, each of which will be discussed individually in this chapter. The electronics configurations and experimental techniques will be discussed at the end of this chapter.

3.1 Low Temperatures

The Dilution Refrigerator. The dilution refrigerator is a Model 75 Oxford Instruments[1] dilution refrigerator with several alterations: (1) The gas handling system has been built into a single, portable cabinet (~3 ft x 4 ft x 6 ft in size) to allow the refrigerator to be set up in either the 23 T Bitter magnets or the 30 T hybrid magnet. (2) The outside diameters of the cryostat have been reduced to provide clearance for the

---

[1]Oxford Instruments, Bedford, MA 01730

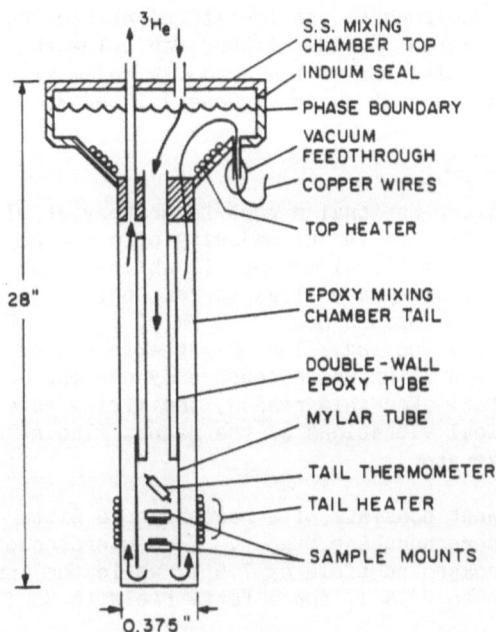

Fig. 3.1. Cross-section of the dilution
refrigerator cryostat.

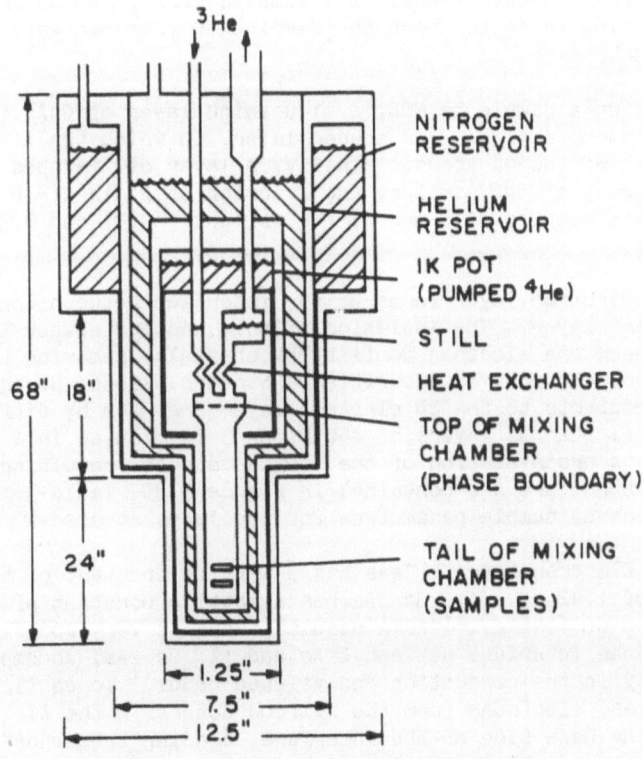

Fig. 3.2. Cross-section of the dilution
refrigerator mixing chamber.

30 T hybrid magnet. (3) The mixing chamber of the dilution refrigerator has been specially designed to provide efficient cooling of the samples, which are immersed in the ^{3}He-^{4}He mixture far below the phase boundary in the mixing chamber. Diagrams of the cryostat and mixing chamber of the dilution refrigerator are given in Figures 3.1 and 3.2.

3.2 High-field Magnets

The Francis Bitter National Magnet Laboratory at MIT has provided the high-field magnets utilized in the majority of our experiments. The 23 T Bitter magnets are conventional magnets in which copper plates and insulators are stacked to provide a helical current path. At peak field, these magnets draw 38 kA and consume nearly 10 MW of power. De-ionized water provides the necessary cooling. The DC-magnetic fields have low amplitude AC-components 6 Hz and 84 Hz which cause eddy current heating in the dilution refrigerator. For this reason, the mixing chamber is entirely epoxy. The mechanical vibrations of the magnet also result in heating in the dilution refrigerator.

The hybrid magnet consists of a reconfigured Bitter magnet insert surrounded by a superconducting magnet. The superconducting magnet provides a constant background field of 7.5 T, while the insert provides an additional field up to 22.5 T, for a total field to 30 T.

3.3 High Mobility Samples

Growth of the Samples. All of the samples were grown by molecular beam epitaxy, MBE, by which the semiconductor crystal is grown one atomic layer at a time in ultra-high vacuum. This allows excellent control in the doping profiles of the crystal, as well as exceedingly small background impurity concentrations. All samples were grown on Cr-doped GaAs substrates. The Cr doping pins the Fermi energy in mid-gap, ensuring an insulating substrate.

In growing a sample by MBE, a ~1 μ thick layer of GaAs is grown, followed by an undoped $Al_xGa_{1-x}As$ spacer layer, in which the Al atoms substitute for some of the Ga atoms. Finally, a layer of Si-doped $Al_xGa_{1-x}As$ is grown. For most of our samples, the fraction of Al is x = 0.3. The exceptions are samples of F and G, for which x = 0.37 and 0.39, respectively.

The 2D electron layer forms at the interface between the GaAs and the undoped spacer layer. The inclusion of this undoped spacer layer significantly enhances the electron mobility by spatially removing the 2D electrons from the ionized impurities. [Stormer, 1983b; Hwang, 1984] Electrical contacts to the 2D electron layer are made by diffusing indium through the $Al_{.3}Ga_{.7}As$ layers at 450°C for 5-10 minutes in a hydrogen atmosphere. A cross-section of the samples and the resulting electron energy band structure are contained in Figure 3.3. Table 3.1 contains a list of important sample parameters for specimens studied.

The 2D Electron Layer. GaAs has a lattice constant of 5.654 A and an energy gap of 1.42 eV. $Al_{.3}Ga_{.7}As$ has a lattice constant of 5.656 A and an energy gap of 1.46 eV.[2,3] As a result, there is very good lattice matching at the interface between GaAs and $Al_{.3}Ga_{.7}As$, accompanied by a discontinuity in the conduction and valence bands.[7] To equilibrate the Fermi energies, electrons from the silicon donors in the $Al_{.3}Ga_{.7}As$ migrate to the GaAs side of the interface, bending the bands[3] to form a triangular energy well at the interface. The electron confinement results

---

[2]Values for $Al_{.3}Ga_{.7}As$ are interpolated from values for GaAs and AlAs.

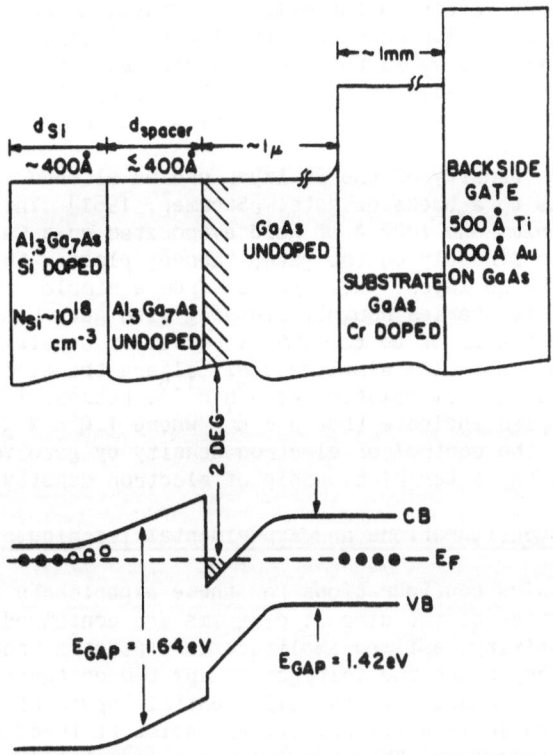

Fig. 3.3. High-mobility samples: (a) cross-section, (b) energy diagram.

Table 3.1. Sample parameters. See Fig. 3.3 for cross-section of the samples and definitions of the parameters.

Sample Names		d_{Si} ($\mathring{A}$)	N_{Si} ($10^{18}cm^{-3}$)	d_{spacer} ($\mathring{A}$)	n_{2D} ($10^{11}cm^{-2}$)	μ_0 ($10^3\frac{cm^2}{Vs}$)
A	TO7060#2	400	·2.0	370	1.5	600
B	TO7060#6	400	2.0	370	1.5	600
C	TO7060 K	400	2.0	370	1.6	600
D	TO7062 S	400	2.0	275	2.5	300
E	TO7060#7	400	2.0	370	1.5	600
F	1368 Z38	360	0.8	300	2.7	1100
G	1401 Z81	290	0.9	370	1.5	500
H	D489 A	800	1.0	425	1.9	79
I	D489 B	800	1.0	425	2.0	72
J	CO7030	—	—	170	2.5	140
K	TO7060#8	400	2.0	370	1.5	600
L	TO7060 α	400	2.0	370	1.5	600
M	TO7062 α	400	2.0	275	2.5	300

in quantized energy levels in the well. A quantum mechanically two-dimensional system results when only the lowest electron level is populated and the thermal energy is too low to excite electrons to higher levels. Calculations yield an energy level spacing of ~10 meV and an electron confinement to a region ~100 Å thick. [Stern, 1984; Hurkx, 1985]

The electron density of the 2D layer can be altered via application of a voltage bias to a backside gate. [Stormer, 1981] The gates consist of 500 Å of titanium and 1000 Å of gold evaporated on a GaAs wafer. The sample is mounted directly on the gate, thereby placing the gate roughly 1 mm away from the 2D layer. As expected from a simple capacitor model, the electron density varies roughly linearly with gate bias for biases up to ±700 V. Gate biases of up to ±1000 V result in density increases as high as 80%. The gate bias simultaneously alters the electron mobility according to an empirical relation of $\mu \propto n^{1.5}$. [Chang, 1983] Measurements on our samples indicate that $\mu \propto n^{\gamma}$, where $1.0 \leq \gamma \leq 1.7$. Figure 3.4 shows the control of electron density by gate voltage. Figure 3.5 shows the interrelationship of electron density and mobility.

3.4 Electronics Configurations and Experimental Techniques

The electronics configurations for these experiments are straight-forward. Schematics of the circuit diagrams are contained in Figure 3.6. To measure resistivity, a fixed amplitude current at a known frequency is applied to the sample and the voltages at any two contacts on a Hall bridge sample are connected to the differential inputs of the PAR Model 124 lock-in.[3] The lock-in filters out all noise at frequencies other than the signal frequency, thus providing very sensitive signal detection. The output of the lock-in is plotted versus magnetic field, yielding plots like those of Figure 1.1. To measure conductivity, a fixed amplitude voltage at a known frequency is applied to a central contact on the sample and the voltage across a current-sensing resistor is input to the lock in, yielding traces such as those in Figure 3.7. The frequencies used are between 2 Hz and 23 Hz typically. The lower frequencies experience fewer problems with phase shifts for small signals, but require longer time constants and are generally noisier.

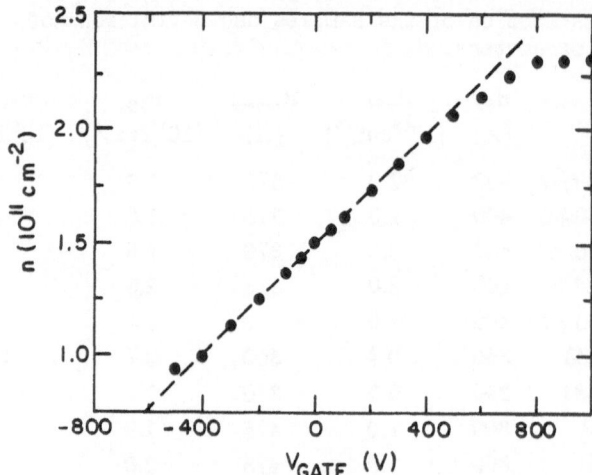

Fig. 3.4. Electron density controlled by gate voltage.

[3]Princeton Applied Research, Inc., Princeton, NJ 08540.

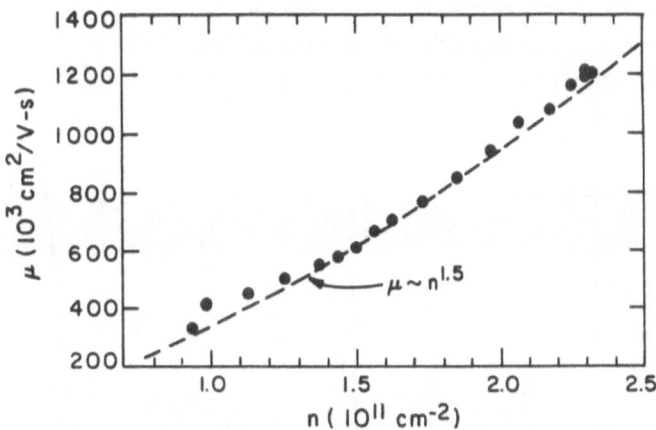

Fig. 3.5. Interrelationship of electron density and mobility.

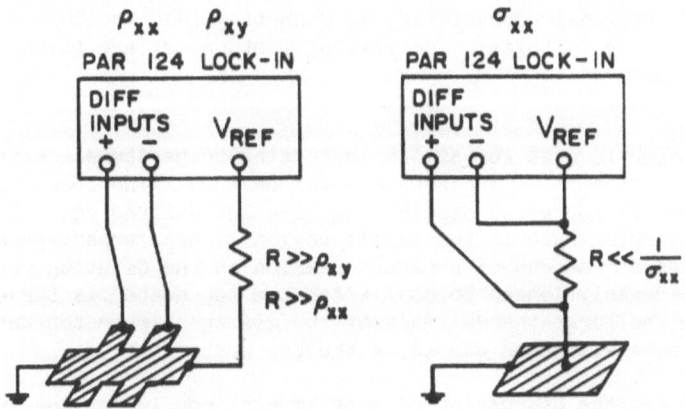

Fig. 3.6. Circuit diagrams for measuring (a) resistivity, and (b) conductivity. The conditions for R ensure that a constant current (voltage) is applied to the sample for measuring resistivity (conductivity).

Many experiments, including all of the activation energy experiments, require the variation of temperature by means of one of the two mixing chamber heaters (see Fig. 3.2). The tail heater is the primary heater for the rapid sample heating and cooling that it allows. For example, by applying 100 μW to the tail heater for two minutes, the samples will warm to 1.5 K and then return to 100 mK in as little as 10 min. The cool-down rate can be controlled by increasing the length of time that the tail heater is on. The best results for activation energy temperature sweeps occur by leaving 80 μW on the tail heater for 12-25 min. The samples then take 60-120 minutes to cool from 1.2 K to 100 mK. Because of the long distance between the phase boundary, where all cooling takes place, and the tail region, temperature gradients between the tail thermometer and the samples are not a problem provided the tail heater is off.

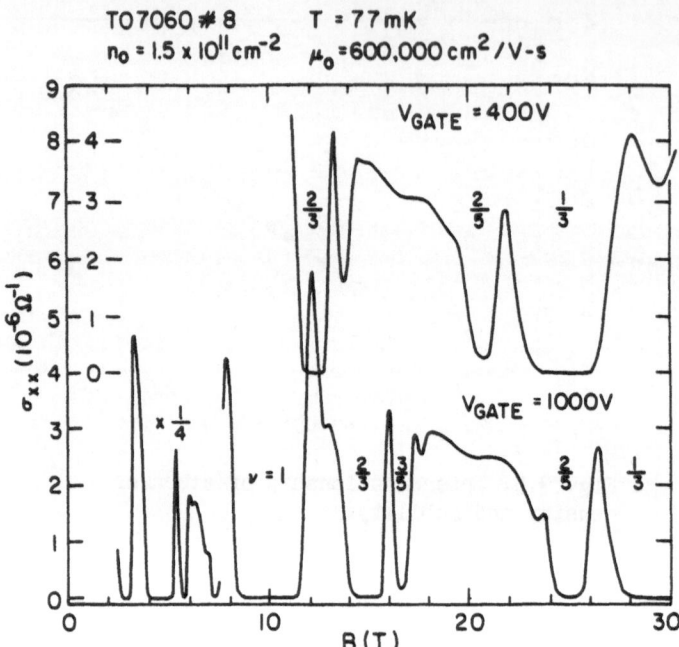

TO7060 #8   T = 77mK
$n_0 = 1.5 \times 10^{11} cm^{-2}$ $\mu_0 = 600,000 \, cm^2/V\text{-}s$

Fig. 3.7. Diagonal conductivity vs magnetic field for two different gate voltages. The filling factors, ν, are given in the diagram.

The top heater is used for smooth increasing-temperature sweeps or accurate stabilization at an intermediate temperature. The power applied to the top heater is increased smoothly to give the desired equilibrium temperature or warming rate in the sample region. These experiments are usually performed at the end of a magnet session as the dilution refrigerator takes considerably longer to cool after the top heater is turned off. This is because the large thermal mass of the mixing chamber top and the heat exchanger have also been warmed by the top heater.

This concludes the discussion of experimental configurations and techniques. In the next chapter, the experimental results will be presented and discussed.

CHAPTER 4: EXPERIMENTAL RESULTS

4.1 New Fractions in the FQHE

After the initial discovery of the FQHE (see Fig. 1.3) at filling factors $\nu = \frac{1}{3}$ and $\frac{2}{3}$, later research revealed additional fractions (see Table 1.1). The latest developments in discovering new fractions are given in Figures 4.1-4.3.

Figures 4.1 and 4.2 contain data from Clark et al. [1986]. The spectacular array of new fractions results from the enhancement of the sample mobilities after illumination by a red L.E.D. The most startling feature of the data is the suggestion of even denominator fractions in the range $2 < \nu < 4$ for which the second Landau level is partially filled.

Interest in high magnetic field limit centers on attempts to observe new fractions of the form $\nu = 1/q$ (q odd), which correspond to the other fundamental Laughlin ground states. Weak features $\nu = \frac{1}{5}$ have been previously reported. [Mendez, 1984; Chang, 1984a] Very recently, however, a strongly developed minimum at $\nu = \frac{1}{3}$ has been observed, as shown in Figure 4.3 [Boebinger, unpublished] The next natural goal is the $\nu = \frac{1}{5}$ minimum; however, the FQHE is expected to be terminated by the onset of Wigner crystallization at $\nu \sim \frac{1}{7}$. [Levesque, 1984; Lam, 1984]

4.2 Competition between Neighboring Minima

Introduction. The early theoretical models of the FQHE are based on an ideal 2D electron system in the absence of disorder and at zero temperature. The actual 2D heterojunction systems, however, are far from ideal model systems. They contain appreciable random potential fluctuations which are expected to have a considerable effect on this novel electronic state. In fact, the observation of a quantized Hall plateau is attributed to the presence of potential fluctuations, which are able to localize excess carriers or quasiparticles at filling factors near a given integral or fractional filling factor. Excessive randomness, however, will destroy the coherence of the ground state, roughly when the strength of the potential fluctuations becomes comparable to the size of the associated quasiparticle gap. Temperature has a similarly destructive influence as kT approaches the gap value.

Several aspects of these phenomena have been observed: (a) The observed temperature dependence of the minima in ρ_{xx} and σ_{xx} suggests quasiparticle localization at the lowest experimental temperatures (see Section 4.2). (2) Earlier publications have revealed a mobility-dependent competition between the IQHE and the FQHE. It is observed that the FQHE requires 2D systems of much higher mobility ($\sim 10^5 \mathrm{cm}^2/\mathrm{Vs}$, compared to $\sim 10^4/\mathrm{cm}^2/\mathrm{Vs}$ for the IQHE). Lower mobility samples exhibit exclusively the IQHE, even at a filling factor of $\nu = \frac{2}{3}$, for example. [Paalanen, 1982; Paalanen, 1984] (3) The FQHE is destroyed by temperatures above ~ 2 K. These phenomena indicate the complexity of the FQHE at finite temperature in the presence of the potential fluctuation of real systems. In this section, we present additional data which indicates the intricacy of the interrelationship of temperature and disorder.

Early work on the FQHE concentrated on the extreme quantum limit ($\nu < 1$) where the higher magnetic fields facilitate observations. For this reason, we first present several sets of data on $\nu < 1$ which illustrate the "normal" development of the FQHE as temperature is decreased. The data were taken on sample B (see Table 3.1) during a single temperature sweep. Figure 4.4 shows three traces of ρ_{xx} versus magnetic field. Note that as the temperature is decreased, the minima at $\nu = 1$, $\frac{2}{3}$, and $\frac{2}{5}$ become better developed. Also, note that the local maximum between the neighboring minima at $\nu = \frac{2}{3}$ and $\frac{2}{5}$ increases as the temperature is decreased, even for the case of a very weakly developed $\frac{2}{5}$ minimum. This behavior has been observed for the weakly developed $\frac{4}{5}$, $\frac{3}{7}$, and $\frac{4}{9}$ minima as well.

Observations at $\nu > 1$. We now turn attention to some particularly interesting data on the FQHE taken at $\nu > 1$. Figure 4.5 shows experimental traces of ρ_{xy} and ρ_{xx} as functions of magnetic field. Note that the magnetic field axes have been scaled so that the filling factors of all traces align. All data were taken on specimen B during a single seven-hour experimental session. Figure 4.5 (a) shows the plateaus in the Hall resistance at $\nu = \frac{4}{3}$ and $\frac{5}{3}$. The Hall resistance at $\nu = \frac{4}{3}$ is quantized to $\rho_{xy} = 3h/4e^2$ to an accuracy of 3 parts in 10^3, while at $\nu = \frac{5}{3}$ it is quantized to $\rho_{xy} = 3h/5e^2$ to an accuracy of 4 parts in 10^3. Both values remain constant within this accuracy over a range of about 2 kG. These results confirm an earlier report on the accuracy of fractionally quantized levels at filling factors $\nu > 1$. [Ebert, 1984] Figures 4.5(b)-(d)

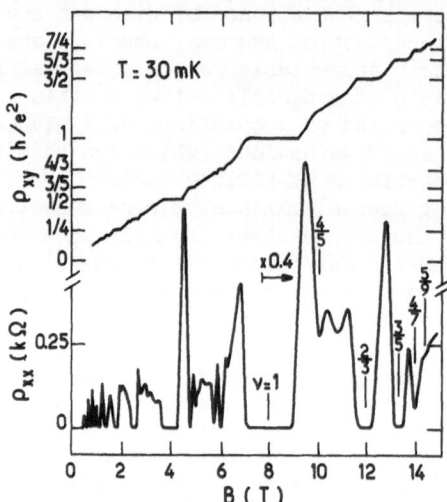

Fig. 4.1. The resistivity and Hall voltage (upper trace) at 30 mK. The fractional occupancies for $\nu < 1$ are indicated on the trace. |Clark, 1986|.

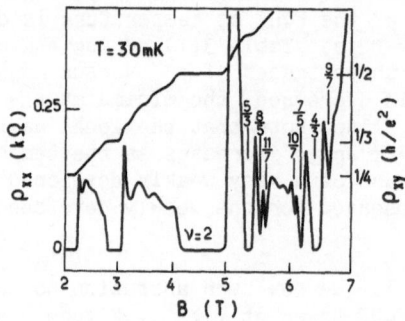

Fig. 4.2. Low-field resistivity and Hall traces. |Clark, 1986|.

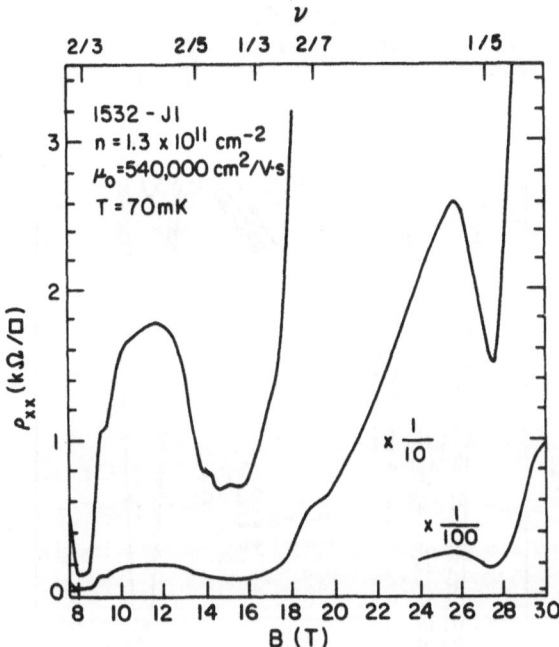

Fig. 4.3. Recent observation of well-developed minimum at $\nu = \frac{1}{5}$.

show the minima in the diagonal resistivity ρ_{xx} at $\nu = \frac{4}{5}$, $\frac{5}{3}$, $\frac{7}{5}$, and $\frac{8}{5}$. The data of Fig. 4.5(b) were taken without backside gate bias. In Figs. 4.5(c) and 4.5(d) the application of a backside gate bias has resulted in an increased electron density. The minima are shifted towards higher fields where they are better developed at comparable temperatures. At a fixed density, the minima become stronger as the temperature is lowered. This behavior is identical to that exhibited by the FQHE for $\nu < 1$. [Stormer, 1983a; Chang, 1984a]

An unexpected feature of the data in Fig. 4.5 is the influence of the $\frac{4}{3}$ minimum on the neighboring $\frac{7}{5}$ minimum. This is apparent in Figs. 4.5(c) and 4.5(d), where the $\frac{7}{5}$ minimum seems to be disappearing into the shoulder of the $\frac{4}{3}$ minimum as the temperature is decreased. This contrasts with the previously discussed behavior of the neighboring $\frac{2}{3}$ and $\frac{4}{5}$ minima (see Fig. 4.4). Figure 4.6 contains five experimental traces of ρ_{xx} in the region of $\nu > 1$ at temperatures ranging from 100 mK - 1.7 K. The traces are displaced in the vertical direction for clarity. Here the competition between neighboring minima is observed for two different pairs: (1) With decreasing temperatures in the $\frac{7}{5}$ minimum disappears into the shoulder of the $\frac{4}{3}$ minimum. (2) The minimum at $\frac{8}{3}$ is only visible at temperatures above ~700 mK. At temperatures below ~700 mK, the $\frac{8}{3}$ minimum disappears completely into the shoulder of the broad $\nu = 3$ minimum. There appears to exist a competition between neighboring minimum in which the stronger minimum suppresses the weaker one as temperature is decreased. A similar mobility-dependent competition between minima and plateaus associated with the FQHE and the IQHE has been reported previously. [Paalanen, 1982; Paalanen, 1984] Only high mobility samples, comparable to those used in our experiments, show the FQHE. Lower mobility samples show broad

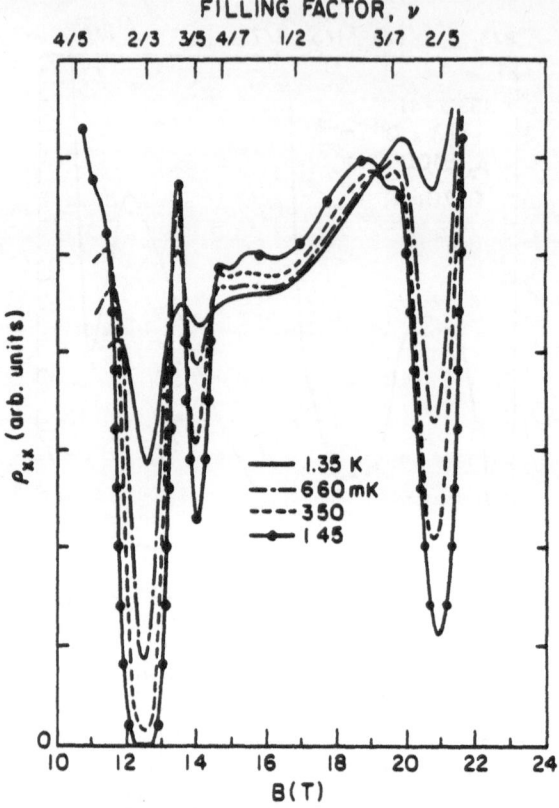

Fig. 4.4. "Normal" development of minima with decreasing temperature: ρ_{xx} versus magnetic field at $\frac{2}{5} < \nu < \frac{4}{5}$.

features at integral ν which suppress the additional structure associated with the FQHE. Our data suggest that such competition does not only exist between the FQHE and the IQHE, but also between neighboring minima possibly belonging to different levels of the hierarchial model of the FQHE. Furthermore, our results suggest that the competition is not solely characterized by the mobility of the samples. There is clearly a peculiar temperature dependence to the competition whereby, with decreasing temperature, the number of minima developed in ρ_{xx} can first increase (Fig. 4.6) and then decrease. The possibility of this type of behavior has not been explicitly dealt with in the existing theories on the FQHE.

Apparently, a given FQHE minimum exists only over a finite temperature range. The upper limit is determined by the thermal energy required for quasiparticle excitations across the energy gap. The lower limit is set by a threshold temperature below which localization precludes correlation between particles to form the appropriate condensed state.

We do not have an explicit model for this behavior, nor does the present literature provide one. However, one might want to speculate as to the origin of our observations following the lines of thought of the hierarchial model. Uncorrelated electrons, which exhibit the IQHE, form

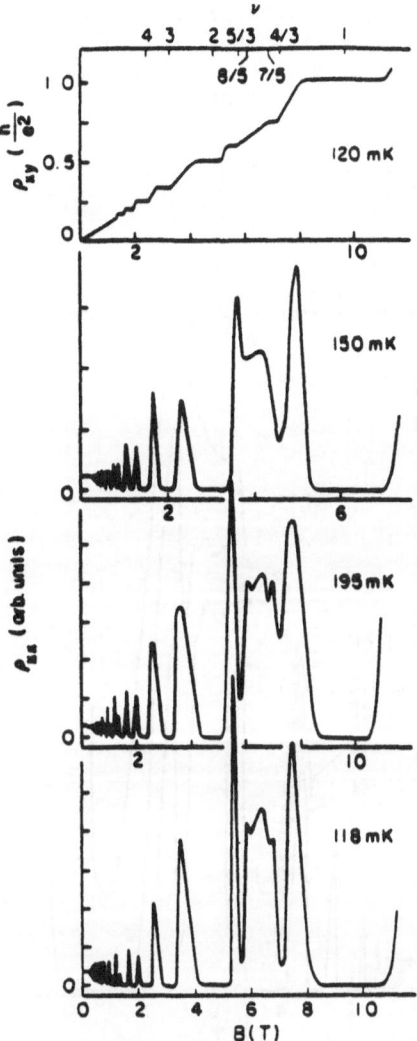

Fig. 4.5. Observations at $\nu > 1$: ρ_{xy} and ρ_{xx} versus magnetic field. The temperature for each trace is given in the figure.

the basis of this hierarchy. At rational filling factor $\nu = 1/q$ (and $1 - 1/q$, the electron-hole symmetric state for $B \to \infty$) electron correlation favors the formation of a condensed state. Deviation from these primitive filling factors leads to the creation of quasiparticles or quasiholes on this condensate. At given rational fractions $\nu = p/q$, the motion of these quasiparticles becomes correlated and forms a new ground state on the first level of the hierarchy. Higher levels are derived analogously from correlation of the quasiparticles of the next lower level ground state. In particular, the $\frac{2}{5}$ state is a daughter state of the $\frac{1}{3}$ state and, hence, the $\frac{7}{5}$ state derives from the $\frac{4}{3}$ state. From a more general point of view, the $\frac{8}{3}$ can be regarded as a daughter of the $\nu = 3$ integral state. Hence, our experimental findings might indicate a general scheme whereby daughter states are successively destroyed by their parental states as the temperature is lowered.

Which state will survive for $T \to 0$ probably depends on the strength of the random potential fluctuations as compared to the gap energy of the fractional state in question. For example, as the filling factor increases from exactly $\nu = \frac{4}{3}$, an increasing number of quasiparticles are created on the $\nu = \frac{4}{3}$ state. They initially localize, leading to the plateau in ρ_{xy} and the wide stretch of zero resistance in ρ_{xx}. As the system drifts further off $\nu = \frac{4}{3}$, at finite temperature, additional quasi-

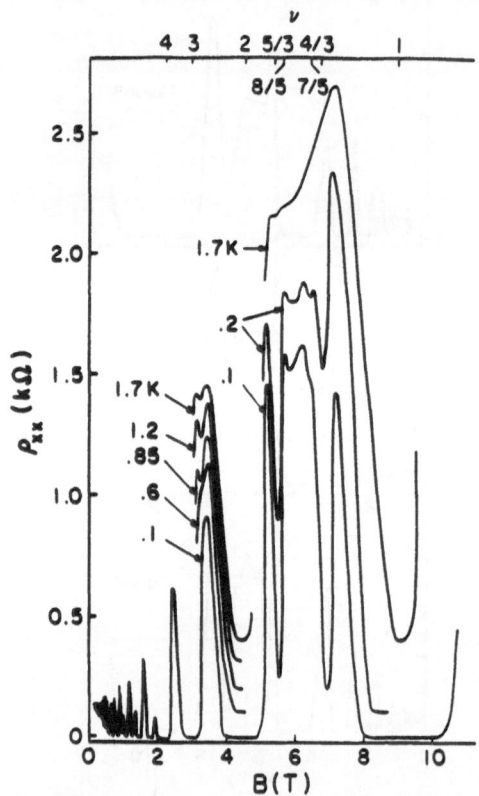

Fig. 4.6. Competition between neighboring minima: ρ_{xx} versus magnetic field for temperatures ranging from 100 mK to 1.7 K.

particles can no longer be localized and the transport coefficients deviate from their ideal value. In the vicinity of $\nu = \frac{7}{5}$, the quasiparticles of the $\frac{4}{3}$ state can be favorably correlated forming a new ground state visible as a dip in ρ_{xx}. At lower temperature and/or stronger potential fluctuations, quasiparticles of the $\frac{4}{3}$ state are localized out to $\nu = \frac{7}{5}$ and beyond, preventing the formation of the $\frac{7}{5}$ ground state. Temperature, therefore, seems to play a complex role in the FQHE: It brings into existence daughter states by delocalizing the quasiparticles of the parent state, thereby allowing for quasiparticle correlation, and it destroys the daughter state once kT surpasses the characteristic quasiparticle gap energy.

Scaling Theory of the FQHE. Currently, attempts are being made to develop a theoretical understanding of transport in the IQHE and FQHE under realistic experimental conditions. An empirical scaling model which incorporates white-noise disorder has emerged. [Laughlin, 1985; Laughlin, 1986] It related the diagonal and off-diagonal components of the conductivity tensor (see Figure 4.7). The inelastic scattering length is the relevant length scale and will vary as a function of electron mobility and temperature. In a typical experimental situation, the electron mobility is fixed. It is plausible to assume that the scattering length is increased monotonically as the temperature is decreased, in which case the flow lines of Figure 4.7 correspond to decreasing temperature. Then, for a low mobility sample which only exhibits the IQHE, the "high tempera-ture" limit is represented by the dashed line marked "A" in Figure 4.7. As temperature is decreased, the transport coefficients change according to the flowlines in the diagrams. At T = 0, such a sample will exhibit zero conductivity states in σ_{xx} and quantized plateaus in σ_{xy} for integral filling factors.

Note that the scaling theory contains two types of fixed points, denoted as circles and squares in Figure 4.7. The circles represent moving the Fermi level through a narrow extended-state band in the quasi-particle density of states. The squares represent the closing of a mobility gap in the quasiparticle density of states.

There are some obvious limitations to the scaling theory: There is no quantitative recipe for the incorporation of temperature into the theory. Additionally, the inability of the Laughlin ground state to screen poten-tial fluctuations implies the existence of non-white-noise disorder. In the spirit of these limitations, our experimental discussion assumes that ρ_{xy} is well-behaved, that is, is linear in magnetic field and forms plateaus at low temperatures corresponding to the well-developed minima in ρ_{xx}. Then, in the minima of ρ_{xx},

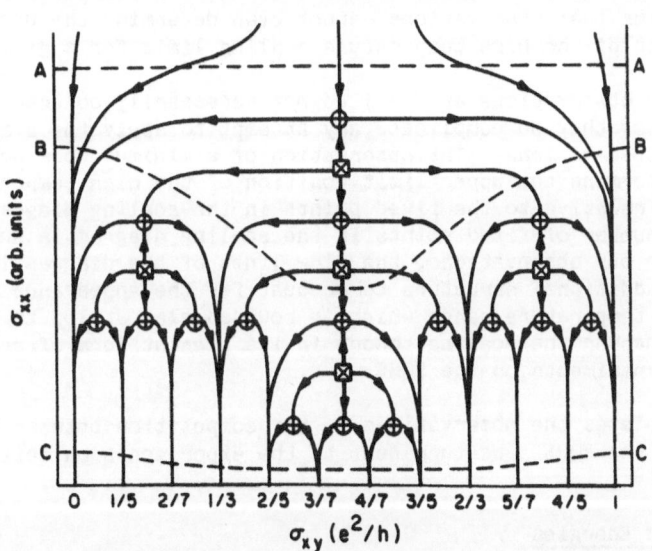

Fig. 4.7. Scaling diagram for the FQHE for $0 < \nu < 1$. The units of conductivity are e^2/h. The diagram is the same for $1 < \nu < 2$. The three dashed lines are discussed in the text.

$$\sigma_{xx} = \frac{\rho_{xx}}{\rho_{xx}^2 + \rho_{xy}^2} \sim \frac{\rho_{xx}}{\rho_{xy}^2} \propto \rho_{xx}$$

and we can speak of ρ_{xx} and σ_{xx} interchangeably.

The "normal" temperature-dependent behavior of the FQHE, illustrated in Fig. 4.4, is incorporated into the scaling model. In particular, the upward flowing lines in Figure. 4.7 correspond to the local maxima which increase with decreasing temperature. The data of Figure 4.4 suggest that the "high temperature" scaling limit for this sample may be given by the dashed line labelled "B" in Figure 4.7. Then, according to the scaling picture, at T = 0, this sample will certainly exhibit zero-conductivity states in σ_{xx} and quantized plateaus in σ_{xy} for ν = 1, $\frac{1}{3}$, $\frac{2}{3}$, $\frac{4}{3}$, and $\frac{1}{3}$. Note that this sample may be properly represented by the dashed line labelled "C," in which case all of the fractions in Figure 4.7 will be observed at zero temperature, several of which will first appear at temperatures below our experimental range. Observations at finite temperature can necessarily only determine an upper limit to the position of the high temperature scaling limit for a given sample.

We now turn our attention to the data taken at ν > 1 which appears to complicate the simple behavior described in the scaling diagram. In light of the behavior observed in Figure 4.4 on the same sample on minima which are on the same levels of the hierarchy, the data taken at T ~ 0.2 K in Figs. 4.5 and 4.6 suggest that the minimum at ν = $\frac{2}{3}$ will continue to develop with decreasing temperature. In this case, the dashed line labelled "B" applies to this sample for 1 < ν < 2, as well. The subsequent disappearance of $\frac{2}{3}$ minimum with decreasing temperature reveals that, in fact, the dashed line "B" is incorrect. The dashed line, which is shown passing immediately below a mobility-gap-closing fixed point (denoted by a square), must pass above this point to describe the observed behavior. The appearance and subsequent disappearance of the $\frac{4}{3}$ minimum in Figure 4.6 represents a similar ambiguity regarding the position of a mobility-gap-closing fixed point relative to the high temperature scaling limit. It seems that observations cannot even determine the upper limit to the position of the high temperature scaling limit for a given sample.

While the observations at ν > 1 do not necessarily contradict the scaling diagram, they do complicate any attempt to apply the diagram to experimental observations. The observation of a minimum does not unerringly determine the upper limit position of the high temperature scaling limit relative to the fixed points in the scaling diagram. While the existing number of fixed points in the scaling diagram is sufficient to account for our observations, the flow lines of the diagram in Fig. 4.7 must acquire additional structure to account for the appearance of minima over a finite temperature range which is bounded also at low temperatures. One wonders whether the scaling theory in its present form offers anything relevant to experiments on the FQHE.

This completes the observations on the competition between neighboring minima in the FQHE. We turn next to the experiments on activation energies.

4.3 Activation Energies

Temperature Dependence of the Thirds. The temperature dependences of ρ_{xx} and σ_{xx} have been previously interpreted as activation energies in the FQHE. [Chang, 1983, 1984a, 1984b; Tsui, 1983; Ebert, 1984] The value of ρ_{xx} or σ_{xx} at the minimum corresponding to a given fractional factor is determined as a function of temperature from ~1.5 K to ~100 mK. Figure 4.8 shows such graphs for $\nu = \frac{2}{3}$ at two different magnetic fields.

The data of Fig. 4.8(a) follow a straight line, indicating activated conduction. The activation energy, $\Delta/2$, is determined from

$$\rho_{xx}(T) = \rho_0 \exp(-\Delta/2T) \text{ and } \sigma_{xx}(T) = \sigma_0 \exp(-\Delta/2T).$$

As defined here, Δ represents the quasiparticle pair-creation energy. (Note that this definition of Δ differs by a factor of two from the Δ's defined in previous experimental work.) We choose this definition because the lowest energy excitation which can clearly be observed by magneto-transport experiments is the pair-creation of charged quasiparticles. In particular, it is not clear that the neutral excitation at the roton mimi-mum could be observed.

All data taken at magnetic fields between 6 and ~10 T indicate simple activated behavior. (Occasionally ρ_{xx} deviates from a simple activated dependence at higher T as a result of the weak minimum riding on a slightly temperature-dependent background.) At B ~ 10 T, a deviation of the lowest T becomes observable and increases as the magnetic field is increased. Figure 4.8(b) shows this substantial deviation at B = 20.8 T.

Low Temperature Behavior. Data like those in Figure 4.8(b) fit very well over the entire temperature range to a sum of activated conduction at higher T and hopping conduction at lower T [solid curve in Fig. 4.8(b)]. This dependence suggests that the quasiparticles in the FQHE become localized at low temperatures, in analogy to the localization of electrons in the IQHE. [Tsui, 1982c; Ebert, 1983] The data could not be fitted if we assumed only hopping conduction over the entire temperature range.

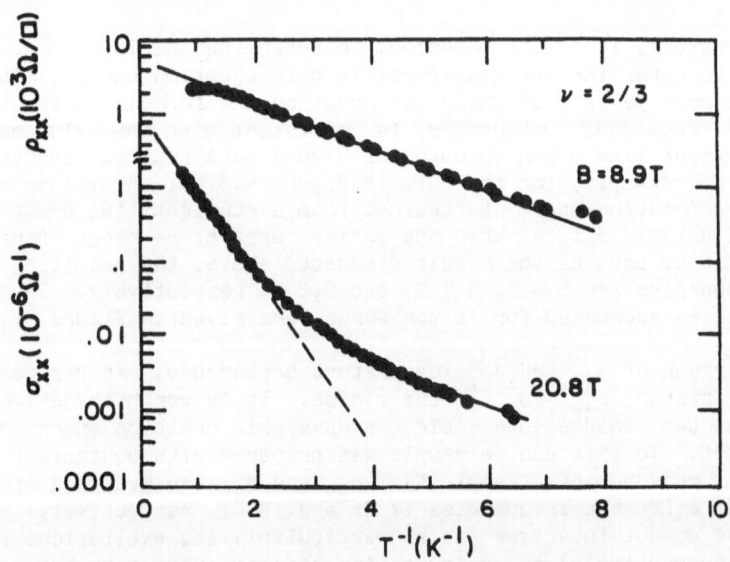

Fig. 4.8. Temperature dependence of the minimum at $\nu = \frac{2}{3}$ (a) at B = 8.9 T, showing simple activated behavior; (b) at B = 20.8 T, showing the smooth, curved deviation from activated behavior at lower temperatures.

The formula used to model the hopping conduction in a magnetic field is from Ono,

$$\sigma_{xx}(T) = \sigma_{2D,B}(T)\exp\left[-(T_0^{Ono}/T)^{\frac{1}{2}}\right],$$

where

$$\sigma_{2D,B}(T) = e^2\gamma_0/k_BT \text{ and } k_BT_0^{Ono} = \eta_0/[D(E_F)\ell_0^2],$$

where γ_0 is some constant depending upon electron-phonon coupling strength, the phonon density of states, and other material constants and η_0 is a constant of order unity which is related to the critical concentration of the percolation problem. [Ono, 1982] This formula results from the Gaussian decay of the localized electron wave function in the presence of a strong magnetic field: $\psi \propto \exp(-r^2/2\ell_0^2)$.

The two-dimensional Mott variable-range hopping formula fits the low-T deviation as well:

$$\sigma_{xx}(T) = \sigma_{2D}(T)\exp\left[-(T_0^{Mott}/T)^{\frac{1}{3}}\right],$$

where

$$\sigma_{2d}(T) = e^2\gamma_0D(E_F)^{\frac{1}{3}}/4(\pi\alpha k_BT)^{\frac{2}{3}} \text{ and } k_BT_0^{Mott} = (27\alpha^2)/\pi D(E_F),$$

where γ_0 depends on the electron-phonon coupling strength and α is determined by the exponential decay of the localized electron wave function $\psi \propto \exp(-\alpha r)$. [Mott, 1979]

Ihm and Phillips have suggested the existence of a second activation energy in the FQHE, due to excitations of electrons which, in the presence of potential fluctuations, have not condensed into the Laughlin quantum liquid. [Ihm, 1985] Attempts to fit the data by use of two activation energies, $\Delta_1/2$ and $\Delta_2/2$, are equally successful:

$$\sigma_{xx}(T) = \sigma_1\exp(-\Delta_1/2T) + \sigma_2\exp(-\Delta_2/2T).$$

Fortunately, it is not necessary to determine the low-T conduction mechanism to determine the quasiparticle pair creation energy. The pair creation energy is only slightly dependent on the formula chosen to fit the lower-T data and, furthermore, is consistent with the value determined from a straight line drawn through the high-T data [dashed line in Fig. 4.8(b)]. For example, for the data in Figure 4.8(b), the values of the high-T pair creation energy determined from a straight line drawn through the high-T data is 5.1 K. When the entire temperature range of data is computer fit to each of the models discussed above, the resulting pair creation energies are 5.5 K, 5.7 K, and 5.2 K, respectively. These variations are accounted for in the error bars given in Figure. 4.11.

Comparison of ρ_{xx} and σ_{xx} Temperature Dependence. As has been previously stated, $\rho_{xx} \propto \sigma_{xx}$ in the minima. It is worth investigating whether the two measurements yield the same pair creation energies, as one would expect. To this end, a sample was prepared with contacts for both ρ_{xx} and σ_{xx} measurements (sample K). ρ_{xx} and σ_{xx} are measured simultaneously, at different frequencies (7 Hz and 13 Hz, respectively) so the two signals do not interfere and at particularly low excitations (10 nA and 1 mV, respectively) to avoid heating effects. The temperature dependences of the two are very similar over the experimental temperature range, as shown in Fig. 4.9. Of particular importance to this study, the pair creation energies determined from the two curves are nearly identical. The slight difference between the two low-T deviations is perhaps due to the unequivalent current paths for the two geometries.

Magnetic Field Dependence of the Thirds. The data taken on the minima at $\nu = \frac{1}{3}, \frac{2}{3}, \frac{4}{3}$, and $\frac{5}{3}$ exhibit a smooth evolution as magnetic field is increased. Figure 4.10 contains data on $\nu = \frac{4}{3}$ at B = 5.9 T; $\nu = \frac{2}{3}$ at B = 8.9 T, 11.5 T, and 20.8 T; and $\nu = \frac{1}{3}$ at B = 18.3 T and 29.0 T. Three aspects of this figure should be noted:

(1) There does not appear to be any difference among data taken on the different filling factors, $\nu = \frac{1}{3}, \frac{2}{3}, \frac{4}{3}$, and $\frac{5}{3}$, suggesting that they all result from the same energy spectrum above the condensed ground state.

(2) The activation energy seems to increase monotonically with magnetic field.

(3) The low-temperature deviations increase with magnetic field.

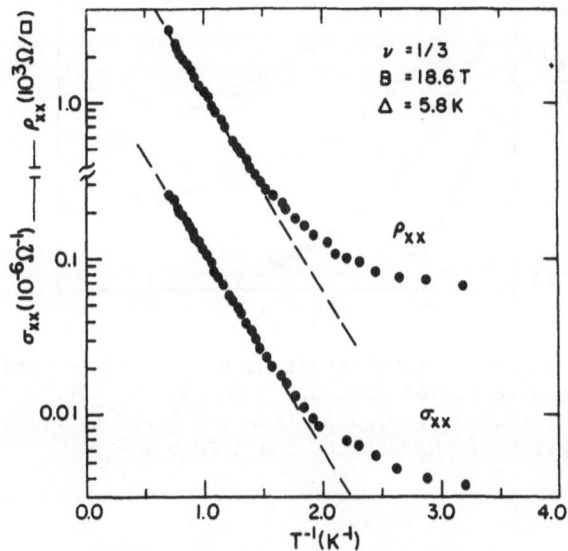

Fig. 4.9. Direct comparison of temperature dependences of ρ_{xx} and σ_{xx} at the $\nu = 1/3$ minimum for B = 18.6 T. The two sets of data were taken simultaneously on sample K. The dashed lines correspond to a pair creation energy of 5.8 K.

The reasons for this last point are straight-forward. At higher magnetic fields, the activation energy is higher. As a result, the magnitude of the conductivity decreases to $\sim 10^{-7} \Omega^{-1}$ at a correspondingly higher temperature, at which point the contribution due to hopping conduction becomes apparent. Indeed, the data in Figure 4.10 indicates that the conduction is activated until the conduction has decreased by 1-1.5 orders of magnitude from the value at ~ 1 K. Also contributing to the increased localization effects at higher magnetic fields will be the magnetic-field-induced compression of the wave function of the charged quasiparticle. The quasiparticles would therefore "see" shorter wavelength fluctuations in the random potential and would have additional opportunity to localize.

Pair Creation Energies. Figure 4.11 presents the pair creation energies from the data on $\nu = \frac{1}{3}$, $\frac{2}{3}$, $\frac{5}{3}$, and $\frac{4}{3}$ from the high mobility samples. The data from sample A in Figure 4.11 are from a study of the $\frac{4}{3}$ minimum by Chang, et al. [unpublished; Chang, 1983]

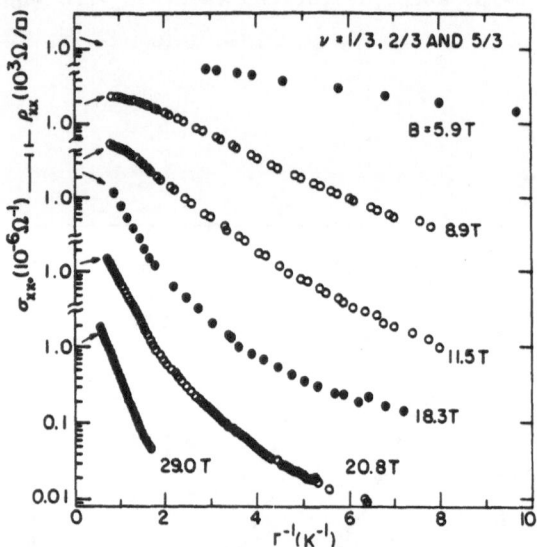

Fig. 4.10. Overview of temperature dependence of the thirds minima for magnetic fields from 5.9 T to 29.0 T. The data represented by open, filled circles is from $\nu = \frac{2}{3}$ and $\frac{1}{3}$, respectively, with the exception of the data at 5.9 T from $\nu = \frac{5}{3}$.

The error bars in the diagram result roughly equally from the sum of the estimated errors from fitting the experimental data and the uncertainties in the correction for magnetoresistance of the mixing chamber tail thermometer at high magnetic fields. The actual reproducibility of the data is better indicated by the variations in the data taken on a given sample at similar magnetic fields, for example, the two points from sample L at 18.5 and 18.8 T.

Four features of Figure 4.11 should be stressed:

(1) There is no apparent sample dependence among these samples of similar mobility. Also, the changes in mobility due to the backside gate bias do not significantly affect the pair creation energies observed on these very high mobility samples, provided that $\mu_{calc} \gtrsim 400,000$ cm^2/V s.

(2) The data for $\nu = \frac{1}{3}$ and $\frac{4}{3}$ overlap at B ~ 15-20 T. Also, the data for $\nu = \frac{2}{3}$ and $\frac{5}{3}$ are consistent with the data for $\nu = \frac{4}{3}$ at similar magnetic fields. This suggests a very similar pair creation energy, $^3\Delta$, for all of the filling factors: $\nu = \frac{1}{3}$, $\frac{2}{3}$, $\frac{5}{3}$, and $\frac{4}{3}$.

(3) The observed pair creation energies are much smaller than theo-

retically predicted. As discussed in chapter 2, these theories all yield quasiparticle pair-creation energies for $\nu = \frac{1}{3}$ and $\frac{2}{3}$ of the form $\Delta = Ce^2/\varepsilon\ell_0$, where the constant of proportionality, C, is model dependent. In our experimental units, this corresponds to $\Delta = 50.95\ C\ B^{1/2}$, where Δ is in kelvin and B is in tesla. The consensus among theorists is that the most accurate calculations based upon Laughlin's theory yield C ~ 0.10. To compare this result with our experimental data, Figure 4.10 contains a curve of $^3\Delta$ vs B for C = 0.030, over a factor of three smaller than the theoretical value for an ideal two-dimensional system.

(4) $^3\Delta$ does not follow the predicted $B^{1/2}$ magnetic field dependence. Rather, the phenomenon has a finite threshold at B ~ 5.5 T. For higher magnetic fields, there is a roughly linear increase in $^3\Delta$ up to B ~ 18 T, followed by a weaker magnetic field dependence of $^3\Delta$ ~ 5.2 K for B $\geq$ 18 T.

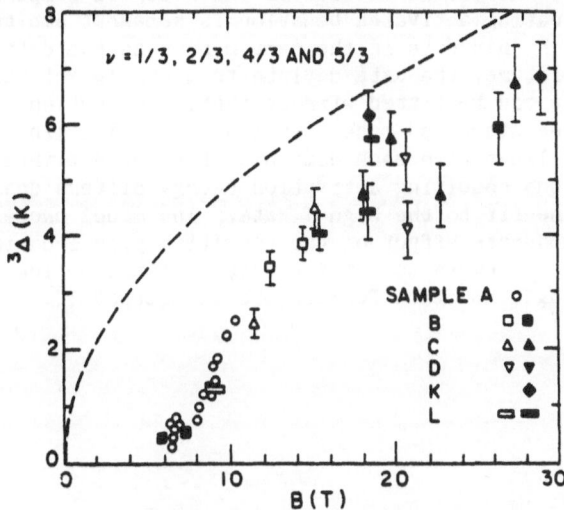

Fig. 4.11. The pair creation energy for the thirds versus magnetic field. The open, filled symbols correspond to $\nu = \frac{2}{3}$ and $\frac{1}{3}$, respectively, except for the two filled squares at 5.9 T, 7.4 T, which are on $\nu = \frac{4}{3}$ and $\frac{5}{3}$, respectively. The dashed line corresponds to $^3\Delta = 0.030\ e^2/\varepsilon\ell_0$, about a factor of three smaller than the calculated energy gap for an ideal two-dimensional system.

Electrically Neutral Excitations. The dispersion relation resulting from the Laughlin model indicates that the quasielectron-quasihole interactions can result in electrically neutral excited states (see Fig. 2.2). The neutral excitation at the "roton minimum" in the dispersion relation of the quasiparticles corresponds to an energy gap of ~ 0.075 $e^2/\varepsilon\ell_0$ and occurs at $k\ell_0$ ~ 1.4, where k is the wave vector of the quasiexciton and is proportional to the interparticle distance. [Haldane, 1985; Girvin, 1985] Although it is not immediately clear that the electrically neutral excitations would be observed in magnetotransport measurements, a calculation of the dc conductivity in the FQHE regime suggests that the thermally activated magneto-rotons can provide a channel for dissipation. [Platzman, 1985] The resulting temperature dependence of the transport coefficients would be activated, with an activation energy, $\Delta/2$, equal to the magneto-roton gap. These electrically neutral excitations when compared to the observed activation energies, $^3\Delta/2$, also lie well above the observed values. Note that for the magneto-roton excitation, the parameter Δ corresponds to twice the gap energy. Thus the data in Figure 4.11 must be compared to a theoretical calculation of $C \sim 0.15$, which is a factor of five larger than the dashed line in Figure 4.11.

Temperature Dependence of the Fifths. In the interest of completing a comprehensive study of activation energies in the FQHE, we now turn attention toward filling factors other than the thirds. Unfortunately, the only other minima which are sufficiently well developed to allow a meaningful study are the minima at $\nu = \frac{2}{5}$ and $\frac{3}{5}$. As such, we have studied the temperature dependence of ρ_{xx} and σ_{xx} at $\nu = \frac{2}{5}$ and $\frac{3}{5}$ for 14 T $< B <$ 29 T. The data in Figure 4.4 indicate that the minima at $\nu = \frac{2}{5}$ and $\frac{3}{5}$ do not exhibit as large a temperature dependence as the minima at the thirds. Within our temperature range, ρ_{xx} and σ_{xx} change by 1-2 orders of magnitude. Figure 4.12 shows the temperature dependence of the minimum at $\nu = \frac{2}{5}$ at $B = 25.2$ T. The interpretation of the observed temperature dependence as indicating activated behavior is somewhat ambiguous; however, we approach this data in the same manner as the data on the thirds. In this picture, the data deviate from simple activated behavior at low-T. The data can be fitted with activated conduction at higher T and any of the three discussed models at lower T (solid line in Fig. 4.12). The dashed lines give each component for the Activated-Ono hopping model. Note that the resulting activation energy differs dramatically from a straight-line-fit to the high-T data. The model chosen to fit the low-T data has a moderate effect on the resulting pair creation energy. As with the thirds, attempts to fit the data over the entire temperature range with a hopping conduction formula are unsuccessful.

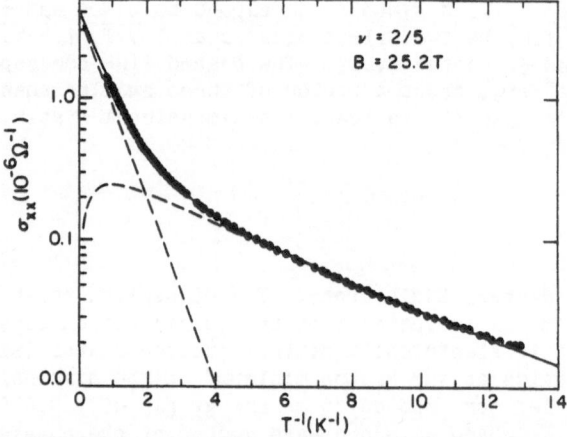

Fig. 4.12. Temperature dependence of the minimum at $\nu = \frac{2}{5}$.

Magnetic Field Dependence on the Fifths. The sets of data in Figure 4.13 show a magnetic field evolution for the fifths similar to that of the thirds. The activation energy increases monotonically with magnetic field. However, the onset of the low-T deviation occurs at higher temperatures for the fifths, despite the smaller activation energies. The fifths seem to be more affected by the potential fluctuations, resulting in a more significant low-T deviation due to localization.

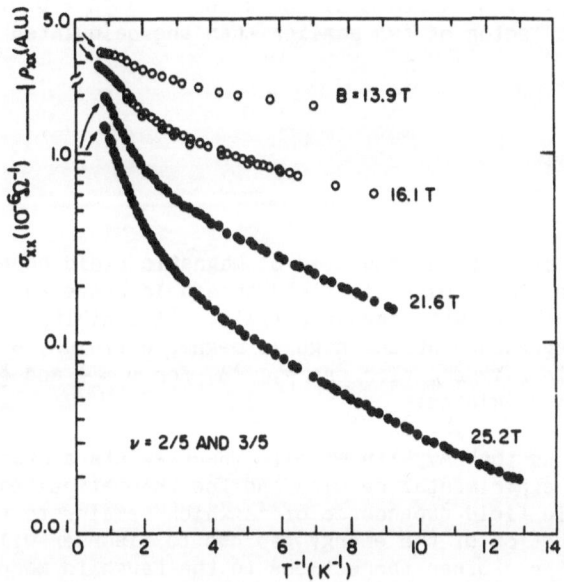

Fig. 4.13. Overview of temperature dependence of the fifths minima for magnetic fields from 13.9 T to 25.2 T. The open (filled) circles represent data on the $\frac{2}{5}$ ($\frac{3}{5}$) minimum.

Pair Creation Energies. The nineteen sets of experimental data again suggest a single pair creation energy, $^5\Delta$ for $\nu = \frac{2}{5}$ and $\frac{3}{5}$, which is plotted versus magnetic field in Figure 4.14. As previously stated, the pair-creation energies at $\nu = p/q$ will scale as $q^{-\frac{5}{2}}$ if the quasi-particles can be approximated as point charges. This yeilds $^5\Delta = 0.28\,^3\Delta$ and C ~ 0.030 for the pair creation energy at $\nu = \frac{2}{5}$ and $\frac{3}{5}$. Figure 4.14 contains a curve of $^5\Delta$ vs B for C = 0.015 for comparison of theory and experiment. Experimentally, $^5\Delta \sim 0.4\,^3\Delta$.

In conclusion, we find a single pair creation energy, $^3\Delta$ for $\nu = \frac{1}{3}$, $\frac{2}{3}$, $\frac{4}{3}$, and $\frac{5}{3}$ in magnetic fields up to 28 T. $^3\Delta$ is much smaller than expec-

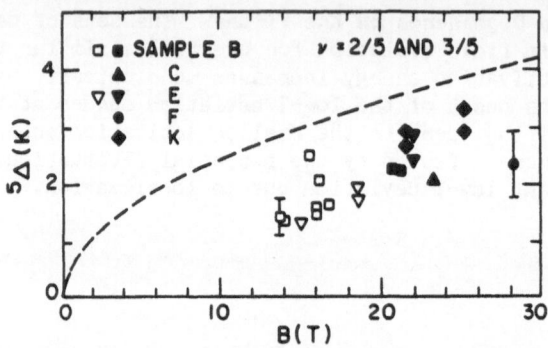

Fig. 4.14. The pair creation energy for the fifths versus magnetic field. The open, filled symbols correspond to $\nu = \frac{3}{5}$ and $\frac{2}{5}$, respectively. The dashed line corresponds to $^5\Delta = 0.015 \ e^2/\epsilon\ell_0$, about a factor of two smaller than the calculated energy gap.

ted and does not exhibit the expected $B^{\frac{1}{2}}$ magnetic field dependence. Instead, $^3\Delta$ has a finite magnetic field threshold above which it has a roughly linear increase with magnetic field. $^3\Delta$ exhibits a weaker magnetic field dependence at the highest magnetic fields, $B \geq 18$ T. We also find a single pair creation energy, $^5\Delta$, for $\nu = \frac{2}{5}$ and $\frac{3}{5}$. $^5\Delta$ is also much smaller than predicted.

Corrections to the Laughlin Model. There exists a startling discrepancy between the experimental results and the theoretical calculations of C and the magnetic field dependence of $^3\Delta$ which remains to be explained. The possible reduction of the energy gap due to disorder will be discussed in the next chapter. Other corrections to the Laughlin model due to admixture of higher Landau levels, finite layer thickness of the two-dimensional system, and the possibility of spin-reversed quasiparticle excitations are considered here.

The admixture of higher Landau levels is found to essentially preserve a Laughlin-like quantum liquid ground state. [Yoshioka, 1984, 1986] The admixture also reduces the energy gaps in the FQHE; however, the reduction is calculated for a four electron system to be ~17% at B = 6.7 T. As magnetic field increases, the calculated reduction of the energy gap decreases. Thus, the effect of Landau mixing along cannot account for the observed discrepancies between theory and experiment.

The finite thickness of the two-dimensional system weakens the short-range Coulomb interaction between electrons. This effect becomes important when $\ell_0 \sim z_0$, where z_0 is the layer thickness and increases as magnetic field is increased. Calculations on finite number systems by two groups have found substantial reductions from the ideal two-dimensional case. [Zhang, 1986; Yoshioka, 1986] The calculations of Zhang and Das Sarma can qualitatively account for the observed weaker magnetic field dependence of $^3\Delta$ at the higher magnetic fields. However, the calculated magnitude of the energy gap is about a factor of three higher than experiments.

Yoshioka considers finite layer thickness and admixture of higher Landau levels together. His calculations find a reduced energy gap which is about a factor of two larger than experiment. The energy gaps at $\nu = \frac{1}{3}$ and $\frac{2}{3}$ are slightly different, although the effect is approximately the same magnitude as the uncertainties in the experiments. Furthermore, Yoshioka notes that the electron system accomodates a small amount of disorder by creating the lowest energy excitations which can alter the local charge density, that is, magneto-rotons. If these magneto-rotons exist in sufficient numbers, then activation energy experiments may detect the thermal activation from the magneto-roton minimum to the quasiparticle pair creation energy at infinite separation. He calculates this energy difference with the corrections included and finds it to be fifty percent higher than the experimental results at the highest magnetic fields, where the agreement is the best. Neither Zhang and Das Sarma nor Yoshioka find a magnetic field threshold resulting from the effects of finite layer thickness or admixture of higher Landau levels.

The elementary excitations above the Laughlin ground state may consist of spin-reversed quasiparticles. [Chakraborty, 1986a,b] From calculations on 4- and 5-electron systems, the spin-reversed quasielectron and spin-polarized quasihole production is found to be energetically favorable for magnetic fields less than ~12 T. In this range of magnetic field, the Zeeman energy is dominant and, thus, the energy gap scales linearly with magnetic field. Although this calculation cannot account for the finite threshold magnetic field, the calculation does find a slope for the linear dependence which is in very close agreement with the roughly linear dependence of $^3\Delta$ observed near the finite threshold magnetic field. The observed magnetic field dependence of $^3\Delta$ is therefore interpreted as a linear dependence crossing over to a square root dependence at ~12 T.

This concludes the present discussion on corrections to the Laughlin model. The next section will present experimental and theoretical evidence that the most important correction probably results from the finite mobility of the samples.

4.4 Effects of Mobility and Localization in the FQHE

Introduction. There exists a startling discrepancy between the experimental results and the theoretical calculations of C and the magnetic field dpendence of $^3\Delta$ which remains to be explained. A reduction of the many-particle gap due to disorder and subsequent thermal excitation to a mobility edge provides a qualitative explanation for the reduced values of C as well as for the finite threshold field. [Chang, 1983] Recent theoretical work is attempting to assess quantitatively the effects of disorder and finite thickness of the two-dimensional electron system on the energy gaps in the FQHE. From the discussion in the previous section, the disorder effects might be expected to account for the bulk of the deviations between experiment and theory. This section presents experimental and theoretical evidence that this is the case.

Negative Biased High-Mobility Samples. Negative gate biases were applied to two samples with $\mu_0 = 300,000$ cm^2/V s, to achieve reduced mobilities of 100,000 to 250,000 cm^2/v s (see section 3.3). Five resulting pair creation energies, plotted in Figure 4.15, are consistent with a reduction of the energy gap with decreasing mobility. The five data are consistent with a simple phenomenological model in which the theoretical considerations are assumed to yield the correct magnetic field dependence of the energy gap in the infinite mobility limit. [Chang, 1983] In the presence of disorder, the excitation energy level would be expected to broaden. The broadened energy level will contain extended quasiparticle states in the center of the level separated from localized quasiparticle states in the tails of the level by a mobility edge. Transport experiments will detect excitations to the mobility edge. The observed

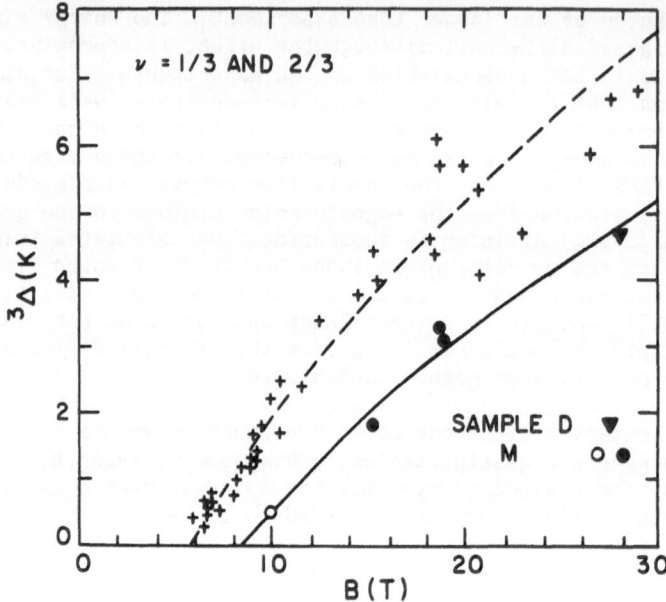

Fig. 4.15. Pair creation energies of the thirds minima for inter-mediate mobilities, achieved by negatively biasing two high-mobility samples. The open (filled) circles are from the $\frac{2}{3}$ $\left(\frac{1}{3}\right)$ minima.

pair production energy will thus be

$$^3\Delta(\mu) = \Delta_0(\mu,B) - \Gamma(\mu) = C(\mu)\left(\frac{e^2}{\varepsilon\ell_0}\right) - \Gamma(\mu),$$

where $\Gamma(\mu)$ is the half-width of the extended quasiparticle states, assumed to be magnetic-field-independent. We emphasize that this model is not suggested as a rigorous analysis of our pair creation data. It is merely offered as a "guide to the mind's eye."

The data from Figure 4.11 are shown as crosses in Figure 4.15. A least-squares fit to this high-mobility data yields C(400-1000) = 0.049 and Γ(400-1000) = 6.0 K. It should be noted that this curve does not adequately model the data at the highest magnetic fields; the data describe a curve with higher curvature. The five data of intermediate mobilities are consistent with C(100-250) = 0.040 and Γ(100-250) = 5.9 K.

Finite Mobility Corrections to the Laughlin Model. Recently, consid-erable theoretical effort has been directed toward understanding the effect of disorder on the excitation spectrum of the FQHE. Independently, Zhang et al. and Rezayi and Haldane have considered the effect of a single impurity in a finite system, for an attractive Coulombic impurity and delta function impurities, respectively. [Zhang, 1985; Rezayi, 1985] Both find an interesting oscillation in the charge density surrounding an impurity. Furthermore, there is a reduction of the energy gap in the vicinity of the impurity, although the magnitude of this reduction is dependent on the size of the finite system. [Zhang, 1985]

To date, there have been two direct attempts to quantitatively fit the experimental activation energy results. [MacDonald, 1986; Gold, 1986]

Both determine the effect of disorder within the single mode approximation of Girvin, 1985 and they reach similar results. Of greatest importance with regard to the experiments, these calculations establish the existence of a finite threshold magnetic field in the presence of disorder. It is found that disorder results in a reduction of the activation energy which is enhanced at the higher magnetic fields by the effects of finite thickness of the electron layer.

MacDonald, et al. show that the calculated corrections to the magneto-roton energy gap, given the sample parameters, are of the order required to agree with the experimental results. Upon adjusting a single parame-ter to fit the experimental threshold magnetic field, they achieve notable agreement with the data of Figure 4.11 over the entire range of magnetic field. Their fitted curve, while resembling the curve from the naive model shown in Figure 4.15, fits the data more accurately. In particular, the high curvature of the data in Figure 4.11 is reproduced, as is the magnitude of the activation energy at the highest magnetic fields. On the other hand, Gold derives equations describing the effect of disorder on the energy gap. He then performs a two-parameter fit, and achieves similarly good agreement with the experimental results.

The theoretical work described above may well represent as good agreement between theoretical calculations and activation energy experiments as can be achieved for some time, given the dominant effects of disorder.

Localization in Low Mobility Samples. Returning to the experiments, we consider three samples of low mobility. The temperature dependences of the $\frac{1}{3}$ and $\frac{2}{3}$ minima on samples J and I (μ_o = 140,000 and 72,000 cm^2/Vs, respectively) are contained in Figures 4.16 and 4.17. These results from samples with μ_o < 200,000 cm^2/Vs are more difficult to interpret than the results from the high-mobility samples. Meaningful fits to the data of the sum of two conduction mechanisms, as discussed in section 4.3, are not possible here. This is due to the reduced curvature and dynamic range of the data from lower mobility samples. However, it is fruitful to consider our attempts to fit the low-mobility data with individual conduction formulae. From observations on the high-mobility samples, we expect that the effects due to hopping conduction will be largest at high magnetic fields. Indeed, eight out of nine of the sets of data for B > 23 T fit the Ono hopping formula over the entire temperature range. (The sole exception is the data on sample J at 17.6 T, but Fig. 4.16 reveals this to be a pretty rotten set of data.) To indicate the quality of the experimental fit to the Ono formula, Figure 4.18 contains the set of data on $\nu = \frac{1}{3}$ at B = 25.2 T. This is the set with the greatest dynamic range, chosen to best distinguish among the three candidate low-T conduction formulae. While the data in Figure 4.18 seems sufficient to clearly distinguish among the three formula, it is conceivable that the magnetic field noise causes local heating in either the 2D layer or the carbon resistor thermometer. This could introduce errors in our thermometry below ~100 mK. However, even if all data below 125 mK are rejected as completely unreliable (believed to be conservative), the remaining data are sufficient to distinguish among the three conduction formulae. We find, therefore, that at high magnetic fields the observed conduction is consistent with the functional form of the Ono hopping formula.

At lower magnetic fields, we speculate that the observed conduction will include components from both hopping and activated conduction. This "crossover regime" has a somewhat increased curvature, as observed on the high mobility samples. At the lowest magnetic fields, we may reach a regime in which only the activated conduction is observed. This proposed sequence is consistent with what is observed at lower magnetic fields.

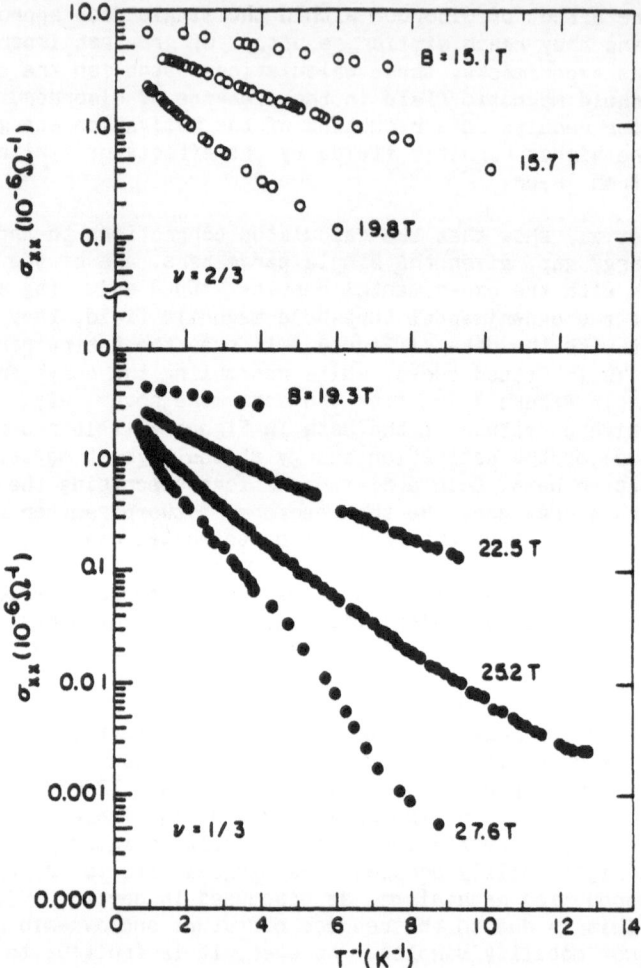

Fig. 4.16. Temperature dependence of the $\frac{1}{3}$ and $\frac{2}{3}$ minima for sample J, for which μ_0 = 140,000 cm²/Vs.

The implications of this interpretation are several. Firstly, the Ono formula, which was derived for two-dimensional hopping in the presence of a magnetic field, is substantiated by our experiments on the charged quasiparticles in the FQHE. In turn, the fit to the Ono formula suggests that the quasiparticles are localized by short wavelength fluctuations and that the localized quasiparticle states have gaussian tails in high magnetic fields. Also, we now have support for the attribution of the low-T deviations observed on the high-mobility samples to the effects of quasiparticle localization. Furthermore, it is reasonable to favor the Ono formula to describe the low-T deviations in the high-mobility data.

In light of this, the fitted values of the parameters T_0^{Ono} from the Ono formula deserve to be examined further.

The fitted values of T_0^{Ono} from the high mobility data are plotted versus magnetic field in Figure 4.19. From the Ono formula

$$k_B T_0^{Ono} = \eta_0 / [D(E_F)\ell_0^2],$$

where η_0 is a constant of order unity which we set equal to one. [Ono,

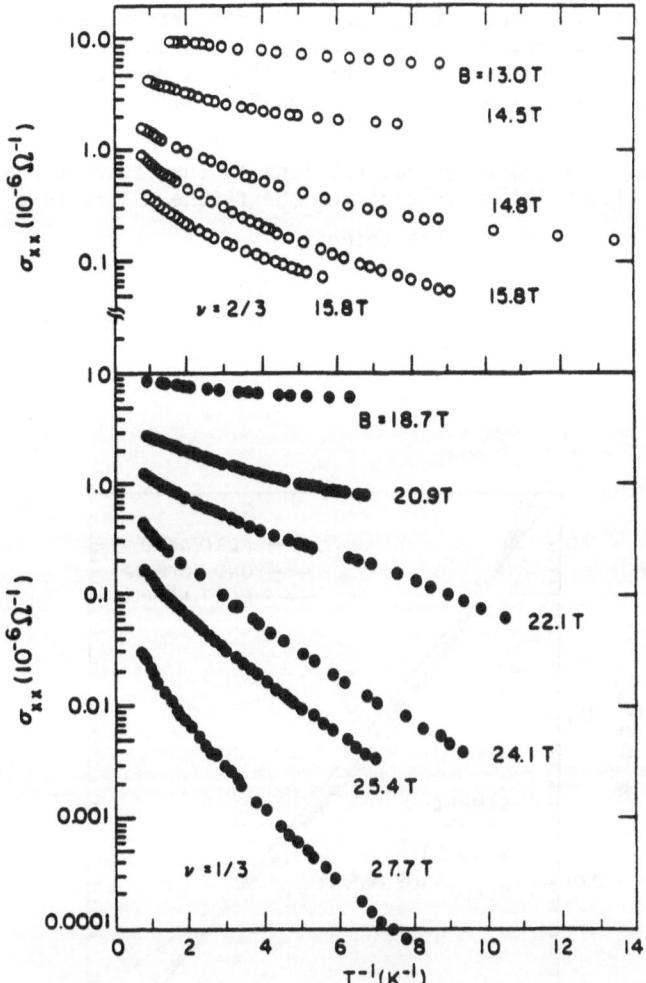

Fig. 4.17 Temperature dependence of the $\frac{1}{3}$ and $\frac{2}{3}$ minima for sample I, for which $\mu_0 = 72,000$ cm²/Vs.

1982] If we assume that the quasiparticle density of states at a given fractional fitting is independent of magnetic field, we find that $k_B T_0$ should be linear in magnetic field. While the data of Figure 4.19(a) does not even begin to suggest a linear dependence, we fit a straight line to the data for an order-of-magnitude determination of the density of quasiparticle states. From the slope of this line, we can determine the order of magnitude of the quasiparticle density of states at the Fermi energy. The line in Figure 4.20(a) corresponds to $D_{1/3,2/3}(E_F) \sim 2.8 \times 10^{12}$ meV⁻¹ cm⁻² for the thirds. This is a factor of 2.7 larger than reported for the electron $D(E_F)$ at a lower magnetic field from a lower mobility GaAs/Al$_{.3}$Ga$_{.7}$As heterostructure in the quantum Hall regime. [Ebert, 1983] In light of the many different experimental parameters, we attach no great significance to the number 2.7, except to say that our results on quasiparticles are found not to differ greatly from the results on electrons. However, the calculated magnitude for the quasiparticle density of states is a factor of 98 *larger* than the electron density of states in zero magnetic field, which should serve as an approximate *upper limit* [see Fig. 1.2(c)]. This discrepancy is also noted by Ebert, et al. for the electrons.

The plot of T_0^{Ono} versus B for the fifths in Figure 4.19(b) shows significantly less scatter, due most probably to the increased influence of localization in the observed data, which allows a more accurate determination of the localization parameters (see Fig. 4.12). The magnitude of T_0^{Ono} for the fifths is observed to be about a factor of seven smaller than for the thirds. The least-squares fit line in the figure corresponds to $D_{2/5,3/5}(E_F) \sim 1.9 \times 10^{13}$ meV^{-1}cm^{-2} for the fifths. This is a factor of 6.8 larger than observed for the thirds.

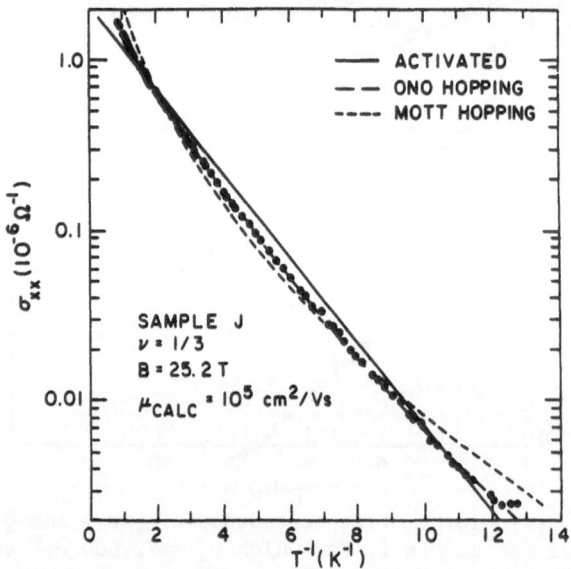

Fig. 4.18 Temperature dependence of σ_{xx} at $\nu = \frac{1}{3}$ from a low-mobility sample. The three lines show least-squares fits to the three conduction formulae.

To ostensibly remove the effects of the difference in charge for the quasiparticles, we can express the density of states in natural units: ℓ_0 for length and $(e^*)^2/\varepsilon\ell_0$ for energy. Then,

$$\frac{D^*_{\frac{2}{5},\frac{3}{5}}(E_F)}{D^*_{\frac{1}{3},\frac{2}{3}}(E_F)} \sim 6.8 \left(\frac{e^*_{\frac{1}{3}}}{e^*_{\frac{2}{5}}}\right)^{\frac{3}{2}} = 3.2.$$

This completes the discussion of our data on localization in the FQHE. In the next chapter we compare our results to those of other researchers and review the conclusions of the lecture notes.

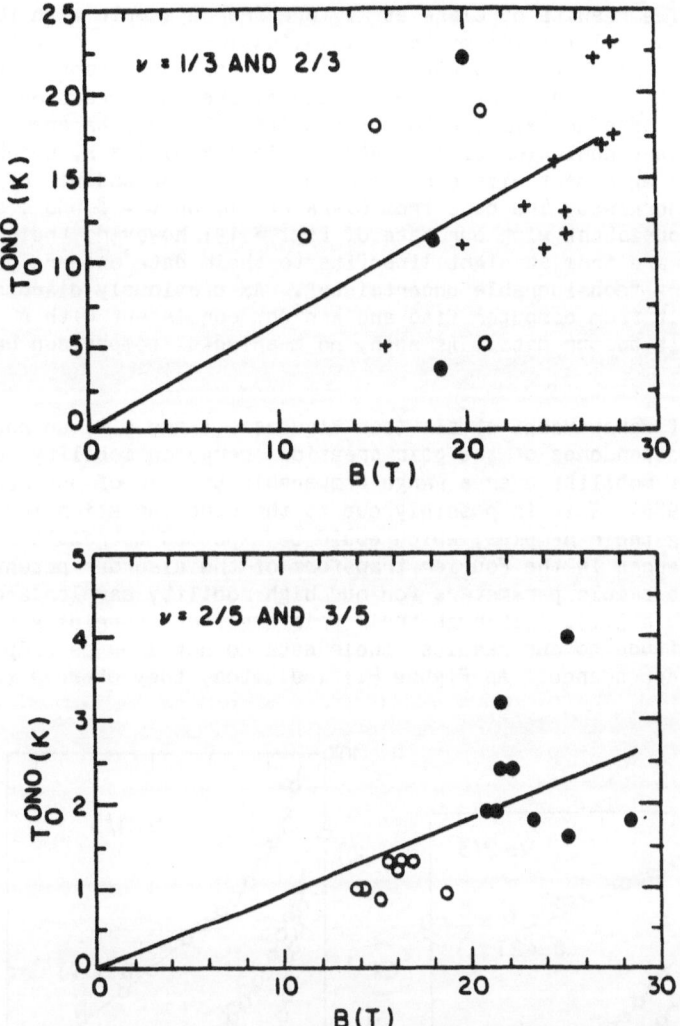

Fig. 4.19. The open (filled) circles represent T_0^{Ono} for (a) the $\frac{2}{3}$ $\left(\frac{1}{3}\right)$ minimum and (b) the $\frac{3}{5}$ $\left(\frac{2}{5}\right)$ minimum versus magnetic field from the high-mobility samples. The solid lines are discussed in the text. [The "+" data in (a) are from the low-mobility samples. The lower scatter is due most likely to the increased accuracy in determining the localization parameters from these samples.]

5.1 Comparison with Results of Other Researchers

Several other groups have reported activation energies on GaAs/Al$_{.3}$Ga$_{.7}$As heterostructures. The results are summarized in Table 5.1. The results of Ebert el al., which are from samples of comparable mobility, are perfectly consistent with our data on $^3\Delta$ vs B of Fig. 4.11. [Ebert, 1984] The results of Clark et al. are from a sample with the mobility enhanced to 2.1 x 10^{6}cm^2/Vs by illumination from a red LED. [Clark, 1986] Although they do not explicitly report the temperature dependence that they observe at $\nu = \frac{2}{3}$, $\frac{1}{3}$, and $\frac{5}{3}$, the pair creation energies that they report are similar to our results. Their data are roughly consistent with a translation of our data vertically by 1.4 K, which is qualitatively in agreement with our observed mobility dependence of the pair creation energies. The data from Clark et al. on $\nu = \frac{4}{5}$ and $\frac{4}{7}$ are superficially consistent with our data of Fig. 4.14; however, their reported values are from straight line fits to their data and are described as having "considerable uncertainty". As previously discussed, our values result from computer fits and are not consistent with a straight line fit to our data. As such, no meaningful comparison can be made.

The data of Wakabayashi et al. (see Figures 5.1 and 5.2) do not show any systematic dependence of the pair creation energy on mobility, despite the variation of mobility over a range comparable to that of our study. [Wakabayashi, 1986] This is possibly due to the wide variation in sample parameters among their samples, which could be expected to result in significant differences in the Fourier transform of the disorder potential. In contrast, the sample parameters for our high-mobility samples are quite similar (see Table 3.1). Although their pair creation energies are similar in magnitude to our results, their data do not show as simple a magnetic field dependence. As Figure 5.1 indicates, they observe the

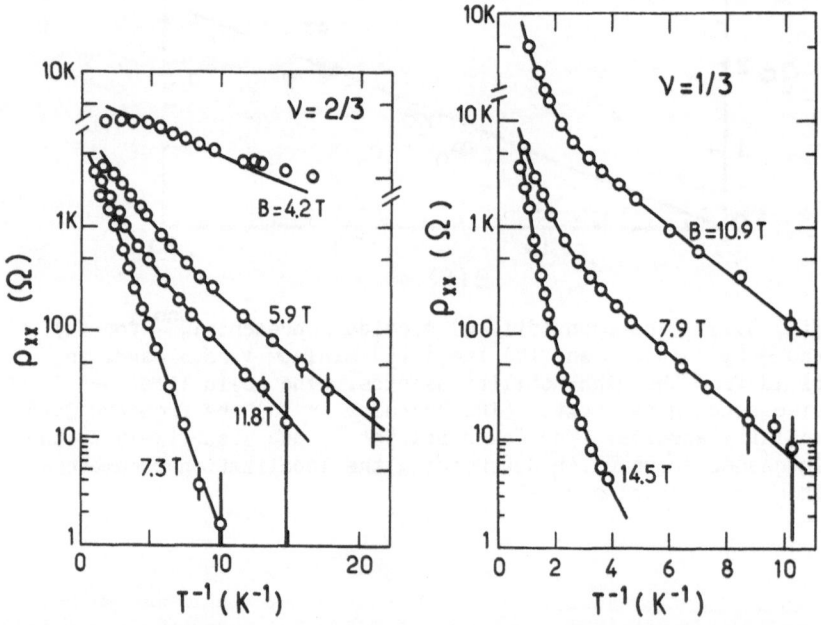

Fig. 5.1. Examples of the temperature dependence of ρ_{xx} at the ν = 2/3 and 1/3 minima. The solid lines represent fits to the model of two activation energies. |Wakabayashi, 1986|.

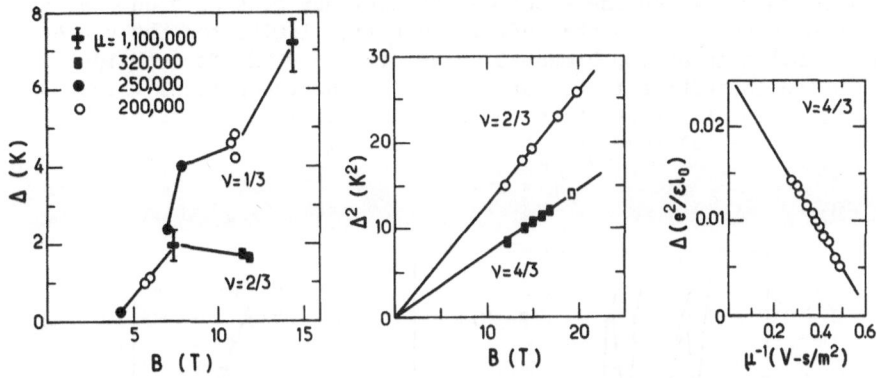

Fig. 5.2. The higher activation energy
versus magnetic field for $\nu = 2/3$ and $1/3$.
The sample mobilities are given in the inset.
|Wakabayashi, 1986|.

deviations from activated behavior at low temperatures, although they
interpret these deviations as a second activation energy.

The FQHE is more difficult to observe in silicon MOSFETs, due to
decreased electron mobilities from surface roughness and scattering from
impurities near the interface. Nonetheless, a study of activation
energies in the FQHE has been completed on silicon MOSFETs, (Kukushkin,
1986). The data from this study are given in Figures 5.3-5.6. The major
conclusions from this study are the observation of the theoretically
predicted scaling of the energy gaps at $\nu = 2/3$ and $4/3$ with the square
root of magnetic field (see Fig. 5.5). Additionally, the dependence of

Table 5.1. Pair creation energies from other
researchers.

ν	B (T)	Δ (K)	μ_0 ($10^3 \frac{cm^2}{V_s}$)	Reference
5/3	5.5	.17	1050	Ebert, 1984
4/3	6.8	.50	1050	"
4/3	6.8	.61	1050	"
5/3	4.8	1.4	2100	Clark, 1986
4/3	6.0	1.0	2100	"
2/3	12.0	4.4	2100	"
7/5	5.7	0.4	2100	"
3/5	13.3	1.4	2100	"
2/3	7.5	1.94	1100	Wakabayashi, 1986
1/3	14.5	7.16	1100	"
2/3	11.7	1.66	320	"
1/3	7.1	2.34	230	"
2/3	5.8	1.04	200	"
1/3	11.1	4.40	200	"

the activation energy on the electron mobility has been studied (see Fig. 5.6). An extrapolation to the infinite mobility limit, for the $\nu = 4/3$ minimum, results in an experimental value of C ~ 0.026 for the ideal system, which is about a factor of four below the theoretical value. The major limitation of the study is the extremely limited variation of ρ_{xx} over the experimental temperature range (see Fig. 5.4).

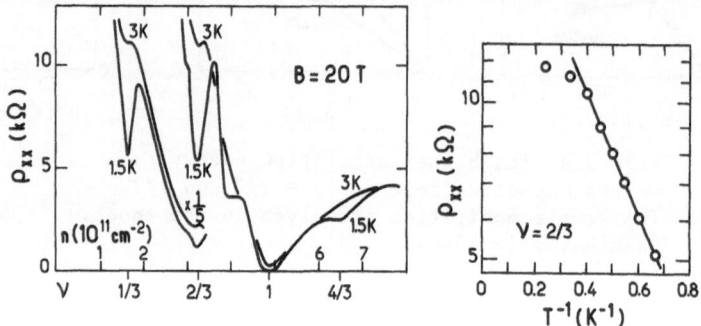

Fig. 5.3. The behavior of the diagonal components of the magnetoresistivity tensor ρ_{xx} in the region of fractional filling factor $\nu = 2/3$. |Kukushkin, 1986|.

Fig. 5.4. The temperature dependence of ρ_{xx} at $\nu = 2/3$. |Kukushkin, 1986|.

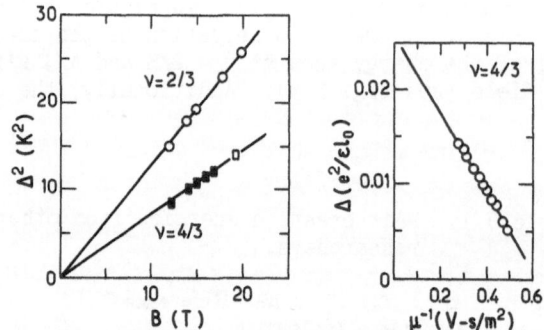

Fig. 5.5. The dependence of the activation gap Δ on B at $\nu = 2/3$ and 4/3 measured for two Si-MOSFETs (dark and open symbols) with fixed mobility: $\mu_e = (3.5\pm0.1).10^4$ cm^2/Vs (circles), $\mu_e = (2.7\pm0.1).10^4$ cm^2/Vs (squares). |Kukushkin, 1986|.

Fig. 5.6. The dependence of the activation energy in the units of Coulomb energy on the reciprocal electron mobility at $\nu = 4/3$. |Kukushkin, 1986|.

5.2 Conclusions

We have performed a range of experiments on the FQHE. Firstly, we observe a temperature-dependent competition between neighboring minima in ρ_{xx} which is similar to the previously observed mobility-dependent competition between the IQHE and the FQHE. These observations require a

complication of the scaling picture of the FQHE which makes more difficult the application of this theoretical picture to experiments.

The bulk of the experiments in these notes concern the energy gaps in
the FQHE. Temperature-dependent studies of the minima in ρ_{xx} and σ_{xx}
reveal activated conduction with deviations from activated conduction at
the lower temperatures. These deviations, which are attributed to the
effects of quasiparticle localization, are found to increase at higher
magnetic fields. Activation energy experiments on similar high-mobility
samples provide an empirical determination of the magnitude of the energy
gap for $\nu = \frac{1}{3}, \frac{2}{3}, \frac{4}{3}$, and $\frac{5}{3}$ for magnetic fields to 29 T. These data, which
constitute the single most important result of this research, are
contained in Fig. 4.11. Within the experimental accuracy and for the
finite magnetic field range covered for the different fractions, a single
magnetic-field-dependent activation energy is observed for all of the
thirds filling factors. Above a magnetic field threshold at ~5.5 T, this
activation energy increases roughly linearly and then rolls over to a
weaker magnetic field dependence at magnetic fields above ~18 T. The
observed sample dependence of the activation energies is smaller than the
variations observed on any single sample and no difference is found
between activation energies from ρ_{xx} and σ_{xx}.

Activation energy experiments on the relatively weaker minima at $\nu =$
$\frac{2}{5}$, though more difficult to interpret, yield activation energies which are
approximately 2.5 times smaller than the activation energies observed on
the thirds. This data, which is contained in Fig. 4.14, is necessarily
over a more limited magnetic field range.

The reduction of the energy gap with decreasing mobility is experimentally established through observations on two intermediate-mobility
samples. Attempts to further study this effect on low-mobility samples
instead provide information on quasiparticle localization. Conduction
consistent with the functional form of the Ono hopping formula is observed
at high magnetic fields, while the data at lower magnetic fields is
interpreted as an admixture of activated and hopping conduction. The
observation of hopping conduction lends experimental support to the claim
that the low temperature deviations from activated behavior result from
the effects of localization. Estimates of the quasiparticle density of
states are roughly consistent with the observations by other researchers
on electrons in the IQHE, although both are much higher than expected from
theoretical considerations. The increased localization effects observed
on the fifths results in an estimated density of states approximately
seven times larger than that derived from observations on the thirds.

ACKNOWLEDGEMENTS

The author gratefully acknowledges the substantial contributions of
H.L. Stormer, D.C. Tsui, and A.M. Chang to the research presented herein.
Special thanks to J.C.M. Hwang, A. Weimann, A. Cho and C. Tu for providing
the samples. The author is supported by a fellowship from the Hertz
Foundation. The Francis Bitter National Magnet Laboratory is supported
by the National Science Foundation through its Division of Materials
Research. A final thanks to M. O'Meara for duty above and beyond in
preparing this manuscript.

REFERENCES

Arovas, Daniel, J.R. Schrieffer, and Frank Wilczek, 1984, Fractional
statistics and the quantum Hall effect, Phys. Rev. Lett. 53:722.

Boebinger, G.S., A.M. Chang, H.L. Stormer, and D.C. Tsui, 1985a, Competition between neighboring minima in the fractional quantum Hall effect, Phys. Rev. B32:4268.

Boebinger, G.S., A.M. Chang, H.L. Stormer, and D.C. Tsui, 1985b, Magnetic field dependence of activation energies in the fractional quantum Hall effect, Phys. Rev. Lett. 55:1606.

Boebinger, G.S., A.M. Chang, H.L. Stormer, D.C. Tsui, J.C.M. Hwang, G. Weimann, A. Cho, and C. Tu, 1986, Activation energies of fundamental and higher order states in the fractional quantum Hall effect, Surf. Sci. 170:129.

Boebinger, G.S., D.C. Tsui, H.L. Stormer, G. Weimann, J.C.M. Hwang, A. Cho, and C. Tu, unpublished results on the $\nu = 1/5$ minimum in ρ_{xx} and studies on low-mobility samples.

Chakraborty, T., 1985, Elementary excitations in the fractional quantum Hall effect, Phys. Rev. B31:4026.

Chakraborty, T., P. Pietiläinen, and F.C. Zhang, 1986a, Elementary excitations in the fractional quantum Hall effect and the spin-reversed quasiparticles, Phys. Rev. Lett. 57:130.

Chakraborty, T., 1986b, Spin-reversed quasiparticles in the fractional quantum Hall effect - many body approach, Phys. Rev. B34:2926.

Chang, A.M., M.A. Paalanen, D.C. Tsui, H.L. Stormer, and J.C.M. Hwang, 1983, Fractional quantum Hall effect at low temperatures. Phys. Rev. B28:6133.

Chang, A.M., P. Berglund, D.C. Tsui, H.L. Stormer, and J.C.M. Hwang, 1984a, Higher-order states in the multiple-series, fractional, quantum Hall effect, Phys. Rev. Lett. 53:997.

Chang, A.M., M.A. Paalanen, H.L. Stormer, J.C.M. Hwang, and D.C. Tsui, 1984b, Fractional quantum Hall effect at low temperatures, Surf. Sci. 142:173.

Clark, R.G., R.J. Nicholas, A. Usher, C.T. Foxon, and J.J. Harris, 1986, Odd and even fractionally quantized states in GaAs-GaAlAs heterojunctions, Surf. Sci. 170:141.

Ebert, G., K. von Klitzing, C. Probst, E. Schuberth, K. Ploog, and G. Weimann, 1983, Hopping conduction in the Landau level tails in $GaAs-Al_xGa_{1-x}As$ heterostructures at low temperatures, Solid State Commun. 45:625.

Ebert, G., K. von Klitzing, J.C. Maan, G. Remenyi, C. Probst, G. Weimann, and W. Schlapp, 1984, Fractional quantum Hall effect at filling factors up to $\nu = 3$, J. Phys. C17:L775.

Fano, G., F. Ortolani, and E. Colombo, 1986, Configuration-interaction calculations on the fractional quantum Hall effect, Phys. Rev. B34:2670.

Girvin, S.M., A.H. MacDonald, and P.M. Platzman, 1985, Collective-excitation gap in the fractional quantum Hall effect, Phys. Rev. Lett. 54:581.

Girvin, S.M., A.H. MacDonald, and P.M. Platzman, 1986, Magneto-roton theory of collective excitations in the fractional quantum Hall effect, Phys. Rev. B33:2481.

Gold, A., 1986, Disorder effects on the transport properties in the fractional quantized Hall regime, Europhysics Lett. 1:241.

Haldane, F.D.M., 1983, Fractional quantization of the Hall effect: a hierarchy of incompressible quantum fluid states. Phys. Rev. Lett. 51:605.

Haldane, F.D.M. and E.H. Rezayi, 1985, Finite-size studies of the incompressible state of the fractionally quantized Hall effect and its excitations, Phys. Rev. Lett. 54:237.

Halperin, B.I., 1984, Statistics of quasiparticles and the hierarchy of fractional quantized Hall states, Phys. Rev. Lett. 52:1583.

Halperin, B.I., Z. Tesanovic, and F. Axel, 1986, Compatibility of crystalline order and the quantized Hall effect, Phys. Rev. Lett. 57:922(c).

Hurkx, G.A.M. and W. van Haeringern, 1985, Self-consistent calculations on GaAs-Al$_x$Ga$_{1-x}$As heterojunctions, J. Phys. C18:5617.

Hwang, J.C.M., A. Kastalsky, H.L. Stormer, and V.G. Keramidas, 1984, Transport properties of selectively doped GaAs-(AlGa)As hetero-structures grown by molecular beam epitaxy, Appl. Phys. Lett. 44:802.

Ihm, J. and J.C. Phillips, 1985, Activation energies and localization in the fractional quantized Hall effect, J. Phys. Soc. Jpn. 54:1506.

Kittel, C., 1976, "Introduction to Solid State Physics," 5th ed., John Wiley & Sons, Inc., New York.

Kivelson, S., C. Kallin, D.P. Arovas, and J.R. Schrieffer, 1986, Cooperative-ring-exchange theory of the fractional quantized Hall effect, Phys. Rev. Lett. 56:873.

Kukushkin, I.V. and V.B. Timofeev, 1986, Activation gaps in the energy spectrum and influence of disorder on the fractional quantum Hall effect in silicon MOSFETS, Surf. Sci. 170:148.

Lam, Pui K., and S.M. Girvin, 1984, Liquid-solid transition and the fractional quantum-Hall effect, Phys. Rev. B30:473.

Laughlin, R.B., 1981, Quantized Hall conductivity in two dimensions, Phys. Rev. B23:5632.

Laughlin, R.B., 1983, Anomalous quantum Hall effect: an incompressible quantum fluid with fractionally charged excitations. Phys. Rev. Lett. 50:1395.

Laughlin, R.B. 1984a, The gauge argument for accurate quantization of the Hall conductance, in "Two-Dimensional Systems, Heterostructures, and Superlattices," G. Bauer, F. Kuchar, H. Heinrich, eds., Springer-Verlag, Berlin, p.272.

Laughlin, R.B., 1984b, Primitive and composite ground states in the fractional quantum Hall effect, Surf. Sci 142:163.

Laughlin, R.B., 1984c, Excitons in the fractional quantum Hall effect, Physica (Amsterdam) 126B:254.

Laughlin, R.B., M.L. Cohen, J.M. Kosterlitz, H. Levine, S.B. Libby, and A.M.M. Pruisken, 1985, Scaling of conductivities in the fractional quantum Hall effect, Phys. Rev. B32:1311.

Laughlin, R.B., 1986, Destruction of the fractional quantum Hall effect by disorder, Surf. Sci. 170:167.

Levesque, D., J.J. Weis, and A.H. MacDonald, 1984, Crystallization of the incompressible quantum fluid state of a two-dimensional electron gas in a strong magnetic field, Phys. Rev. B30:1056.

MacDonald, A.H., K.L. Liu, S.M. Girvin, and P.M. Platzman, 1986, Disorder and the fractional quantum Hall effect: activation energies and the collapse of the gap, Phys. Rev. B33:4014.

Mendez, E.E., L.L. Chang, M. Heiblum, L. Esaki, M. Naughton, K. Martin, and J. Brooks, 1984, Fractionally quantized Hall effect in two-dimensional systems of extreme electron concentration, Phys. Rev. B30:7310.

Morf, R., and B.I. Halperin, 1986, Monte-Carlo evaluation of trial wave functions for the fractional quantized Hall effect: disk geometry, Phys. Rev. B33:2221.

Mott, N.F., and E.A. Davis, 1979, "Electronic Properties in Non-Crystalline Materials," 2nd ed., Clarendon, Oxford.

Ono, Y., 1982, Localization of electrons under strong magnetic fields in a two-dimensional system, J. Phys. Soc. Jpn. 51:237.

Paalanen, M.A., D.C. Tsui, and A.C. Gossard, 1982, Quantized Hall effect at low temperatures, Phys. Rev. B25:5566.

Paalanen, M.A., D.C. Tsui, A.C. Gossard, and J.C.M. Hwang, 1984, Disorder and the fractional quantum Hall effect, Solid State Commun. 50:841.

Platzman, P.M., S.M. Girvin, and A.H. MacDonald, 1985, Conductivity in the fractionally quantized Hall effect, Phys. Rev. B32:8458.

Rezayi, E.H., and F.D.M. Haldane, 1985, Incompressible states of the fractionally quantized Hall effect in the presence of impurities: a finite-size study, Phys. Rev. B32:6924.

Stern, F., and S. Das Sarma, 1984, Electron energy levels in GaAs-$Ga_{1-x}Al_xAs$ heterojunctions, Phys. Rev. B30:840.

Stormer, H.L., A.C. Gossard, and W. Wiegmann, 1981, Backside-gated modulation-doped GaAs-(AlGa)As heterojunction interface, Appl. Phys. Lett. 39:493.

Stormer, H.L., A. Chang, D.C. Tsui, J.C.M. Hwang, A.C. Gossard, and W. Wiegmann, 1983a, Fractional quantization of the Hall effect, Phys. Rev. Lett. 50:1953.

Stormer, H.L., 1983b, Electron mobilities in modulation-doped GaAs-(AlGa)As heterostructures, Surf. Sci. 132:519.

Tsui, D.C., A.C. Gossard, B.F. Field, M.E. Cage, and R.F. Dziuba, 1982a, Determination of the fine-structure constant using GaAs-$Al_xGa_{1-x}As$ heterostructures, Phys. Rev. Lett. 48:3.

Tsui, D.C., H.L. Stormer, and A.C. Gossard, 1982b, Two-dimensional magne-totransport in the extreme quantum limit, Phys. Rev. Lett. 48:1559.

Tsui, D.C., H.L. Stormer, and A.C. Gossard, 1982c, Zero-resistance state of two-dimensional electrons in a quantizing magnetic field, Phys. Rev. B25:1405.

Tsui, D.C., H.L. Stormer, J.C.M. Hwang, J.S. Brooks, and M.J. Naughton, 1983, Observation of a fractional quantum number, Phys. Rev. B28:2274.

von Klitzing, K, G. Dorda, and M. Pepper, 1980, New method for high-accuracy determination of the fine-structure constant based on quantized Hall resistance, Phys. Rev. Lett. 45:494.

Wakabayashi, J., S. Kawaji, J. Yoshino, and H. Sakaki, 1986, Activation energies of the fractional quantum Hall effect in GaAs-AlGaAs heterostructures, J. Phys. Soc. Jpn. 55:1319.

Yoshioka, D., B.I. Haperin, and P.A. Lee, 1983, Ground state of two-dimensional electrons in strong magnetic fields and 1/3 quantized Hall effect, Phys. Rev. Lett. 50:1219.

Yoshioka, D., 1984, Effect of the Landau level mixing on the ground state of two-dimensional electrons, J. Phys. Soc. Jpn. 53:3740.

Yoshioka, D., 1986, Excitation energies of the fractional quantum Hall effect, J. Phys. Soc. Jpn. 55;885.

Zhang, F.C., V.Z. Vulovic, Y. Guo, and S. Das Sarma, 1985, Effect of charged impurity on the fractional quantum Hall effect: exact numerical treatment of finite system, Phys. Rev. B32:6920.

Zhang, F.C. and S. Das Sarma, 1986, Excitation gap in the fractional quantum Hall effect: finite layer thickness corrections, Phys. Rev. B33:2903.

INTERACTIONS IN 2-D ELECTRON SYSTEMS

Philip M. Platzman

AT&T Bell laboratories, 600 Mountain Ave.
Murray Hill, New Jersey   07974

I. Some Qualitative Aspects of Itinerant 2-D Systems

INTRODUCTION

In recent years there has been a great deal of interest in the properties of itinerant two-dimensional (2-D) electron gases.[1-4] Electrons localized in a heterostructure i.e. at the interface between GaAs and GaAlAs,[1] (Fig. 1.1) or at the interface of a metal-oxide semiconductor sandwich (MOS)[2] or at the surface of a liquid helium metal-helium-vacuum system (MHV) (Fig. 1.2)[3] are examples of such systems. In such cases the electrons are trapped in a potential well with a well defined set of energy levels (subbands) for motion perpendicular to the interface shown in Figs. 1.1-1.2. At low temperatures, if the density is not too high they are often confined to the lowest subband but are free to move parallel to the interface. A finite concentration, n per unit area, of them is placed in this subband either by doping, as in the case of a heterostructure, (Fig. 1.1), or by the application of a voltage, i.e. a parallel plate capacitor arrangement as shown in Fig. 1.2.

The reason for the interest in such systems stems from several key experimental characteristics:

1. It is possible at low temperatures to insure, that they are rather "ideal" 2-D systems. More precisely the excited states energies, for motion perpendicular to the interface, can be made large relative to the in plane energies.

2. The density of electrons and thus the strength of the Columb interaction among electrons may be varied over many orders of magnitude. For the MHV system $10^6 < n < 10^9 \, cm^{-2}$ has been studied. The lower limit is set by the fact that it is difficult to see fewer electrons while the upper limit is fixed by a rather interesting macroscopic instability of the liquid surface.[4] For thin films of helium on insulating substrates higher densities, $n \cong 10^{10} \, cm^{-2}$, have been achieved.[5] For the MOS and GaAlAs systems $3 \times 10^{11} < n < 10^{12} \, cm^{-2}$. The range of densities here is fixed by the microscopic nature of the interface between Si and SiO_2 in MOS and by

the doping technique for the heterostructure. This range of six orders of magnitude in density corresponds to a range of about eight orders of magnitude in the Coulomb energies since the dielectric constant and mass of the carriers are different in these systems.

3. The energy scale set by the Fermi and Columb energies, is typically smaller than 100 $^\circ K$ so that external perturbations which are easily applied (magnetic field, temperature, electric fields, etc.) significantly influence the behavior of the system.

4. The effect of impurities may be made small, i.e. very high mobilities have been achieved (MOS $\mu \lesssim 2\times10^4$, $cm^2 V/sec$ GaAs and MHV $\mu \lesssim 10^6$ $cm^2 V/sec$)

5. For the MHV system, coupling to the ripplons on the surface (analogous to electron phonon interaction) may be made significant and interesting by the application of electric fields.[6]

6. For the MHV system, disorder (Helium gas vapor above the surface) may be introduced in a controlled manner.[5]

Despite the very important progress made in understanding the physics of such systems we suggest here that both experimentally and theoretically many of the important phenomenon remain to be explored. In this lecture, we will qualitatively discuss some general aspects of the many body physics of such systems.

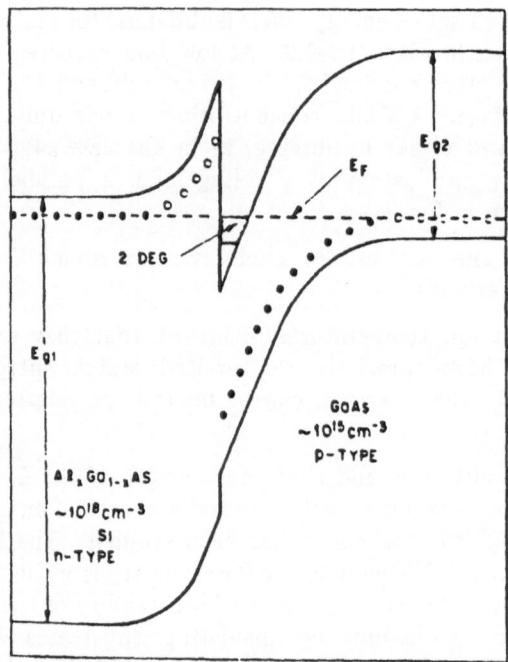

Fig. 1.1 Energy scheme at a modulation-doped GaAs-(AlGa)As heterojunction interface showing the position of the two-dimensional electron gas (2DEG), Fermi energy E_F and energy gaps E_{g1} and E_{g2} (not to scale).

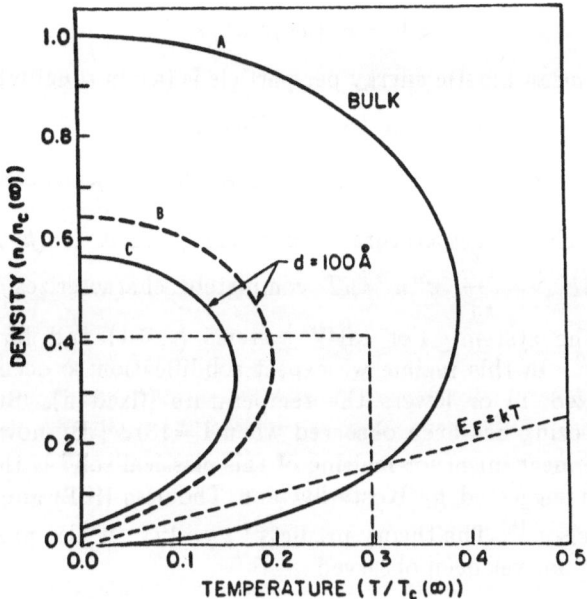

Fig. 1.3 Qualitative phase diagram for 2D electron system. Curve A
is for bulk He, B and C are for 100 A films on dielectric substrates
with ε = 10 and -∞ (metal) respectively.

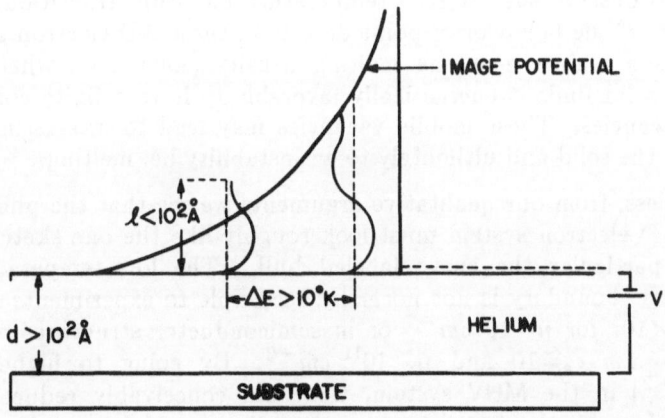

Fig. 1.2 Schematic of configuration for electrons trapped in an MHV
system. The wave functions perpendicular to the interface are drawn
in. The static voltage V fixes the average charge density.

COLUMB INTERACTIONS (no magnetic field)

In those cases where, to a good approximation, the electrons interact in the plane via a $1/r$ potential (where r is the distance measured in the plane) the mean strength of the interaction energy per particle is roughly

$$<V> \cong (n\pi)^{1/2} e^2/\epsilon_0 . \tag{1.1}$$

In this case the mean kinetic energy per particle is (again roughly)

$$<K.E.> = k_B T \qquad T \rightarrow \infty$$

$$= \pi \hbar^2 n/m_e \equiv E_F \qquad T \rightarrow 0 \tag{1.2}$$

In the high temperature classical limit $(k_B T \gg \pi \hbar^2 n/m_e)$ $\Gamma \equiv <V>/<K.E.> = \dfrac{e^2}{\epsilon_0} \pi^{1/2} n^{1/2} k_B T$ completely characterizes the statistical mechanics of the system. For MHV systems $(\epsilon_0 \cong 1)$ and for $n \cong 10^9 cm^{-2}$, $T \cong 1\,°K$, $\Gamma \cong 10^2$. In this regime we expect solidification to occur as one raises the density (fixed T) or lowers the temperature (fixed n). Such a *classical* $(k_B T \gg E_F)$ freezing has been observed when $\Gamma \cong 131$.[7] By now it is generally agreed that the mechanism for melting of this classical solid is the unbinding of pairs of defects suggested by Kosterlitz and Thouless (KT) and elaborated by Halperin and Nelson.[8] The theory predicts $\Gamma_{KT} = 125$. It also predicts a hexatic phase which has not yet been observed.

There is, however, another region of the phase diagram, i.e. the low temperature region where $E_F \gg k_B T$. In this regime $\Gamma = (me^2)(\dfrac{1}{\pi n})^{1/2}$ is simply the parameter r_s (the mean interparticle separation measured in effective Bohr radii). In this case we expect melting to occur as one raises the density, (the Wigner transition).[9] No real theory of this transition exists although estimates[10] and a more reliable numerical calculation has been made. The numerical calculations suggest an $r_s \cong 37$.[11] Needless to say it would be interesting to observe such a zero temperature quantum transition. A recent calculation[12] of the behavior of point detects in these 2-D electron gases shows that there is a regime near the melting density (solid side) where the zero temperature solid finds it energetically favorable to have a finite concentration of mobile vacancies. These mobile vacancies may lead to interesting magnetic properties of the solid and ultimately to an instability i.e. melting.

Nevertheless, from our qualitative arguments we see that the phase diagram for such a 2-D electron system must look roughly like the one sketched in Fig. (1.3);[10] in particular the curve labeled bulk. The low temperature (T=0) quantum phase boundary is not normally accessible to experiments on bulk He where $r_s \cong 3 \times 10^3$ for $n = 10^9 cm^{-2}$ or in semiconductor structures where $r_s \lesssim 3$ because $m_e \cong .1m$ $\epsilon_0 \cong 10$ and $n \geq 10^{11}$ cm^{-2}. By going to higher densities $(n \sim 10^{12} cm^{-2})$ in the MHV system, we could conceivably reduce r_s to one hundred, and make the fermi energy a few degrees Kelvin so that quantum effects would become evident. However, getting to the regimes $r_s \sim 10$ would be difficult without some new experimental breakthrough.

The full quantum regime becomes accessible if one is willing to work with films of helium on materials with large effective dielectric constants (see Fig. 1.4).[13] In such cases, it is possible to change the effective interaction between electrons due to the presence of image charges in the underlying substrate when the film thickness d is comparable to the interparticle spacing. As an example we consider a thin film of dielectric on a metal. The metal can, for the purpose of these screening arguments, be considered a dielectric with an $\epsilon = -\infty$. In this case the law of force between electrons, at large distances, looks dipolar, i.e.,

$$V(r) \approx \frac{e^2}{r} \left(\frac{d}{r} \right)^2 \qquad (1.3)$$

While no quantitative theory exists for such situations it is clear that the screening of the Columb interaction by the substrate pushes the quantum region of the phase diagram to lower densities and temperatures. This fact and the fact that the films can support higher densities makes the quantum region experimentally accessible. One simple estimate for the modified phase boundary for a film of 100 Å thickness on a metal with $\epsilon = -\infty$ (solid curve) and an insulator with $\epsilon = 10$ (dashed curve) is shown in Fig. 1.3.[13] It displays a

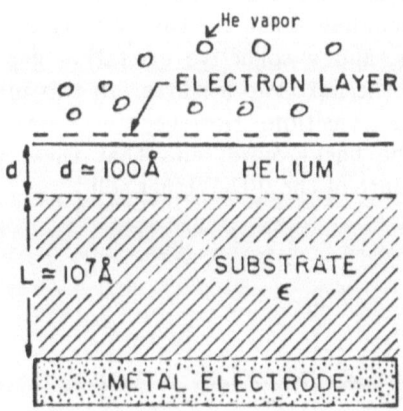

Fig. 1.4 Geometric configuration for charged He film. The circles above the electron layer are Helium vapor.

new feature i.e. a fluidlike region which persists as $n \rightarrow 0$, $T \rightarrow 0$. This new interesting feature arises because the force is dipolar so that $<V> \sim n^{3/2}$. Investigations of the fluid-like phase down to very low temperatures in the presence of ripplons, excitations of the He film, might also reveal a possible superfluid transition.

COLUMB INTERACTIONS (large perpendicular magnetic field)

When a large field B $(\hbar\omega_c > E_F > k_B T)$ is applied perpendicular to the interface the kinetic energy of the electrons in the plane is quenched (see Fig. 1.5) i.e. the electrons all reside in the lowest Landau level. In this case carrier carrier interactions dominate the behavior at low T. Magneto-resistance and Hall resistance measurements conducted over the past few years on a variety of modulation doped GaAs-(AlGa)As heterostructures at temperatures below 1 °K have shown that these quantities (unlike their non-interacting versions) behave in a remarkably rich fashion.[1,14,15] While the GaAs system is important and interesting, it is limited. Primarily one cannot go below densities of a few times 10^{11}. This means that with the highest DC fields available (250 kG) it is not possible to decrease the fractional filling ν of the lowest Landau level $(\nu \equiv E_F/\hbar\omega_c)$ much below 1/5. In addition the carrier concentration cannot be changed by large factors in the same sample. Finally, since the behavior of this unique systems is so striking it seems important to verify the behavior in, for example, MHV systems under similar experimental conditions. This would help elucidate the connection between the theory of the 2-D no impurity 1/r electron gas and the actual experiment.

Recent theoretical calculations[16,17] strongly suggest that the striking behavior of GaAs heterostructures near $\nu = 1/3$ is due to the existence of a novel ground state of the electron gas at rational fractions (odd denominators) of Landau level filling. This new electronic ground state has a number of remarkable properties including a gap for the excitation of fractionally charged quasi-particles and holes, and a collective excitation gap (Magneto-Roton).[18] This will be the topic of the next two lectures. In this picture the electron gas does *not* undergo a phase transition. However we "know" that there are phase transitions lurking in the background; and that these phase transitions are connected with the existence of the 2-D Wigner solid.

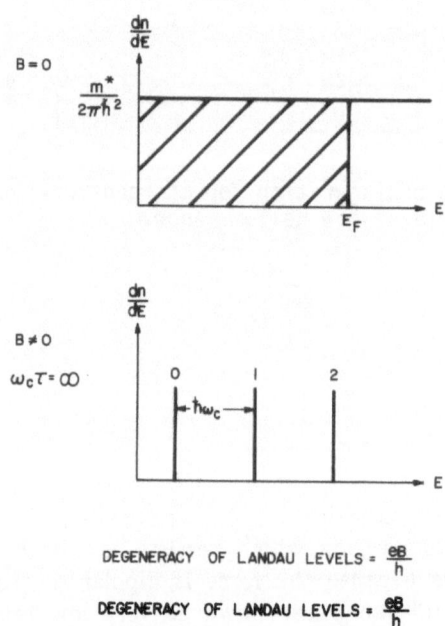

Fig. 1.5 Density of 2-D states for a non-interacting electron gas with or without a magnetic field.

The Hamiltonian describing a 2-D electron gas in the extreme quantum limit (no Landau Level mixing) is,

$$H = 1/2\sum_q V(q)\left[\bar{\rho}(q)\bar{\rho}(-q) - \rho e^{-q^2\ell^2}\right] \tag{1.4}$$

where $\bar{\rho}(q)$ is the projected density operator. The absence of a kinetic energy term would suggest that a solid is the ground state of such a system, i.e. it seems that Eq. (1.4) describes a set of interacting classical particles. However, the operators $\bar{\rho}(q)$ do not commute with one another. Nevertheless, in the limit $B\rightarrow\infty^{[19]}$ $(\ell\equiv\hbar c/eB)^{1/2}$.

$$[\bar{\rho}_q,\bar{\rho}_{-q}] \rightarrow \ell \tag{1.5}$$

and we do have a solid probably triangular in the limit of infinite magnetic field.

Early on it was suggested that in the experimental regime $(r_s\cong 1, \nu\geq 1/5)$ that there was a transition to solid and that the correlated ground state for the observed experimental anomalies was a Wigner crystal.[19,20] Several variational calculations of the ground state energy of such a system showed that it was indeed favored over the uniform gas like ground state,[20,21] however unlike the experiments there were no special values of the filling factors. In addition the experiments show vanishing dissipation in the limit of zero temperature and no discernible non-linear conductivity. Both of these observations are difficult to reconcile with a solid in the presence of impurities.

R. Laughlin[16] using a trial wave function of the form,

$$\Psi_m(z_1\cdots z_n) = \prod_{i<j}(z_i-z_j)^m\, e^{-1/4\sum_k|z_k|^2} \tag{1.6}$$

where $z_i = x_i + iy_i$ $(\ell = 1)$ and m odd showed:

1. The density of such a state is uniform and $\nu=1/m$.

2. The energy is about 10% lower than the solid at $\nu=1/3$.

3. The energy required to add a particle to such a state is finite.

All of these results are in good qualitative agreement with the experiment. The current theory of the liquid neglects coupling to other Landau levels. This is equivalent to assuming,

$$\gamma = (e^2/\epsilon_0)\pi n^{1/2}/\hbar\omega_c \equiv \nu r_s < 1. \tag{1.7}$$

The experimental value of r_s for GaAs at $n\cong 5\times 10^{11}$ is about two so that $\gamma\cong .6$ for $\nu=1/3$ and a theory assuming $\gamma=0$ is probably okay. The larger r_s for fixed ν, the more the solid is favored. At $\nu=1/3$ we guess that the solid is lower in energy for $r_s\cong 10$. At small ν for any r_s the solid is also favored. Lam and Girvin suggest[22] that the solid is the lowest state for $\nu<1/7$ even for $r_s=0$. At the current time there is considerable debate and uncertainty regarding the exact value of these numbers. Nevertheless the qualitative aspects of the statements made here are correct.

All of these rough arguments about the general nature of the phase diagram is summarized in Figs. (1.6-1.7). In Fig. (1.6) the T=0 situation is sketched as a function of r_s and ν.[23] The solid dashed vertical lines indicate the Laughlin incompressible fluid which runs into the solid liquid phase boundary probably with a cusp at an $r_s \cong 10$ for $\nu = 1/3$. The shaded region around $\nu = 1/3$ and $1/5$ is meant to show some type of Wigner solidification of excited quasi-particle (holes) in a narrow region around the fluid state.[23] The asymptote in the boundary curve shown as $r_s \cong 10$ is the zero field point on the curve in Fig. (1.3).

At finite temperatures we can characterize the *phase diagram* as shown in Fig. (1.7). For $r_s = \infty$ the magnetic field has no effect on the classical Kosterlitz Thouless melting curve $\Gamma = 131$. The curves drawn in for different r_s are only meant to show trends although a crude estimate of such curves based on a Lindeman like melting criterion was made in some unpublished work by Fukuyama and Platzman. In essence quantum fluctuations (finite r_s) decrease the solid regime, while a big field enhances the solid region. Needless to say no real theory of any of this exists and an experimental investigation of these quantum transitions would be exciting.

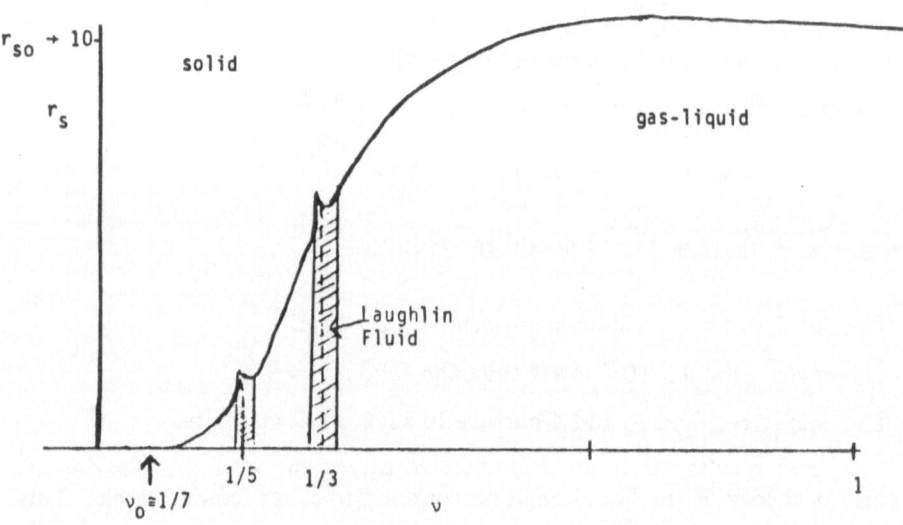

Phase Diagram T=0

Fig. 1.6 Qualitative phase diagram at T = 0 for a 2-D electron gas in the lowest Landau level ($\nu \equiv E_F/\hbar\omega_c$).

ONE ELECTRON COUPLED TO PHONONS

When the 2-D electron system is trapped at the Helium liquid vacuum interface it has been recently demonstrated that there are striking effects due to the coupling of the electron to the quantized excitations of the liquid surface ripplons.[6,24] The basic form of the problem is simply elucidated when we remember that an electron on the surface is always subject to some (fairly uniform) electric field $E_\perp$,

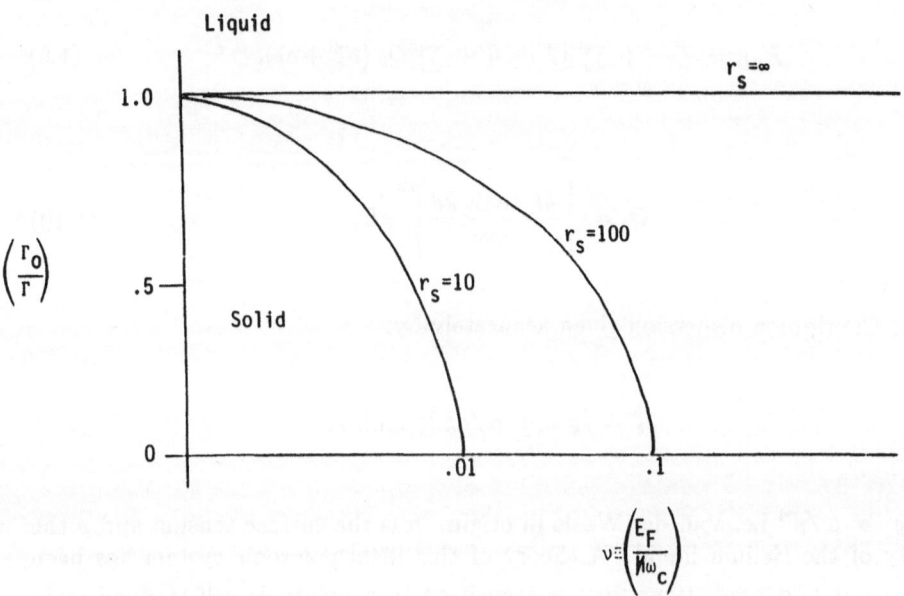

Fig. 1.7 The qualitative effects of r_s and magnetic field on the solidification boundary of a 2-D electron gas.

$$eE_\perp = eE_{ext} + \frac{e^2}{4d^2} \frac{(\epsilon-1)}{(\epsilon+1)} . \tag{1.8}$$

Here E_{ext} is essentially determined by the charge density n_e per unit area at the surface, i.e., $E_{ext}=4\pi n_e$. The second term comes from the image charge induced in the dielectric substrate (dielectric constant ϵ) supporting the He film. As the film gets thin the total field must get large independent of the charge density. Since the coupling to the surface deformation is quite accurately given by $V = eE_\perp \cdot \delta$ where δ is the displacement of the surface we immediately see that thin films means strong 2-D electron ripplon coupling. In fact several authors have shown that this is an almost ideal 2-D polaron problem as perhaps first envisioned by Frohlich with a short range and a variable coupling strength.[6,25] The Hamiltonian describing this one electron problem is

$$H_{2D} = \frac{p^2}{2m} + \sum_k a_k^+ a_k \omega_k + \sum Q_k (a_k^+ + a_k) e^{i\vec{k}\cdot\vec{r}} \tag{1.9}$$

with

$$Q_k = \left(\frac{\hbar k \tanh kd}{2\rho\omega} \right)^{12} eE_\perp \tag{1.10}$$

and ω_k the ripplon dispersion given accurately by,

$$\omega_k^2 = (g' k + \sigma/\rho k^3) \tanh kd \tag{1.11}$$

with $g' = \alpha/d^4$ i.e. Van der Waals in origin. σ is the surface tension and ρ the density of the Helium liquid. A theory of this ideal polaronic system has been worked out and predictions for the transition to a polaronic self-trapped state were made. Fig. (1.8) is a plot of mass versus coupling constant

$$\alpha \equiv (eE_\perp)^2 / 8\pi\sigma k_c^2 \tag{1.12}$$

at T=0. The recent preliminary experiments of Andrei[24] show that there is indeed a dramatic change in the properties of the electrons on the surface. In this very beautiful experiment bulk helium was charged with a density of electrons $n \cong 10^8 cm^{-2}$. The thickness d of the helium was then continuously decreased. A plot of the mobility versus d is shown in Fig. 1.9. Several of the features of this experiment are consistent with the one electron ripplon theory.[6] However a great deal more work on this problem must be done, density temperature, frequency etc., before one has a good feel for all the phenomenon involved. There surely will be a regime (low density, and low temperature) where the one electron theory works well. However, there will also be a regime (high density) where many electrons coupled to retarded phonons and by instantaneous Columb interactions will be a necessary part of the description. It is not hard to conceive of a new 2-D superfluid phase in such a system.

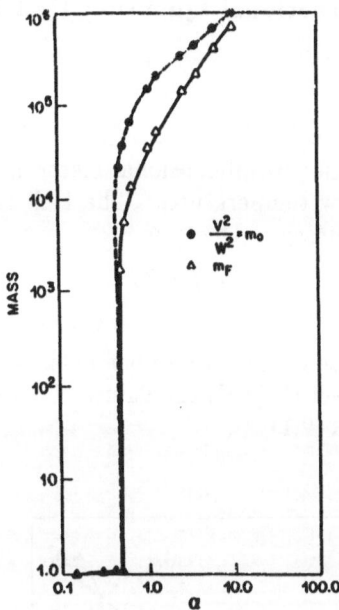

Fig. 1.8 Effective Feynman mass for model 2-D polaron system.

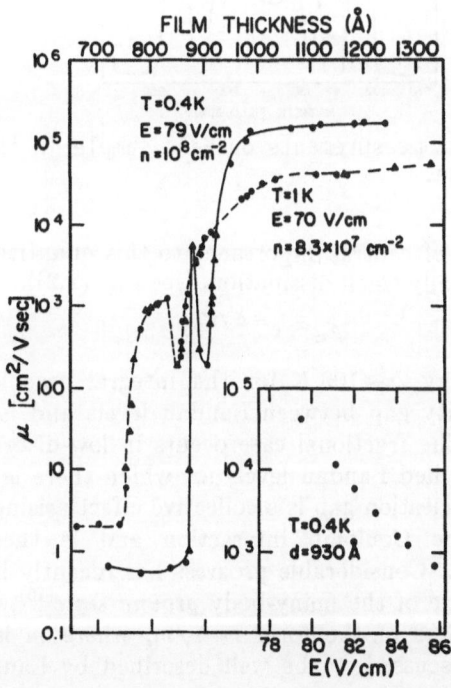

Fig. 1.9 Measurement of the mobility μ^* as a function of the film thickness d at two different temperatures. Triangles denote increasing d; circles, decreasing d. The lines are guides for the eye. E is the applied field. The inset shows the E dependence of μ^* at a fixed film thickness, and density n = 10^8 cm^{-2}.

II. Excitations in the Fractional Quantum Hall Effect - MAGNETO ROTONS

INTRODUCTION

The quantum Hall effect[1,16,26] is a remarkable macroscopic quantum phenomenon occurring in the two-dimensional electron gas (inversion layer) at high magnetic fields and low temperatures. The Hall resistivity is found to be quantized with extreme accuracy in the form

$$\rho_{xy} = h/e^2 i. \tag{2.1}$$

In the integral Hall effect the quantum number i take on integral values. In the fractional quantum Hall effect (FQHE) the values of i are rational fractions with odd denominators. (see Fig. (2.1))

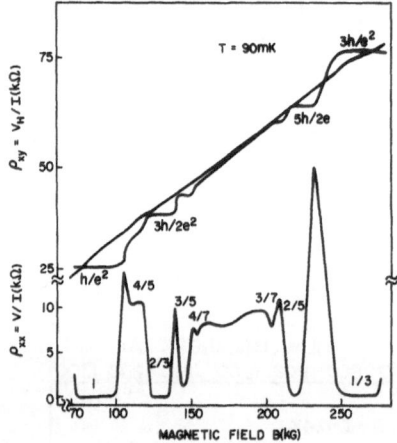

Fig. 2.1 Transport measurements on GaAs sample in the Fractional Quantum Hall Regime.

Associated with and of central importance to this quantization of ρ_{xy} is the appearance of exponentially small dissipation, (see Fig. (2.2)).

$$\rho_{xx} \sim e^{-\Delta/2T}. \tag{2.2}$$

The activation energy $\Delta \approx 100\ K$ for the integral case is associated with disorder and the mobility gap between Landau levels and is thus primarily a single-particle effect. The fractional case occurs in low-disorder, high-mobility samples with partially filled Landau levels for which there is no single-particle gap. In this case the excitation gap is a collective effect arising from many-body correlations due to the Coulomb interaction and is therefore smaller in magnitude ($\Delta_{1/3} \approx 6\ K$). Considerable progress has recently been made toward understanding the nature of the many-body *ground state*[16], which at least for Landau-level-filling factors of the form $\nu = 1/m$, where m is an odd integer, appears, as we have discussed, to be well described by Laughlin's variational wave function.

In this lecture we would like to describe a microscopic theory of the excitation spectrum of such systems. The theory is closely analogous to Feynman's theory of the excitation spectrum of superfluid Helium[27]. It goes under the assumption that all significant excitations are collective in character. We begin by discussing Feynman's theory and then discuss the application of these ideas to Fermion systems in a magnetic field.

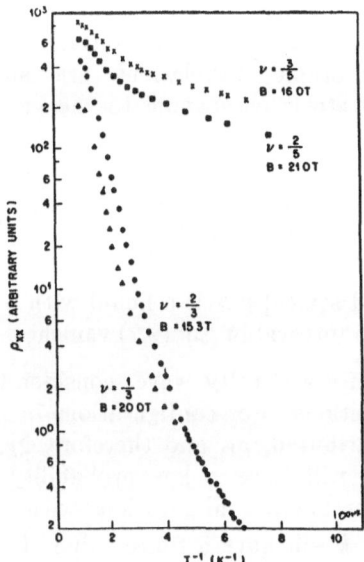

Fig. 2.2 Measured activation in the FQHE.

REVIEW OF FEYNMAN'S THEORY OF ^{4}He

In a beautiful series of papers, Feynman has laid out an elegantly simple theory of the collective excitation spectrum in superfluid ^{4}He[27]. Even though the underlying ideas were developed for a neutral Bose system in three dimensions, they are sufficiently general that they can be applied *mutatis mutandis* to a charged Fermi system in two dimensions in a high magnetic field. Let us therefore now review Feynman's arguments.

Because ^{4}He is a Bose system, the ground-state wave function is symmetric under particle exchange and has no nodes. Using these facts Feynman argues that there can be no low-lying single-particle excitations, so that the only low-energy excitations are long-wavelength density oscillations (phonons). Now suppose that somehow one knew the exact ground-state wave function, Ψ. Then one could write the following variational wave function to describe a density-wave-excited state which still contains most of the favorable correlations occurring in the ground state:

$$\Phi_k(r_1,...,r_n) = N^{-1/2}\rho_k\Psi(r_1,...,r_N), \tag{2.3}$$

where ρ_k is the Fourier transform of the density

$$\rho_k = \int d^2R \exp(-ik\cdot R)\rho(R), \tag{2.4}$$

$$\rho(R) = \sum_{j=1}^{N} \delta^2(R-r_j), \tag{2.5}$$

so that

$$\rho_k = \sum_{j=1}^{N} \exp(-ik \cdot r_j). \tag{2.6}$$

Note that since the ansatz wave function contains the ground-state wave function as a factor, favorable correlations are automatically built in. Nevertheless, the excited state is orthogonal to the ground state as required. This may be seen by writing

$$<\Psi | \Phi_k > = N^{-1/2} \int d^2R \exp(-ik \cdot R) <\Psi | \rho(R) | \Psi >. \tag{2.7}$$

By hypothesis is the ground state $|\Psi>$ is a liquid with a homogeneous density. Hence (for $k \neq 0$) the overlap integral in Eq. (2.7) vanishes.

To see that Φ_k represents a density wave, consider the following. Φ_k is a function of the particle positions. For configurations in which the particles are more or less uniformly distributed, ρ_k, and therefore Φ_k, will be close to zero. Hence such configurations will have a low probability of occurrence in the excited state. On the other hand, configurations with some degree of density modulation at wave vector k will have a finite value of ρ_k, and Φ_k, will match the phase of the density wave. All phases are equally likely if Ψ corresponds to a liquid. Hence the average density is still uniform in the excited state. A simple way of interpreting all of this to a view ρ_k as one of a set of collective coordinates describing the particle configuration. We can then make an analogy with the simple harmonic oscillator in which the exact excited-state wave function Ψ_0 by multiplication by the coordinate.

$$\Psi_1(x) = x \Psi_0(x). \tag{2.8}$$

Ψ_1 is orthogonal to Ψ_0, and Ψ_1 vanishes for the configuration ($x=0$) at which Ψ_0 is peaked ("uniform density"). Ψ_1 is nonzero when the coordinate is displaced ("density wave"), but has a phase that varies with displacement so that the average value of the coordinate is zero ("uniform average density") even in the excited state.

Having established an ansatz variational wave function, we need to evaluate the energy. This requires a knowledge of the norm of the wave function

$$s(k) = <\Phi_k | \Phi_k > = N^{-1} <\Psi | \rho_k^\dagger \rho_k | \Psi >, \tag{2.9}$$

but this is nothing more than the static structure factor for the ground state, a quantity which can be directly measured using neutron scattering. For later purposes it will be convenient to note that $s(k)$ is also related to the Fourier transform of the radial distribution function for the ground state, $g(r)$:

$$s(k) = 1 + \rho \int d^2R \exp(-ik \cdot R)[g(R) - 1]$$

$$+ \rho(2\pi)^2 \delta^2(k), \tag{2.10}$$

where ρ is the average density.

The variational estimate for the excitation energy is the usual expression

$$\Delta(k) = f(k)/s(k), \tag{2.11}$$

where

$$f(k) = \langle \Phi_k | H - E_0 | \Phi_k \rangle, \tag{2.12}$$

and H is the Hamiltonian and E_0 is the ground-state energy. Using Eq. (2.3) we may rewrite Eq. (2.12) as

$$f(k) = N^{-1} \langle \Psi | \rho_k^{\dagger} [H, \rho_k] | \Psi \rangle, \tag{2.12a}$$

which will be recognized as the oscillator strength. Because the potential energy and the density are both simply functions of position, they commute with each other. The kinetic energy, on the other hand, contains derivatives which do not commute with the density. This yields the universal result

$$f(k) = \frac{\hbar^2 k^2}{2m}, \tag{2.13}$$

making the oscillator-strength sum independent of the interaction potential. We emphasize this point because a rather different result will be obtained for the case of the FQHE.

Combining Eqs. (2.13) and (2.12) yields the Feynman-Bijl formula for the excitation energy,

$$\Delta(k) = \frac{\hbar^2 k^2}{2ms(k)}. \tag{2.14}$$

We can interpret this as saying that the collective-mode energy is just the single-particle energy $\hbar^2 k^2 / 2m$ renormalized by the factor $1/s(k)$ which represents the effect of correlated motion of the particles. We emphasize that since we have invoked the variational principle, $\Delta(k)$ is a rigorous upper bound to the lowest excitation energy at wave vector k.

In order to gain a better understanding of the meaning of $\Delta(k)$ and the underlying assumptions that have been used, let us rederive Eq. (2.14) by a different method. Consider the dynamic structure factor defined by ($\hbar = 1$ throughout)

$$S(k, \omega) = N^{-1} \sum_n \langle 0 | \rho_k^{\dagger} | n \rangle \delta(\omega - E_n + E_0) \langle n | \rho_k | 0 \rangle, \tag{2.15}$$

where the sum is over the complete set of exact eigenstates. Using Eq. (2.15) the static structure factor is related to $S(k, \omega)$ by

$$s(k) = \int_0^{\infty} d\omega \, S(k, \omega), \tag{2.16}$$

and using Eq. (2.12a) we see that the oscillator strength is related to $S(k,\omega)$ by

$$f(k) = \int_0^\infty d\omega\, \omega S(k,\omega). \tag{2.17}$$

Substitution of these results into Eq. (2.11) shows that the Feynman-Bijl expression for $\Delta(k)$ is actually the exact first moment of the dynamic structure factor. That is, $\Delta(k)$ represents the average energy of the excitations which couple to the ground state through the density. This is consistent with the idea that $\Delta(k)$ is a variational bound on the collective-mode energy, since the average energy necessarily exceeds the minimum excitation energy. Note that if the oscillator-strength sum is saturated by a single mode, then the mean excitation energy and the minimum will be the same and the Feynman-Bijl expression will be exact. This idea is consistent with Feynman's argument that there are no low-lying single-particle excitations. The assumption that Φ_k in Eq. (2.3) is a good variational wave function is equivalent to assuming that the density-wave saturates the oscillator-strength sum rule.

How well does the Feynman-Bijl formula work? Evaluating $s(k)$ in Eq. (2.14) from the experimental neutron-scattering cross section and speed-of-sound data yields a collective-mode frequency which vanishes linearly at small k (with a slope corresponding exactly to the speed of sound) and exhibits the famous "roton minimum" near $k = 2\text{Å}^{-1}$. The roton minimum is due to a peak in $s(k)$ associated with the short-range order in the liquid. The predicted dispersion curve (see Fig. (2.3)) is in good qualitative agreement with experiment, but the predicted frequency at the roton minimum is about a factor of 2 too high. Feynman and Cohen[28] have shown that this problem can be remedied by including back-flow corrections which guarantee that the continuity equation $\Delta \cdot <J> = 0$ is satisfied by the variational wave function. These corrections bring the predicted mode energy into excellent quantitative agreement with experiment.

These considerations of Feynman's arguments lead us to the following conclusions. One expects on quite general grounds that (at least for long wavelengths) the low-lying excited states of any system will include density waves. If because of special circumstances (such as those occurring in superfluid ^{4}He) there are no low-lying single-particle states, then the Feynman-Bijl expression for the collective-mode energy will be valid. The expression is also qualitatively correct even at short wavelengths and yields quantitative agreement with experiment if explicit back-flow corrections are included.

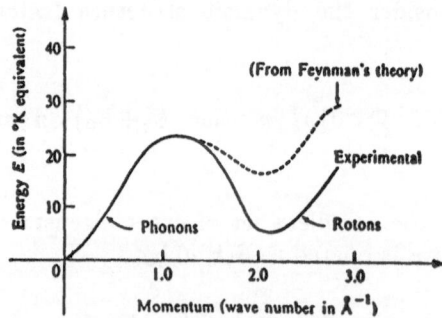

Fig. 2.3 The Phonon-Roton Spectrum in Superfluid Helium.

APPLICATION TO FERMIONS

The single-mode approximation seems less likely to succeed for Fermi systems. For instance, examination of $S(k,\omega)$ for the three-dimensional electron gas (jellium) shows the existence of not only a collective mode (the plasmon) but also a large continuum of single-particle excitations. (see Fig. 2.4) The low-lying continuum is due to the small kinetic energy of excitations across the Fermi surface. Despite this problem Lundqvist and Overhauser[29] have shown that very useful results can be obtained from the single-mode approximation (SMA). It is straightforward to prove from the compressibility sum rule that the plasmon mode saturates the oscillator-strength sum rule in the limit $k \to 0$. Hence the SMA gives the exact plasma frequency $\Delta(0) = \hbar\omega_p$. For finite k the SMA breaks down, but as noted above $\Delta(k)$ is the exact first moment of the excitation spectrum. Thus for large k, $\Delta(k)$ lies at the centroid of the continuum. Hence the SMA is exact at long wavelengths and gives a reasonable fit to the single-particle continuum even at short wavelengths. The same statement is true for the two-dimensional case, where the plasmon dispersion is $\omega_p \sim k^{1/2}$ at long wavelengths.

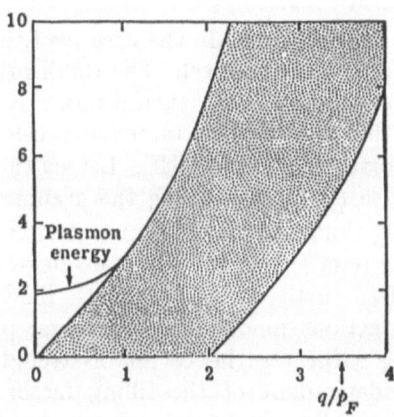

Fig. 2.4 The excitation spectrum for the 3-D electron gas in the Random Phase Approximation.

We are now in a position to ask what the effect of a large magnetic field is on the Fermi system. It is important to note that the neutral excitations are still characterized by a conserved wave vector k, but that in two dimensions the magnetic field quenches the single-particle continuum of kinetic energy, leaving a series of discrete, highly degenerate Landau levels evenly spaced in energy at intervals of $\hbar\omega_c$, where ω_c is the classical cyclotron frequency. Consider first the case of the filled Landau level ($\nu=1$). Because of Pauli exclusion the lowest excitation is necessarily the cyclotron mode in which particles are excited into the next Landau level. Furthermore, from Kohn's theorem we know that in the limit of zero wave vector, the cyclotron mode occurs at precisely $\hbar\omega_c$ and saturates the oscillation-strength sum rule. Hence, once again the SMA is exact at long wavelengths and yields $\Delta(0) = \hbar\omega_c$. To see explicitly that this is so, note that for $\nu=1$ the radial distribution function is known exactly (neglecting mixing of Landau levels in the ground state),

$$g(r) = 1 - \exp(-r^2/2l^2), \qquad (2.18)$$

where $\ell = (eB/\hbar c)^{-1/2}$ is the magnetic length. Using (10) and $\rho = \nu/(2\pi l^2)$, we have, for $k \neq 0$,

$$s(k) = 1 - \exp(-k^2 l^2/2). \qquad (2.19)$$

The predicted mode energy is

$$\Delta(k) = \frac{\hbar^2 k^2}{2m[1 - \exp(-k^2 l^2/2)]}, \qquad (2.20)$$

but $\hbar^2/ml^2 = \hbar\omega_c$, so that we finally obtain

$$\Delta(k) = \hbar\omega_c \frac{(kl)^2/2}{1 - \exp(-k^2 l^2/2)}, \qquad (2.21)$$

which has the correct limit $\Delta(0) = \hbar\omega_c$. We emphasize that this was derived with a ground-state structure factor calculated by neglecting Landau-level mixing, but that Kohn's theorem requires that the same result be obtained (for $k \rightarrow 0$) for the exact ground state.

For the FQHE we are interested not in the case $\nu = 1$, but rather we need to consider the fractionally filled Landau level. The Pauli principle now no longer excludes low-energy intra-Landau-level excitations. It is these low-lying excitations rather than the high-energy inter-Landau-level cyclotron modes which are of primary importance to the FQHE. Let us therefore consider what the SMA yields for the case $\nu < 1$ by recalling the argument leading to Kohn's theorem. At asymptotically long wavelengths an external perturbation couples only to the center-of-mass (c.m.) motion. The c.m. degree of freedom has the excitation spectrum of a free particle in the magnetic field and is unaffected by the correlations and interactions among the individual particles. Hence once again the cyclotron mode saturates the oscillator-strength sum rule and the SMA yields $\Delta(0) = \hbar\omega_c$ independent of the filling factor and the interaction potential. The SMA thus tells us nothing about the low-energy modes of interest.

The root of this difficulty can be traced back to Eq. (15). For small k the variational excited state has most of its weight in the next-higher Landau level. We can greatly improve the variational bound on the excitation energy by insisting that the excited state lie entirely within the lowest Landau level. This can be enforced by replacing Eq. (2.3) by

$$\Phi_k = N^{-1/2}\bar{\rho}_k\Psi, \qquad (2.22)$$

where $\bar{\rho}_k$ is the projection of the density operator onto the subspace of the lowest Landau level. Providing that this projection can be explicitly carried out, we may derive a new approximation (the projected SMA) for the low-lying collective-mode frequency,

$$\Delta(k) = \bar{f}(k)/\bar{s}(k), \qquad (2.23)$$

where $\bar{f}$ and $\bar{s}$ are, respectively, the projected analogs of the oscillator strength and the static structure factor. Formulation of the projection scheme and the derivation of Eq. (2.23) are carried out in the next section.

PROJECTION ONTO THE LOWEST LANDAU LEVEL

The formal development of quantum mechanics within the subspace of the lowest Landau level has been presented elsewhere.[30] We briefly review here the pertinent results. Taking the magnetic length to be unity ($l=1$) and adopting the symmetric gauge (with $B=-B\hat{z}$), the single-particle eigenfunctions of kinetic energy and angular momentum in the lowest Landau level are

$$\Phi_m(z)=\frac{1}{(2\pi 2^m m!)^{1/2}}z^m\exp(-|z|^2 4),\qquad (2.24)$$

The projected density operator, using this basis, is

$$\bar{\rho}_k=\sum_{j=1}^{N}\exp\left[-ik\frac{\partial}{\partial z_j}\right]\exp\left[-i\frac{k^*}{2}z_j\right]\qquad (2.25)$$

where all derivatives operate to the left.

In analogy with Eq. (12a), the projected oscillator strength is

$$\bar{f}(k)=N^{-1}<0|\bar{\rho}_k^{\dagger}[\bar{H},\bar{\rho}_k]|0>,\qquad (2.26)$$

where $|0>$ is the ground state (represented as a member of the Hilbert space of analytic functions). Note that previously it was the kinetic energy which contained derivatives and hence failed to commute with the density. Now the kinetic energy is an irrelevant constant, but both the potential-energy and density operators contain derivatives. Thus Eq. (2.26) becomes

$$\bar{f}(k)=N^{-1}<0|\bar{\rho}_k^{\dagger}[\bar{V},\bar{\rho}_k]|0>.\qquad (2.27)$$

The meaning of this is that since the kinetic energy has been quenched by the magnetic field, the scale of the collective-mode energy is set solely by the scale of the interaction potential, which is, of course, as it should be.

In order to evaluate it is convenient to note that the projected density operators obey

$$[\bar{\rho}_k,\bar{\rho}_q]=(e^{k^*q/2}-e^{kq^*/2})\bar{\rho}_{k+q}.\qquad (2.28)$$

It is convenient to use parity symmetry in k to rewrite Eq. (2.27) as a double commutator,

$$\bar{f}(k)=\frac{1}{2N}<0|[\bar{\rho}_k^{\dagger},[\bar{V},\bar{\rho}_k]]|0>.\qquad (2.29)$$

Using Eq. (1.4), Eq. (2.29) is readily evaluated with the commutation relation given in (2.28),

$$\bar{f}(k)=\frac{1}{2}\sum_q v(q)(e^{q^*k/2}-e^{qk^*/2})$$

$$\times[\bar{s}(q)e^{-k^2/2}(e^{-k^*q/2}-e^{-kq^*/2})\qquad (2.30)$$

$$+\bar{s}(k+q)(e^{k^*q/2}-e^{kq^*/2})],$$

where $\bar{s}(q)$ is the projected static structure factor,

$$\bar{s}(q) = N^{-1} <0|\bar{\rho}_q^\dagger \bar{\rho}_q|0>. \tag{2.31}$$

Using

$$\rho_q^\dagger \rho_q = \bar{\rho}_q^\dagger \bar{\rho}_q + (1 - e^{-|q|^2/2}),$$

one obtains the relation

$$\bar{s}(q) = s(q) - (1 - e^{-|q|^2/2}), \tag{2.32}$$

where $s(q)$ is the ordinary static structure factor given in Eq. (2.10). From Eq. (2.19) we see that $\bar{s}(q)$ vanishes identically for the filled Landau level. This is simply a reflection of the fact that it is not possible to create any excitations within the lowest Landau level when it is completely filled.

Clearly, Eq. (2.30) is more complicated than its analog, (2.13). Nevertheless, it is still true that knowledge of the static structure factor is all that is required to evaluate (2.30) and (2.31) and hence obtain the projected SMA mode energy:

$$\Delta(k) = \bar{f}(k)/\bar{s}(k). \tag{2.33}$$

The essence of the Feynman-Bijl result (2.11) is still maintained - namely that one can express a dynamical quantity, the collective-mode energy, solely in terms of static properties of the ground state.

Let us begin our evaluation of (2.33) by consideration of the small-k limit. We assume throughout that the ground state is an isotropic and homogeneous liquid. Direct expansion of Eq. (2.30) shows that, for small k, $\bar{f}(k)$ vanishes like $|k|^4$. Indeed, one can show that this is true, in general, because Kohn's theorem tells us that the total oscillator-strength sum $f(k) = \hbar^2 k^2/2m$ is saturated by the cyclotron mode (to leading order in $|k|^2$). Hence the intra-Landau-level contribution [which is $\bar{f}(k)$] must quite generally vanish faster than $|k|^2$ for both solid and liquid ground states. Given that

$$\bar{f}(k) \sim |k|^4, \tag{2.34}$$

it follows from (2.33) that a necessary (but not sufficient) condition for the existence of a finite direct ($k=0$) gap is

$$\bar{s}(k) \sim |k|^4. \tag{2.35}$$

Equation (2.35) is a *sufficient* condition only within the SMA, but as the following argument shows, Eq. (2.35) is always a *necessary* condition for a gap. Equation (2.33) gives the exact first moment of the (intra-Landau-level) excitation spectrum. If $\bar{s}(k)$ vanishes slower than $|k|^4$, then the mean excitation energy vanishes as k approaches zero and there can be no gap. If $\bar{s}(k) \sim |k|^4$, then the mean excitation energy is finite for small k. This does not prove that there is a gap; however, it seems plausible that in this system (unlike ordinary jellium) there can be no low-lying single-particle excitations to invalidate the SMA and defeat the gap since the kinetic energy necessary to produce such a continuum has been quenched by the magnetic field.

In order to go beyond the small-k limit in evaluating Eq. (2.33), we need to have $\bar{s}(k)$ for finite k. Lacking the experimental structure factor that was available for the case of ^{4}He, we are forced to adopt a specific model for the ground state. We have chosen to use the Laughlin ground-state wave function Eq. (1.6), since it appears to be quite accurate and because the static structure factor is available through the 2DOCP analogy.

The static structure factor for the 2DOCP has been computed by both Monte Carlo (MC) and hypernetted-chain (HNC) methods. Using these results we have evaluated the collective mode dispersion relation (see Fig. (2.5)). The gap is finite at zero wave vector and exhibits a deep minimum at finite k. This magneto-roton minimum is caused by a peak in $\bar{s}(k)$ and is, in this sense, quite analogous to the roton minimum in helium. We interpret the deepening of the minimum in going from $\nu = \frac{1}{3}$ to $\nu = \frac{1}{7}$ to be a precursor of the collapse of the gap which occurs at the critical density ν_c for Wigner crystallization. From Fig. 2.5 we see that the minimum gap is very small for $\nu < \frac{1}{5}$. This is consistent with a recent estimate of the critical density, $\nu_c = 1/(6.5 \pm 0.5)$. Within mean-field theory, the Wigner crystal transition is weakly first order and hence occurs slightly before the roton mode goes completely soft. Further evidence in favor of this interpretation of the roton minimum is provided by the fact that the magnitude of the primitive reciprocal-lattice vector for the crystal lies close to the position of the magneto-roton minimum, as indicated by the arrows in Fig. 2.5. These ideas suggest the physical picture that the liquid is most susceptible to perturbations whose wavelength matches the crystal lattice vector. This will be illustrated in more detail.

Having provided a physical interpretation of the gap dispersion and the magneto-roton minimum, we now examine how accurate the SMA is. Figure 2.6 shows the excellent agreement between the SMA prediction for the gap and exact numerical results for small ($N = 6, 7$) systems recently obtained by Haldane and Rezayi. Those authors have found by direct computation that the single-mode approximation is quite accurate, particularly near the roton minimum, where the lowest excitation absorbs 90% of the oscillator strength. This means that the overlap between our variational state and the exact lowest excited eigenstate *exceeds* 0.98. We believe this agreement confirms the validity of the SMA and the use of the Laughlin-state static structure factor.

Near $k = 0$ there is a small ($\sim 20\%$) discrepancy between $\Delta_{SMA}(0)$ and the numerical calculations. It is interesting to speculate that the lack of dispersion near the roton minimum may combine with residual interactions to produce a strong pairing of rotons of opposite momenta leading to a two-roton bound state of small total momentum. This is known to occur in helium. For the present case $\Delta_{1/3}(0)$ happens to be approximately twice the minimum roton energy. Hence the two-roton bound state which has zero oscillator strength could lie slightly below the one-phonon state which absorbs all of the oscillator strength. For $\nu < \frac{1}{3}$ the two-roton state will definitely be the lowest-energy state at $k = 0$. It would also be interesting to compare the numerical excitation spectrum with a multiphonon continuum computed using the dispersion curves obtained from the SMA.

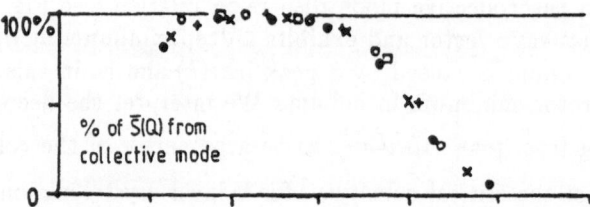

Fig. 2.5 For comparison of the Single Mode Approximation with the numerical experiments ref. (31) circles are from a seven and triangles for a six particle system.

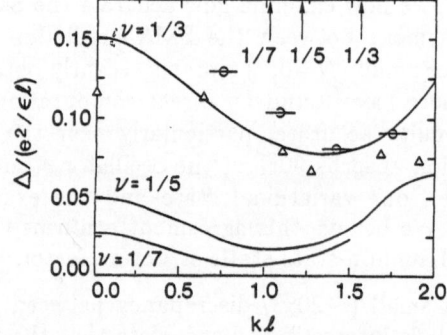

Fig. 2.6 Overlap of SMA wave function with exact numerical wave function.

The calculations discussed in the preceding sections have all used the point Coulomb interaction in 2-D. In order to make a comparison with experiment, it is important to recognize that the finite extent of the electron wave functions perpendicular to the plane cuts off the divergence of the Coulomb interaction at short distances. For both GaAs and Si devices, this thickness is on the order 100 Å, which exceeds the magnetic length at high fields ($l=66$ Å at $B=15$ T).

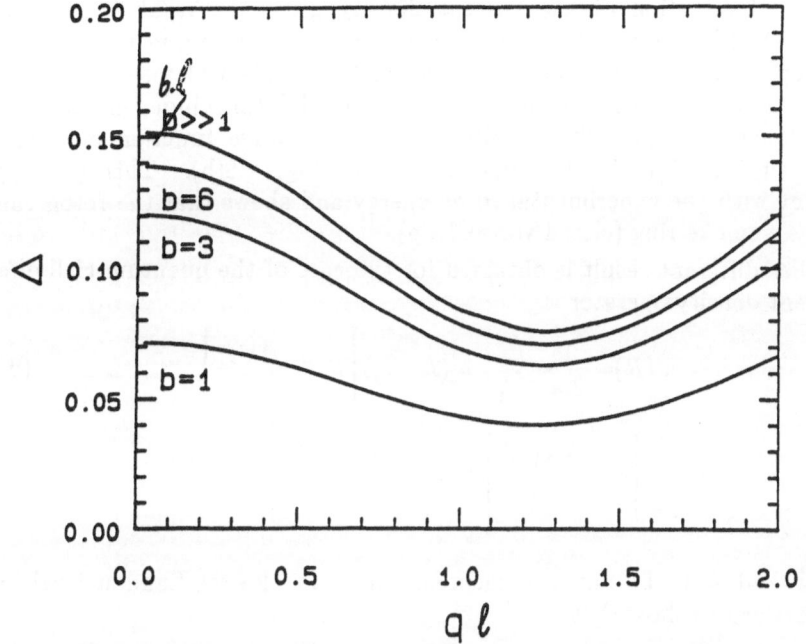

Fig. 2.7 The effect of finite electron extent on the collective excitation spectrum. $bl \gg 1$ corresponds to zero spreading.

Using the Fang-Howard variational form for the charge distribution normal to the plane i.e.,

$$g(z) = \frac{b^3}{2} z^2 e^{-bz}, \qquad (2.36)$$

Figure 2.7 shows the collective-mode dispersion for four different values of the dimensionless thickness parameter bl. We see that experimentally relevant values of bl cause a significant reduction in the size of the gap. Furthermore, this reduction is B-field dependent. Naively, we expect $\Delta \sim B^{1/2}$ (for fixed ν) since the natural unit of energy is $e^2/\epsilon l$. However, $bl \sim B^{-1/2}$ and therefore Δ does not rise as rapidly as might be expected at large B.

BACKFLOW CORRECTIONS

It is apparent from Fig. 2.5 that the SMA works extremely well - better, in fact, than it does for helium. Why is this so? Recall that, for the case of helium, the Feynman-Bijl formula overestimates the roton energy by about a factor of two. Feynman traces this problem to the fact that a roton wave packet made up from the trial wave functions violates the continuity equation

$$\nabla \cdot <J> = 0. \qquad (2.37)$$

To see how this happens, consider a wave packet

$$\Phi(r_1, \ldots, r_N) = \int d^2k \, \xi(k) \rho_k \Psi(r_1, \ldots, r_N), \qquad (2.38)$$

where $\xi(k)$ is some function (say a Gaussian) sharply peaked at a wave vector k located in the roton minimum. It is important to note that this wave packet is quasistationary because the roton group velocity $d\Delta/dk$ vanishes at the roton minimum. Evaluation of the current density gives the result schematically illustrated in Fig. 2.8(a). The current has a fixed direction and is nonzero only in the region localized around the wave packet. This violates the continuity equation (2.37) since the density is (approximately) time independent for the quasistationary packet. The modified variational wave function of Feynman and Cohen includes the backflow shown in Fig. 2.8(b). This gives good agreement with the experimental roton energy and shows that the roton can be viewed as a smoke ring (closed vortex loop).

A rather different result is obtained for the case of the quantum Hall effect. The current density operator is

$$J(R) = \frac{1}{2m} \sum_{j=1}^{N} \delta^2(R - r_j) \left[p_j + \frac{eA(r_j)}{c} \right] \qquad (2.39)$$

$$+ \left[p_j + \frac{eA(r_j)}{c} \right] \delta^2(R - r_j) \, .$$

Taking Φ and Ψ to be any two members of of the lowest Landau level, it is straightforward to show that

$$<\Phi | J(R) | \Psi> = -\frac{1}{2} \nabla \times <\Phi | M(R) | \Psi>, \qquad (2.40)$$

where

$$M(R) = \rho(R) \hat{z}, \qquad (2.41)$$

and $\rho(R)$ is the density and $\hat{z}$ is the unit vector normal to the plane. It follows immediately from (40) that

$$\nabla \cdot <J(R)> = 0 \qquad (2.42)$$

for any state in the lowest Landau level. Hence the backflow condition is automatically satisfied. The current flow for the magneto-roton wave packet calculated from (2.40) is illustrated schematically in Fig. 2.8(c). We see that the magneto-roton circulation is rather different from the smoke ring in bulk helium shown in Fig. 2.8(b).

Equation (2.42) is paradoxical in that it implied $\partial \rho/\partial t = 0$ for every state in the lowest Landau level. This is merely a reflection of the fact that the kinetic energy has been quenched and perturbations can cause particles to move only be means of (virtual) transitions to higher Landau levels. One can resolve this paradox by noting that there are really two different current operators we can consider. The first is the ordinary (instantaneous) current discussed above. The second is the slow (time-averaged) $E \times B$ drift of the particles in the magnetic field. Restriction of the Hilbert space to the lowest Landau level eliminates the fast degrees of freedom associated with the cyclotron motion but retains the slow (drift) coordinates.

For the case of the magneto-roton wave packet discussed above, we note that the excess particle density is circularly symmetric. hence the (mean) electric field is radial and the particle drift is purely circular, as illustrated previously in Fig. 2.8(c). Hence one is once again led to the conclusion that the continuity condition is automatically satisfied by the magneto-roton wave packet. We believe that this accounts for the excellent results obtained using the SMA.

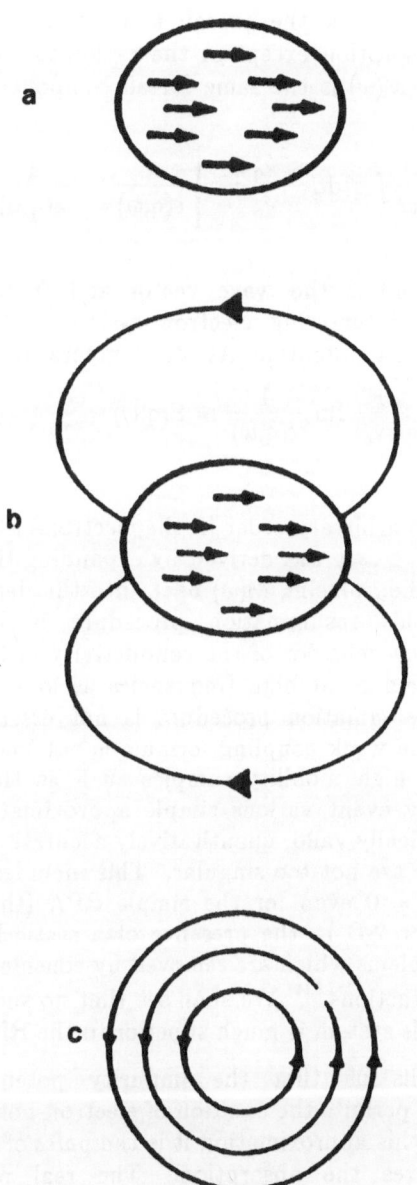

Fig. 2.8 Schematic illustration of the current distribution in a roton wave packet (a) helium with no backflow (b) helium with backflow and (c) lowest Landau level magneto-roton.

III. Transport and Impurity Effects in the Magneto-Roton Picture

Having shown how the SMA leads to a good microscopic description of the lowest lying excited states FQHE it is natural to ask if we can accurately describe the observed transport properties of such systems.[31]

Formally the long wave length AC conductivity for the system may be written in terms of a memory function $M(\omega)$.[32] In particular,

$$\sigma_+ = \sigma_{xx} + i\sigma_{xy} = \frac{ine^2/m}{\omega - \omega_c + M(\omega)} \tag{3.1}$$

where m is the band mass (no Coulomb corrections) of the electron. In the absence of impurities which break the overall translational symmetry, $M(\omega)$ vanishes and there is no absorption except at the cyclotron frequency. In the presence of weak impurities $M(\omega)$ is the same for all components of the tensor $\sigma_{\alpha\beta}$ and is given by,

$$M(\omega) = \frac{1}{4\pi m\omega} \int q^3 dq \frac{\langle \phi_q^2 \rangle}{v_q} \left[\frac{1}{\epsilon(q,\omega)} - \frac{1}{\epsilon(q,o)} \right] \tag{3.2}$$

Here $v_q = 2\pi e^2/q$ and $\epsilon(q,\omega)$ is the wave vector and frequency-dependent dielectric constant for the interacting electron system and $\langle \phi_q^2 \rangle$ is the configuration averaged impurity potential. At zero temperature

$$-\frac{1}{\pi v_q} \operatorname{Im} \frac{1}{\epsilon(q,\omega)} \equiv S(q,\omega) \tag{3.3}$$

Equation (3.2) is exact to arbitrary order in the electron-electron interaction and to leading order in $\langle \phi_q^2 \rangle$. It was derived by expanding the Kubo formula in ϕ_q and then resumming i.e., placing $M(\omega)$ back into the denominator of the expression for $\sigma_{\alpha\beta}(\omega)$. This resummation procedure is important when considering the low frequency behavior of the conductivity in the absence of B. In the presence of a DC field or at high frequencies as long as one is not at cyclotron resonance the resummation procedure is not essential. While the exact range of validity of the weak coupling formula is not known we expect it to be particularly good in high mobility samples such as those used in the FQHE experiments. In any event various simple approximations to Eq. (3.2) give perfectly sensible, physically valid, quantitatively accurate results as long as the expressions for $\epsilon(q,\omega)^{-1}$ are not too singular. This includes all cases in 3-D and the case of 2-D for $B = 0$ even for the simple RPA (the most singular) expression for $\epsilon(q,\omega)$.[33] For 2-D in the presence of a static B field the RPA form for $\epsilon(q,\omega)$ leads to problems which are removed by considering higher order effects in the Coulomb interactions.[34] We shall see that no such problems arise with the SMA, which for this system is much superior to the RPA.

Physically Eq. (3.2) tells us that the impurity potential breaks the translational symmetry and permits the creation of electron-hole pair excitations at finite wave-vector q. In this approximation it is the *poles* of $\epsilon^{-1}(q,\omega)$ i.e., the Im $M(\omega)$ which characterizes the absorption. The real problem for any interacting system is of course how to compute $\operatorname{Im}[1/\epsilon(q,\omega)]$. In the case of interest here, at low temperatures and high frequency, we use

$$S(q,\omega) = n\bar{s}(q)\delta(\omega - \Delta(q)). \tag{3.4}$$

with $\Delta(q)$ given in Eq. (2.23) in the imaginary part of Eq. (3.2) to compute, for example, the AC conductivity at $T = 0$. As a specific model of the impurity potential we choose,[35]

$$<\phi_q^2> = \frac{n_I}{n}\left(\frac{2\pi e^2}{q} e^{-\delta q}\right)^2. \tag{3.5}$$

This $<\phi_q^2>$ characterizes the scattering of electrons from ions with concentration n_I randomly placed at a distance δ from the interface

In Fig. (3.1) we have plotted the frequency dependent $\mathrm{Im}\,M(\omega)$ in units of $1/\tau_0$. This scattering rate is determined from the zero field mobility $\mu_{DC}(B=0) = e\tau_0/m$. This procedure allows us to normalize out n_I as long as we assume that $\mu_{DC}(B=0)$ is dominated by the same set of donors n_I. (Note that for $B=0$ it is necessary to compute τ_0 using RPA screening.) We have set $\nu = 1/3$, $\epsilon_\infty = 10$, (all e^2 are divided by ϵ_∞ the background dielectric constant of the GaAs), $r_s = me^2/\hbar k_F = 1$, $\delta = 2\ell$ and $m = 0.07\,m_e$. In Fig. (3.1), we have also assumed that all of the absorption near $\omega = \Delta_{min}$ comes from the rotons whose dispersion is well approximated by

$$\Delta(k) = \Delta_{min} + (|\vec{k}| - k_R)^2/2m_R \tag{3.6}$$

with $m_R \cong 10m$ and $k_R \cong 1.4/\ell$ for this system. The following qualifications should be noted in connection with the use of Eq. (3.6). For $|\vec{k}| \leq k_R$, Δ_k is fairly well represented by Eq. (3.6) for $\Delta_k < 2\Delta_{min}$. For $|\vec{k}| > k_R$ there is no single mode which alone absorbs the oscillator strength. Nevertheless the *mean* energy of the relevant modes *is* well represented by Eq. (3.6). In any case the impurity potential matrix element most strongly weights the region $|\vec{k}| \leq k_R$ where the SMA and Eq. (3.6) are valid.

As expected, the absorption shows a sharp onset at $\omega = \Delta_{min}$ with a square root singularity which will of course be smeared out by higher order impurity effects. Note that the strength of the square root singularity is quite large (in units of $1/\tau_0$). There have been as yet no reported experiments, but such a strong frequency dependence should be experimentally observable.

One might suppose that the small dispersion in $\Delta(k)$ for large k would cause an additional sharp structure in the absorption at the frequency $\Delta(\infty)$. This does not occur because the exciton mode has negligible oscillator strength. All of the intra-Landau level oscillator strength is concentrated at wave vectors near the roton minimum. This suggests the possibility of interesting experiments using probes of specific wavelength ($\sim 200\text{Å}$) such as phonons or gratings to study the collective mode dispersion in more detail.

We now turn to a discussion of the conductivity at zero frequency but finite temperature. At the present time very little is understood about the origin of dissipation in the fractional regime. The accepted wisdom is that the dissipation is linear in the number of thermally activated quasi-particles.[35,36] The activation energy for this mechanism may be estimated as follows. For large k, the exciton energy $\Delta(k)$ is the energy required to produce a quasi-electron quasi-hole pair separated by a distance $k\ell^2/\nu$. Hence the energy to

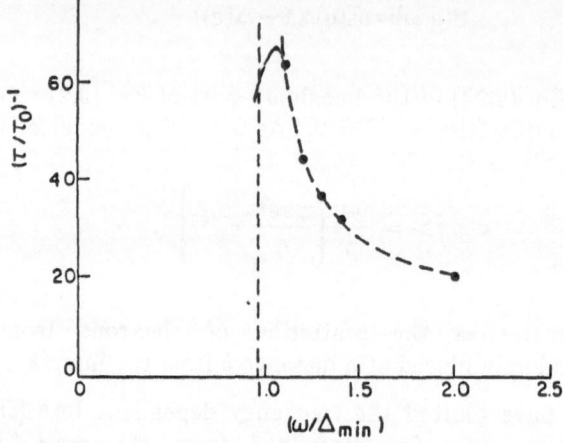

Fig. 3.1 The effective frequency dependent collision rate τ^{-1} in units of the DC B = 0 collision rate τ_0^{-1}.

produce a pair of fully separated quasi-particles of charge $\pm\nu$ is $\Delta(\infty)$. The law of mass action tells us that the activation energy for the creation of quasi-electrons and quasi-holes will be $\Delta(\infty)/2$. Of course such arguments are only qualitative. They imply a gap but do not lead to a quantitative expression for σ_{xx} in the presence of disorder.

We present here a specific calculation of the DC conductivity using the finite temperature version of the weak coupling formulae Eqs. (3.1-3.3) including only the effect of the magneto-roton excitations. We then discuss the intriguing questions raised by these results. For weak coupling and large magnetic field

$$\sigma_{xx} \sim - \sigma_{xy} \left. \frac{\text{Im}M(\omega)}{\omega_c} \right|_{\omega \to 0}. \tag{3.7}$$

Clearly Eqs. 3.2 and 3.7 give $\sigma_{xx} = 0$ at zero temperature because of the excitation gap. At finite temperatures the scattering of thermally activated rotons produces a finite conductivity, which can be evaluated using,

$$\frac{1}{\pi v_q} \text{Im} \frac{1}{\epsilon(q,\omega)} = (1 - e^{-\beta\omega}) Z^{-1} \sum_{n,m} e^{-\beta E_n}$$

$$|<m|\rho_q|n>|^2 \delta(E_m - E_n - \omega), \tag{3.8}$$

where Z is the partition function which for low temperatures is approximated by unity. Taking $|n> \cong \bar{\rho}_k|o>$ and $E_n = \Delta(k)$ we can evaluate $\text{Im}(1/\epsilon(q,\omega))$ directly. We do however need to know matrix elements of the form; $<o|\bar{\rho}_{-k-q} \rho_q \bar{\rho}_k| o>$. Expanding this for q small (but not zero) the exact leading order term can be shown to be[37]

$$<o|\bar{\rho}_{-k-q}\rho_q\bar{\rho}_k|o> = \text{in}(\vec{k} \times \vec{q}) \cdot \hat{z}\ell^2\bar{s}(k), \tag{3.9}$$

where $\hat{z}$ is a unit vector normal to the plane. Because the impurity setback distance is a few times the Landau orbit size, small q is a reasonable approximation. Using Eq. (3.9) and substituting Eq. (3.8) into Eq. (3.2) yields

$$\text{ImM}(\omega) \cong 0.1 \frac{e^4}{\hbar} \left[\frac{m_R}{m} \right] (k_R \ell)^3 \left[\frac{n_I}{n} \right] \frac{\ell}{\delta^3} \left(\frac{k_B T}{\Delta_c} \right)^{1/2} \Delta \beta e^{-\beta \Delta_{\min}} \qquad (3.10)$$

which gives $\Delta_{\min}$ as the thermal activation energy for the dissipation. The last term which diverges as the roton width (cutoff) Δ_c approaches zero comes from the fact that the rotons at the minimum have a singular (one-dimensional like) density of states. Treatment of the scattering in the self-consistent Born approximation will remove this weak divergence. However such considerations are unimportant here. We can assume that $\Delta_c \cong \Delta_{\min}$ for typical mobility samples and $k_B T < \Delta_c$. The physical process described by the weak coupling formula Eq. (3.10) with only magneto-roton states included can be summarized as follows (see DC Fig. 3.2). A low frequency long wave length photon, (wiggly line) virtually excites an electron to the *next* Landau level since this is the only dipole allowed transition at long wave length. This explains why ω_c appears in the energy denominator in Eq. (3.1). The excited electron scatters from an impurity dropping back down to the lowest Landau level creating a virtual excited state within the lowest Landau level. In this case we have *assumed* it is a roton with wave vector q. This roton then decays into a pair of rotons wave vector $\vec{k}$ and $-\vec{k} + \vec{q}$. If one of the rotons ($\vec{k}$) happens to be thermally excited this is a real final state and we can, in this weak impurity limit, have a real absorption. Equation (3.10) has prompted an interesting array of questions. The only approximation we have used in driving Eq. (3.10) is that of working to lowest order in the impurity potential. It has been suggested to us that even though there is no off-diagonal long range order in this system, it might be appropriate to take our analogy with superfluid helium even further and use the Landau two-fluid argument[38] to treat the population of thermally activated rotons as a "normal fluid" which comes to equilibrium in the rest frame of the impurities while the superfluid continues to flow. This effect is not contained within the memory function formalism since it is infinite order in the impurity potential. The analog of the Landau argument for a charged system is the following. In the presence of a uniform Hall field $\vec{E}$ the collective excitation spectrum is modified

$$\Delta'(k) = \Delta(k) + e\ell^2 (\vec{k} \times \hat{z}) \cdot \vec{E} \qquad (3.11)$$

due to the separation of charge along the direction of the electric field. For sufficiently weak fields the roton will still be a metastable local minimum in the excitation spectrum and the gas of rotons could simply equilibrate to its new distribution function. If this picture is correct then the relaxation rate determined from Eq. (3.10) does not represent the steady state dissipation but rather a characteristic relaxation rate for the normal fluid. Such a characteristic time might be visible in the AC conductivity. However, it is fair to say that such an *argument* is *not* a proof since there are additional complications in GaAs heterostructures which could very well negate it. For example, the existence of bulk substrate phonons which can act as an energy and momentum source for rotons is one of them.

The best current estimate for $\Delta(\infty)/2$ and $\Delta_{\min}$ for these samples is about 10 Kelvin at B = 28 Tesla. This is approximately a factor of 4 larger than the observed activation temperature of 2.5-3K. Thus it appears that even in the very highest mobility GaAs devices, disorder has a significant effect on the gap. It seems reasonable to take into account the renormalization of the roton and quasi particle energies by impurities. We have done this for the magneto-rotons by considering a weak coupling theory of the scattering of magneto rotons.[37]

In the presence of disorder the magneto roton states at different wave vectors are coupled. In an external potential $\Phi(\vec{r})$.

$$<\Psi_{k'}|\Phi|\Psi_k> = \phi_q M(k',k) \tag{3.12}$$

where $q \equiv (k'-k)$ and

$$\frac{<0|\bar{\rho}_{-k},\rho_q\rho_k|0>}{(\bar{s}(k')\bar{s}(k))^{1/2}} = (2\pi)^2\delta^2(k'-k-q)M(k',k) \tag{3.13}$$

Because of the extremum in $\Delta_k(k)$, the broadening of these modes must be treated self-consistently even for arbitrarily weak disorder. Since the excited states are labeled by wave vector the problem is formally identical to the broadening of single-particle states by disorder, except that the bare dispersion relation is different and the effective potential is nonlocal. We treat the broadening in a self-consistent Born approximation. Our approach, then, is entirely analogous to that adopted by Kallin and Halperin[35] in discussing the magnetoplasmon excitations which occur near $\hbar\omega_c$, and to facilitate comparison we adopt their notation. The Dyson equation for the configuration averaged magnetoroton propagator is

$$D^{-1}(k,\omega) = \omega - \Delta(k) - \Pi(k,\omega), \tag{3.14}$$

where the magnetoroton self-energy is

$$\Pi(k,\omega) = \frac{1}{L^2}\sum_q <\phi_q^2>_c \tag{3.15}$$

$$\times |M(k+q,k)|^2 D(k+q,\omega)$$

Using Eq. (3.9) for $M(k',k)$ and an impurity potential of the form given by Eq. (3.5) the Dyson equation Eq. (3.14) was solved numerically. The solutions are accurately approximately by setting $D(k+q,\omega) \cong D(k,\omega)$ in Eq. (3.15). The approximate solution may then be written in the form

$$D(k,\omega) = \int \frac{d\omega'}{\pi} \frac{\rho(k,\omega')}{\omega-\omega'-i\eta}, \tag{3.16}$$

where $\rho(k,\omega)$, the spectral density of the magnetoroton propagator at wave vector k. It is given by

$$\rho(k,\omega) = \left[1 - \left[\frac{\omega-\Delta(k)}{\Gamma_\nu(k)}\right]^2\right]^{1/2}, |\omega-\Delta(k)| \lesssim \Gamma_\nu(k) \tag{3.17}$$

$$\Gamma_\nu^2(k) = \frac{4}{L^2}\sum_q <U(-q)U(q)>_c |M(k+q,k)|^2. \tag{3.18}$$

In all the cases we have studied Eqs. (3.16-3.17) provide an excellent approximation to the numerical solutions to the Dyson equation. This was expected because of the importance of small-angle scattering and the remaining discussion will be in terms of these expressions. We should remark that our approach can be formally justified only in the weak scattering limit, i.e., only when $\Gamma_\nu(k) \ll \Delta(k)$. In examining its predictions for the strong scattering limit below, we do not expect numerical accuracy but rather a qualitative indication of the importance of impurity effects on the determination of the experimentally observed activation energies.

When $\Gamma_\nu(k) > \Delta(k)$ in Eqs. (3.17) $\rho(k,\omega)$ remains nonzero for $\omega < 0$. This implies that the incompressible liquid ground state is no longer stable and the true ground state is probably of the nature of a Wigner glass. Since the fractional quantum Hall effect will not occur in that regime the condition $\Gamma_\nu(k) < \Delta(k)$ is a necessary one for the occurrence of the fractional quantum Hall effect. From Eq. (3.18) we have that for $\delta > \ell$.

$$\Gamma_\nu(k) = C \frac{k\ell}{\sqrt{2}} \left[\frac{\ell}{\delta} \right] \tag{3.19}$$

C is a correction factor, which must be smaller than one to account for the weakening of the scattering due to correlations in ion positions, finite donor-layer thickness, and finite electron-layer thickness and an actual uncertainty in the number of donors. Since $k\ell \sim \sqrt{2\pi\nu}$ at the magnetoroton minimum ($k = k_R$), we have, by requiring $\omega_0(k^*) > \Gamma_\nu(k^*)$ the following condition for the occurrence of the fractional Quantum Hall effect

$$\frac{\delta}{\ell} \sim \frac{C\sqrt{\pi\nu}}{\Delta(k^*)}. \tag{3.20}$$

For a given filling factor Eq. (3.20) may be interpreted as providing a minimum setback distance δ or alternately a threshold field ($\ell = 25\mathring{A} / \sqrt{H}$, H in tesla).

The form of Eq. (3.18) suggests that (assuming magneto-rotons lead to dissipation) the experimentally measured gap in the resistivity data should be identified with $\widetilde{\Delta}$

$$\widetilde{\Delta} = \Delta(k_R) - \Gamma_\nu(k_R). \tag{3.21}$$

Detailed calculations assuming a setback distance of 300 $\mathring{A}$, a thickness $t \cong 100\mathring{A}$ for the region over which donors are distributed and a finite electron width $z_0 \cong 150\mathring{A}$ where performed. If the numerical factor expressing our ignorance of the correlation of ion positions and concentration is adjusted by 40% the solid curve shown in Fig. (3.3) is obtained. The curve with its one adjustable parameter is in excellent although possibly fortuitous agreement with experiment.

Similarly any detailed theory of the quasi-particle quasi-hole absorption must consider the effects of disorder. This has not been done. However, one expects that the change in the energy $\Delta(\infty)$ is smaller than the change in Δ_{min} since the charge on the quasi-particle is only e/3 and the coupling is to charge fluctuations of the donor. In addition, such a theory must consider the possible localization (no DC absorption) of the quasi-particles by the disorder. Indeed the finite width of the fractional plateaus presumably arises from quasi-particle localization.

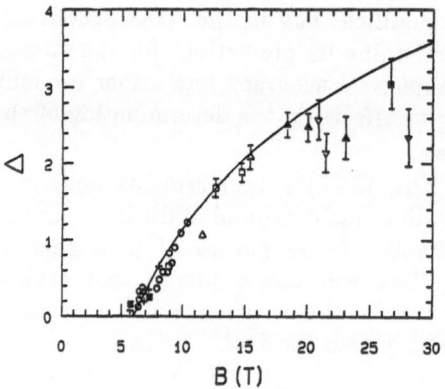

Fig. 3.2 A schematic representation of the process involving "magneto-rotons" which gives rise to 1) the AC conductivity and (b) the temperature activated DC resistivity.

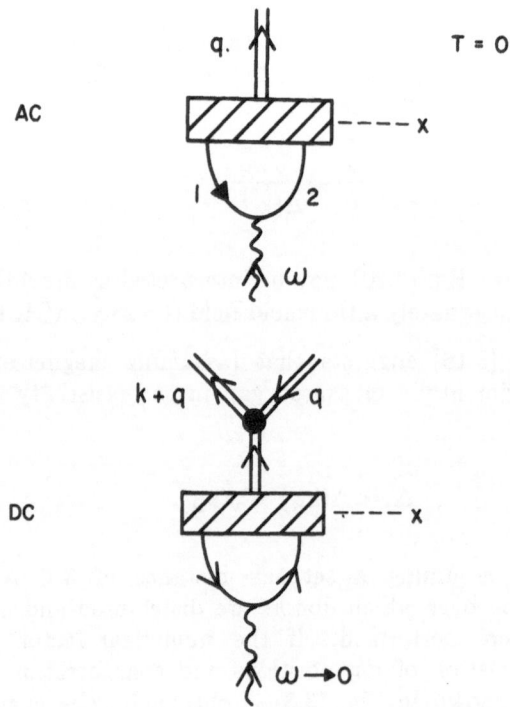

Fig. 3.3 The fit of Eq. (3.21) to the data of ref. (30).

To summarize our arguments, there seem to be two possibilities both of which are interesting. Either the activation energy is Δ_{min} or it is $\Delta(\infty)/2$ (with possible modifications by impurity effects). However, if the later is correct then we must seriously consider the possibility that other effects associated with the two-fluid phenomenology of superfluidity may occur.

In conclusion we have used the recently developed roton picture of collective excitation modes and a weak impurity coupling memory function formulation to compute the high frequency conductivity in the fractional quantum Hall regime. The AC conductivity exhibits a strong square root singularity at the threshold for roton creation. This distinctive signature should be visible in an experiment. We also used the memory function formalism to compute the DC dissipation due to scattering of thermally activated rotons. Possible modification of this result due to superfluid-like effects was discussed.

REFERENCES

[1] H. L. Stormer, *Surf. Science* **132**, 519 (1983); *Adv. in Solid State Phys.* **XXIV**, 25 (1981).

[2] T. Ando, A. Fowler and F. Stern, *Rev. Mod. Phys.* **54**, 437 (1982).

[3] Y. P. Monarkha and V. Shikin, *Fiz. Nitk. Temp.* **8**, 563 (1982) *[Sov. J. Low Temp. Phys.* **8**, 279 (1982)].

[4] M. Wannier and P. Leiderer, *Phys. Rev. Lett.* **42**, 315 (1984); H. Ikezi, R. W. Giannetta and P. M. Platzman, *Phys. Rev.* **25**, 4488 (1982).

[5] K. Kajita, in Proceedings of Electronic Properties of 2-D Systems (1983, Oxford), published in *Surf. Science* **142**, (1984).

[6] S. A. Jackson and P. M. Platzman, *Phys. Rev.* **B24**, 499 (1981).

[7] C. C. Grimes and G. Adams, *Phys. Rev. Lett.* **42**, 795 (1979); D. S. Fisher, B. I. Halperin and P. M. Platzman, *Phys. Rev. Lett.* **42**, 798 (1979).

[8] M. Kosterlitz and D. Thouless, *J. Phys.* **C6**, 1181 (1973); B. Halperin and D. Nelson, *Phys. Rev.* **B19**, 2457 (1979).

[9] E. P. Wigner, *Phys. Rev.* **46**, 1002 (1934).

[10] P. M. Platzman and H. Fukuyama, *Phys. Rev.* **B10**, 3150 (1974).

[11] D. Ceperley, *Phys. Rev.* **B18**, 3126 (1979).

[12] V. Elser, private communication.

[13] F. M. Peeters and P. M. Platzman, *Phys. Rev. Lett.* **50**, 2021 (1983).

[14] D. C. Tsui, H. L. Stormer and A. C. Gossard, *Phys. Rev. Lett.* **48**, 1559 (1982).

[15] D. C. Tsui, H. L. Stormer, J. C. M. Huang, J. S. Brooks and M. J. Naughton, *Phys. Rev.* **B28**, 2274 91983); E. E. Mendey, M. Heiblum, L. L. Chang and L. Esaki, *Phys. Rev.* **B28**, 4886 (1983).

[16] R. B. Laughlin, *Phys. Rev. Lett.* **50**, 1395 (1983).

[17] F. D. M. Haldane, *Phys. Rev. Lett.* **51**, 605 (1983).

[18] S. M. Girvin, A. H. MacDonald, and P. M. Platzman *Phys. Rev. Letters*, **54**, 581 (1985), *Phys. Rev. B* **33** 2481 (1986).

[19] H. Fukuyama, P. M. Platzman and P. W. Anderson, *Phys. Rev.* **B19**, 5211 (1979).

[20] D. Yoshioka and H. Fukuyama, *J. Phys. Soc. Japan* **47**, 394 (1979).

[21]   K. Maki and Y. Zotos, *Phys. Rev.* **B28,** 4349 (1983).

[22]   P. Lam and S. Girvin, *Phys. Rev.* **B30,** (1984).

[23]   B. I. Halperin (private discussions).

[24]   E. Andrei, *Phys. Rev. Lett.* **52,** 1449 (1984).

[25]   L. M. Sander, *Phys. Rev.* **B** 4350 (1975).

[26]   H. L. Stormer in *Festkorperpobleme (Advances in Solid State Physics),* ed. P. Grosse (Perjanon/Vieweg, Braumshweig 1984) Vol. 24 p. 25.

[27]   R. P. Feynman, *Statistical Mechanics,* (Benjamin 1972).

[28]   R. P. Feynman, M. Cohen *Phys. Rev.* **107,** 1189 (1956).

[29]   B. Lundqvist, *Phys. Konders Matter* **6,** 206 (1967); A. W. Overhauser, *Phys. Rev. B* **3,** 1888 (1971).

[30]   S. M. Girvin and T. Jack, *Phys. Rev. B* **29** 5617 (1984).

[31]   F. D. M. Haldane and E. H. Rezayi, *Phys. Rev. Lett.* **54,** 237 (1985).

[32]   G. S. Boebinger, A. M. Chang, H. L. Stormer, and D. C. Tsui, *Phys. Rev. Letters* **55,** 1606 (1986).

[33]   A. Ron and N. Tzoar, *Phys. Rev.* **131,** 12 (1963); W. Götze and P. Wolfle, *Phys. Rev. B* **6,** 1226 (1972).

[34]   N. Tzoar, P. M. Platzman, and A. Simons, *Phys. Rev. Lett.* **36,** 1200 (1976).

[35]   C. S. Ting, S. C. Ying, and J. J. Quinn, *Phys. Rev. B* **16,** 4980 (1979).

[36]   C. Kallin and B. I. Halperin, *Phys. Rev. B* **6,** 3635 (1985).

[37]   R. B. Laughlin, *Physica* **126B,** 254 (1984).

[38]   A. H. MacDonald, K. L. Liu, S. M. Girvin and P. M. Platzman, *Phys. Rev. B* **33,** 4014 (1986).

[39]   E. M. Lifshitz and L. P. Pitawvskii, *Statistical Physics,* (Pergamon, New York, 1981).

ELECTRON-PHONON INTERACTION IN TWO-DIMENSIONAL SYSTEMS:

POLARON EFFECTS AND SCREENING

J.T. Devreese and F.M. Peeters

Physics Department, University of Antwerp (UIA & RUCA)
Antwerpen, Belgium; and University of Technology, Eindhoven
The Netherlands

INTRODUCTION

The polaron concept, as originally proposed by Landau to describe an electron (or hole) dressed with a cloud of virtual LO-phonons, has been useful both to describe the physical properties of charge carriers in ionic crystals and in polar semiconductors and as a field theoretical model.

The cyclotron resonance of electrons and holes in alkali halides is influenced considerably by polaron effects; the polaron mass can be more than doubled with respect to the band mass. In some polar semiconductors the resonance condition $\omega_{LO} = \omega_c$ ($\frac{\omega_{LO}}{2\pi}$ is the long wavelength frequency of the longitudinal optical phonons and $\frac{\omega_c}{2\pi}$ is the cyclotron frequency) can be realised experimentally and the Landau levels are then observed to be modified by the polaron coupling. Also to interpret mobility measurements in alkali halides and polar semiconductors the Fröhlich polaron interaction is needed. Sophisticated theoretical models have been proposed to analyse the physical properties of polarons: Green's function techniques, canonical transformations, path integral technique.

The polaron effects mentioned above relate to bulk materials ("3D"). Also in the two-dimensional electron gas, as realised in heterojunctions and in inversion layers, built from polar semiconductors, the polaron interaction plays a role and must be taken into account to interpret observations of cyclotron resonance, mobility, optical absorption, etc. .... . In the two-dimensional electron systems realised so far (like GaAs-$Al_xGa_{1-x}As$) the polaron coupling strength (Fröhlich α) is "weak" ($\alpha < 1$); therefore some form of perturbation theory will be sufficient to evaluate the polaron effects in those systems. Characteristic of the 2DEG in heterostructures is also a relatively high electron density, so that many-body effects become significant. The competition between polaron effects and many-body screening will turn out to be a central characteristic of the 2DEG in heterostructures like GaAs-$Al_xGa_{1-x}As$. Like in three dimensions, polaron effects are enhanced near the resonance condition $\omega_c = \omega_{LO}$. Therefore cyclotron resonance experiments will be of central importance for the study of polaron effects in the 2DEG.

131

In the present series of lectures, I review the theoretical studies on the physics of polarons in two dimensions performed over the last three years in Antwerpen in collaboration between F.M. Peeters, Wu Xiaoguang and J.T. Devreese. The first chapter is devoted to the study of the free polaron in two dimensions in the single particle approximation. First, the self-energy and the effective mass of the polaron in two dimensions are calculated. The self-energy is evaluated to fourth order of perturbation theory. Approximate scaling relations are obtained at all coupling. These scaling relations suggest qualitative similarities between some physical properties (here the self-energy and the effective mass) of polarons in two dimensions and in three dimensions. These similarities are contrasted to the dramatic differences which are found to occur between the polaron energy-momentum dispersion in 2D respectively 3D. In the second chapter the "many-polaron problem" in two dimensions is analysed. This study is greatly facilitated by the use of a "many-polaron canonical transformation" which is a generalisation of the well-known Lee-Low-Pines polaron transformation(s) to the case of many polarons (cfr. ref. [20]). Both the self-energy and the effective mass per polaron are treated dynamically (i.e. using time-dependent many-body response functions). The analysis confirms that screening tends to reduce polaron effects. It is worth noting that the study of the dynamical screening of the polaron mass was facilitated by defining the mass from the optical absorption spectrum. Subsequently, the plasmon-phonon coupled modes and their dispersion for the 2DEG are calculated; it is suggested that these modes can be observed experimentally. In the third chapter the cyclotron resonance of polarons in 2D is investigated. First the Landau levels for polarons in two dimensions are calculated with considerable precision. Next the magneto-optical absorption spectrum for polarons in 2D is derived using the memory function formalism. The motivation is that this is the quantity which is determined experimentally. If the one-polaron approximation is made, and if the non-parabolicity of the conduction band is taken into account, cyclotron resonance experiments for p-type InSb inversion layers can be explained quantitatively but this is not the case for GaAs-Al$_x$Ga$_{1-x}$As heterostructures. It is shown that the incorporation of many-body effects allows for an understanding of the GaAs-Al$_x$Ga$_{1-x}$As cyclotron resonance data in the polaron resonance region, although a discrepancy persists at small magnetic fields.

I. THE FREE POLARON IN TWO DIMENSIONS: SINGLE PARTICLE APPROXIMATION

Ia. The Hamiltonian

If the Fröhlich interaction, describing the coupling between an electron and the LO-phonons of a 3D polar crystal, is adapted to the special case where the electron is confined to motion in a two-dimensional space ("ideal" 2D case) the following Hamiltonian results

$$H = \frac{p^2}{2m} + \sum_k \hbar\omega_{LO} \, a_k^+ a_k + \sum_k (V_k a_k \, e^{i\vec{k}\cdot\vec{r}} + V_k^* a_k^+ \, e^{-i\vec{k}\cdot\vec{r}}) \tag{1a}$$

Formally this expression is identical to the 3D-Fröhlich Hamiltonian. The two-dimensional character of the electron motion (say in the x-y plane) is accounted for i) by the fact that $\vec{p}$ and $\vec{r}$, the momentum and position operator of the electron, have a x- and a y-component only (i.e. no z-component), ii) by the expression for V_k which takes the following form for the two-dimensional case:

$$V_k = -i\hbar\omega_{LO} \, (\frac{\sqrt{2}\pi\alpha}{Ak})^{\frac{1}{2}} \, (\frac{\hbar}{m_b\omega_{LO}})^{\frac{1}{4}} \tag{1b}$$

m_b is the band mass of the electron, ω_{LO} is the optical phonon frequency, k the wave number of the LO-phonon, A the surface of the 2D system. $a_{\vec{k}}^+$, $a_{\vec{k}}$ are respectively the creation and annihilation operators for the LO-phonons of wave vector $\vec{k}$. For the ideal 2D case the vector $\vec{k}$ has x- and y-components only. α is the standard (-3D) Fröhlich coupling constant.

Notice that in the Hamiltonian (1a) the LO-phonon frequency is assumed to be that of the bulk phonons in the Debye approximation.

The Hamiltonian (1) for the ideal two-dimensional polaron mass was first derived from the standard Fröhlich Hamiltonian by J. Sak [1].

It is useful in actual experimental circumstances to take into account the non-zero width of the electron layer, which leads to a deviation from the "ideal" 2D model. This is done most conveniently by describing the motion of the electron in the z-direction, i.e. normal to the quasi-2D electron layer, by a variational wave function

$$\psi_0(z) = (\frac{b^3}{2})^{\frac{1}{2}} \, z \, e^{-bz/2} \tag{2a}$$

The following form for V_k is then derived:

$$V_k = i\hbar\omega_{LO} \, (\frac{4\pi\alpha}{Vk^2})^{\frac{1}{2}} \, (\frac{\hbar}{2m_b\omega_{LO}})^{\frac{1}{4}} \, <\psi_0|e^{ik_z z}|\psi_0> \tag{2b}$$

The parameter b is determined approximately, before including the electron-phonon interaction [2]. One obtains

$$b = (\frac{48\pi \, N \, m_b \, e^2}{\hbar^2 \, \varepsilon_0})^{\frac{1}{3}} \tag{2c}$$

with

$$N = n_d + (\frac{11}{32}) \, n_e \tag{2d}$$

n_d and n_e are the depletion and the charge carrier density respectively.

The Hamiltonian (1a) with (1b) or (2b) for V_k is the starting point of what follows. Derivations of this Hamiltonian can be found in many publications and it suffices here to refer to the literature for that purpose.

Ib. Groundstate Energy and Effective Mass of a Free Polaron in Two Dimensions [3]

The first step in any polaron theory consists of calculating the groundstate energy and the effective mass.

In ref. [3] we derived exact and approximate results for the groundstate energy of a polaron in two dimensions.

For the weak coupling expansion, we first used the Feynman path integral representation for the polaron partition function in which the phonon variables can be eliminated rigorously. The free energy F is then obtained from

$$e^{-\beta F} = e^{-\beta F_0} \, <e^{S_I[\vec{r}(t)]}>_{S_0} \tag{3a}$$

where F_0 is the free energy for a free polaron which is described by the action

$$S_0 = \frac{1}{2m} \int_0^\beta [\dot{\vec{r}}(t)] \tag{3b}$$

and

$$S_I = \sum_{\vec{k}} |V_k|^2 \int_0^\beta du \int_0^\beta du' \, G_{\omega_S}(u-u') \, e^{i\vec{k}.[\vec{r}(u)-\vec{r}(u')]} \tag{3c}$$

with

$$G_\omega(u) = \frac{1}{2} \, n(\omega) \, (e^{\omega|u|} + e^{\omega(\beta-|u|)}) \tag{3d}$$

$< \, >_{S_0}$ is a path integral average with weight e^{S_0}.

To derive the groundstate energy to order α^2 we needed to calculate

$$<S_I[\vec{r}(t)]>_{S_0} \tag{4a}$$

and

$$<S_I[\vec{r}(t)] \, S_I[\vec{r}(t)]>_{S_0} \tag{4b}$$

For the technical details, the reader is referred to [3].

Our result for the groundstate energy of the Fröhlich polaron in two dimensions is

$$E = -A\alpha - B\alpha^2 + O(\alpha^3) \tag{5a}$$

with the following expressions for A and B:

$$A = \frac{\pi}{2} \tag{5b}$$

$$B = \frac{\pi^2}{8} - G + 2 \int_0^1 \frac{K(x)}{(1+x^2)^2} \, dx = 0.06397 \tag{5c}$$

and G, Catalan's constant, is:

$$G = \sum_{m=0}^{\infty} \frac{(-1)^m}{(2m+1)^2} = 0.91596 \tag{5d}$$

The groundstate energy to order α had been obtained in ref. [1], the result to order α^2 was first obtained in our treatment of ref. [3]. A previous attempt to calculate the coefficient B of the fourth order perturbation theory had produced an erroneous result [4].

For strong coupling the adiabatic approximation, which becomes exact in that limit, was also worked out in ref. [3]. We adapted the scheme of Miyake to two dimensions and obtained

$$E = - 0.4047 \, \alpha^2 \tag{6}$$

a numerical result which is expected to be exact for the given digits.

Although the strong coupling limit is mainly of academic interest, it never-theless provides a checking ground to test the accuracy of "all-coupling" theories.

The adaptation of the variational Feynman path integral treatment of the 3D polaron [5] to the 2D case is straightforward [6]. The following upper bound for the groundstate energy of a 2D polaron is then obtained

$$E = \frac{(v-w)^2}{2v} - \frac{\alpha}{2} \left(\frac{\pi}{2}\right)^{\frac{1}{2}} \int_0^\infty dt \; \frac{e^{-t}}{\sqrt{D(t)}} \tag{7a}$$

with

$$D(t) = \frac{w^2}{2v^2} t + \frac{v^2-w^2}{2v^3} (1-e^{-vt}) \tag{7b}$$

where v and w are variational parameters.

It is worthwhile to write down the weak and strong coupling expressions of the result (7a) for the Feynman model:
for the weak coupling limit, i.e. $\alpha \to 0$:

$$E = -\frac{\pi}{2} \alpha - \frac{\pi^2}{216} \alpha^2 + O(\alpha^3) \tag{7c}$$

in the strong coupling limit

$$E = -\frac{\pi}{8} \alpha^2 + O(\alpha^0) \tag{7d}$$

Consequently the leading weak coupling term is obtained rigorously from the variational path integral treatment, the coefficient of the asymptotic limit $\alpha \to \infty$ is off by only 3%.

Ic. Approximate Scaling Relation for the Groundstate Energy and the Effective Mass of Polarons in 2D and 3D

The main result of our work in ref. [3], apart from the rigorous cal-culation of the fourth order perturbation result for the groundstate energy of the 2D polaron is the observation that a very simple scaling relation exists which links the 2D and 3D polaron groundstate energies and which is valid for the Feynman polaron model:

$$E_{2D}(\alpha) = \frac{2}{3} E_{3D} \left(\frac{3\pi}{4} \alpha\right) \tag{8}$$

This scaling relation (8) is not valid for the exact polaron groundstate energy as can be seen directly by considering the weak and strong coupling expansions of the groundstate energies. For weak coupling (8) is satisfied rigorously to order α but only approximately for higher order terms in α.

In the asymptotic limit of strong coupling (8) leads to $E_{2D}(\alpha) = -0.4016 \, \alpha^2$ if the r.h.s. of (8) is filled out with Miyake's result. This should be compared with our rigorous (numerical) 2D result obtained in ref. [3] $E_{2D} = 0.4047 \, \alpha^2$ from which it follows that for strong coupling, (8) is valid at the 1% level. Scaling relations like (8) allow to extrapolate physical results from 3D to 2D; e.g. as the Feynman polaron has no "insta-bility" in 3D (see ref. [7]) it follows that there is no instability for the Feynman polaron in 2D either. It should be emphasized once more that (8) is not a property of the rigorous polaron groundstate energy.

The fact that also some other approximations satisfy the relation (8) is less significant because they are of limited validity compared to the Feynman polaron model.

For the polaron effective mass a similar scaling relation exists for the Feynman model:

$$\frac{m_{2D}^*(\alpha)}{m_{2D}} = \frac{m_{3D}^*(\frac{3\pi}{4}\alpha)}{m_{3D}} \tag{9}$$

In the presence of an applied static magnetic field H the scaling relations like (8) or (9) are no longer satisfied, as discussed in ref. [8].

With respect to the scaling relations (8) or (9) it is probably worthwhile to refer to ref. [9] where rigorous scaling relations were obtained for the kinetic energy of the electron, the total polaron energy and the interaction energy in 3D. Starting from arguments like those in ref. [9] it might be possible to obtain extensions of (8) and (9).

Id. Dispersion Relation of a 2D Polaron [10]

The energy-momentum relation of Fröhlich polarons in 3D was studied in the early polaron theories [11]. Self-energy corrections and modifications to the polaron dispersion due to the phonon scattering states (von Neumann's non-crossing theorem) were analysed. Although the non-parabolicity in the polaron dispersion of the Fröhlich polaron was occasionally invoked to explain experimental data (e.g. as related to non-Ohmic transport), these studies remained largely formal.

The dispersion relation of the polaron in 2D turns out to be of interest because it behaves in an entirely different way from its 3D counterpart. The typical 2D dispersion relation could be of significance for the physical properties of 2D electrons (or quasi-2D electrons) in heterojunctions.

In this section the analysis is limited to the weak coupling case and we choose zero thickness for the electron layer. The basic expression for the energy-momentum relation at weak coupling takes the form:

$$E(k) = \frac{\hbar^2 k^2}{2m} - P \sum_{\vec{q}} \frac{|V_q|^2}{\hbar\omega_{LO} + \frac{\hbar^2(\vec{k}-\vec{q})^2}{2m} - \frac{\hbar^2 k^2}{2m} - \Delta(k)} \tag{10}$$

P denotes the principal value. In this expression $\hbar\vec{q}$ is used for the phonon momentum and $\hbar\vec{k}$ for the electron momentum. With the expression (1b) for V_q the polaron correction in the r.h.s. of (10) becomes

$$\text{Re } \Delta E(\vec{k}) = -\text{Re } \frac{\alpha}{\sqrt{1-\Delta(k)}} K \left(\frac{k}{\sqrt{1-\Delta(k)}}\right) \tag{11}$$

K(x) is the complete elliptic integral. $\Delta(k)$ can take different forms depending on the approximation which is chosen.

i) Rayleigh-Schrödinger perturbation theory (RSPT)

The RSPT corresponds to $\Delta(k) = 0$ in (10) and is valid if $k \ll k_{LO}$. In that case we obtain the following momentum dependent polaron mass (2D)

$$\frac{m^*}{m} = \frac{1}{1 - \alpha \frac{\pi}{8} (1 + \frac{27}{8} k^2 + \frac{375}{64} k^4 + \ldots)} \tag{12}$$

The divergence in $\Delta E(k)$ at $k = k_{LO}$ arises because the RSPT is no longer valid in that case and degenerate perturbation theory should be used.

ii) Tamm-Dancoff approximation

It is possible to extend the validity of the approximation (10) to the energy range $E(k) - E(o) < 2\hbar\omega_{LO}$ by making the "improved" Tamm-Dancoff approximation [12] which leads to (11) with $\Delta(k) = \Delta E(k) - \Delta E(o)$, with $\Delta E(o) = -\alpha \frac{\pi}{2}$.

The drastic difference between the energy-momentum relation for a 2D and a 3D polaron is illustrated in Fig. 1. The value $\alpha = 0.07$, chosen in this figure, corresponds to a GaAs-Al$_x$Ga$_{1-x}$As heterostructure. While in 3D the "improved" Tamm-Dancoff approximation leads to a continuous $E(k)$ dispersion for all k the $E(k)$ relation in 2D exhibits a (zero width) gap in the spectrum and is a three valued function for $k \gg k_{LO}$.

The difference between the 2D and 3D cases is most clearly manifested by a plot of the density of states (D.O.S.) as shown in Fig. 1. The density of states for a free electron behaves as $N(E) = $ constant in 2D, and as $N(E) = \sqrt{E}$ in 3D. Fig. 1 reveals a singular behaviour around $E(k) = E(o) + \hbar\omega_{LO}$ in the D.O.S., qualitatively different in 2D from 3D.

For a GaAs-Al$_x$Ga$_{1-x}$As heterostructure, taking into account a realistic broadening (replace $\frac{1}{E}$ by $\frac{1}{E+i\Gamma}$ in eq. (10), with a Γ value taken from experiment), leads to a structure in the density of states around $E(k) = E(o) + \hbar\omega_{LO}$ which is an order of magnitude larger than that of the corresponding structure in 3D.

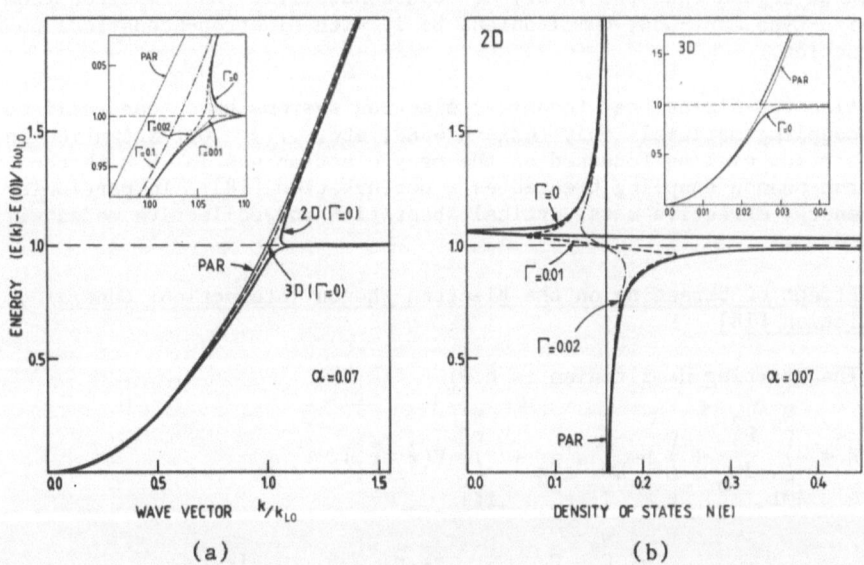

Fig. 1. Energy-momentum relation (a) and the density of states (b) for the 2D and the 3D polaron. For the 2D case the results with different values of the broadening parameter Γ are also shown.

The present section therefore reveals a physical property of a 2D polaron (its density of states), which is dramatically different from the corresponding 3D property. Such behaviour should be contrasted with the behaviour of the polaron groundstate energy and effective mass, which is very similar in 2D and 3D. In ref. [10] we have also discussed the 2D-D.O.S. for polarons in the context of the oscillations in the derivative of the current-voltage characteristics of GaAs-Al$_x$Ga$_{1-x}$As tunnel junctions discovered by Hickmott et al.

Our results on the 2D-D.O.S. of polarons, if combined with Ihm's [13] explanation of the Hickmott oscillations [14], suggest that if a similar tunnel experiment as the one proposed by Hickmott et al. were performed in 2D one would expect current oscillations which are much more pronounced than those observed by Hickmott et al.

II. MANY-POLARON PROBLEM IN TWO DIMENSIONS. EFFECT OF SCREENING OF THE ELECTRON-PHONON INTERACTION

IIa. Introduction

It is hardly possible to excite "many electrons" in an ionic crystal with large coupling constants like an alkali halide, a silver halide or ionic crystals like SiO_2, Al_2O_3, etc. ... . The Fröhlich electron-LO phonon interaction is therefore only of importance in the context of many-body systems characterized by a weak electron-phonon coupling (GaAs, CdO and some other III-V and II-VI compounds).

Many-polaron effects have been observed in 3D in GaAs where plasmon-phonon modes coupled through the Fröhlich interaction (ω^+, ω^-, see ref.[15]) occur [16]. The theoretical predictions that the ω^--mode always satisfies $\omega^- < \omega_{TO}$ (k=0) (see ref. [17]) seems to be confirmed experimentally. Also in the experiments on "droplets" in GaAs and other polar semiconductors the Fröhlich electron (hole) - LO phonon interaction plays a role. Of course it can be argued that the theory of superconductivity has resulted from a Fröhlich-type many-body Hamiltonian, be it with electron-acoustical phonon interaction.

Also two-dimensional (quasi -) electron systems have been realised for weak coupling materials only (GaAs-AlGaAs, etc. ...). Our attention in this section will be focussed on the many-electron gas in 2D with the electron-phonon coupling treated as a perturbation [18]. In particular the self-energy effective mass, optical absorption and collective modes will be studied.

IIb. Effect of Screening on the Electron-Phonon Interaction: General Formulation [18]

The starting Hamiltonian is now:

$$H = \sum_{j=1}^{N_s} \frac{p_j^2}{2m_b} + \sum_k \hbar\omega_{LO}\, a_k^+ a_k + \sum_{i<j}^{N_s} V(\vec{r}_i - \vec{r}_j)$$

$$+ \sum_{j=1}^{N_s} \sum_k (V_k a_k\, e^{i\vec{k}.\vec{r}_j} + V_k^* a_k^+\, e^{-i\vec{k}.\vec{r}_j}) \tag{13a}$$

$\vec{p}_j$ is the momentum operator of electron j, $\vec{r}_j$ its position operator. N_s is the total number of electrons. $V(\vec{r}-\vec{r})$ represents the Coulomb interaction between two electrons in the quasi-2D electron layer. V_k is given by Eq. (2b). The other symbols in (13a) were mentioned before.

The Hamiltonian (13a) describes the quasi-2D electron gas interacting with the system of LO-phonons. The choice of b allows us to adapt the width of the 2D electron layer; as $b \to \infty$ the "exact" 2D limit arises. The quasi-2D characteristics of the system enter the problem through the modification of the basic electron-electron interaction and of the electron-phonon interaction. $V(\vec{r}_i - \vec{r}_j)$ is obtained by averaging over the electron wave function in the z-direction. The Fourier transform of $V(\vec{r})$ is $V(k_\parallel)$:

$$V(k_\parallel) = \frac{2\pi e^2}{k_\parallel \epsilon_\infty} f(k_\parallel, b) \tag{13b}$$

with

$$f(k,b) = \frac{(8b^3 + 9kb^2 + 3k^2 b)}{8(k+b)^3} \tag{13c}$$

($k_\parallel$ is the component of $\vec{k}$ parallel to the 2D electron layer). (13b) can be considered as a modification of $\frac{2\pi e^2}{k_\parallel \epsilon_\infty}$ by introducing a wave-number and layer-dependent dielectric function. The model discussed here does only take into account the lowest subband. The validity of this model is discussed in ref. [19].

IIc. The Many-Polaron Canonical Transformation and Dynamical Screening of the Electron-Phonon Interaction. Effect on the Self-Energy

The study of the many-polaron problem at relatively weak coupling is greatly facilitated by the introduction of a many-particle canonical transformation for polarons, which is a generalisation of the so-called Lee-Low-Pines transformation:

$$U = \exp \sum_{j=1}^{N_j} s(r_j) \tag{14a}$$

with

$$s(r_j) = \sum_{\vec{k}} (f_k a_k e^{i\vec{k}\cdot\vec{r}_j} - f_k^* a_k^+ e^{-i\vec{k}\cdot\vec{r}_j}) \tag{14b}$$

f_k are variational parameters.

This generalisation (14a-b) of the Lee-Low-Pines transformation was introduced in ref. [20] by Lemmens, Devreese and Brosens. The unitary transformation (14) "dresses" each polaron in the many-polaron system with its self-induced cloud of virtual polarisation.

The many-polaron canonical transformation (14) leads to an expression for the groundstate energy of the many polaron system which is of the same structure as for the single polaron problem, except that the many-body effects are incorporated via the structure factor $S(k)$ of the 2D electron gas:

$$E = -\alpha \hbar \omega_{LO} \, k_{LO}^{-1} \int_0^\infty dk \, \frac{S^2(k)}{S(k) + k^2 k_{LO}^{-2}} \, f(k,b) \tag{15}$$

Several standard approximations for $S(k)$ in the case of a 2D electron gas can now be used, e.g. the Random Phase Approximation (RPA) or the

Hartree-Fock approximation (HF). Our approximation is called "dynamical screening approximation" because $S(k)$ is defined as the zeroth moment of the dynamical structure factor $S(k,\omega)$ of the 2D electron system:

$$S(k) = \int_0^\infty d\omega \, S(k,\omega) \qquad (16a)$$

$$S(k,\omega) = \frac{1}{\pi \, V(k)} \, \text{Im} \, \frac{-1}{\varepsilon(k,\omega)} \qquad (16b)$$

$\varepsilon(k,\omega)$ is the dielectric function. It is important to remark that $S(k,o) \neq S(k)$. The effect of the non-zero width of the electron layer is taken into account by the choice of $V(k)$ in the expression for the dielectric function

$$\varepsilon(k,\omega) = 1 - V(k) \, \chi(k,\omega) \qquad (16c)$$

where $\chi(k,\omega)$ is the polarisation of the (2D) electron gas. Indeed $V(k) = \frac{2\pi e^2}{k \, \varepsilon_\infty} \, f(k,b)$ and $f(k,b)$ depends on the magnitude of the width of the electron layer.

IId. Static Approximation

In ref. [21] the influence of the electron-phonon screening and the non-zero width of the 2D electron layer on the polaron effects in (polar) semiconductor heterojunctions was studied by Das Sarma, using a simple perturbation scheme. As stated in [18] the approach of [21] implies that a one-polaron picture is used where screening is merely taken into account by the replacement:

$$|V|_k^2 \rightarrow |V_k| \, \varepsilon^{-2}(k,o) \qquad (17a)$$

where $\varepsilon(k,o)$ is the <u>static</u> dielectric function of the 2D electron gas. This leads to the following result for the groundstate energy:

$$E = -\alpha \hbar \omega_{LO} \, k_{LO}^{-1} \int_0^\infty dk \, \frac{1}{\varepsilon^2(k,o)} \, \frac{f(k,b)}{(1+k \, k_{LO}^{-2})} \qquad (17b)$$

(17b) was first studied within the Thomas-Fermi (TF) approximation [21] and later [22] with $\varepsilon(k,o) = \varepsilon_{RPA}(k,o)$. (17b) is called the static approximation.

A typical result of our study of the screening of the electron-phonon interaction in GaAs-Al$_x$Ga$_{1-x}$As heterostructures is shown in Fig. 2. In Fig. 2 the groundstate energy of the many polaron system (for zero layer thickness (a) and for non-zero width of the electron layer (b)) (expressed per polaron) is shown, as a function of the electron density, for the different approximations discussed here. For a detailed discussion the reader is referred to ref. [18]; the main conclusions regarding the screening of the electron-phonon interaction, in particular in GaAs-Al$_x$Ga$_{1-x}$As heterostructures, are the following:
- our treatment confirms that screening leads to a reduction of the effective electron-phonon interaction for a 2D polaron gas, as it occurs in GaAs-Al$_x$Ga$_{1-x}$As heterostructures. However our results indicate that the <u>dynamical</u> treatment of the screening effects on the electron-phonon interaction results in larger polaron effects for electron densities $n_e < 10^{12} \text{cm}^{-2}$ in comparison with the static approach.
- our results are not in contradiction with those of ref. [23], where it was found - in cyclotron resonance experiments on InSb inversion layers

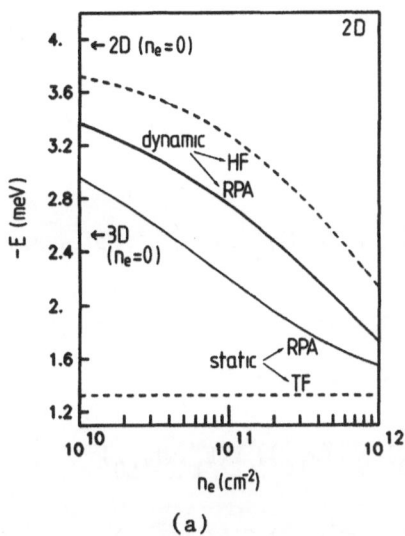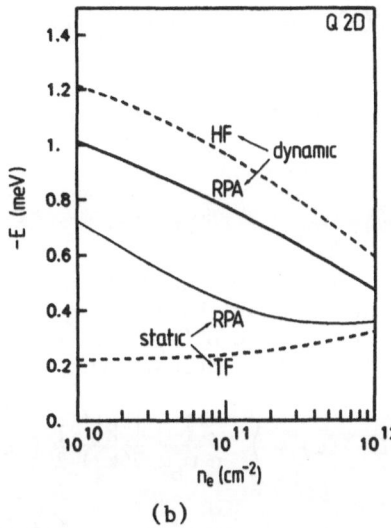

Fig. 2. (a) The contribution of the electron-phonon interaction to the groundstate energy as a function of the electron density. The results for i) the LLP approximation (static structure factor taken from the RPA (thick solid curve) and the Hartree-Fock (dash-dotted curve) approximation), ii) the static RPA approximation (thin solid curve) and iii) the static TF approximation (dashed curve) are shown. The energy correction is calculated for the ideal 2D system.
(b) Same as (a) but now including the effect of the non-zero width of the electron layer.

that at the resonance condition for the cyclotron resonance frequency ω_c ($\omega_c \approx \omega_{LO}$) the polaron effects were not reduced by the many-particle effects. There is no contradiction because the results of ref. [18] are obtained for the non-resonant situation.

IIe. Effects of Screening on the Polaron Mass and the Optical Absorption in 2D [24]

In ref. [18] no unitary transformation approach, equivalent to (14), but applicable to calculate the polaron mass, was developed. However, our subsequent investigations of the optical absorption spectrum for a 2D electron gas interacting with LO-phonons, allowed us to obtain this mass renormalisation, including the effect of dynamical screening in a different way. This is briefly discussed in what follows.

To evaluate the optical absorption for a 2D system of electrons interacting with LO-phonons it is convenient to use the following expression for the a.c. conductivity [24]

$$\sigma(\omega) = \frac{i \, n_e \, e^2/m_b}{\omega - \Sigma(\omega)} \qquad (18a)$$

Here ω is the frequency of the incident radiation. $\Sigma(\omega)$ is often called the memory function, it depends on the relevant physical parameters describing the system (here including α, n_e, m_b, ω). For the 2D electron system interacting with LO-phonons, we obtain

141

$$\Sigma(\omega) = \frac{1}{\omega} \int_0^\infty dt \; (1 - e^{i\omega t}) \; \text{Im} \; F(t) \tag{18b}$$

$$F(t) = - \sum_{\vec{k}} \frac{k_\parallel}{n_e m_b \hbar} |V_k|^2 \; (i \; D(k_\parallel, t) + i \; n(\omega_{LO}) \; D^R(k_\parallel, t)) \; e^{-i\omega_{LO}t} \tag{18c}$$

where $n(\omega_{LO}) = \left(e^{\beta \hbar \omega_{LO}} - 1 \right)^{-1}$, $\beta = \frac{1}{kT}$; $D(k_\parallel, t)$ $(D^R(k_\parallel, t))$ are the electron
(-retarded) density-density correlation functions. To obtain $\sigma(\omega)$ to
dominant order in the coupling at weak coupling, it is sufficient to evaluate $D(k_\parallel, t)$ and $D^R(k_\parallel, t)$ with $\alpha = 0$.
Explicitly, the imaginary part of $\Sigma(\omega)$ takes the form:

$$\text{Im} \; \Sigma(\omega) = \sum_{\vec{k}} \frac{k_\parallel^2}{n_e m_b \omega} \frac{|V_k|^2}{V(k_\parallel)} \frac{1}{2} \; [(n(\omega + \omega_{LO}) - n(\omega_{LO})) \; S(k_\parallel, \omega + \omega_{LO})$$

$$- (1 + n(\omega - \omega_{LO}) + n(\omega_{LO})) \; S(k_\parallel, \omega - \omega_{LO})] \tag{19a}$$

where

$$S(k, \omega) = - \text{Im} \; \epsilon^{-1}(k, \omega) \tag{19b}$$

is the structure factor of the many-electron systems, which is related to
the dielectric function. $\text{Re} \; \Sigma(\omega)$ is related to (19a) via a Kramers-Kronig
relation. Expressions of $\epsilon(k, \omega)$ for a 2D electron gas are available within the RPA [25].

By casting (18a) in the Drude form for the a.c. conductivity one

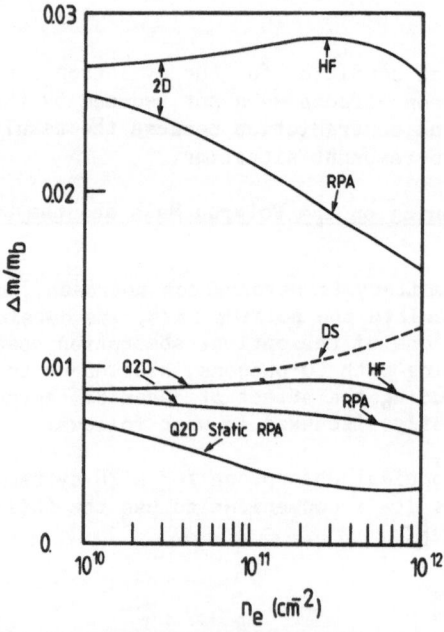

Fig. 3. Mass renormalisation at zero temperature is plotted as a
function of the electron density for both the ideal two-
dimensional (two upper curves) and the quasi-two-dimensional (lower part of the figure) electron system. The
results from different approximations are shown.

obtains for the mass-renormalisation:

$$\frac{\Delta m}{m_b} = - \frac{Re \ \Sigma(\omega)}{\omega} \tag{20}$$

The expression (20) provides one with the mass of a polaron in a 2D electron gas as influenced by dynamical screening.

In Fig. 3 the resulting $\frac{\Delta m}{m_b}$ is plotted as a function of the electron density. The dielectric function was calculated both with the RPA and with the Hartree-Fock approximation. Also the quasi-2D system was considered and the static approximation was made. The main conclusion is that the screening of the electron-phonon interaction reduces the mass renormalisation. The same is true for non-zero width of the electron layer. The combined effect of dynamical screening and non-zero width of the 2D electron layer results in a slow decrease of the mass renormalisation as a function of the density of the electrons.

It would be interesting to examine whether eq. (20) with Re $\Sigma(\omega)$ connected to (19b) via a Kramers-Kronig relation can also be derived using the canonical transformation (14).

The optical absorption for the 2D polaron is now obtained using:

$$Re \ \Sigma(\omega) \propto \frac{Im \ \Sigma(\omega)}{(\omega - Re \ \Sigma(\omega))^2 + (Im \ \Sigma(\omega))^2} \tag{21}$$

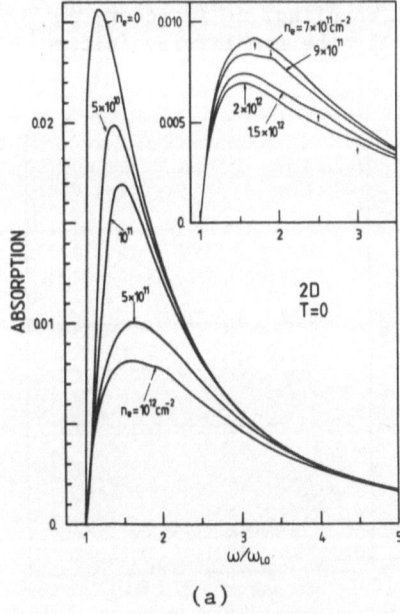

(a)

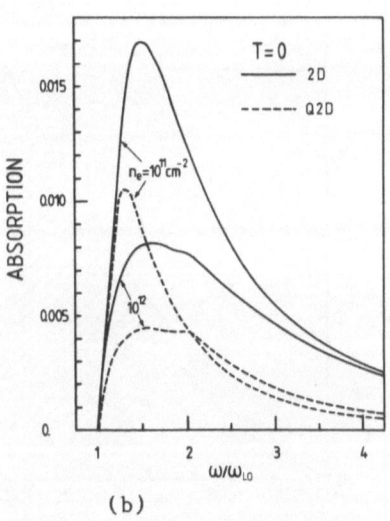

(b)

Fig. 4. (a) Optical absorption spectrum of an ideal two-dimensional electron gas interacting with polar optical phonons at zero temperature is plotted for different values of the electron density. In the insert the arrows indicate the position of the frequency $\omega_{LO} + E_F/\hbar$.
(b) Zero temperature absorption spectrum is shown for an ideal two-dimensional and a quasi-two-dimensional electron system.

a formula which was obtained in ref. [26] for the case of one polaron as described by the Feynman model.

In ref. [24] the right hand side of eq. (21) was evaluated for the case of a GaAs-Al$_x$Ga$_{1-x}$As heterostructure. A typical result is shown in Fig. 4. It is seen that the optical absorption for the 2D polaron gas:
i) is reduced by the screening of the Fröhlich interaction;
ii) is reduced by the non-zero width of the electron layer;
iii) shows a shoulder-structure for electron densities around 10^{12}cm^{-2}.
It would be interesting to study these effects experimentally.

IIf. Plasmon-Phonon Interaction [28]

Plasmon–phonon coupled modes result from the structure (poles, imaginary parts, ...) of the dielectric function f the many-polaron system. Their study therefore fits in this chapter on the effects of screening on the properties of the (weak-coupling) polaron gas. Of aprticular interest in the dispersion of these coupled modes.

In 3D the dispersion of the coupled plasmon-phonon modes for all wave vectors q was first investigated in ref. [17]. The q→0 limits were studied before [27]. In ref. [17] it is shown that, within the RPA, the dielectric function of the LO-phonon bath and of the electron gas is additive for weak coupling. The rather complex behaviour of the coupled plasmon-LO-phonon modes for weak coupling polarons was observed in GaAs using Raman scattering techniques [16].

In ref. 28 we have applied the studies 17 of coupled plasmon-phonon modes to analyse the 2D polaron gas at weak coupling, as it occurs in GaAS-Al$_x$Ga$_{1-x}$As heterostructures. Technically the simplifying feature,

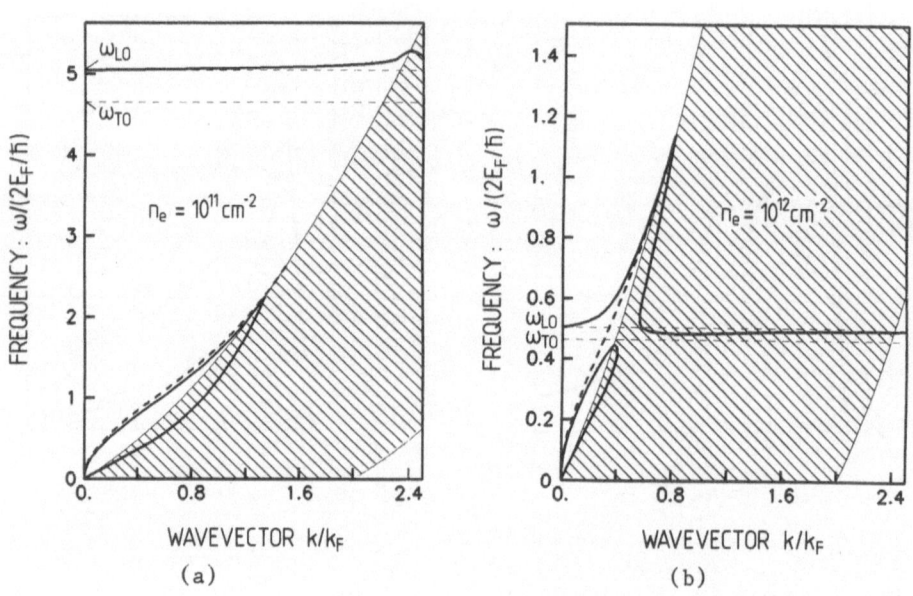

Fig. 5. The collective excitations of the 2DEG in a GaAs-Al$_x$Ga$_{1-x}$As heterostructure (heavy solid curves) are shown as a function of the wave vector k. The plasmon branch for the unperturbed 2DEG is given by the heavy dashed curve. The shaded area corresponds to the pair-excitation region ("Landau damping"). The electron density is (a) $n_e = 10^{11}$cm^{-2} and (b) $n_e = 10^{12}$cm^{-2}.

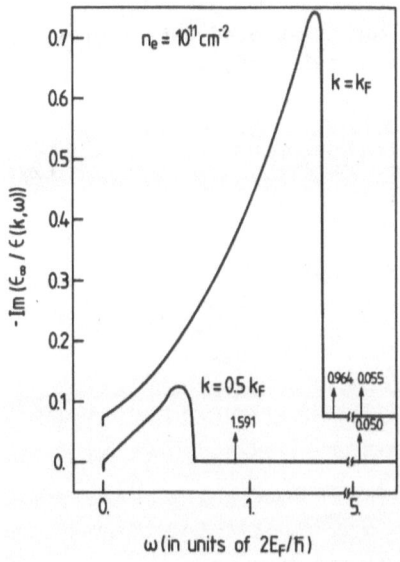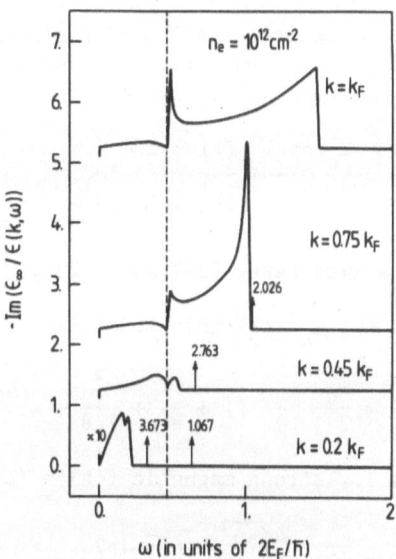

Fig. 6. The energy-loss function as a function of frequency for
different values of the wave vector k. The arrows correspond to delta peaks with an oscillator strength indicated
by the number above the arrows. The electron density is
(a) $n_e = 10^{11} cm^{-2}$ and (b) $n_e = 10^{12} cm^{-2}$. The dashed line
in (b) indicates the position of the TO frequency.

just as for the 3D system resides in the additivity of the polarisation of
the electrons and that of the phonons in the RPA. For the details of the
straightforward calculation I refer to ref. [28].

Typical results are shown in Figs. 5 and 6 for two different electron
densities. Especially for sufficiently large electron densities
($n_e - 10^{12} cm^{-2}$ in GaAs-Al$_x$Ga$_{1-x}$As heterostructures) the plasmon-phonon
coupling modifies the 2D unperturbed excitation spectrum quite drastically.
Two main effects are found:
i) a shift in the phonon and the plasmon frequency (for $\omega_p < \omega_{TO}$ the plasmon
   frequency is lowered, while for $\omega_p > \omega_{TO}$ the frequency is shifted to
   higher values due to the plasmon-phonon coupling);
ii) a splitting of the plasmon branch (manifest as $n_e > 10^{12} cm^{-2}$) occurs as
   ω_p tends to ω_{TO}

It would be very interesting to obtain experimental data for the
plasmon-phonon dispersion in 2D electron layers and heterostructures. The
oscillator strengths in Fig. 6 provide interesting data which can guide
the experiment in searching for the most pronounced structures.

III. THE 2D POLARON SYSTEM SUBJECTED TO A STATIC MAGNETIC FIELD

IIIa. A Groundstate Energy for the 2D Polaron in a Magnetic Field

In ref. [29] the treatment of Peeters and Devreese [30] of the physical properties of a Feynman polaron in a magnetic field was adapted to the
2D case. The groundstate energy, effective mass and several thermodynamic

quantities were evaluated. The only caveat concerns the variational nature of the method which can only be based on Feynman's conjecture.

Some interesting limiting expressions for the groundstate energy of a 2D polaron in a magnetic field were derived in ref. [29]:
a) for small coupling limit $\alpha \ll 1$:

$$E = \frac{\omega_c}{2} \frac{\alpha\pi\sqrt{\omega_c}}{2} \frac{\Gamma(1 + 1/\omega_c)}{\Gamma(\frac{1}{2} + 1/\omega_c)} \tag{22a}$$

(see also refs. [31] and [32]).

a1) For $\omega_c \ll 1$ this leads to:

$$E = \frac{\omega_c}{2} - \frac{\pi\alpha}{2} [1 + \frac{\omega_c}{8} + \frac{\omega_c^2}{128} + \ldots] \tag{22b}$$

a2) for strong magnetic fields ($\omega_c \gg 1$):

$$E = \frac{\omega_c}{2} - \frac{\alpha\sqrt{\pi\omega_c}}{2} - \frac{\alpha\sqrt{\pi}}{\sqrt{\omega_c}} \frac{\ln 2}{} \tag{22c}$$

b) for strong coupling ($\alpha \to \infty$) one obtains

$$E = -\frac{\pi}{8} \alpha^2 - 2\ln 2 - \frac{1}{2} + \frac{\omega_c}{2v^2} \tag{22d}$$

$$v = \frac{\pi}{4} \alpha^2 - (4\ln 2 - 1) \qquad \text{as } \alpha \to \infty. \tag{22e}$$

It is shown in ref. [29] that the scaling relations (8) and (9) are no longer valid - even for the Feynman model - when $\omega_c \neq 0$.

The study of ref. [29] then mostly concentrates on the sudden transition from a dressed state to a stripped state of the polaron, which arises if $\alpha > 1.6$ for sufficiently large ω_c. Although the discontinuous character of this transition might be an artefact of the method, a strong - (but possibly) - continuous modification of the polaron characteristics should be expected for increasing ω_c and in a limited ω_c interval.

IIIb. Energy Levels of 2D Polarons in a Magnetic Field. Position of the Landau Levels

Before discussing the magneto-absorption of weak coupling polarons in 2D I will first discuss the influence of the electron-LO-phonon interaction on the Landau levels of 2D electrons within second order perturbation theory. Evidently the starting expression is:

$$\Delta E_n = \sum_{m=0}^{\infty} \sum_{q} \frac{|M_{nm}(q)|^2}{D_{nm}} \tag{23a}$$

where M_{nm} is the Fröhlich coupling matrix element

$$M_{nm}(\vec{q}) = \langle m, k_z; \vec{q} | H_I | n, 0, \bar{0} \rangle \tag{23b}$$

and, in 2D,

$$D_{nm} = \hbar\omega_{LO} - \Delta_n + \hbar\omega_c(m-n) \tag{23c}$$

The choice of Δ_n determines the type of perturbation theory which is used:
i) $\Delta_n = 0$ corresponds to Rayleigh-Schrödinger perturbation theory (RSPT);
ii) $\Delta_n = \Delta E_n$ results in Wigner-Brillouin perturbation theory (WBPT);

iii) $\Delta_n = \Delta E_n - \Delta E_0^{RSPT}$ gives the so-called "improved" Wigner-Brillouin perturbation theory (IWBPT) which leads to the correct pinning behaviour for small α.

We have found it very convenient to introduce the following integral representation for $\frac{1}{D_{nm}}$:

$$\frac{1}{D_{nm}} = \int_o^\infty du\ e^{-D_{nm}u} \tag{24a}$$

This representation transforms (23a) into:

$$E_n = - \sum_q |V_q|^2 \int_o^\infty du\ e^{-(1-\Delta_n)u} \langle n,o|e^{i\vec{q}\cdot\vec{r}(u)}\ e^{-i\vec{q}\cdot\vec{r}(o)}|n,o\rangle \tag{24b}$$

The attractive feature of (24b) is that it expresses E_n in terms of the density-density correlation function which was the basis of much of our previous work on polarons (see e.g. [26], [33]). The correlation function in (24b) is most easily expressed using the equations of motion for $\vec{r}(t)$. In the 2D case (24b) then leads to the following expression:

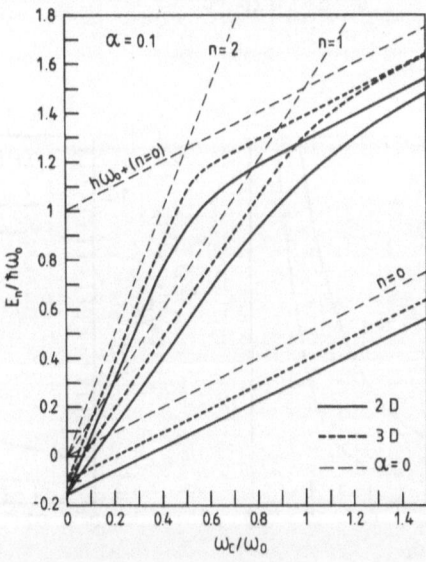

Fig. 7. The first three Landau levels as function of the magnetic field for the unperturbed energy levels (thin dashed curves), the 2D polaron (thick solid curves) and the 3D polaron (thick dashed curves). The electron-phonon coupling constant is $\alpha = 0.1$.

$$\Delta E_n = -\alpha \frac{\pi}{2} \frac{1}{\sqrt{\omega_c}} \sum_{m=0}^{n} \binom{n}{m} \left[\frac{(2m-1)!!}{2m}\right] \frac{1}{m!} \frac{\Gamma((1-\Delta_n)/\omega_c - m)}{\Gamma((1-\Delta_n)/\omega_c + 1/2)} \tag{25}$$

where $\Gamma(x)$ is the delta function. (25) had been obtained by D. Larsen [31] for $\Delta_n = 0$ and $n = 0,1,2$.

In Fig. 7 the first three Landau levels are plotted as a function of magnetic field for $\alpha = 0.1$. Both the 2D and the 3D case are shown. Fig. 7 reveals the well-known features for weak coupling polaron Landau levels. The polaron levels are i) shifted to lower energies, ii) at weak magnetic fields they are bending downward due to mass renormalisation, iii) at $n\omega_c \approx \omega_{LO}$ the n-th Landau level does not cross the LO-phonon + (n-1)-th Landau level, and iv) for $\omega_c \to \infty$ all Landau levels become pinned to $\hbar\omega_{LO} + \Delta E_0^{RSPT}$. Note from Fig. 7 that the polaron effects (energy shift, mass renormalisation, level splitting at $n\omega_c \approx \omega_0$) are more pronounced in 2D than in 3D. The present results refer to one polaron. As suggested by the results of Section II however, screening can reduce the polaron effects. Interesting expansions can now be obtained from eq. (25) for limiting strengths of the magnetic field. E.g. for weak magnetic fields ($\frac{\omega_c}{\omega_{LO}} \ll 1$) we obtain, to first order in α, (i.e. with $\Delta_n = 0$):

$$E_n = -\alpha \frac{\pi}{2} \left[1 + \frac{2n+1}{8} \omega_c + \frac{18n(n+1)+1}{128} \omega_c^2 + \frac{5}{1024}(2n+1)(10n^2+10n-1)\omega_c^3 \right. $$
$$\left. + \ldots\right] \tag{26a}$$

(see also (22b)).
Defining the cyclotron mass m^* from $\Delta E_{n+1} - \Delta E_n$ one finds

$$\frac{m^*}{m} = \frac{1}{1 - \alpha \frac{\pi}{8}(1 + \frac{9}{8}\omega_c + \frac{145}{126}\omega_c^2 + \ldots)} \tag{26b}$$

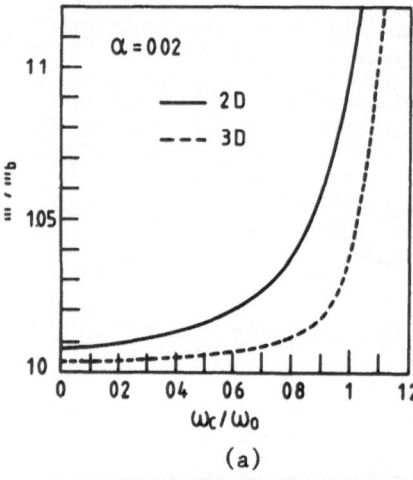
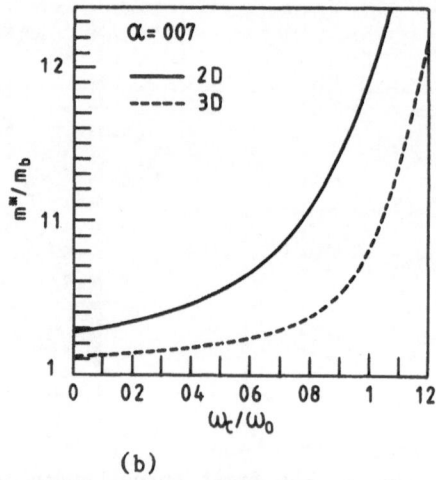

(a)   (b)

Fig. 8. Cyclotron resonance mass as function of the magnetic field for the 2D (solid curve) and the 3D (dashed curve) polaron and for (a) $\alpha = 0.02$ and (b) $\alpha = 0.07$.

In Fig. 8 the polaron cyclotron mass, calculated with IWBPT, is plotted and compared to the 3D case. This figure clearly reveals an enhancement of the mass renormalisation for a polaron in a system with reduced dimensionality. Eq. (25) also provides results for the splitting and pinning of the Landau levels. For the splitting of the Landau level n at $n\omega_c = \omega_0$ (i.e. vv) one obtains 2 ΔE_n with:

$$(\Delta E_n) = \alpha^{\frac{1}{2}} \left((\frac{-}{n})^{\frac{1}{2}} \frac{(2n-1)!!}{2^{n+1}n!} \right)^{\frac{1}{2}} \tag{27}$$

a result obtained for n = 1, in ref. [21].

In the asymptotic limit of high magnetic fields, one obtains for the energy of the n-th Landau level (IWBPT):

$$E_n = 1 + E_0^{RSPT} + \frac{1}{2} \omega_c - \frac{\alpha}{\sqrt{\omega_c}} \frac{\sqrt{\pi}(2n-1)!!}{2^{n+1}n!} \tag{28}$$

see also (22c).

The results in this Section IIIb are only applicable if $E_n-E_0 < \hbar\omega_{LO}$. In ref. [34] these results were generalised to incorporate the energy range $2\hbar\omega_{LO} > E_n-E_0 > \hbar\omega_{LO}$. The generalisation is possible by using degenerate perturbation theory and imposing the correct pinning behaviour via IWBPT. As we are primarily interested in the position of the energy levels it is sufficient to calculate the real part of the energy; this allows to avoid the difficulties associated with the continuous nature of the energy spectrum as $2\hbar\omega_{LO} > E_n-E_0 > \hbar\omega_{LO}$. The following result was obtained:

$$\Delta E_n = \frac{\alpha\sqrt{\omega_c}}{2} \sum_{m=0}^{\infty} \frac{\delta(m,n)}{\mu_n} \sum_{\ell,\ell'=0}^{S} \binom{S}{\ell} \binom{S}{\ell'} \frac{(-)^{\ell+\ell'}\Gamma(|n-m|+\ell+\ell'+1/2)}{(|n-m|+\ell)!(|n-m|+\ell')!} \tag{29}$$

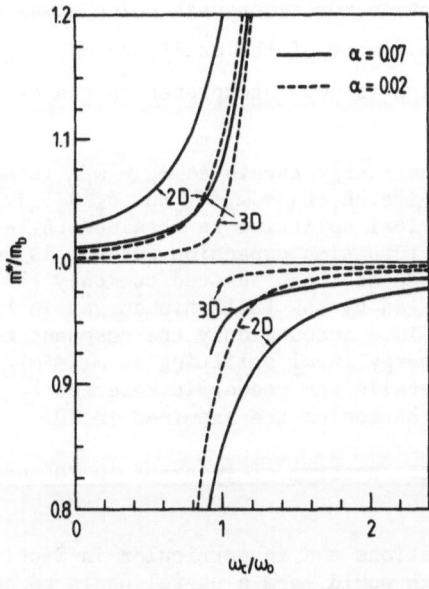

Fig. 9. Cyclotron polaron mass in 2D and 3D for two values of the electron-phonon coupling constant $\alpha = 0.02$ and $\alpha = 0.07$ which correspond respectively to InSb and GaAs.

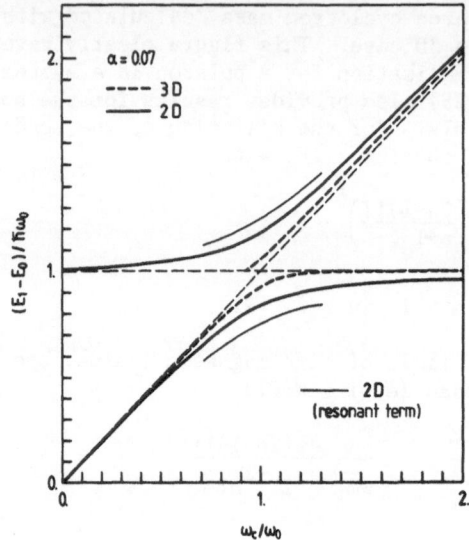

Fig. 10. Energy of the first Landau level (E_1) shifted with the
groundstate energy (E_0) as a function of the magnetic
field for the 2D (solid curve) and the 3D (dashed curve)
case. If only the resonant term is included within WBPT
the thin solid curve is obtained.

One now relates $E_1-E_0 = \hbar\omega_c^*$ to the position of the cyclotron resonance
frequency ω_c from which the cyclotron mass is derived $\frac{m^*}{m} = \frac{\omega_c}{\omega_c^*}$. This cyclotron mass is plotted in Fig. 9 and compared to the 3D case. The result
below the continuum is obtained from Section IIIb.

From Fig. 9 it is seen that for any magnetic field two branches exist.
If $\omega_c/\omega_0 < 1$ the lowest branch has the dominant oscillator strength and
corresponds to the cyclotron resonance peak which gives a polaron cyclotron
mass $m^* > m_b$. For $\frac{\omega_c}{\omega_{LO}} > 1$ most of the oscillator strength is transferred
to the upper branch, which is now interpreted as the cyclotron resonance
peak with $m^* < m_b$.

The splitting of the energy levels at $\omega_c = \omega_{LO}$ is not symmetric around
the unperturbed result (i.e. $E_1-E_0 = \omega_c \pm \delta_\pm$ but $\delta_+ \neq \delta_-$). It is important
to remark that a symmetrical splitting is obtained ($\delta_+ = \delta_-$) if only the
resonant term in the perturbation expansion for ΔE_n is retained, within
WBPT. The latter approach has been adapted commonly [35] and it leads to
the energy levels indicated by the full thin curves in Fig. 10. It should
be realised that taking into account only the resonant term leads to a 48%
overestimation of the energy level splitting at $\omega_c = \omega_{LO}$ for the 2D case
($\alpha = 0.07$). For more details the reader is referred to ref. [34] where
also LO-phonon assisted harmonics are examined in 2D.

IIIc. Cyclotron Resonance Spectrum of Polarons in 2D and Comparison to
Experiment

In the previous sections and in particular in Section IIIb., we have
introduced material which would form a useful basis to analyse cyclotron
resonance data on electrons in heterojunctions composed of polar semicon-

ductors like GaAs-Al$_x$Ga$_{1-x}$As. We would have to add non-parabolicity and possibly screening effects to the results of IIIb. before comparing them to experiment.

Although such an approach in which the cyclotron mass is related to the position of the Landau levels might be sufficient for weak electron-LO phonon coupling we prefer to calculate directly the cyclotron resonance absorption spectrum itself. The theoretical cyclotron resonance frequency is then determined from the position of the calculated cyclotron resonance peak. I.e. we prefer to calculate directly the measured quantity. This procedure is well adapted if one wishes to generalise the treatment to include many-body effects, non-parabolicity, etc. ... .

What one has to do is to generalise the expression (18a) for the a.c. conductivity to include the case where a static magnetic field is applied. This leads to the expression:

$$\sigma(\omega) = \frac{i \; n_e \; e^2/m_b}{\omega - \omega_c - \Sigma(\omega)} \tag{30a}$$

The magneto-optical absorption (to within a factor) is then given by

$$\mathrm{Re} \; \Sigma(\omega) \propto \frac{-\mathrm{Im} \; \Sigma(\omega)}{(\omega - \omega - \mathrm{Re} \; \Sigma(\omega))^2 + (\mathrm{Im} \; \Sigma(\omega))^2} \tag{30b}$$

This expression is a generalisation of that obtained for zero magnetic field in ref. [26] (see also ref. [ ]).

The physical approximations (perturbation theory, Feynman model for the polarons, RPA, Hartree-Fock for the many-body system) go into the calculation of the memory function $\Sigma(\omega)$. The basic quantity entering into $\Sigma(\omega)$ is the space Fourier transform of the density-density correlation function:

$$I(k,t) = \langle e^{i\vec{k}.\vec{r}(t)} \; e^{-i\vec{k}.\vec{r}(o)} \rangle \tag{30c}$$

which was already encountered above in studying the optical absorption of polarons in 2D. Of course, $I(k,t)$ depends on ω_c because the equation of motion for $\vec{r}(t)$ involves the magnetic field; (30c) has to be generalised to include many-body effects. Detailed expressions for $I(k,t)$ are presented in Appendix A and B.

As $T \to 0$ the expression $\mathrm{Im} \; \Sigma(\omega)$ tends to zero and the cyclotron resonance peak becomes a delta function whose position is given by the zero of the non-linear equation:

$$\omega - \omega_c - \mathrm{Re} \; \Sigma(\omega) = 0 \tag{31}$$

The solution of (31) is denoted as $\omega = \omega_c$. (31) also shows that $\mathrm{Re} \; \Sigma(\omega)$ is the shift in the cyclotron resonance peak position due to the Fröhlich electron-phonon interaction.

The resulting cyclotron resonance spectrum is shown in Fig. 6 of Appendix A for several broadening parameters. The non-zero width of the electron layer is taken into account in Fig. 7 of Appendix B. It is apparent from these figures that i) for $\omega_c \sim \omega_{LO}$ the cyclotron resonance peak is split into two peaks, ii) with increasing magnetic field strength the oscillator strength of the peak at the lowest frequency is transferred to the peak at the frequency above the LO-phonon frequency, ii) the peak below ω_{LO} is shifted to lower frequency in comparison to ω_c while the peak above ω_{LO} is shifted to higher frequencies as compared to ω_c.

We have compared to experiment our polaron cyclotron resonance results for InSb inversion layers and for GaAs-Al$_x$Ga$_{1-x}$ heterostructures.

As explained in Appendix A and B, it was crucial to incorporate the non-parabolicity of the electron bands before comparing to experiment (see eq. (12) of Appendix A). First a comparison with experiment was made within a one-particle theory including the theory sketched above. For the InSb inversion layer theory and experiment fit quite closely (see Fig. 11, or Fig. 12 of Appendix A). The dashed line in Fig. 11 corresponds to the cyclotron resonance mass as due to non-parabolicity without polaron effects. Below the resonant magnetic field (B = 2.9 T) no polaron effects have to be invoked in order to explain the experimental data. However, around the resonant field polaron effects are important as seen from Fig. 11 (see ref. [39].

A one-polaron picture with inclusion of non-parabolicity and the non-zero width of the electron layer is apparently sufficient to explain the InSb inversion layer data. No screening effects have to be included.

The situation regarding the cyclotron resonance data for the GaAs-Al$_x$Ga$_{1-x}$As heterostructures is less clear, both experimentally and theoretically. Experimentally the cyclotron mass data on very similar samples, obtained by two groups [36][37], are drastically different. Recent data [38] tend to confirm the data of Merkt and Horst. The theoretical analysis, based on the same one-polaron theory as applied to InSb inversion layers, only provides reasonable agreement to the experiment of ref. [36] if the coupling constant is effectively lowered (see discussion in Appendix A). Therefore we were led to study the influence of many-body effects on the cyclotron resonance mass of 2D polarons. Fermi-Dirac statistics and static

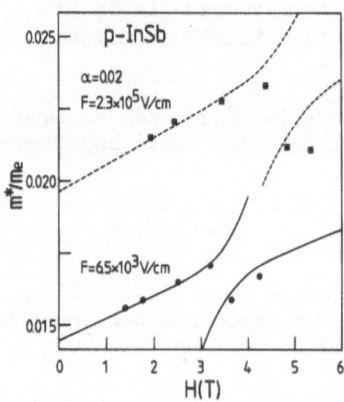

Fig. 11. The experimental data for InSb inversion layers (Ref. 23) are compared with the present theoretical results (solid and dashed curves). The solid circle and square points correspond to the experimental electron densities n_e = 2×10^{11} and 10^{12}cm^{-2}, respectively.

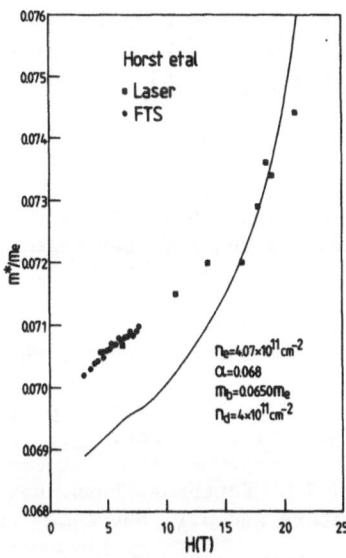

Fig. 12. The theoretical results (solid curve) are compared to
the experimental data of Horst et al. [36] for a
GaAs-Al$_{1-x}$Ga$_x$As heterostructure which are given by the
full circles and the square points.

screening were incorporated. The treatment allows us to get reasonable
agreement with the data of ref. [36] using the established α-value for
GaAs, i.e. without artificially reducing it. The result is shown in Fig.
12 (Fig. 6 of Appendix B).

The agreement between theory and experiment is now good when H > 15T.
For weaker magnetic fields the present theory does not explain the data.

The interested reader is referred to Appendix A and B where the
details of our study of the cyclotron resonance of polarons in 2D are
presented.

REFERENCES

[1] J. Sak, Phys. Rev. B6, 3981 (1972).
[2] T. Ando, B. Fowler and F. Stern, Rev. Mod. Phys. 54, 437 (1982).
[3] Wu Xiaoguang, F.M. Peeters and J.T. Devreese, Phys. Rev. B31, 3420 (1985).
[4] S. Das Sarma and B.A. Mason, Ann. Phys. (N.Y.) 163, 78 (1985).
[5] R.P. Feynman, Phys. Rev. 97, 660 (1955).
[6] W. Huybrechts, Solid State Commun. 28, 95 (1978); M. Matsuura, ibid. 44, 1471 (1982).
[7] F.M. Peeters and J.T. Devreese, Phys. Stat. Sol. (b) 110, 631 (1982).
[8] Wu Xiaoguang, F.M. Peeters and J.T. Devreese, Phys. Rev. B32, 7964 (1985).
[9] L.F. Lemmens, J. De Sitter and J.T. Devreese, Phys. Rev. B8, 2717 (1973).
[10] F.M. Peeters, P. Warmenbol and J.T. Devreese, to be published (1987).
[11] See e.g. D.M. Larsen, Phys. Rev. 144, 697 (1966).
[12] See e.g. G. Whitfield and R. Puff, Phys. Rev. 139, A338 (1965), or:
A.A. Klochikhin, Sov. Phys. Solid State 21, 1770 (1979).

[13] J. Ihm, Phys. Rev. Lett. 55 , 999 (1985).

[14] T.W. Hickmott, P.M. Solomon, F.F. Fang, F. Stern, R. Fisher and H. Morkoc, Phys. Rev. Lett. 52, 2053 (1984).

[15] A. Mooradian and G.B. Wright, Phys. Rev. Lett. 16, 999 (1966).

[16] W. Richter, in "Polarons and Excitons in Polar Semiconductors and Ionic Crystals", J.T. Devreese and F.M. Peeters, eds., Plenum Publishing Corp., New York (1985), p. 209.

[17] L.F. Lemmens and J.T. Devreese, Solid State Commun. 14, 1339 (1974); L.F. Lemmens, F. Brosens and J.T. Devreese, ibid. 17, 337 (1975).

[18] Wu Xiaoguang, F.M. Peeters and J.T. Devreese, Phys. Stat. Sol. (b) 133, 229 (1986).

[19] T.S. Rahman and D.L. Mills, Phys. Rev. B23, 4081 (1981).

[20] L.F. Lemmens, J.T. Devreese and F. Brosens, Phys. Stat. Sol. (b) 82, 439 (1977).

[21] S. Das Sarma, Phys. Rev. B27, 2590 (1983); B32, 4034(E) (1985).

[22] S. Das Sarma and A.B. Mason, Phys. Rev. B31, 5536 (1985); B32, 2656(E) (1985).

[23] M. Horst, U. Merkt and J.D. Kotthaus, Phys. Rev. Lett. 50, 754 (1983).

[24] Wu Xiaoguang, F.M. Peeters and J.T. Devreese, Phys. Rev. B34, 2621 (1986).

[25] F. Stern, Phys. Rev. Lett. 18, 546 (1967).

[26] J.T. Devreese, in "Polarons in Ionic Crystals and Polar Semiconductors", North-Holland Publishing Co., Amsterdam (1972), p. 113.

[27] B.B. Varga, Phys. Rev. A137, 1896 (1965).

[28] Wu Xiaoguang, F.M. Peeters and J.T. Devreese, Phys. Rev. B32, 6982 (1985).

[29] Wu Xiaoguang, F.M. Peeters and J.T. Devreese, Phys. Rev. B32, (1985).

[30] F.M. Peeters and J.T. Devreese, Phys. Rev. B25, 7281 (1982); B25, 7302 (1982).

[31] D.M. Larsen, Phys. Rev. B30, 4807 (1984).

[32] F.M. Peeters and J.T. Devreese, Phys. Rev. B31, 3689 (1985).

[33] F.M. Peeters and J.T. Devreese, Phys. Rev. B25, 7281 (1982).

[34] F.M. Peeters, Wu Xiaoguang and J.T. Devreese, Phys. Rev. B33, 4338 (1986).

[35] See e.g. W. Zawadzki, P. Pfeffer and H. Sigg, Solid State Commun. 53, 777 (1985).

[36] M. Horst, U. Merkt, W. Zawadzki, J.C. Maan and K. Ploog, Solid State Commun. 53, 403 (1985).

[37] H. Sigg, P. Wyder and J. Perenboom, Phys. Rev. B31, 5253 (1985).

[38]   Nicholas, private communication.

[39] F.M. Peeters, Wu Xiaoguang and J.T. Devreese, Physica Scripta 34 (1986).

APPENDIX A

THEORY OF THE CYCLOTRON RESONANCE SPECTRUM
OF A POLARON IN TWO DIMENSIONS

by: Wu Xiaoguang, F. M. Peeters and J. T. Devreese

published in Phys. Rev. B**34**, 8800 (1986)

ABSTRACT

The magneto-optical absorption spectrum of a two-dimensional polaron is calculated by using a memory function approach. The cyclotron resonance frequency and the cyclotron resonance mass of the polaron are obtained for weak electron-phonon coupling. The absorption spectrum exhibits peaks around the cyclotron frequency ω_c and the LO-phonon assisted harmonics $\omega_{LO} + n\omega_c$ ($n = 1, 2...$). The oscillator strength and the position of the peaks are investigated as a function of the magnetic field strength. A Landau level broadening parameter is introduced phenomenologically in order to remove the divergencies in the magneto-optical absorption spectrum. The effect of the finite width of the two-dimensional electron layer is also investigated. After taking into account the effect of the non-parabolic energy band of the electron, the calculated cyclotron resonance masses are compared to the experimental data for GaAs-AlGaAs heterostructures and InSb inversion layers. In order to explain the experimental results for GaAs-AlGaAs heterostructures with our one polaron theory an effective electron-phonon coupling constant has to be used which is smaller than generally accepted.

I. INTRODUCTION

In polar semiconductors and ionic crystals an electron interacts with longitudinal optical (LO) phonons. In the presence of a magnetic field there will be so-called *resonant polaron effects* when ω_c, the unperturbed cyclotron frequency, approaches ω_{LO}, the LO-phonon frequency. The main evidence for the existence of resonant polaron effects is provided by a cyclotron resonance experiment. In such an experiment the mass renormalization of the electron due to the polaron effect is observed clearly. Over the last few decades three-dimensional (3D) polarons have been extensively studied. For a review of the theoretical and experimental progress in this field we refer to Ref.1.

Recently, due to technological progress in material growth (e.g. the advent of MBE), quasi-two-dimensional (Q2D) electron systems have been created in polar semiconductors. Examples of such systems are GaAs-AlGaAs heterostructures, GaInAs heterojunctions, InSb inversion layers *etc.*[2]. Only very recently polaron effects have been studied in these 2D systems[3-12]. The main emphasis was on the investigation of the peak position of the cyclotron resonance line and on the splitting of the line around the resonance frequency $\omega_c = \omega_{LO}$. In Ref.8, by using a Green's function approach, Das Sarma and Madhukar made a formal calculation to investigate the Landau level correction and the magneto-optical anomalies in the resonant region. They showed that the influence of the electron-phonon coupling in 2D systems can lead to a splitting of the cyclotron resonance line when $\omega_c \sim \omega_{LO}$ which is similar to that for the 3D polaron (see, for instance, Ref.13 and references therein). In their calculation off-resonance terms in the perturbation theory were neglected. Subsequently Larsen studied the cyclotron resonance of a 2D polaron using the Rayleigh-Schrödinger perturbation theory (RSPT)[9]. In Ref.9 the effect due to the non-zero density of the electron gas on the polaron Landau levels was also investigated by summing the most divergent terms in the perturbation theory to all orders. The electron-phonon interaction correction to the Landau levels of the 2D polaron has also been studied by the present authors using the so-called Improved-Wigner-Brillouin perturbation theory (IWBPT)[10]. Most of the above mentioned studies are based on a perturbation calculation of the position of the Landau levels of the 2D polaron. The cyclotron resonance mass of the electron is then obtained from the difference in energy between two adjacent Landau levels.

In this paper we present a calculation of the 2D polaron cyclotron resonance spectrum which is based on a memory function approach[13−15]. Instead of calculating polaron energy levels we calculate the magneto-optical absorption spectrum which is expressed in terms of a memory function. Our motivation is that the magneto-optical absorption is the physical measured quantity. The cyclotron resonance frequency and the cyclotron resonance mass of the electron are obtained from the position of certain peaks in the magneto-optical absorption spectrum. The present paper is divided into two parts: the first part contains our theoretical calculation and in the second part our results are compared with recent experimental data[4−6].

In the theoretical part of this paper we limit ourselves to a single electron with a parabolic energy band, interacting with bulk LO-phonons. The effect of the finite width of the 2D electron layer is included by considering the lowest subband where the standard variational wave function is used[2]. We also introduce a Landau level broadening parameter[2,16] in order to remove the divergencies in the absorption spectrum. This is equivalent to a standard procedure in which the unperturbed density of states, which consists of a series of delta functions, is replaced by a set of Gaussian functions. The electron-phonon interaction is treated as a perturbation and the memory function is calculated to first-order in the electron-phonon coupling constant. The spin, the effect resulting from Fermi-Dirac statistics and the screening effect of the electron gas are neglected in the present study.

The present paper is organized as follows: in Sec.II we outline the calculation of the memory function including the effect of the finite width of the 2D electron layer and of the Landau level broadening. The magneto-optical absorption spectra of the 2D polaron are then calculated. Sec.III contains our numerical results and discussion. In Sec.IV we take into account the band nonparabolicity and compare our calculations with recent experimental data. Our conclusion is presented in Sec.V.

II. FORMULATION AND CALCULATION

The electron-phonon system is described by the Fröhlich Hamiltonian

$$H = \frac{1}{2m_b}\left(\vec{p} + \frac{e\vec{A}}{c}\right)^2 + \sum_{\vec{k}} \hbar\omega_{LO}\, a_{\vec{k}}^\dagger a_{\vec{k}}$$
$$+ \sum_{\vec{k}}\left(V_{\vec{k}}\, a_{\vec{k}}\, e^{i\vec{k}\cdot\vec{r}} + V_{\vec{k}}^*\, a_{\vec{k}}^\dagger\, e^{-i\vec{k}\cdot\vec{r}}\right)\ ,$$

(1)

where $\vec{p}\,(\vec{r})$ is the momentum (position) operator of an electron. $a_{\vec{k}}^{\dagger}\,(a_{\vec{k}})$ is the creation (annihilation) operator of a bulk LO-phonon with wave vector $\vec{k}$ and energy $\hbar\omega_{\mathrm{LO}}$. The magnetic field H is taken perpendicular to the 2D electron layer and the z-axis is chosen along the direction of the magnetic field. In Eq.(1) $V_{\vec{k}}$ is given by

$$V_{\vec{k}} = i\hbar\omega_{\mathrm{LO}} \left(\frac{4\pi\alpha}{V k^2}\right)^{1/2} \left(\frac{\hbar}{2m_b\omega_{\mathrm{LO}}}\right)^{1/4} \langle\psi_0|e^{ik_z z}|\psi_0\rangle \quad , \tag{2}$$

where $\psi_0(z) = (b^3/2)^{1/2} z e^{-bz/2}$ is the variational wave function describing the motion of the electron in the direction normal to the 2D electron layer. b is given by $b = (48\pi N m_b e^2/\hbar^2\epsilon_0)^{1/3}$, where $N = n_d + (11/32)n_e$ and n_d and n_e are the depletion and carrier charge density respectively. For simplicity we take into account only the lowest subband and neglect all higher subbands. Such an approximation results in Eq.(2).

Within linear response theory the dynamical conductivity of the system can be written as

$$\sigma(\omega) = \frac{in_e e^2/m_b}{\omega - \omega_c - \Sigma(\omega)} \quad , \tag{3}$$

where $\Sigma(\omega) = \Sigma(\alpha, \omega_c, b; \omega)$ is the memory function[13-15]. $\omega_c = eH/m_b c$ is the cyclotron resonance frequency when no polaron effects are present. The magneto-optical absorption is defined as the real part of Eq.(3) (within a factor)

$$\frac{-\mathrm{Im}\Sigma(\omega)}{[\omega - \omega_c - \mathrm{Re}\Sigma(\omega)]^2 + [\mathrm{Im}\Sigma(\omega)]^2} \quad . \tag{4}$$

The zero magnetic field limit of Eq.(4) leads to a similar expression as was obtained in Ref.17. In the present paper the memory function will be calculated within a similar approximation as in Ref.15. This amounts to a perturbational calculation of $\Sigma(\omega)$. To first-order in the electron-phonon coupling constant the memory function has the form (see also Refs.17 and 18 for the zero magnetic field case)

$$\Sigma(\omega) = \frac{1}{\omega}\int_0^\infty dt(1 - e^{i\omega t})\mathrm{Im}F(t) \quad , \tag{5a}$$

and

$$F(t) = -\sum_{\vec{k}} \frac{k_\parallel^2}{m_b\hbar}|V_{\vec{k}}|^2 [\,(1 + n(\omega_{\mathrm{LO}}))\langle e^{i\vec{k}\cdot\vec{r}(t)}e^{-i\vec{k}\cdot\vec{r}(0)}\rangle$$
$$- n(\omega_{\mathrm{LO}})\langle e^{-i\vec{k}\cdot\vec{r}(0)}e^{i\vec{k}\cdot\vec{r}(t)}\rangle]e^{-i\omega_{\mathrm{LO}}t} \quad , \tag{5b}$$

where $n(\omega_{\mathrm{LO}}) = (e^{\beta\hbar\omega_{\mathrm{LO}}} - 1)^{-1}$ is the number of phonons and $k_\parallel^2 = k_x^2 + k_y^2$. $\langle\cdot\rangle$ stands for the average and must be calculated without electron-phonon interaction in order to be consistent with the memory function calculation.

The problem is now reduced to the calculation of a density-density correlation function

$$I(\vec{k}, t) = \langle e^{i\vec{k}\cdot\vec{r}(t)}e^{-i\vec{k}\cdot\vec{r}(0)}\rangle \quad . \tag{6}$$

In the following we will concentrate on the zero temperature case and calculate $I(\vec{k}, t)$ to zero order in the electron-phonon interaction which amounts to a replacement of the hamiltonian H by H_0 (H_0 is the hamiltonian of a free electron in a magnetic field). The calculation of $I(\vec{k}, t)$ is analogous to a similar calculation given in Ref.10. Here we give the main steps

$$I(\vec{k}, t) = \sum_n \langle 0|e^{iH_0t/\hbar}|0\rangle\langle 0|e^{i\vec{k}\cdot\vec{r}}|n\rangle\langle n|e^{-iH_0t/\hbar}|n\rangle\langle n|e^{-i\vec{k}\cdot\vec{r}}|0\rangle$$

$$= \sum_n e^{iE_0t/\hbar - \Gamma^2t^2/8}\langle 0|e^{i\vec{k}\cdot\vec{r}}|n\rangle e^{-iE_nt/\hbar - \Gamma^2t^2/8}\langle n|e^{-i\vec{k}\cdot\vec{r}}|0\rangle$$

$$= e^{-\Gamma^2t^2/4}\sum_n e^{iE_0t/\hbar}\langle 0|e^{i\vec{k}\cdot\vec{r}}|n\rangle e^{-iE_nt/\hbar}\langle n|e^{-i\vec{k}\cdot\vec{r}}|0\rangle \tag{7}$$

$$= e^{-\Gamma^2t^2/4}\exp\left(-\frac{\hbar k^2}{2m_b\omega_c}(1 - e^{-i\omega_c t})\right) \quad,$$

where $|n\rangle$ is the wave function of the unperturbed n-th Landau level. In Eq.(7) we introduce the Landau level broadening parameter Γ by assuming that the propagator of a free electron is $\exp(-iE_nt/\hbar - \Gamma^2t^2/8)$ (see Refs.2 and 16). For convenience we choose Γ independent of the Landau level number.

After some algebra we arrive at

$$F(t) = \eta \int_0^\infty dx\, x^2 f(x, b_0)\exp\left(-x^2(1 - e^{-i\omega_c t}) - \frac{\Gamma^2t^2}{4} - i\omega_{\mathrm{LO}}t\right) \quad, \tag{8}$$

where $\eta = 2\alpha\omega_{\mathrm{LO}}{}^3(\omega_c/\omega_{\mathrm{LO}})^{3/2}$ and $b_0 = b(\hbar/2m_b\omega_c)^{1/2}$. The form factor $f(k, b)$ is given by $f(k, b) = (8b^3 + 9b^2k + 3bk^2)/(8(b + k)^3)$ which expresses the finite width of the 2D electron layer.

The explicit form of the memory function can be obtained from Eqs.(5). In the present paper we will give numerical results for the case of zero lattice temperature. In this limit the real part of the memory function becomes

$$\mathrm{Re}\Sigma(\omega) = \sum_{n=0}^\infty \frac{B_n}{\omega\Gamma}\left[2\,\mathrm{D}\left(\frac{\varepsilon_n}{\Gamma}\right) - \mathrm{D}\left(\frac{\varepsilon_n + \omega}{\Gamma}\right) - \mathrm{D}\left(\frac{\varepsilon_n - \omega}{\Gamma}\right)\right] \quad, \tag{9a}$$

and the imaginary part

$$\mathrm{Im}\Sigma(\omega) = \sum_{n=0}^\infty \frac{\sqrt{\pi}B_n}{2\omega\Gamma}\left[\exp\left(-\frac{(\varepsilon_n + \omega)^2}{\Gamma^2}\right) - \exp\left(-\frac{(\varepsilon_n - \omega)^2}{\Gamma^2}\right)\right] \quad, \tag{9b}$$

with

$$B_n = \frac{\eta}{n!}\int_0^\infty dx\, x^2 f(x, b_0)x^{2n}e^{-x^2} \quad, \tag{9c}$$

where $\varepsilon_n = \omega_{\mathrm{LO}} + n\omega_c$ and $\mathrm{D}(x)$ is the Dawson integral. The real and imaginary part of the memory function given by Eqs.(9a) and (9b) satisfy the Kramers-Kroning relation

$$\mathrm{Re}\Sigma(\omega) = -\frac{1}{\pi}\int_{-\infty}^{+\infty} dx\,\frac{\mathrm{Im}\Sigma(x)}{\omega - x} \quad, \tag{10}$$

where the integral is interpreted as a principal integration.

III. NUMERICAL RESULTS AND DISCUSSION

We have performed the numerical calculation of the memory function and of the magneto-optical absorption spectrum. First, we study the case with zero Landau level broadening, i.e. $\Gamma = 0$. In this case the imaginary part of the memory function consists of a series of delta functions at the frequencies $\omega = \omega_{\mathrm{LO}} + n\omega_c$ $(n = 0, 1, 2...)$. The real part of the memory function diverges at $\omega = \omega_{\mathrm{LO}} + n\omega_c$ as $1/(\omega - \omega_{\mathrm{LO}} - n\omega_c)$ (see Eq.(9a)). Due to this special structure of the imaginary part of the memory function the magneto-optical absorption spectrum consists of a series of delta function peaks. The position of these peaks are determined by the equation $\omega - \omega_c - \mathrm{Re}\Sigma(\omega) = 0$ and are thus not influenced by the imaginary part of the memory function. The delta function peaks in the absorption spectrum have oscillator strength $\pi(1 - \frac{\partial}{\partial\omega}\text{Re}\Sigma(\omega))^{-1}$.

In Fig.1 the frequencies (ω^*) of the first four peaks in the magneto-optical absorption spectrum are plotted as a function of the magnetic field for an ideal 2D system. These peaks correspond to the transitions of the polaron from the ground state $(n = 0)$ to the n-th $(n = 1, 2, 3)$ Landau level. The splitting of ω_1^* and ω_2^* around $\omega_c = \omega_{LO}$, and the pinning behaviour of ω_2^* for $\omega_c \gg \omega_{LO}$, are clearly seen from this figure. In Ref.19 a detailed comparison was made between the results of the IWBPT calculation and the present results. It was suggested in Ref.19 that the present approach is a rather good approximation in calculating the polaron cyclotron resonance mass even for $\alpha \sim 0.1$.

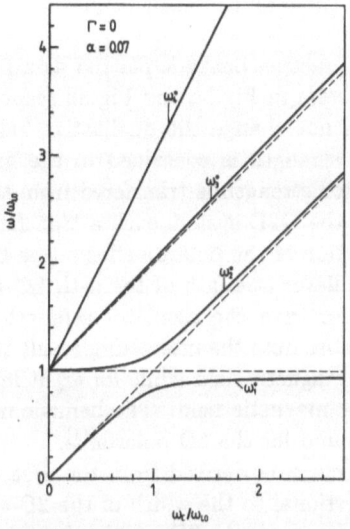

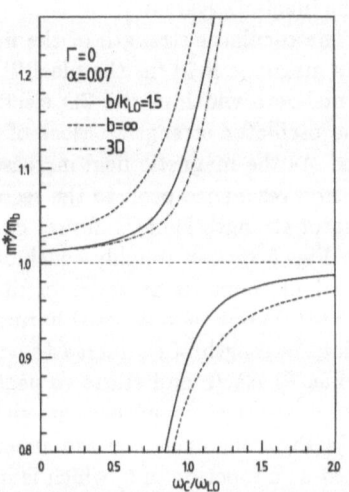

Fig.1 The position of the first four peaks in the magneto-optical absorption spectrum are plotted as a function of the magnetic field strength for an ideal 2D system.

Fig.2 The cyclotron resonance mass derived from the absorption spectrum is shown as a function of the magnetic field for the 2D and the Q2D system. The 3D result is given by the dash-dotted curve.

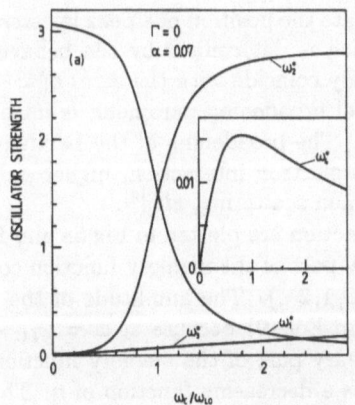

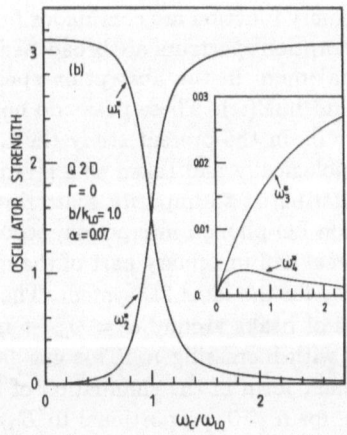

Fig.3 The oscillator strength of the first four delta function peaks in the absorption spectrum is shown as a function of the magnetic field for the ideal 2D system (Fig.3a) and for the Q2D system (Fig.3b).

We compare the ideal 2D and Q2D results in Fig.2. The effective mass of the electron, which is derived from ω_1^* and ω_2^* by $m^* = \omega_c/\omega_1^*$ for $\omega_c < \omega_{LO}$ and $m^* = \omega_c/\omega_2^*$ for $\omega_c > \omega_{LO}$, is plotted as a function of the magnetic field strength. Similar as for the zero magnetic field case[20] the finite width of the electron layer reduces the polaron effects. This reduction is a consequence of the fact that the form factor $f(k,b) \leq 1$ in Eq.(9) while for the ideal 2D system $f(k,b = \infty) = 1$. Notice that the splitting of the first two Landau levels is not symmetrical about the LO-phonon energy, i.e. $\omega_{LO} - \omega_1^* \neq \omega_2^* - \omega_{LO}$, at $\omega_c = \omega_{LO}$. A similar result was recently found for the IWBPT theory[10]. In Fig.2 we also plot the corresponding 3D results for $\omega^* < \omega_{LO}$ (for $\omega^* > \omega_{LO}$ the situation is more complicated since the imaginary part of the memory function is different from zero[13]). For the ideal 2D system the polaron effect is enhanced compared to the 3D case, i.e., we have a larger polaron mass correction. For the zero magnetic field case a similar result was found earlier[20]. In the zero magnetic field limit we find the familiar result $m^*/m_b = 1 + \pi\alpha/8$ for the ideal 2D system.

The oscillator strength of the first four delta function peaks is plotted as a function of the magnetic field for the ideal 2D and Q2D system in Fig.3a and Fig.3b respectively. The non-zero width of the 2D electron layer does not change the qualitative behaviour of the oscillator strength. Most of the oscillator strength is contained in the first two peaks. As the magnetic field increases, the oscillator strength is transferred from the first cyclotron resonance peak to the second peak. For the Q2D system such a transfer of the oscillator strength is more abrupt due to the reduction of the polaron effects (see Fig.3b). From Figs.3 we can roughly conclude that the oscillator strength of the n-th LO-phonon assisted harmonic is an order of magnitude smaller than the oscillator strength of the $(n-1)$-th LO-phonon assisted harmonic. Furthermore note the interesting result that the oscillator strength of ω_3^* increases with increasing magnetic field while for ω_4^* it increases up to $\omega_c \sim \omega_{LO}/2$ and starts to decrease for larger magnetic field. The behaviour of the oscillator strength found here are similar to that found for the 3D polaron[1].

In Fig.4 the cyclotron resonance mass of the electron derived from the first peak is plotted as a function of b, which is inversely proportional to the width of the 2D electron layer, for different values of the magnetic field. As b increases the 2D electron layer becomes narrower and closer to the ideal 2D system. Consequently the effective electron-phonon coupling strength is enhanced and the polaron correction to the cyclotron resonance mass of the electron increases as shown in Fig.4. We point out that in the limit of $b \to 0$ we do not recover the 3D results. This is due to the fact that in the present study we only consider the lowest subband.

In the case $\Gamma > 0$, the Landau levels have a finite width and the real and imaginary part of the memory function are continuous functions of the frequency. All delta function peaks in the absorption spectrum are broadened. In this case the position of a peak is determined by the maximum in the absorption spectrum which is determined by the behaviour of $\text{Re}\Sigma(\omega)$ and $\text{Im}\Sigma(\omega)$. These peaks do not necessarily coincide with the zeros of $\omega - \omega_c - \text{Re}\Sigma(\omega) = 0$. In the present study the Landau level broadening parameter is introduced phenomenologically and taken as a given constant. The broadening of the Landau levels may be attributed to impurity scattering, electron-electron interaction, higher orders of the electron LO-phonon interaction, acoustical phonon scattering, etc.[2].

The real and imaginary part of the memory function are plotted in Fig.5a and Fig.5b respectively for the ideal 2D system. The imaginary part of the memory function consists of a series of peaks around $\omega = \omega_{LO} + n\omega_c$ ($n = 0, 1, 2...$). The amplitude of the peaks decreases with increasing n. This can be seen from Eqs.(9) because at $\omega = \omega_{LO} + n\omega_c$ the dominant term in the summation of the imaginary part of the memory function (see Eq.(9)) is, for $n \neq 0$, proportional to B_n/n which is a decreasing function of n. The real part of the memory function, which can be obtained from the imaginary part by using the Kramers-Kroning relation, is an oscillating function. Approximately we have $\text{Re}\Sigma(\omega) = 0$ when $\text{Im}\Sigma(\omega)$ attains its maximum values.

The magneto-optical absorption spectrum of the ideal 2D system for ω near ω_{LO} is

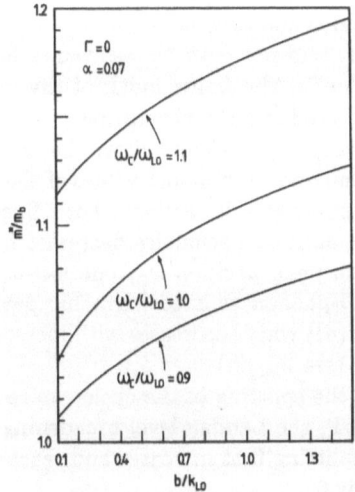

Fig.4 The cyclotron resonance mass derived from the first peak of the absorption spectrum is shown as a function of b for different values of the magnetic field.

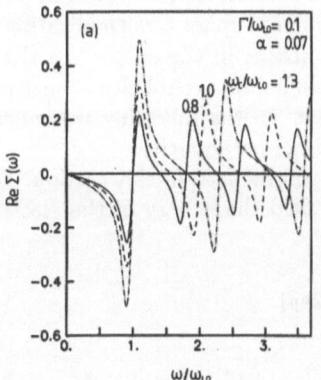 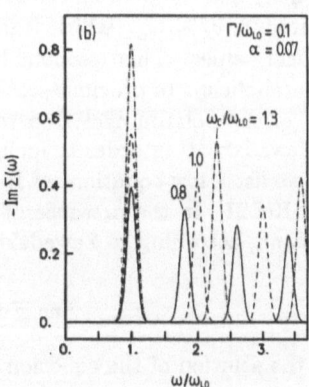

Fig.5 The real (Fig.5a) and imaginary (Fig.5b) part of the memory function is plotted as function of the frequency for an ideal 2D system and for three different values of the magnetic field.

plotted in Fig.6 for different values of Γ and three values of the magnetic field strength which are chosen in such a way that they are near the resonance condition $\omega_c \sim \omega_{LO}$. Similar to the case $\Gamma = 0$, we observe that the amplitude of the first peak decreases for increasing ω_c while the amplitude of the second peak increases. The Landau level broadening parameter not only affects the amplitude of the peaks but also their position. As Γ increases the peaks broaden and become less pronounced. For sufficient large Γ values the cyclotron resonance peaks will not even be resolved. The absorption spectrum of the Q2D system is plotted in Fig.7. The finite width of the 2D electron layer reduces the splitting of the peaks and also reduces the absorption.

The magneto-optical absorption spectrum above the LO-phonon continuum is plotted in Fig.8 for the ideal 2D system and for different values of the magnetic field strength. The LO-phonon assisted harmonics are clearly resolved. For a fixed magnetic field strength the amplitude of the LO-phonon assisted harmonics decreases for higher harmonics. This is due to the fact that near the peak position ω_n, one has $\omega_n \approx \omega_{LO} + n\omega_c$ $(n = 1, 2...)$ and the amplitude of the LO-phonon assisted harmonic peak is approximately given by $\text{Im}\Sigma(\omega_n)/(\omega_n - \omega_c)^2$ (see Eq.(4)) which decreases with increasing n because also $\text{Im}\Sigma(\omega_n)$ decreases with increasing ω_n (see Fig.5b).

Finally in Fig.9 we plot the splitting of the cyclotron resonance peak $(\Delta = \omega_2^* - \omega_1^*)$ at $\omega_c = \omega_{LO}$ as a function of Γ, the Landau level broadening parameter, for the ideal 2D system. As Γ increases the splitting first increases and reaches a maximum after which it decreases rapidly (see also Fig.6).

IV. COMPARISON WITH EXPERIMENT

In this section we apply the theory, developed in the previous sections, to analyse the experimental polaron cyclotron resonance data of Refs.4 and 5 for GaAs-AlGaAs heterostructures and of Ref.6 for InSb inversion layers. To make a realistic comparison with the experiments we have to take into account the nonparabolicity of the electron energy band which leads to an effective electron mass which increases with increasing magnetic field strength. In this aspect it has a similar effect on the electron effective mass as polaron effects have, at least when $\omega_c \ll \omega_{LO}$. The essential difference between the effect of band nonparabolicity and the polaron contribution is that polaron effects induce a resonant contribution around $\omega_c \sim \omega_{LO}$ which leads to a pronounced mass renormalization for these magnetic field values. This resonant behaviour is absent in the case when there is only band nonparabolicity. In previous sections energy band nonparabolicity is neglected in the calculation of the polaron cyclotron resonance mass. In the following we will apply the theory of Zawadzki[21] in order to include the band nonparabolicity.

First, we list a few equations of Ref.21, which will be used in this section. For details we refer to Ref.21. In the presence of a magnetic field the energy of the electron in the lowest subband, according to Zawadzki, is given by

$$\varepsilon_{np} = \varepsilon_\| + z(\varepsilon_g + 2\varepsilon_\|) \quad , \tag{11a}$$

where z is the solution of the equation

$$\frac{8}{3}z^{3/2} + \frac{4}{5}z^{5/2} = \left(\frac{\varepsilon_g}{2m_b}\right)^{1/2} \frac{4\pi eF\hbar}{(\varepsilon_g + 2\varepsilon_\|)^2} \left(\frac{3}{4}\right) \quad , \tag{11b}$$

and

$$\varepsilon_\| = -\frac{\varepsilon_g}{2} + \left[\left(\frac{\varepsilon_g}{2}\right)^2 + \varepsilon_g\hbar\omega_c\left(n + \frac{1}{2}\right)\right]^{1/2} \tag{11c}$$

Here ε_g is the energy band gap. F is an electric field which determines the electron confining triangular well potential and which is treated here as a fitting parameter. In

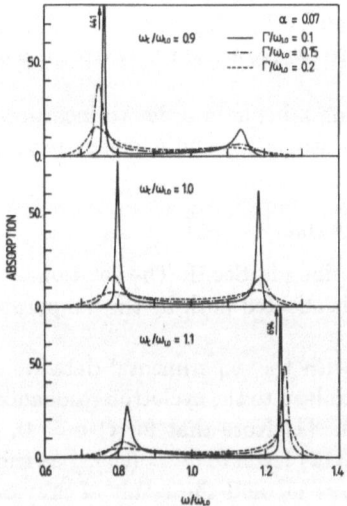

Fig.6 The absorption spectrum of an ideal 2D system is shown for different values of Γ and for magnetic field values around the resonance condition.

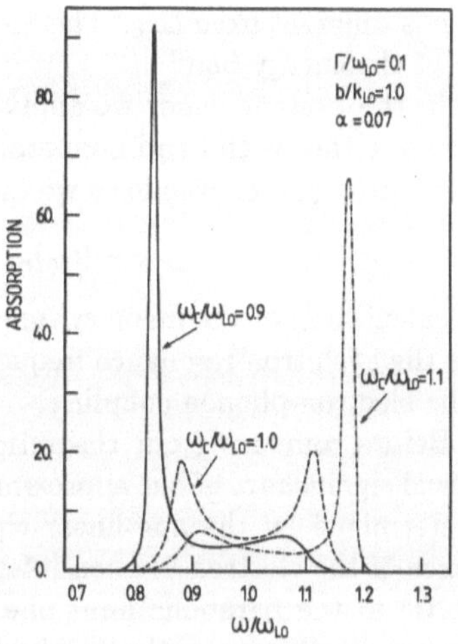

Fig.7 The absorption spectrum of the Q2D system.

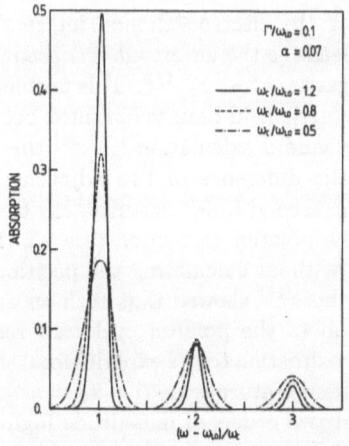

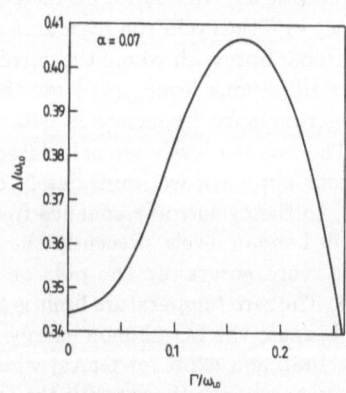

Fig.8 The LO-phonon assisted harmonics of the absorption spectrum shown for the ideal 2D system and for different values of the magnetic field.

Fig.9 The splitting of the first two cyclotron resonance peaks at $\omega_c = \omega_{LO}$ is shown as a function of Γ the Landau level broadening parameter for the ideal 2D system.

this way the electron density does not directly enter the theory. The spatial extent of the subband wave function is determined by the variational parameter b which is given by $b = 2(3eFm_b/2\hbar^2)^{1/3}$ and is consequently entirely determined by the electric field F.

Let us first consider band nonparabolicity and disregard momentarily polaron effects. The cyclotron resonance frequency is then given by $\hbar(\omega_c)_{np} = \varepsilon_{np}(n = 1) - \varepsilon_{np}(n = 0)$ which is different from $\hbar\omega_c$. This shift from ω_c to $(\omega_c)_{np}$ arises solely from the nonparabolicity of the energy band.

In the present paper we apply the following scheme in order to incorporate polaron effects together with band nonparabolicity. We use $(\omega_c)_{np}$ and b, given above, as input to the equation (for convenience we take $\Gamma = 0$)

$$\omega - (\omega_c)_{np} - \mathrm{Re}\Sigma(\alpha, (\omega_c)_{np}, b; \omega) = 0 \quad . \qquad (12)$$

The calculation of the memory function is described in Sec.II. The solution ω^* of Eq.(12) gives the cyclotron resonance frequency which is affected both by the nonparabolicity and by the electron-phonon coupling.

Before comparing our theoretical results with the experimental data we discuss the physical significance of our approximations which lead to the cyclotron resonance frequency as determined by the non-linear equation (12). (1) Note that for i) $\alpha = 0$, i.e., in the absence of the electron-phonon interaction, Eq.(12) leads to $\omega^* = (\omega_c)_{np}$ as should be the case; ii) in the parabolic limit one has $(\omega_c)_{np} = \omega_c$ and the result of previous sections is recovered. Consequently the right limiting behaviour is obtained if either the electron-phonon interaction or the nonparabolicity are switched off. (2) The correction to ω^* due to the band nonparabolicity and the polaron effect are not considered to be additive in Eq.(12) because i) $\mathrm{Re}\Sigma$ also contains $(\omega_c)_{np}$ and ii) the solution of Eq.(12) denoted by ω^* results as the solution of a non-linear equation. (3) Our approximation for the combined incorporation of the band nonparabolicity and the electron-phonon interaction corresponds to a local parabolic approximation to the band nonparabolicity. In calculating the polaron effect *all Landau levels are incorporated* in the calculation of $\mathrm{Re}\Sigma$ within this local parabolic approximation. This is in contrast to earlier work of other investigators[4,5,7,21] where only the resonance term in the perturbation theory for the electron-phonon interaction was considered (see also Ref.19). This latter approach leads to the unfortunate consequence[21] that for $\omega_c \to 0$ the cyclotron resonance mass diverges as $m^* \sim \omega_c^{-1/2}$. This problem is not present in our approach where the correct zero magnetic field limit is obtained because we sum over all Landau levels. (4) Note that in the standard calculation[4,5,7,21] the polaron cyclotron resonance frequency is determined as the difference of two adjacent Landau levels. The Landau levels are calculated within e.g. second-order perturbation theory. In the present approach we immediately calculate the polaron correction (see e.g. Ref.15), i.e. $\mathrm{Re}\Sigma$, to the cyclotron resonance frequency ω^* without calculating the position of the individual Landau levels. Recently the present authors[19] showed that such an approach describes more accurately the polaron contribution to the polaron cyclotron resonance mass. (5) The zero temperature limit is a good approximation to the experimental situation $T \sim 4K$ because the LO-phonon energy induces a temperature scale (i.e. $T_0 = \hbar\omega_{LO}/k_B \approx$ 283K for InSb and 426K for GaAs) which is almost two orders of magnitude higher.

In comparing our theory with the experimental data we are confronted with the problem that the value of F, or equivalently the width of the 2D electron layer, is not known experimentally. Therefore we have to take F as a fitting parameter. The nonparabolicity of the electron energy band (without polaron effect) results in a electron effective mass m^*_{np} which is almost a linear function of the magnetic field strength[4]. The incorporation of the polaron effect leads to a strong increase of the cyclotron resonance mass m^* when $(\omega_c)_{np} \sim \omega_{LO}$ and consequently m^* deviates from the linear behaviour of m^*_{np}. The stronger the electron-phonon coupling strength, the bigger the deviation of m^* from m^*_{np}. The band mass m_b also affects m^* but has little influence on the slope of m^*.

It is worth noticing that in the experiment of Horst *et al.* and of Sigg *et al.* the samples have almost the same electron density, i.e. $n_e = 4.07 \times 10^{11}\,\mathrm{cm}^{-2}$ and $4 \times 10^{11}\,\mathrm{cm}^{-2}$,

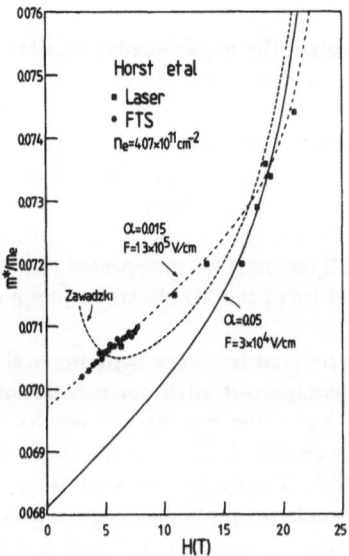

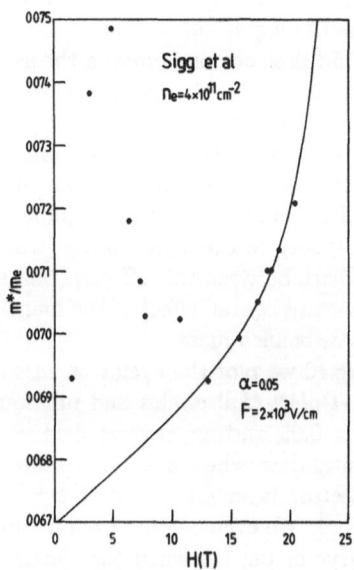

Fig.10 The experimental data of Horst *et al.* [4] (full circle and square points) are compared with the present theoretical results (solid and dash-dotted curves). The dashed curve is the results of Ref.19.

Fig.11 The experimental data of Sigg *et al.* [5] (full circle points) are compared with the present theoretical results (solid curve).

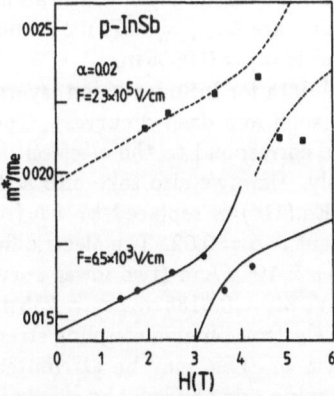

Fig.12 The experimental data for InSb inversion layers[6] are compared with the present theoretical results (solid and dashed curves). The full circle and square points correspond to the experimental electron densities $n_e = 2 \times 10^{11} \text{cm}^{-2}$ and $n_e = 10^{12} \text{cm}^{-2}$, respectively.

respectively. However the polaron cyclotron resonance mass in both experiments shows quite different behaviour, particularly in the lower magnetic field region (see Fig.10 and Fig.11). This is a puzzling fact which needs more experimental attention. Therefore in the present study we will concentrate on the higher magnetic field region where the polaron effect is believed to be dominant. Even in the high magnetic field region (i.e. $H > 12\,\text{T}$) the experimental results of Horst et al. for the cyclotron resonance mass are slightly higher than those of Sigg et al..

We find that our one polaron theory can not explain the experimental results quantitatively in the case of the GaAs-AlGaAs heterostructures if we use the generally accepted value for the electron-phonon coupling constant, i.e. $\alpha = 0.068$. For any choice for F we find that with $\alpha = 0.068$ the calculated cyclotron resonance mass shows a rather large increase compared with the experimental data when $(\omega_c)_{np} \sim \omega_{\text{LO}}$. Therefore we will consider α as an effective electron-phonon coupling constant and adapt its value in order to fit our theory to the experiments (without confusion we will use the same notation α). The difference between this effective value for α and 0.068 may be interpreted as resulting from a constant overall effect of the many-particle nature of the 2D electron system on the cyclotron resonance mass.

In Fig.10 we plot the cyclotron resonance mass derived from the experimental result of Horst et al.[4] (full circles and full squares) and compare it with our calculation. By using $\alpha = 0.05$ and an electric field $F = 3 \times 10^4\,\text{V/cm}$ we can fit our results to the experimental data when $H > 15\text{T}$ (solid curve). The electric field F results in a width of the 2D electron layer $< (z - <z>)^2 >^{1/2} = \sqrt{3}/b \approx 55\text{Å}$. The dashed curve is the result of Zawadzki[21]. We even can obtain a fitting over the whole magnetic field region (see dash-dotted curve in Fig.10) when the electron-phonon coupling constant is taken $\alpha = 0.015$, which is about 4.5 times smaller than the well accepted value $\alpha = 0.068$, and an electric field $F = 1.3 \times 10^5\,\text{V/cm}$ which results in a very thin 2D electron layer of 34Å.

Next we analyse the experimental data of Sigg et al.[5] in Fig.11 where we have to use $\alpha = 0.05$ in order to fit the experiment in the region $H > 12\text{T}$. This value for α is consistent with our previous one. Notice that the cyclotron resonance mass derived from the experimental data of Ref.5 exhibits a pronounced peak structure at $H \sim 5\,\text{T}$ (see Fig.11). This is not caused by polaron effects and was attributed in Ref.5 to a change of the filling factor from $\nu > 1$ to $\nu < 1$. Here we use a smaller electric field value $F = 2 \times 10^3\,\text{V/cm}$ which leads to a electron layer width about 135Å. The other physical parameters used in the calculation are $\hbar\omega_{\text{LO}} = 36.75\,\text{meV}$ and $\varepsilon_g = 1520\,\text{meV}$. The band mass m_b is taken from Ref.4 and is $m_b = 0.0665\,m_e$.

In Fig.12 The experimental data for InSb inversion layers[6] (for the lowest subband) are compared with our results (solid and dashed curves). The experimental data plotted as full circles and square points correspond to the electron densities $n_e = 2 \times 10^{11}\text{cm}^{-2}$ and $n_e = 10^{12}\text{cm}^{-2}$, respectively. Here we also take into account the effect of spin. For this purpose $\hbar\omega_c(n + 1/2)$ in Eq.(11c) is replaced by $\hbar\omega_c(n + 1/2) - |g_0^*|\mu_{\text{B}}H/2$. The electron-phonon coupling constant is $\alpha = 0.02$. The electric fields are $F = 2.3 \times 10^5\,\text{V/cm}$ (two upper curves) and $F = 6.5 \times 10^3\,\text{V/cm}$ (two lower curves) respectively. The other parameters are $\hbar\omega_{\text{LO}} = 24.4\,\text{meV}$, $m_b = 0.0135\,m_e$, $\varepsilon_g = 235\,\text{meV}$ and $|g_0^*| = 51$.

It is clear that the effective electron-phonon coupling strength is reduced considerably in the experiments of Refs.4 and 5. This may be attributed to occupation effects (i.e. Fermi-Dirac statistics) and screening arising from the electron-electron interaction, which are neglected in the present study. It has been suggested in Refs.3 and 5 that electron screening may play an important role in the electron-phonon interaction in the cyclotron resonance experiment and may modify the polaron effects considerably. In the experiment of Ref.3 no significant polaron effects could be identified. This can be understood by noticing the fact that the largest probe laser energy used in the experiment is only about $\hbar\omega_{\text{LO}}/3$, corresponding to a wave-length $\lambda = 96\mu m$ (see Ref.3), which is far from the polaron resonance condition. Therefore the experimental results could be explained by band nonparabolicity only.

In Ref.6 the cyclotron resonance frequency ω^* is measured below and above the LO-phonon energy. For $\omega^* < \omega_{LO}$ the cyclotron resonance mass m^* is almost a linear function of the magnetic field (see Fig.12). Such a linear behaviour of m^* can also be explained by band nonparabolicity. However band nonparabolicity can not explain the splitting of m^* for $\omega^* \sim \omega_{LO}$. The splitting of m^* in the experiment of Ref.6 is so large that it is necessary to take $\alpha = 0.02$ to explain the experimental data. $\alpha = 0.02$ is the well accepted value for the electron-phonon coupling constant in InSb. This is in contrast to the above analysis of the experimental results of Refs.4 and 5 for the GaAs-AlGaAs heterostructures where we need a smaller electron-phonon coupling constant to explain the data with our theory. For the moment, it is not clear why the electron-phonon coupling is not reduced in the experiment of Ref.6 by the many-particle effects. A possibility is that occupation effects and screening are not important at the resonant condition itself[22]. Note also that in order to explain the $\omega_c < \omega_{LO}$ results of Ref.6 no polaron effects have to be invoked.

V. CONCLUSION

In the present paper we have calculated the magneto-optical absorption spectrum of a single 2D polaron. In order to remove the divergencies in the absorption spectrum we have introduced a Landau level broadening parameter in a phenomenological way. The finite width of the 2D electron layer is incorporated into the memory function calculation by considering the lowest subband. It is found that the finite width of the 2D electron layer reduces considerably the polaron effects. In order to make a realistic comparison between our results and the experimental data, we have taken into account the effect resulting from the non-parabolic electron energy band. When we use an effective electron-phonon coupling constant which is smaller than the real values, we are able to explain quantitatively the experimental data for GaAs-AlGaAs heterostructures in the high magnetic field region. Effects from Fermi-Dirac statistics and the electron screening are probably responsible for the reduction of the electron-phonon coupling. But this needs further theoretical consideration before we can be definite about it. We have also compared our theoretical results with the experimental data of Ref.6 for a InSb inversion layer. We find that in this case the electron-phonon coupling constant is not reduced and that the one polaron theory could explain the experimental results.

In the present study we have neglected the many-body effects. We have studied the problem of one polaron. Therefore the present theory is expected to be valid in the limit of low electron density only. Recently the optical absorption spectrum of a 2D polaron has been calculated for the zero magnetic field case by the present authors[23]. In Ref.23 the dynamical screening effects due to the electron-electron interaction on the electron LO-phonon interaction have been found to be important. For the Q2D electron system like the GaAs heterostructures in Refs.4 and 5 the electron density is not very high: $n_e \approx 4 \times 10^{11} \text{cm}^{-2}$. Near the resonance condition $\omega_c \approx \omega_{LO}$ the filling factor $\nu \leq 1$ and electrons will be in the lowest Landau level. This probably is the reason why the present one polaron theory provides such a close agreement with the experimental results.

ACKNOWLEDGEMENTS

This work is partially sponsored by F.K.F.O. (Fonds voor Kollektief Fundamenteel Onderzoek, Belgium), project No. 2.0072.80. One of the authors (F.M.P.) acknowledges financial support from the Belgian National Fund for Scientific Research. Wu. X. wishes to thank The International Culture Co-operation of Belgium for financial support. We are pleased to thank M. Horst, U. Merkt and W. Seidenbusch for providing us with their experimental data and H. Sigg for interesting discussions.

REFERENCES

[1] *Polarons and Excitons*, edited by C. Kuper and G. Whitfield (Oliver and Boyd, Edinburgh, 1963); *Polarons in Ionic Crystals and Polar Semiconductors*, edited by J. T. Devreese (North-Holland, Amsterdam, 1972); *Polarons and Excitons in Polar Semiconductors and Ionic Crystals*, edited by J. T. Devreese and F. M. Peeters (Plenum, New York, 1984).

[2] T. Ando, A. B. Fowler, and F. Stern, Rev. Mod. Phys. **54**, 437 (1982).

[3] W. Seidenbusch, G. Lindemann, R. Lassnig, J. Edlinger, and G. Gornik, Surf. Sci. **142**, 375 (1984).

[4] M. Horst, U. Merkt, W. Zawadzki, J. C. Maan, and K. Ploog, Solid State Commun. **53**, 403 (1985).

[5] H. Sigg, P. Wyder, and J. A. A. J. Perenboom, Phys. Rev. **B31**, 5253 (1985).

[6] M. Horst, U. Merkt, and J. P. Kotthaus, Phys. Rev. Lett. **50**, 754 (1983).

[7] R. Lassnig and W. Zawadzki, Surf. Sci. **142**, 388 (1984).

[8] S. Das Sarma and A. Madhukar, Phys. Rev. **B22**, 2823 (1980).

[9] D. Larsen, Phys. Rev. **B30**, 4595 (1984).

[10] F. M. Peeters and J. T. Devreese, Phys. Rev. **B31**, 3689 (1985); F. M. Peeters, Wu Xiaoguang, and J. T. Devreese, Phys. Rev. **B33**, 4338 (1986).

[11] S. Das Sarma, Phys. Rev. Lett. **52**, 859 (1984); Phys. Rev. Lett. **52**, 1570 (E) (1984).

[12] D. Larsen, Phys. Rev. **B30**, 4807 (1984).

[13] J. Van Royen, J. De Sitter, L. F. Lemmens, and J. T. Devreese, Phys. Physica **81B**, 101 (1977); J. P. Vigneron, R. Evrard, and E. Kartheuser, Phys. Rev. **B18**, 6930 (1978); J. Van Royen, J. De Sitter, and J. T. Devreese, Phys. Rev. **B30**, 7154 (1984); F. M. Peeters and J. T. Devreese, Physica **127B**, 408 (1984).

[14] W. Götze and P. Wölfe, Phys. Rev. **B6**, 1226 (1972).

[15] F. M. Peeters and J. T. Devreese, Phys. Rev. **B28**, 6051 (1983).

[16] R. R. Gerhardts, Z. Phys. **B21**, 275 (1975); **B21**, 285 (1975).

[17] J. T. Devreese, J. De Sitter, and M. Goovaerts, Phys. Rev. **B5**, 2367 (1972).

[18] R. P. Feynman, R. W. Hellwarth, C. K. Iddings, and P. M. Platzman, Phys. Rev. **127**, 1004 (1962).

[19] F. M. Peeters, Wu Xiaoguang, and J. T. Devreese, Phys. Rev. **B34**, 1160 (1986).

[20] S. Das Sarma, Phys. Rev. **B27**, 2590 (1983); Phys. Rev. **B31**, 4034 (E) (1985); Wu Xiaoguang, F. M. Peeters, and J. T. Devreese, Phys. Status Solidi. **B133**, 229 (1986).

[21] W. Zawadzki, Solid State Commun. **56**, 43 (1985).

[22] M. Horst, U. Merkt, and K. G. Germanova, J. Phys. **C18**, 1025 (1985).

[23] Wu Xiaoguang, F. M. Peeters, and J. T. Devreese, Phys. Rev. **B34**, 2621 (1986).

APPENDIX B

INFLUENCE OF MANY-BODY EFFECTS ON THE CYCLOTRON RESONANCE MASS OF TWO-DIMENSIONAL POLARONS WITH APPLICATION TO GaAs-AlGaAs HETEROSTRUCTURES

Wu Xiaoguang, F. M. Peeters and J. T. Devreese

ABSTRACT

The influence of many-body effects on the cyclotron resonance mass of two-dimensional polarons is investigated. First, *the occupation effect* is considered, which arises from the Fermi-Dirac statistics. In the calculation the Landau levels are considered to be perfectly sharp. We find that (1) the occupation effect reduces the polaron effect, i.e. the cyclotron resonance mass renormalization due to the polaron effect is reduced over the whole magnetic field region. (2) for a fixed electron density the cyclotron resonance mass shows an oscillating behaviour as a function of the magnetic field. The polaron cyclotron resonance mass attains minimal values when the filling factor passes through integer values. Secondly, the effect of *static screening* on the polaron cyclotron resonance mass is examined. The electron screening is found to reduce further the cyclotron resonance mass renormalization. The effect of the finite width of the two-dimensional electron layer is included into the present calculations by considering the lowest subband. After taking into account the nonparabolicity of the electron energy band, the calculated cyclotron resonance mass is compared to the recent experimental data for GaAs-$Al_{1-x}Ga_xAs$ heterostructures. Good agreement is found in the high magnetic field region, i.e. for magnetic fields $H > 12T$.

I. INTRODUCTION

Two-dimensional electron systems, such as formed in inversion layers and semiconductor heterostructures[1] (e.g. InSb inversion layers, GaAs-$Al_{1-x}Ga_xAs$ heterostructures,...), are particularly interesting systems for the study of many-body effects like Fermi-Dirac statistics and screening. In these systems the electron density, and therefore the strength of the electron-electron interaction, can be varied experimentally over a wide range. The aim of the present paper is to investigate the influence of the many-body effects on the cyclotron resonance mass of a two-dimensional (2D) electron gas in which the electrons are coupled to longitudinal optical (LO) phonons.

Recently, polaron effects have been observed in cyclotron resonance experiments[2-4]. Due to polaron effects the cyclotron resonance frequency, and consequently also the cyclotron resonance mass, are shifted. Qualitatively this shift can be explained by a single-polaron picture theory. In Ref.3 the analysis of the experimental data suggested that the electron screening effect may play an important role in modifying the electron-phonon interaction. A considerable number of theoretical studies have been carried out to investigate the cyclotron resonance of the 2D polaron[5-14]. However in most of the studies the occupation effect resulting from the Fermi-Dirac statistics and the screening effect arising from the electron-electron interaction were neglected, with the exception of Refs.6 and 12. These authors[6,12] calculated, within second order perturbation theory, the correction to the position of a Landau level as due to the electron-phonon interaction. Larsen[6] found that the correction to the cyclotron resonance frequency is reduced by the occupation effect. He also pointed out that for integer fillings (an integer number of Landau levels are fully filled) perturbation theroy should be carried out to all order in the electron-phonon coupling strength in order to obtain the correct result. In Ref.12 the occupation effect was treated in a similar fashion as in Ref.6. Lassnig[12] also discussed the coupling between phonons and plasmons by calculating the dynamical polarizibility of the electron gas but its consequence on the polaorn cyclotron resonance mass is not shown explicitly.

In a previous study[14] the cyclotron resonance spectrum of *a single 2D polaron* was calculated using a memory function approach[15−17]. In Ref.14 the theoretical results were compared to the experimental data of Refs.2 and 3 for electrons in a GaAs-Al$_{1-x}$Ga$_x$As heterostructure and of Ref.4 for electrons in a InSb inversion layer. For the GaAs-Al$_{1-x}$Ga$_x$As heterostructures it is found that in order to explain the experimental results with the single-polaron theory an effective electron-phonon coupling constant had to be introduced which is smaller than the well-known value. This was then identified as a reduction of the effective electron-phonon coupling strength. This reduction was then attributed to the many-body effects which were neglected in Ref.14. Of course this procedure is artificial, but nevertheless it gives an idea of the importance of the many-body effects on the electron-phonon interaction in the system. In this paper we will generalize the study of Ref.14 by taking into account the many-body effects in the calculation of the memory function. This memory function is used when we evaluate the cyclotron resonance frequency and the polaron cyclotron resonance mass.

In the presence of a magnetic field, which is applied perpendicular to the 2D electron layer, the energy levels of a 2D non-interacting electron gas are completely quantized into discrete Landau levels. The density of states of this system consists of a series of delta functions. Due to this singular nature of the density of states the broadening of the Landau levels, which certainly exists in real systems, must be treated self-consistently. Such calculations are quite involved[18]. In the present paper the broadening of the Landau levels will be neglected. In doing so, the memory function exhibits singularities at certain frequencies. However, the main concern here is the calculation of the cyclotron resonance mass and these divergences of the memory function do not cause any inconvenience. It is expected that this approximation will be valid as long as the broadening of the Landau levels are small.

In the first part of this paper we will limit ourselves to electrons with a parabolic energy band, interacting with bulk LO-phonons (see Ref.19). The effect of the finite width of the 2D electron layer is included by considering the lowest subband where the standard variational wave function is used[1]. The electron-phonon interaction is treated as a perturbation and the memory function is calculated to first-order in the electron-phonon coupling constant. Spin splitting effects are neglected.

In the present paper the many-body effects are incorporated into the calculation of the memory function in two steps. First, the occupation effect is considered without electron screening. Secondly, the screening effect will be treated within a static screening approximation, i.e. the full dielectric function is replaced by the zero frequency dielectric function. At present a full dynamical screening approach goes beyond our numerical capabilities.

Recently, we calculated the optical absorption spectrum for a 2D interacting polaron gas in the absence of a magnetic field[20]. In Ref.20 many-body effects were investigated within the Hartree-Fock approximation (HF), where only the occupation effect was included, and within the dynamical and static random-phase approximations (RPA) where the electron screening effect was also included. It was found that for the quasi-two-dimensional (Q2D) electron systems, as far as the effective mass of the electron is concerned, the dynamical screening does not lead to a large deviation of the polaron mass from the result of a HF calculation. Of particular interest is that Ref.20 showed that the correction to the polaron mass, calculated within the dynamical screening approximation is, over a wide range of the electron densities, between the results of the static RPA screening approximation and the HF approximation. Here we suggest that this will also be true qualitatively in the presence of a magnetic field. This gives some qualitative justification for the use of the static screening approximation in the present paper.

The present paper is organized as follows: in Sec.II the calculation of the memory function and of the cyclotron resonance frequency is outlined. Section III contains the numerical results and discussion. In Sec.IV, after taking into account the effect of the band nonparabolicity, the theoretical results are compared to the recent experimental data. Our conclusions are presented in Sec.V.

II. FORMULATION AND CALCULATION

The electron-phonon system is described by the many-particle Fröhlich Hamiltonian

$$
H = \sum_{j} \frac{1}{2m_b} (\vec{p}_j + \frac{e\vec{A}(\vec{r}_j)}{c})^2 + \sum_{\vec{k}} \hbar\omega_{LO}\, a_{\vec{k}}^{\dagger} a_{\vec{k}}
$$
$$
+ \sum_{i<j} V(\vec{r}_i - \vec{r}_j) + \sum_{j}\sum_{\vec{k}} \left(V_{\vec{k}}\, a_{\vec{k}}\, e^{i\vec{k}\cdot\vec{r}_j} + V_{\vec{k}}^{*}\, a_{\vec{k}}^{\dagger}\, e^{-i\vec{k}\cdot\vec{r}_j} \right) \quad, \tag{1}
$$

where $\vec{p}_j$ ($\vec{r}_j$) is the 2D momentum (position) operator of the j-th electron. $a_{\vec{k}}^{\dagger}$ ($a_{\vec{k}}$) is the creation (annihilation) operator of a bulk LO-phonon with 3D wave vector $\vec{k}$ and energy $\hbar\omega_{LO}$. $\vec{A}$ is the vector potential describing the magnetic field which is taken perpendicular to the 2D electron layer. The z-axis is chosen along the direction of the magnetic field. In Eq.(1) $V_{\vec{k}}$ represents the interaction of the electron and phonons and is given by

$$
V_{\vec{k}} = i\hbar\omega_{LO} \left(\frac{4\pi\alpha}{Vk^2} \right)^{1/2} \left(\frac{\hbar}{2m_b\omega_{LO}} \right)^{1/4} \langle \psi_0 | e^{ik_z z} | \psi_0 \rangle \quad . \tag{2a}
$$

Here $\psi_0(z) = (b^3/2)^{1/2} z e^{-bz/2}$ is the variational wave function of the electron in the direction normal to the 2D electron layer[1]. The variational parameter b is taken as $b = (48\pi N m_b e^2 / \hbar^2 \epsilon_0)^{1/3}$, where $N = n_d + (11/32)n_e$ and n_d and n_e are the depletion and carrier charge densities respectively. In Eqs.(2) only the lowest subband is included and all higher subbands are neglected. $V(\vec{r} - \vec{r}')$ represents the electron-electron interaction in the Q2D electron layer. The Fourier transform of $V(\vec{r})$ is $V(k_\parallel) = (2\pi e^2 / k_\parallel \epsilon_\infty) f(k_\parallel, b)$ with the standard form factor[1] $f(k,b) = (8b^3 + 9b^2 k + 3bk^2)/(8(b+k)^3)$ and $k_\parallel^2 = k_x^2 + k_y^2$. Note also that[14]

$$
\sum_{k_z} |V_{\vec{k}}|^2 = (\hbar\omega_{LO})^2 \left(\frac{2\pi\alpha}{Ak_\parallel} \right) \left(\frac{\hbar}{2m_b\omega_{LO}} \right)^{1/2} f(k_\parallel, b) \quad, \tag{2b}
$$

where A is the surface area.

The dynamical conductivity of the system in the presence of a magnetic field $\vec{H}$ can be written as[15,16]

$$
\sigma_{\pm}(\omega) = \frac{in_e e^2 / m_b}{\omega \pm \omega_c - \Sigma(\omega)} \quad, \tag{3}
$$

where $\Sigma(\omega) = \Sigma(\alpha, \omega_c, b; \omega)$ is the memory function and $\omega_c = eH/m_b c$ is the unperturbed cyclotron resonance frequency. To first-order in the electron-phonon coupling constant α the memory function has the form

$$
\Sigma(\omega) = \frac{1}{\omega} \int_0^{\infty} dt (1 - e^{i\omega t}) \mathrm{Im} F(t) \quad, \tag{4a}
$$

with

$$
F(t) = -\sum_{\vec{k}} \frac{k_\parallel^2}{n_e m_b \hbar} |V_{\vec{k}}|^2 \left(i\, D(k_\parallel, t) + i\, n(\omega_{LO}) D^{R}(k_\parallel, t) \right) e^{-i\omega_{LO} t} \quad, \tag{4b}
$$

where $n(\omega) = (e^{\beta\hbar\omega} - 1)^{-1}$ is the occupation number. $D(k_\parallel, t)$ $(D^R(k_\parallel, t))$ is the electron (retarded) density-density correlation function[21]. In order to be consistent with the calculation of the memory function, D (D^R) must be calculated without electron-phonon interaction.

By substituting Eq.(4b) into Eq.(4a) the real part of the memory function becomes

$$
\mathrm{Re}\Sigma(\omega) = \sum_{\vec{k}} \frac{k_\parallel^2}{n_e m_b \omega} |V_{\vec{k}}|^2 \frac{\omega^2}{\pi} \int_{-\infty}^{+\infty} dx \frac{(1 + n(x))\mathrm{Im}\Pi^R(k_\parallel, x)}{((x + \omega_{\mathrm{LO}})^2 - \omega^2)(x + \omega_{\mathrm{LO}})}
$$
$$
+ n(\omega_{\mathrm{LO}}) \sum_{\vec{k}} \frac{k_\parallel^2}{n_e m_b \omega} |V_{\vec{k}}|^2 \frac{1}{2} \left[\mathrm{Re}\Pi^R(k_\parallel, \omega + \omega_{\mathrm{LO}}) \right.
$$
$$
\left. + \mathrm{Re}\Pi^R(k_\parallel, \omega - \omega_{\mathrm{LO}}) - 2\mathrm{Re}\Pi^R(k_\parallel, \omega_{\mathrm{LO}}) \right] \quad ,
$$
(5a)

and the imaginary part has the form

$$
\mathrm{Im}\Sigma(\omega) = \sum_{\vec{k}} \frac{k_\parallel^2}{n_e m_b \omega} |V_{\vec{k}}|^2 \frac{1}{2} \left[(n(\omega_{\mathrm{LO}}) - n(\omega_+))\mathrm{Im}\Pi^R(k_\parallel, \omega_+) \right.
$$
$$
\left. + (1 + n(\omega_-) + n(\omega_{\mathrm{LO}}))\mathrm{Im}\Pi^R(k_\parallel, \omega_-) \right] \quad ,
$$
(5b)

where $\omega_\pm = \omega \pm \omega_{\mathrm{LO}}$ and $\Pi = \hbar D$ is the polarization function of the electron gas[21]. The real and imaginary parts of the memory function are related to each other by the Kramers-Kronig relation. Remark that in the calculation of the memory function we need the *frequency dependent* polarization function.

The polarization function of a 2D electron gas has been extensively studied (for instance, see Refs.22 to 24). In the present paper the occupation effect will be included by approximating Π by Π^0 which is the polarization of a non-interacting 2D electron gas. In the following we will concentrate on the zero temperature case, where we only need the imaginary part of the polarization function which is given by

$$
\mathrm{Im}\Pi^0(k, \omega) = \frac{m_b \omega_c}{2\pi\hbar} \sum_{n,n'} C_{n,n'}(k) \, \mathrm{Im}\pi_{n,n'}(\omega) \quad ,
$$
(6a)

with

$$
\mathrm{Im}\pi_{n,n'}(\omega) = -\frac{2}{\hbar}\pi f(\varepsilon_n)(1 - f(\varepsilon_{n'}))(1 - e^{-\beta\hbar\omega})\delta(\omega - (n' - n)\omega_c) \quad ,
$$
(6b)

and

$$
C_{n,n+l}(k) = \frac{n!}{(n+l)!} e^{-x^2} x^{2l} [\mathrm{L}_n^l(x^2)]^2 \quad ,
$$
(6c)

where $x = k(2m_b\omega_c/\hbar)^{-1/2}$ and $\mathrm{L}_n^l(x)$ is the generalized Laguerre polynomial. $\varepsilon_n = (n+1/2)\hbar\omega_c - \mu$ and μ is the chemical potential of the 2D electron gas. $f(x) = (e^{\beta x} + 1)^{-1}$ is the Fermi-Dirac distribution function.

The chemical potential μ depends on the electron density and the temperature of the system. In the present case, μ satisfies the equation $n_e = 2(m_b\omega_c/2\pi\hbar) \sum_{n=0}^{\infty} f(\varepsilon_n)$. By defining the filling factor[1] $\nu = (n_e/2)(2\pi\hbar/m_b\omega_c)$, this equation can be rewritten as $\nu = \sum_{n=0}^{\infty} f(\varepsilon_n)$. At zero temperature the chemical potential is a discrete function of the electron density. In order to get the correct behaviour the limit $\lim_{\beta\to\infty} f_n = f(\varepsilon_n)$ should be treated carefully. Namely, at zero temperature the above equation results in $f_n = 1$ $(n < N_0)$, $f_n = \nu - N_0$ $(n = N_0)$ and $f_n = 0$ $(n > N_0)$ with $N_0 = [\nu]$ the integer part of the filling factor ν.

The imaginary part of the memory function consists of a series of delta functions. The real part of the memory function becomes

$$\text{Re}\Sigma(\omega) = \sum_{n=0}^{N_0} \sum_{n'=N_0}^{\infty} \sum_{\vec{k}} \frac{k_\parallel^2}{2\nu m_b \hbar \omega} \left| V_{\vec{k}} \right|^2 C_{n,n'}(k_\parallel) \Big[\frac{2}{(n'-n)\omega_c + \omega_{\text{LO}}}$$
$$- \frac{1}{(n'-n)\omega_c + \omega_{\text{LO}} + \omega} - \frac{1}{(n'-n)\omega_c + \omega_{\text{LO}} - \omega} \Big] f_n(1 - f_{n'}) \quad . \tag{7}$$

In the zero electron density limit, i.e. $\nu \to 0$, we recover our previous result of Ref.14.

To include the effect of the static screening into the memory function calculation we replace $\left| V_{\vec{k}} \right|^2$ in Eq.(7) by $\left| V_{\vec{k}} \right|^2 / \epsilon^2(k_\parallel)$ where $\epsilon(k) = 1 - V(k)\Pi^0(k, \omega = 0)$. Here the RPA approximation is made for the static dielectric function $\epsilon(k)$ which depends on the magnetic field strength.

III. RESULTS AND DISCUSSION

For the numerical calculations we have taken physical parameters which are typical for the GaAs-$\text{Al}_{1-x}\text{Ga}_x\text{As}$ heterostructure. The filling factor becomes $\nu \approx n_e \times 10^{-12} \omega_{\text{LO}}/\omega_c$ with the electron density n_e in units of cm^{-2}. The electron-phonon coupling constant is $\alpha \sim 0.07$. The cyclotron resonance frequency can be obtained from the magneto-optical absorption spectrum which is proportional to[15]

$$\frac{-\text{Im}\Sigma(\omega)}{[\omega - \omega_c - \text{Re}\Sigma(\omega)]^2 + [\text{Im}\Sigma(\omega)]^2} \quad . \tag{8}$$

Because of the singular nature of the density of states the imaginary part of the memory function is a series of delta functions and consequently the absorption spectrum consists of a series of delta function peaks. The position of these peaks is determined by the equation $\omega - \omega_c - \text{Re}\Sigma(\omega) = 0$ whose solution will be denoted by ω^*. The polaron cyclotron resonance mass is obtained by $m^*/m_b = \omega_c/\omega^*$.

Before presenting the numerical results, first, the physical significance of the memory function, which is given by Eq.(7), will be disscussed. In Eq.(7) there is a double summation over the Landau level number indexes n and n'. Each term in the summation (n, n') corresponds to a specific transition of the polaron. The electron initially in the Landau level n absorbs a photon with energy $\hbar\omega$ and then emits a LO-phonon while the electron makes a transition to the final state which is the Landau level n'. This is different from our earlier single-polaron picture calculations[14] where the initial state of the electron is the $n = 0$ Landau level. In the present approach the initial state of the electron, besides the lowest Landau level, can also be a higher Landau level ($n \geq 0$), but where n is constrained by $n \leq N_0 = [\nu]$ the filling factor.

The transitions of the polaron are restricted by the fact that electrons obey Fermi-Dirac statistics. Therefore there is a factor $f_n(1 - f_{n'})$ for each term in the summation in $\text{Re}\Sigma(\omega)$. Note that this factor $f_n(1 - f_{n'}) \leq 1$ and thus one may conclude that the effective electron-phonon coupling strength is reduced in comparison with the single-polaron picture calculations. Notice also that in Eq.(7) $n \leq n'$, $n' \geq N_0$ and $n \leq N_0$ which is due to the fact that an electron can not make a transition to a state which is already occupied. The resonant term in the summation in $\text{Re}\Sigma(\omega)$ is given by $n = n' = N_0$ and N_0 is determined by the filling factor. N_0 increases when electrons fill up higher Landau levels.

As the filling factor approaches an integer value, the factor $f_n(1 - f_{n'})$ for $n = n' = N_0$ becomes zero. Consequently the resonant term in the summation of $\text{Re}\Sigma(\omega)$ approaches zero. Therefore there will be a further reduction of the polaron effects since the resonant term gives the dominant contribution to the memory function $\text{Re}\Sigma(\omega)$. The vanishing of the resonant term is due to the fact that the Fermi-Dirac statistics prohibit the electron to make a transition to a state which is already occupied.

173

In Fig.1 the polaron correction to the cyclotron resonance mass is plotted as a function of the magnetic field strength at fixed electron densities. The calculations are done for an ideal 2D system, i.e. $b = \infty$ in Eq.(7) (see Eq.(2b)) for the evaluation of the real part of the memory function. Here only the occupation effect has been considered and the electron screening is neglected. The results for $\nu = 0.4\omega_{LO}/\omega_c$ and $\nu = 0.8\omega_{LO}/\omega_c$ correspond to the electron densities $n_e = 4 \times 10^{11} \text{cm}^{-2}$ and $n_e = 8 \times 10^{11} \text{cm}^{-2}$ respectively. The zero electron density result ($\nu = 0$) is from Ref.14.

From Fig.1 it is apparent that for the many-particle system the polaron correction to the cyclotron resonance mass is reduced compared to the single polaron system. The reduction becomes larger with increasing electron density. The correction of the polaron cyclotron resonance mass is further reduced as the filling factor approaches integer values which leads to an oscillating behaviour of the electron effective mass. We found that for integer values of the filling factor the mass renormalization is approximately equal to the zero magnetic field mass renormalization. Of particular interest is the fact that the amplitude of this oscillation of the cyclotron resonance mass decreases as the magnetic field strength decreases (the filling factor increases). In the zero magnetic field limit the polaron cyclotron resonance mass seems to approach the corresponding HF approximation result given in Ref.20 as it should be. In Ref.23 Glasser proved that the electron polarization Π^0 converges to the well-known result as $\omega_c \to 0$. Therefore the present calculation is expected to have the correct limiting behaviour which is difficult to prove numerically. In Fig.2 the oscillations in the polaron cyclotron resonance mass and the cyclotron resonance frequency is shown in more detail for the case of $\nu = 0.4\omega_{LO}/\omega_c$.

In Fig.3 the polaron correction to the cyclotron resonance mass is shown for the Q2D system for two values of the electron density, namely $n_e = 0$ and $n_e = 4 \times 10^{11} \text{cm}^{-2}$. The finite width of the 2D electron layer reduces the polaron corrections considerably (compare Fig.1 and Fig.3). This is in agreement with the zero magnetic field result for the polaron ground state energy and its mass renormalization as found previously[20,25]. The amplitude of the oscillation of the cyclotron resonance mass is also reduced.

Similar as for the ideal 2D case, the occupation effect reduces the polaron correction to the cyclotron resonance mass over the whole magnetic field region. Surprisingly we find that the polaron cyclotron resonance mass, apart from some small oscillations, is almost independent of the magnetic field strength when $\nu \geq 1$. The oscillating behaviour of the effective mass can still be seen but is less pronounced. In the present study only the lowest subband has been considered which is described by the variational wave function. The variational parameter b is chosen in such a way that it corresponds to the electron density used in the calculation. In Fig.3 b is taken as $b = 1.3k_{LO}$ with $k_{LO} = (2m_b\omega_{LO}/\hbar)^{1/2}$ which for a GaAs-Al$_{1-x}$Ga$_x$As heterostructure corresponds to $n_e \sim 4 \times 10^{11} \text{cm}^{-2}$ and $k_{LO} \sim 2.5 \times 10^6 \text{cm}^{-1}$.

Next the effect of static screening on the polaron cyclotron resonance mass will be studied. It is worth noticing that the static dielectric function in the RPA approximation can be obtained in closed form (see also Ref.23). This will benefit the numerical calculations. For $\nu \leq 1$ we have

$$\epsilon(k) = 1 + \frac{4m_b e^2}{\hbar^2 \epsilon_\infty} \frac{f(k,b)}{k} \nu F_0\left(\frac{\hbar k^2}{2m_b\omega_c}\right) \quad , \tag{9a}$$

with

$$F_0(a) = e^{-a} \int_0^a dx \frac{(e^x - 1)}{x} \quad . \tag{9b}$$

For $1 \leq \nu \leq 2$ the dielectric function becomes

$$\epsilon(k) = 1 + \frac{4m_b e^2}{\hbar^2 \epsilon_\infty} \frac{f(k,b)}{k} \Big[F_0\left(\frac{\hbar k^2}{2m_b\omega_c}\right) \\
+ (\nu - 1)F_1\left(\frac{\hbar k^2}{2m_b\omega_c}\right) - (\nu - 1)\left(\frac{\hbar k^2}{2m_b\omega_c}\right)e^{-\hbar k^2/2m_b\omega_c} \Big] \quad , \tag{10a}$$

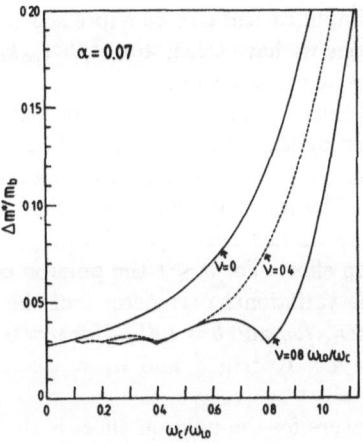

Fig.1 The correction of the cyclotron resonance mass due to polaron effects is plotted as a function of the magnetic field strength for an ideal 2D system. Only the occupation effect is included into the calculations. The solid curve ($\nu = 0.8\omega_{LO}/\omega_c$) and the dashed curve ($\nu = 0.4\omega_{LO}/\omega_c$) correspond to the electron density $n_e = 8 \times 10^{11} \text{cm}^{-2}$ and $4 \times 10^{11} \text{cm}^{-2}$ respectively for a GaAs-Al$_{1-x}$Ga$_x$As heterostructure. The zero electron density result is given by the $\nu = 0$ solid curve.

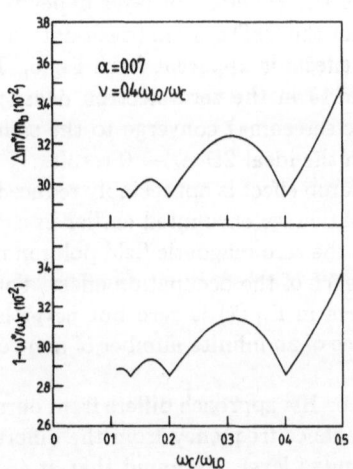

Fig.2 The $\nu = 0.4\omega_{LO}/\omega_c$ curve of Fig.1 in more detail (upper figure) together with the shift in the polaron cyclotron resonance frequency (lower figure).

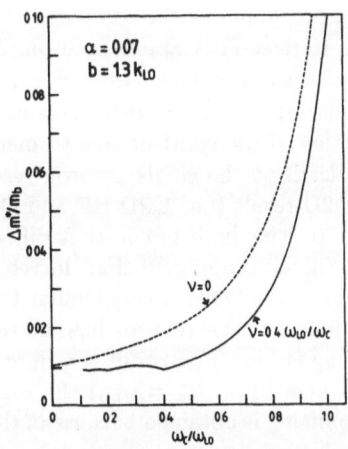

Fig.3 The electron effective mass correction is plotted as a function of the magnetic field strength for a fixed electron density ($n_e = 4 \times 10^{11} \text{cm}^{-2}$) for a GaAs-Al$_{1-x}Ga_x$As heterostructure in the case of a Q2D system. Only the occupation effect is included. The zero electron density results ($\nu = 0$) is given by the dashed curve.

where F_1 is given by

$$F_1(a) = (1 - e^{-a}) + (1 - 2a)F_0(a)$$
$$+ a^2 e^{-a} \int_0^a dx \frac{(e^x - 1 - x)}{x^2} \quad . \tag{10b}$$

The integrals $F_0(a)$ and $F_1(a)$ can further be simplified and can be expressed in terms of the exponential integral. In numerical calculations we have taken $4m_b e^2/\hbar^2 \epsilon_\infty k_{LO} = 1.81$ which is a typical value for a GaAs-Al$_{1-x}$Ga$_x$As heterostructure.

In Fig.4 the effect of the static screening of the electron gas on the polaron cyclotron resonance mass is analysed. The results are given by the solid curve ($\nu = 0.4\omega_{LO}/\omega_c$) and dash-dotted curve ($\nu = 0.6\omega_{LO}/\omega_c$) which correspond to two different electron densities. The calculations are done for the Q2D system. The result in which only the occupation effect is included is also plotted in Fig.4 as the dashed curve ($\nu = 0.4\omega_{LO}/\omega_c$). The static screening effect further reduces the polaron effect. For $\nu \leq 1$ the polaron correction decreases as the electron density increases. The variational parameter b of the subband wave function is chosen as $b = 1.3k_{LO}$ for $\nu = 0.4\omega_{LO}/\omega_c$ and $b = 1.4k_{LO}$ for $\nu = 0.6\omega_{LO}/\omega_c$ which correspond to the electron density $n_e = 4 \times 10^{11} \text{cm}^{-2}$ and $n_e = 6 \times 10^{11} \text{cm}^{-2}$ respectively.

Another interesting quantity which is a measure for the polaron effect is the splitting of the cyclotron resonance line at the optical phonon frequency, i.e. at the resonance condition $\omega_c = \omega_{LO}$. This splitting is given in Fig.5 as function of the electron density and as function of the filling factor ν. The electron density is given in units of k_{LO}^2/π which is about $2 \times 10^{12} \text{cm}^{-2}$ for a GaAs-Al$_{1-x}$Ga$_x$As heterostructure. The results from different approximations are compared in Fig.5: 1) 2D HF: ideal two-dimensional system within the Hartree-Fock approximation; 2) Q2D HF: the same as 1) but for a quasi-two-dimensional system in which the non-zero width of the 2D electron layer depends on the electron density; and 3) the static screening result for the Q2D system (dash-dotted curve). The reduction of the splitting due to many-body effects is apparent from Fig.5. The 2D HF result leads to the single polaron result of Ref.14 in the zero electron density limit. The two Q2D result (i.e. Q2D HF and Q2D static screening) converge to the same limit when $n_e \to 0$. This limit of course is different from the ideal 2D $n_e \to 0$ result.

From Fig.5 it is apparent that the resonant polaron effect is appreciably reduced at the filling factor $\nu = 1$. This is very similar to the situation we encounted earlier in e.g. Fig.1 where the polaron mass renormalization reduces to the zero magnetic field polaron mass at integer filling factor. This result is a pure consequence of the occupation effect. For $\nu = 1$ and at resonance (i.e. $\omega_c = \omega_{LO}$) the resonant term in Eq.(7) is zero but nevertheless a non-zero splitting is obtained because of the presence of an infinite number of non-resonant terms.

A similar result was obtained by Larsen in Ref.6. His approach differs from ours in the sense that he calculated the polaron cyclotron resonance frequency from the difference in energy between the lowest and the first excited Landau level. He found that at $\omega_c = \omega_{LO}$ and $\nu = 1$ the resonant term as calculated by second order perturbation theory gives a zero contribution to the polaron correction to the cyclotron resonance frequency which is consistent with our findings. This led him to argue that at the given condition of $\omega_c = \omega_{LO}$ and $\nu = 1$ every order of perturbation theory contributes and that if these contributions are summed to *infinite order* a non-zero resonant term would be obtained for $\nu \to 1$. Whether or not this is also the case in our approach where the cyclotron resonance frequency is obtained from a dynamical correlation function is not clear at the moment. An experiment where the splitting is followed as function of the electron density could give us very valuable information in this direction. It seems that the GaAs-Al$_{1-x}$Ga$_x$As heterostructure is not directly the best candidate because it is not easy to change the electron density in this system and further more, the condition $\omega_c = \omega_{LO}$ and $\nu = 1$ would imply on electron density $n_e \sim 10^{12} \text{cm}^{-2}$ which is much higher than the electron density of the present

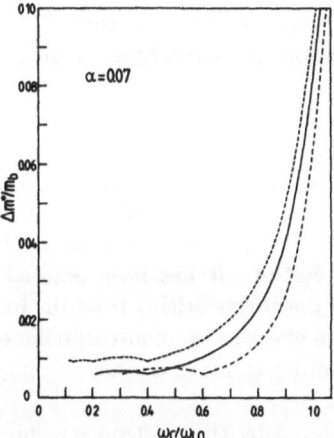

Fig.4 The polaron correction to the electron effective mass is plotted as a function of the magnetic field strength for the Q2D system. The dashed curve ($\nu = 0.4\omega_{LO}/\omega_c$) gives the result where only the occupation effect is included. The solid ($\nu = 0.4\omega_{LO}/\omega_c$) and dash-dotted ($\nu = 0.6\omega_{LO}/\omega_c$) curves give the results where static screening effect (within the RPA) is included.

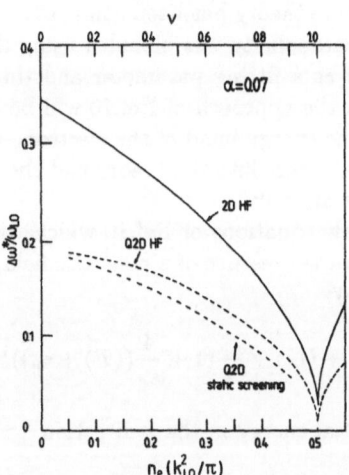

Fig.5 Splitting of the cyclotron resonance peak at the optical phonon frequency as function of the electron density within different approximations (see text). The density is in units of $k_{LO}^2/\pi \sim 2 \times 10^{12} \text{cm}^{-2}$ for GaAs-Al$_{1-x}$Ga$_x$As heterostructures.

day GaAs-Al$_{1-x}$Ga$_x$As heterostructures. For electrons in InSb inversion layers an electron density $n_e \sim 1.4 \times 10^{11}$cm^{-2} is required which can be easily achieved experimentally.

Finally, we want to point out that it was shown in Refs.22 to 24 (see also Eqs.(9) and (10)) that the static dielectric function in the RPA approximation for non-zero magnetic field behaves quite differently from the zero magnetic field result. In the long wavelength limit, where the electron-phonon coupling is believed to be most effective, the RPA static dielectric function for the zero magnetic field diverges[1,22-24] as $1/k$ while for non-zero magnetic field the dielectric function $\epsilon(k)$ has a finite value.

IV. COMPARISON WITH EXPERIMENT

In this section the theory developed in the previous sections will be applied to analyse the experimental data of Refs.2 and 3 for the GaAs-Al$_{1-x}$Ga$_x$As heterostructures (with $x = 0.3$). The cyclotron resonance data for the InSb inversion layers[4] were well explained by the single-polaron theory of Ref.14. It has been pointed out that in these polaron cyclotron resonance experiments the effects arising from the band nonparabolicity play an important role[2,3]. Therefore the effect of the non-parabolic energy band of the electron has to be taken into account in order to make a realistic comparison between the theory and the experimental data.

At this moment, the role played by the electron screening in the polaron cyclotron resonance experiment is not very clear yet. In Ref.3 Sigg et al. employed a perturbative calculation of the polaron correction to the Landau levels where they took the static magnetic field independent Thomas-Fermi screening in analysing their experimental data. Such a treatment for the screening is clearly a very rough approximation. We feel that the foregoing static screening approach gives an indication of the effect of screening on the polaron cyclotron resonance mass but that a full dynamical approach to screening, as was done in Ref.20 for the polaron optical absorption, would be more appropriate. In this section we will consider only the occupation effect and forget about the electron screening. In Ref.14 the single-polaron picture theory has been compared with the experimental data of Refs.2 and 3. The band nonparabolicity was included using the theory of Ref.11 where an electric field was introduced as a fitting parameter and the electron density did not enter the theory. In this section the approach of Ref.10 will be used to treat the effect of the non-parabolic character of the energy band of the electron. In the approach of Ref.10, in principle, no fitting parameter enters into the theory and the band nonparabolicity will be a function of the electron density only.

In the following we list a few equations of Ref.10 which will be used in this section. For details we refer to Ref.10. In the presence of a magnetic field, the energy of an electron in the lowest subband is given by

$$\varepsilon_{np} = -\frac{\varepsilon_g}{2} + \langle U \rangle + \frac{\varepsilon_g}{2}\left(1 + \frac{4}{\varepsilon_g}(\langle T \rangle + \varepsilon_\parallel)\right)^{1/2} \quad , \tag{11a}$$

with the average electron potential energy in the z-direction

$$\langle U \rangle = \frac{12\pi e^2}{\epsilon_0 b}\left(n_d + \frac{11}{16}n_e\right) \quad , \tag{11b}$$

and the average electron kinetic energy in the z-direction

$$\langle T \rangle = \frac{\hbar^2 b^2}{8m_b} \quad , \tag{11c}$$

where ε_g is the band gap and $\varepsilon_\parallel = \hbar\omega_c(n+1/2)$. $b^2 = b_0^2 x$ and x is determined by the equation $x^3 - px - q = 0$ with $p = \hbar^2 b_0^2/2m_b\varepsilon_g$ and $q = (1 + 4\varepsilon_\parallel/\varepsilon_g)$. $b_0 = (48\pi N m_b e^2/\hbar^2\epsilon_0)^{1/3}$ where $N = n_d + (11/32)n_e$.

Due to the band nonparabolicity the cyclotron resonance frequency will be shifted, even without polaron effects. The shifted cyclotron resonance frequency is given by $\hbar(\omega_c)_{np} = \varepsilon_{np}(n = 1) - \varepsilon_{np}(n = 0)$. In order to incorporate polaron effects together with the band nonparabolicity, we insert $(\omega_c)_{np}$ and b_0, as obtained from above, into the memory function which results in the equation

$$\omega - (\omega_c)_{np} - \mathrm{Re}\Sigma(\alpha, (\omega_c)_{np}, b_0; \omega) = 0 \quad . \tag{12}$$

The calculation of the memory function is described in Sec. II. After solving Eq. (12) the cyclotron resonance frequency ω^* is obtained which is affected both by band nonparabolicity and by polaron effects.

Note that in the present approximation the correct limiting behaviour for either vanishing electron-phonon coupling or the parabolic energy band limit (i.e. $\varepsilon_g \to \infty$) is obtained. The correct zero magnetic field limit is also obtained. Further, the polaron effect and the band nonparabolicity are not considered to be additive in Eq. (12). All Landau levels, within a local parabolic energy band approximation, are taken into account in evaluating the real part of the memory function.

The difference between the present calculation and that in Ref. 14 is that, besides the inclusion of the many-body effects, a different approach is used to handle the band nonparabolicity effect. Here the electron density directly enters into the calculation of the memory function and will be taken equal to the experimentally determined value. Although in principle in Eqs. (11) all quantities are known it is found in general very hard to measure the depletion charge density n_d. Therefore n_d will be used as a parameter which is determined in such a way that the theoretical results fit the experimental data as close as possible. In the single-polaron picture theory of Ref. 14 it was found that in order to understand the experimental results of Refs. 2 and 3 quantitatively it was necessary to introduce an effective electron-phonon coupling constant which was 26% lower than the generally accepted value of $\alpha = 0.068$ (for GaAs-Al$_{1-x}$Ga$_x$As heterostructures) which will be used in the present study.

Before presenting the results, it must be pointed out that the experimental data of Ref. 2 behave quite differently from the experimental results of Ref. 3; especially in the low magnetic field region the behaviour is very different. On the other hand the electron densities of the two samples are quite close: $n_e = 4.07 \times 10^{11} \mathrm{cm}^{-2}$ for the experiment of Ref. 2 and $n_e = 4.0 \times 10^{11} \mathrm{cm}^{-2}$ for Ref. 3. It is impossible to fit the present calculations to the experiment data over the whole magnetic field region. Therefore we will concentrate on the high magnetic field region where the polaron effects are believed to be dominant.

In Fig. 6 the cyclotron resonance mass derived from the experimental result of Horst et al.[2] is plotted (full circles and square points) and compared with our theoretical calculation (solid curve). The electron-phonon coupling constant is taken equal to the well-known value $\alpha = 0.068$, the electron band mass $m_b = 0.0650m_e$, and the depletion charge density $n_d = 4 \times 10^{11} \mathrm{cm}^{-2}$ which results in a width for the 2D electron layer $<(z- <z>)^2>^{1/2} = \sqrt{3}/b \approx 40\text{Å}$. Good agreement is found for $H > 15\mathrm{T}$.

In Fig. 7 the experimental data of Sigg et al.[3] is analysed. The cyclotron resonance masses derived from the experimental data of Ref. 3 are plotted as full circles. The theoretical result is given by the solid curve. For $H \geq 12\mathrm{T}$ they agree well. Again we take $\alpha = 0.068$. A slightly smaller band mass $m_b = 0.0648m_e$ is used which gives a better agreement between the calculation and the experimental data. The depletion charge density is much lower $n_d = 10^9 \mathrm{cm}^{-2}$ and corresponds to a width for the 2D electron layer of about 64Å. Notice that the cyclotron resonance mass derived from the experimental data of Ref. 3 changes dramatically at $H \sim 12\mathrm{T}$ and in the region $H < 12\mathrm{T}$ (see Fig. 7). This is probably not caused by polaron effects and was attributed in Ref. 3 to a change of filling factor from $\nu > 1$ to $\nu < 1$ in combination with an anomaly related to possible magneto-plasmon excitations (see also Ref. 26). At this magnetic field value we obtain a small kink in our theoretical result. The other physical parameters used in the calculation

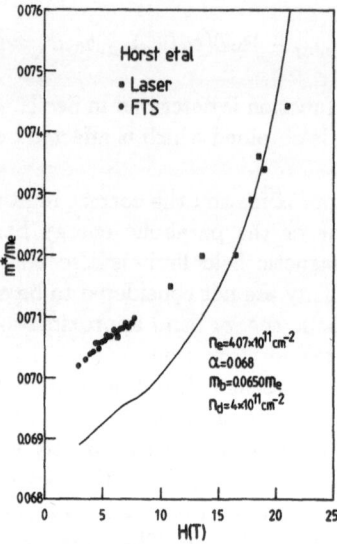

Fig.6 The theoretical results (solid curve) are compared to the experimental data of Horst *et al.*[2] for a GaAs-Al$_{1-x}$Ga$_x$As heterostructure which are given by the full circles and the square points.

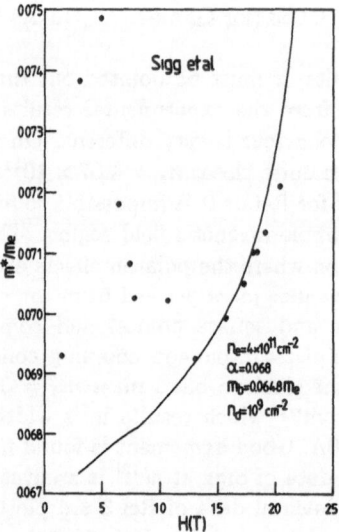

Fig.7 The calculated polaron cyclotron resonance masses (solid curve) are compared to the experimental results of Sigg *et al.*[3] for a GaAs-Al$_{1-x}$Ga$_x$As heterostructure which are given by the full circle points.

are the LO-phonon energy $\hbar\omega_{LO} = 36.75\,\mathrm{meV}$, the static dielectric constant $\epsilon_0 = 12.83$ and the band gap $\varepsilon_g = 1520\,\mathrm{meV}$.

V. CONCLUSION

In the present paper we have investigated the influence of the occupation effect on the polaron cyclotron resonance mass, which results from the Fermi-Dirac statistics. It is found that the occupation effect reduces the polaron effects, i.e., the renormalization of the polaron cyclotron resonance mass is reduced. The electron screening effect is considered within the static RPA approximation. The polaron effects are further reduced by the screening effect. As the filling factor passes through integer values the polaron effects is found to reduce to the zero magnetic field result and this leads to an oscillating behaviour of the polaron cyclotron resonance mass as a function of the magnetic field strength. The non-zero width of the 2D electron layer is incorporated into the memory function calculation by considering the lowest subband. It is found that the non-zero width of the 2D electron layer also reduces the polaron effects. In order to make a realistic comparison between the theoretical results and the experimental data, the effect resulting from the non-parabolic electron energy band has been taken into account. Calculated cyclotron resonance masses have been compared with the experimental data of Refs.2 and 3 for GaAs-Al$_{1-x}$Ga$_x$As heterostructures. Good agreement is found in the high magnetic field region. In making comparison between the theory and the experimental data only the occupation effect is included and the electron screening effect is neglectd.

ACKNOWLEDGEMENTS

This work is partially sponsored by F.K.F.O. (Fonds voor Kollektief Fundamenteel Onderzoek, Belgium), project No. 2.0072.80. One of the authors (F.M.P.) acknowledges financial support from the Belgian National Fund for Scientific Research. One of us (X.Wu) wishes to thank The International Culture Co-operation of Belgium for financial support. We are pleased to thank M. Horst and U. Merkt for providing us with their experimental data and for interesting discussions.

REFERENCES

[1] T. Ando, A. B. Fowler, and F. Stern, Rev. Mod. Phys. **54**, 437 (1982).

[2] M. Horst, U. Merkt, W. Zawadzki, J. C. Maan, and K. Ploog, Solid State Commun. **53**, 403 (1985).

[3] H. Sigg, P. Wyder, and J. A. A. J. Perenboom, Phys. Rev. **B31**, 5253 (1985).

[4] M. Horst, U. Merkt, and J. P. Kotthaus, Phys. Rev. Lett. **50**, 754 (1983).

[5] S. Das Sarma and A. Madhukar, Phys. Rev. **B22**, 2823 (1980).

[6] D. Larsen, Phys. Rev. **B30**, 4595 (1984).

[7] F. M. Peeters and J. T. Devreese, Phys. Rev. **B31**, 3689 (1985); F. M. Peeters, Xiaoguang Wu, and J. T. Devreese, Phys. Rev. **B33**, 4338 (1986).

[8] S. Das Sarma, Phys. Rev. Lett. **52**, 859 (1984); Phys. Rev. Lett. **52**, 1570 (E) (1984).

[9] D. Larsen, Phys. Rev. **B30**, 4807 (1984).

[10] R. Lassnig and W. Zawadzki, Surf. Sci. **142**, 388 (1984).

[11] W. Zawadzki, Solid State Commun. **56**, 43 (1985).

[12] R. Lassnig, Surf. Sci. **170**, 549 (1986). The magnetoplasmon-phonon coupling is also discussed by H. C. A. Oji and A. H. MacDonald, Phys. Rev. **B34**, 1371 (1986).

[13] F. M. Peeters, Xiaoguang Wu, and J. T. Devreese, Phys. Rev. **B34**, 1160 (1986).

[14] Xiaoguang Wu, F. M. Peeters, and J. T. Devreese, Phys. Rev. **B34**, 8800 (1986); F. M. Peeters, Xiaoguang Wu, and J. T. Devreese, Physica Scripta **34** (1986).

[15] J. T. Devreese, J. De Sitter, and M. Goovaerts, Phys. Rev. B5, 2367 (1972); J. Van Royen, J. De Sitter, L. F. Lemmens, and J. T. Devreese, Phys. Physica 81B, 101 (1977); J. P. Vigneron, R. Evrard, and E. Kartheuser, Phys. Rev. B18, 6930 (1978); J. Van Royen, J. De Sitter, and J. T. Devreese, Phys. Rev. B30, 7154 (1984); F. M. Peeters and J. T. Devreese, Physica 127B, 408 (1984); Phys. Rev. B34 (1986).

[16] W. Götze and P. Wölfe, Phys. Rev. B6, 1226 (1972).

[17] F. M. Peeters and J. T. Devreese, Phys. Rev. B28, 6051 (1983).

[18] S. Das Sarma, Phys. Rev. B23, 4592 (1981); T. Ando and Y. Murayama, J. Phys. Soc. Jpn. 54, 1519 (1985); W. Cai and C. S. Ting, Phys. Rev. B33, 3967 (1986).

[19] D. C. Tsui, Th. Englert, A. Y. Cho, and A. C. Gossard, Phys. Rev. Lett. 44, 341 (1980).

[20] Xiaoguang Wu, F. M. Peeters, and J. T. Devreese, Phys. Rev. B34, 2621 (1986).

[21] A. L. Fetter and J. D. Walecka, *Quantum Theory of Many-Particle Systems* (McGraw-Hill, New York, 1971), p.190.

[22] N. J. Horing and M. M. Yildiz, Ann. Phys. (N.Y.) 97, 216 (1976).

[23] M. L. Glasser, Phys. Rev. B28, 4387 (1983).

[24] A. H. MacDonald, J. Phys. C18, 1003 (1985).

[25] S. Das Sarma, Phys. Rev. B27, 2590 (1983); Phys. Rev. B31, 4034 (E) (1985); X. L. Lei, J. Phys. C18, L731 (1985); Xiaoguang Wu, F. M. Peeters, and J. T. Devreese, Phys. Status Solidi. B133, 229 (1986).

[26] Z. Schlesinger, S. J. Allen, J. C. Hwang, P. M. Platzman, and N. Tzoar, Phys. Rev. B30, 435 (1984).

HOT CARRIER EFFECTS IN QUASI-2D POLAR SEMICONDUCTORS

Jagdeep Shah

AT&T Bell Laboratories
Holmdel, N. J. 07733

ABSTRACT

In this series of lectures, I review hot carrier effects in quasi-2D polar semiconductors (quantum wells and heterostructures), with special emphasis on optical techniques for investigating hot carrier effects. After briefly introducing the basic concepts in hot carrier physics, we discuss theoretical calculations of carrier-phonon interactions and hot carrier energy loss rates to the lattice in 3D and quasi-2D systems. We then discuss how these quantities are affected by degeneracy, plasma effects and hot phonons. The bulk of the article is devoted to a discussion of experimental results and their analysis. Three kinds of experiments are discussed: I-V and related transport measurements, direct time-of-flight measurement of velocity-field characteristics, and measurements which use optical spectroscopy to provide direct information about the carrier distribution function in the presence of external perturbations. The optical studies have given valuable new insight into the behavior of hot carrier relaxation processes in quasi-2D systems from femtosecond to steady-state conditions.

I. INTRODUCTION

Investigation of hot carrier effects in semiconductors has played a central role in modern semiconductor physics. Hot carrier phenomena are determined primarily by band structures and carrier-phonon scattering processes, and can therefore provide important information about carrier-phonon interactions, of fundamental interest in semiconductor physics. On the other hand, hot carrier physics determines the behavior of ultrafast, ultrasmall semiconductor devices operating at high electric fields. Therefore, a study of hot carrier effects is also extremely valuable for a thorough understanding of such devices. For these reasons, high field transport and hot carrier effects in bulk semiconductors have received considerable attention over the past several decades [1,2].

The interest in hot carriers in quasi-2D systems arises for the same reasons. From the fundamental point of view, one would like to know how the carrier-phonon interactions in quasi-2D systems such as heterojunctions and quantum wells differ from the interactions in the bulk. The interest in hot carriers in quasi-2D systems is also driven by the intense current activities in high speed devices fabricated from quasi-2D heterostructures. The best example of such devices is the selectively doped heterostructure transistor for which switching times of less than 10 psec have been reported recently [3]. Hot electron physics governs the behavior of such devices.

Investigation of the nonlinear current-voltage characteristics and related transport measurements is the traditional way of studying hot carrier phenomena in semiconductors. The first studies of hot carrier effects in quasi-2D systems were carried out by using this technique. Considerable information about transport properties are obtained from such measurements. However, transport quantities are determined by an average over the carrier distribution function (DF) and do not provide direct information about the microscopic scattering rates. Such information can be obtained by a direct measurement of DF. Optical spectroscopy provides the best means of determining DF in the presence of an applied electric field or an intense optical excitation [4,5] and such studies have provided information of fundamental interest in recent years. Finally, the direct measurement of velocity-field characteristics with a time-of-flight technique, in combination with optical spectroscopy, has also provided new insights into the behavior of hot carriers in quasi-2D systems.

In this review, I will discuss what we have learn about carrier-phonon interactions and hot carrier effects in quasi-2D systems from these three kinds of experiments. The review article is organized as follows. After covering some basic hot electron concepts in Section II, I will review the theoretical work on the electron-phonon interactions in 3D and quasi-2D systems in Sections III and IV. The rest of the review will be devoted to a discussion of the experimental results, and their interpretations and implications (Section V). The discussion of experimental results will be primarily for the GaAs/AlGaAs system since most of the experiments have been performed on this system. I will conclude with a summary in Section VI.

II. BASIC CONCEPTS

At sufficiently low electric fields, the energy gained by the carriers from the applied field is small compared to the average energy of the carriers in the absence of the field. Under these conditions, the distribution function of the carriers is unaffected by the electric field except for a small shift in momentum space that leads to the drift velocity. Since the scattering rates determining the mobility of the carriers depend on DF, the carrier mobility remains independent of the applied electric field and Ohm's law is obeyed at these low fields.

This situation changes considerably as the electric field is increased to a point that the energy gained from the field by the carriers is no longer negligible compared to the average carrier energy. This is the regime of "hot carriers" where DF has changed significantly from its equilibrium value. Therefore, the mobility of the carriers is no longer independent of the electric field and one observes departure from Ohm's law. In the direct gap III-V semiconductors, with subsidiary conduction band minima at the L and X points in the Brillouin zone, electrons are transferred to these subsidiary valleys at high electric fields. Since the mobility of electrons in these valleys is much lower, such a transfer leads to the well-known negative differential resistance, an extreme departure from Ohm's law, in these semiconductors [1].

Under steady state conditions, hot electron phenomena are governed by the balance between the power input into the electronic system from an external perturbation such as an applied electric field and the power loss from the carriers, which is usually dominated by the inelastic collisions between the carriers and the phonons. In transient conditions, the time evolution of the perturbed DF is also determined by carrier- carrier and carrier-phonon collisions. Thus scattering of carriers by phonons and other carriers play a central role in hot carrier effects in semiconductors.

A key to understanding the scattering processes, and thus hot carrier phenomena in semiconductors, is provided by the carrier distribution function. In the simplest cases, when carrier-carrier collisions are more effective compared to other scattering processes, DF can be described in terms of a drifted Maxwellian or Fermi-Dirac distribution function characterized by a carrier temperature T_c larger than the lattice temperature T_L. However, this is not always the case and the determination of DF under various conditions is the central problem in the experimental and theoretical hot electron investigations.

Although we have talked only about electric field heating of the carriers, it is by now well-known [4,5] that photoexcitation of a semiconductor also leads to a heating of the carrier system. The carriers are created with a certain excess kinetic energy immediately after photoexcitation and they share a fraction of this energy with the other carriers in the semiconductor by carrier-carrier collisions, leading to a heating of the carrier system. The demonstration that photoexcitation acts as a source of carrier heating [6] has had important implications in the study of hot carriers. An important example is the use of recently developed pico- and femtosecond laser sources [7] for studying ultrafast hot carrier relaxation phenomena in semiconductors.

Hot carrier effects in quasi-2D systems are governed by the same basic phenomena as in 3D, except that the carrier-phonon interactions are more complicated to evaluate because of complexities arising out of the nature of electronic wavefunctions, the existence of several subbands and phonon modes in quasi-2D. Furthermore, the thickness of the quasi-2D layer enters as an additional parameter in the problem. Finally, in the case of quantum wells, the possibility of the transfer of carriers into the barrier layers (i.e. real-space transfer [8]) must be considered.

III. ELECTRON-PHONON INTERACTIONS IN 3D

The electron-phonon interaction in 3D systems has been investigated extensively in the past. In this section, we review what is known with special emphasis on calculating electron energy loss rate to the lattice. We will consider the simple case of single, spherical, parabolic valley with effective mass m_2. This situation applies quite well to electrons in the Γ valley of III-V semiconductors. The valence band structure is quite complicated but it has been shown that the discussion that follows also applies to holes (with appropriate heavy hole effective mass) if the rate is reduced by a factor of 2.

There are two equivalent ways of calculating the average energy loss rate of carriers to the lattice. In the first, one calculates the energy loss rate for an electron of energy $E(k)$

$$\frac{dE_{\vec{k}}}{dt} = \frac{2\pi}{\hbar} \sum_q \left[|(\vec{k}+\vec{q},N_q-1|H'|\vec{k},N_q)|^2 \hbar\omega_q(1-f(E_{\vec{k}}+\hbar\omega_q)) \right.$$

$$\times \; \delta\left(E_{\vec{k}+q,N_q-1}-E_{\vec{k},N_q}\right)$$

$$- \; |(\vec{k}-\vec{q},N_q-1|H'|E_{\vec{k}},N_q)|^2 \hbar\omega_q$$

$$\left. \times \; (1-f(E_{\vec{k}}-\hbar\omega_q)) \; \delta \; (E_{\vec{k}-q,N_q+1}-E_{\vec{k},N_q}) \right] \tag{1}$$

$$E_{\vec{k}} = \frac{\hbar^2 k^2}{2m} \equiv E \tag{2}$$

$$E_{\vec{k},N_q} = \frac{\hbar^2 k^2}{2m} + N_q \cdot \hbar\omega_q \tag{3}$$

$$f(E_{\vec{k}}) = \frac{1}{1 + \exp\left(\dfrac{E-E_F}{kT_c}\right)} = f(\vec{k}) \tag{4}$$

where T_c is the carrier temperature, and the carrier energy is measured from the band extreme. The first term in Eq. (1) corresponds to phonon absorption and the second to phonon emission. The average rate of change of energy per carrier is then given by averaging over the distribution function,

$$\left<\frac{dE}{dt}\right> = \frac{\displaystyle\int_o^\infty \frac{dE_{\vec{k}}}{dt} \cdot g(E) \, f(E) \, dE}{\displaystyle\int_o^\infty g(E) \, f(E) \, dE} \tag{5}$$

where $g(E)$ is the density of states in the appropriate band.

The second method for calculating the average energy loss rate involves calculating the rate of generation of phonons of a given wavevector q,

$$\frac{\partial N_q}{\partial t} = \frac{2\pi}{\hbar} \left[\sum_{\vec{k}} |(\vec{k},N_q+1|H'|\vec{k}+\vec{q},N_q)|^2 f(\vec{k}+\vec{q})(1-f(\vec{k})) \right.$$

$$\times \; \delta\left(E_{\vec{k}+q,N_q}-E_{\vec{k},N_q+1}\right)$$

$$- \sum_k |(\vec{k}+\vec{q},N_q-1|H'|\vec{k},N_q)|^2 f(k)(1-f(\vec{k}+\vec{q}))$$

$$\left. \times \; \delta\left(E_{\vec{k}+q,N_q-1} - E_{\vec{k},N_q}\right) \right] \tag{6}$$

where $f(\bar{k})$ describes the Fermi-Dirar distribution function (Eq. 4) at energy $E_{\bar{k}}$ given by Eq. (2).

The average energy loss rate per carrier is then given by multiplying $\dfrac{\partial N_q}{\partial t}$ by $\hbar\omega_q$, summing over all $\bar{q}$ and dividing by the total number of electrons

$$< \frac{dE}{dt} > = - \frac{1}{nV} \sum_q \hbar\omega_q \cdot \frac{\partial N_q}{\partial t} \tag{7}$$

Both the methods, of course, give the same result because the order of summation over $\bar{k}$ and $\bar{q}$ should not matter. Later in this lecture, we will be interested in the case in which the phonon population is sufficiently disturbed by hot carrier effects. In this case, we will be interested in determining the phonon population at various wavevectors. For this reason, we will follow the second approach here.

We use the equality

$$\sum_{\bar{k}} F(\bar{k}) \to \int\limits_{k=0}^{\infty} dk \int\limits_{0}^{\pi} d\theta \int\limits_{0}^{2\pi} d\phi \ F(\bar{k}) \ k^2 \sin\theta \cdot 2 \cdot \frac{V}{(2\pi)^3}$$

to evaluate $\dfrac{\partial N_q}{\partial t}$. The first term of $\dfrac{\partial N_q}{\partial t}$ in Eq. (8) is due to emission of phonons and is given by

$$\left(\frac{\partial N_q}{\partial t}\right)_e = \frac{V}{\hbar\pi} \int\limits_0^{\infty} dk \ k^2 |M_e|^2 \left(1 - f(\bar{k})\right) \int\limits_0^{\pi} d\theta \ \sin\theta \ f(\bar{k}+\bar{q}) \tag{9}$$

$$\times \ \delta\left(\frac{\hbar^2 q^2}{2m} + \frac{\hbar^2 kq}{m} \cos\theta - \hbar\omega_q\right)$$

where

$$|M_e|^2 = |(\bar{k}, N_q+1|H'|\bar{k}+\bar{q}, N_q)|^2 \tag{10}$$

$$\left(\frac{\partial N_q}{\partial t}\right)_e = \frac{V}{\hbar\pi} \cdot \frac{m}{\hbar^2 q} \int\limits_0^{\infty} dk \cdot k |M_e|^2 \ (1 - f(\bar{k})) \int\limits_{x_{min}}^{x_{max}} dx \ \delta(x) \ f(\bar{k}+\bar{q}) \tag{11}$$

where

$$x = \frac{\hbar^2 q^2}{2m} + \frac{\hbar^2 kq \cos\theta}{m} - \hbar\omega_q \tag{12}$$

$$x_{min} = \frac{\hbar^2 q^2}{2m} - \frac{\hbar^2 kq}{m} - \hbar\omega_q \tag{13}$$

187

$$x_{max} = \frac{\hbar^2 q^2}{2m} + \frac{\hbar^2 kq}{m} + \hbar\omega_q \tag{14}$$

In order to have non-zero integral x_{min} must be <0 and x_{max} must be >0. This determines the lower limit on k integration to be

$$k_{min} = \frac{m}{\hbar^2 q} \left| \frac{\hbar^2 q^2}{2m} - \hbar\omega_q \right| \tag{15}$$

Changing the integration variable to $\epsilon = \frac{\hbar^2 k^2}{2m} \cdot \frac{1}{kT_c}$ and using

$$\epsilon_{min} = \frac{\hbar^2}{8mkT_c} \left(q - \frac{2m\omega_q}{\hbar q} \right)^2 \tag{16}$$

we get

$$\left(\frac{\partial N_q}{\partial t} \right)_e = \frac{m^2}{\pi\hbar^5 q} kT_c V \int_{\epsilon_{min}}^{\infty} f(\epsilon + x_q)\,(1 - f(\epsilon))\,|M_e|^2 d\epsilon \tag{17}$$

where

$$x_q = \frac{\hbar\omega_q}{kT_c} \tag{18}$$

The second term in Eq. (6) is due to absorption of phonons, can be evaluated in a similar manner

$$\left(\frac{\partial N_q}{\partial t} \right)_a = -\frac{m^2 kT_c V}{\pi\hbar^5 q} \int_{\epsilon_{min}}^{\infty} f(\epsilon)\,(1 - f(\epsilon + x_q))\,|M_a|^2 d\epsilon \tag{19}$$

where

$$|M_a|^2 = |(\vec{k} + \vec{q}, N_q - 1|H'|\vec{k}, N_q)|^2 \tag{20}$$

The sum of Eq. (17) and (19) gives the net rate of generation of phonons of wavevector q

$$\left(\frac{\partial N_q}{\partial t} \right) = \frac{m^2 kT_c V}{\pi\hbar^5 q} \int_{\epsilon_{min}}^{\infty} \Bigg[|M_e|^2 f(\epsilon + x_q)\,(1 - f(\epsilon))$$

$$- |M_a|^2 f(\epsilon)\,(1 - f(\epsilon + x_q)) \Bigg] d\epsilon \tag{21}$$

We now recall the identity

$$f(\epsilon+x_q)\,(1-f(\epsilon)) \equiv (f(\epsilon)-f(\epsilon+x_q))\,N_q(T_c) \qquad (22)$$

where $N_q(T_c)$ is given by

$$N_q(T) = \frac{1}{\exp\left(\dfrac{\hbar\omega_q}{kT}\right)-1} \qquad (23)$$

at $T = T_c$. The matrix elements for phonon emission and absorption may be written as

$$|M_e|^2 = |M_q|^2\,(N_q+1) \qquad (24)$$

$$|M_a|^2 = |M_q|^2\,N_q \qquad (25)$$

Using Eqs. (22), (24) and (25), Eq. (21) may be simplified as

$$\left(\frac{\partial N_q}{\partial t}\right) = \frac{m^2 kT_c V |M_q|^2}{\pi\hbar^5 q}\left(N_q(T_c)-N_q\right)\int_{\epsilon_{min}}^{\infty}(f(\epsilon)-f(\epsilon+x_q))d\epsilon \qquad (26)$$

Since

$$\int_{\epsilon_{min}}^{\infty} f(\epsilon)d\epsilon = \ln\left[1+\exp(\eta-\epsilon_{min})\right] \qquad (27)$$

where $\quad \eta = \dfrac{E_F}{kT_c} \qquad (28)$

we get, using Eq. (16) to eliminate ϵ_{min},

$$\left(\frac{\partial N_q}{\partial t}\right) = \frac{m^2 kT_c V |M_q|^2}{\pi\hbar^5 q}\left(N_q(T_c)-N_q\right)$$

$$\times \ln\left\{\frac{1+\exp\left[\eta-\dfrac{\hbar^2}{8mkT_c}\left(q-\dfrac{2m\omega_q}{\hbar q}\right)^2\right]}{1+\exp\left[\eta-\dfrac{\hbar^2}{8mkT_c}\left(q+\dfrac{2m\omega_q}{\hbar q}\right)^2\right]}\right\} \qquad (29)$$

The average energy loss rate is given by Eq. (7), which may be re-written as

$$
<\frac{dE}{dt}> = -\frac{1}{n} \cdot \frac{1}{2\pi^2} \int_0^\infty \hbar\omega_q q^2 \left(\frac{\partial N_q}{\partial t}\right) dq
$$

$$
= -\frac{m^2 k T_c V}{2\pi^3 \hbar^5 n} \int_0^\infty dq \left[q |M_q|^2 \hbar\omega_q (N_q(T_c) - N_q) \right.
$$

$$
\left. \times \ln \left\{ \frac{1+\exp\left[\eta - \frac{\hbar^2}{8mkT_c}\left(q - \frac{2m\omega_q}{\hbar q}\right)^2\right]}{1+\exp\left[\eta - \frac{\hbar^2}{8mkT_c}\left(q + \frac{2m\omega_q}{\hbar q}\right)^2\right]} \right\} \right] \tag{30}
$$

Equation (30) allows for the possibility that N_q is different from $N_q(T_L)$ i.e. the phonon population is different from that expected at the lattice temperature. This situation $(N_q \neq N_q(T_L))$ corresponds to the existence of non-equilibrium or hot phonons. The simpliest way to evaluate the non-equilibrium phonon population is to assume that excess phonons decay with a characteristic decay time τ_{ph}, i.e. rate of decay is given by $(N_q - N_q(T_L))/\tau_{ph}$. In steady state, one has

$$
\frac{\partial N_q}{\partial t} = \frac{N_q - N_q(T_L)}{\tau_{ph}} \tag{31}
$$

From (29) and (31) we get

$$
\frac{N_q - N_q(T_L)}{\tau_{ph}} = \frac{m^2 k T_c V |M_q|^2}{\pi \hbar^5 q} \left(N_q(T_c) - N_q\right)
$$

$$
\times \ln \left\{ \frac{1+\exp\left[\eta - \frac{\hbar^2}{8mkT_c}\left(q - \frac{2m\omega_q}{\hbar q}\right)^2\right]}{1+\exp\left[\eta - \frac{\hbar^2}{8mkT_c}\left(q + \frac{2m\omega_q}{\hbar q}\right)^2\right]} \right\} \tag{32}
$$

Eq. (32) allows a calculation of N_q for a given set of parameters. Substituting this N_q in (30) allows a calculation of the energy loss rate under these conditions. In general, the integration in Eq. (30) has to be carried out numerically. We will show an example of such a calculation in a subsequent section.

Equations (30) and (32) provide a general formalism for calculating the energy loss reates in the presence of non-equilibrium phonon populations for a degenerate carrier distribution in a simple, parabolic board. We will now apply these to specific cases of carrier phonon interactions.

$$|M_q|^2 = \frac{1}{V} \cdot \frac{2\pi\hbar^2 eE_0}{mq^2} \tag{33}$$

$$eE_0 = \frac{me^2\hbar\omega_{LO}}{4\pi\epsilon_0\hbar^2} \left[\frac{1}{K_\infty} - \frac{1}{K_S} \right] \tag{34}$$

where we have assumed that $\hbar\omega_q = \hbar\omega_{LO}$ and ignored screening. ϵ_0 is the free space permitivity and K_∞ and K_S are optical and static dielectric constants. Substituting (33) into (30), we get

$$\left\langle \frac{dE}{dt} \right\rangle_{po} = -\left(\frac{2\hbar\omega_{LO}}{\pi m} \right)^{\frac{1}{2}} \left(\frac{\hbar\omega_{LO}}{kT_c} \right)^{\frac{1}{2}} \frac{eE_0}{F_{\frac{1}{2}}(\eta)}$$

$$\times \int \frac{dq}{q} \, (N_q(T_c) - N_q) \, \ln\left\{ \frac{1 + \exp\left[\eta - \frac{\hbar^2}{8mkT_c} \left(q - \frac{2m\omega_{LO}}{\hbar_q} \right)^2 \right]}{1 + \exp\left[\eta - \frac{\hbar^2}{8mkT_c} \left(q + \frac{2m\omega_{LO}}{\hbar_q} \right)^2 \right]} \right\} \tag{35}$$

where

$$F_j(\eta) = \frac{1}{\Gamma(j+1)} \int E^j f(\epsilon) \, d\epsilon \tag{36}$$

and we have used the relations

$$f(\epsilon) = \frac{1}{1 + \exp(\epsilon - \eta)} \tag{37}$$

$$n = N_C F_{\frac{1}{2}}(\eta) \tag{38}$$

$$N_c = 2 \left(\frac{2\pi mkT_c}{h^2} \right)^{\frac{3}{2}} \tag{39}$$

Substituting (33), into (32) gives

$$
\frac{2mkT_c eE_0}{\hbar^3 q^3} \left(N_q(T_c) - N_q\right) \ln \left[\frac{1 + \exp\left[\eta - \frac{\hbar^2}{8mkT_c} \left(q - \frac{2m\omega_{LO}}{\hbar q}\right)^2\right]}{1 + \exp\left[\eta - \frac{\hbar^2}{8mkT_c} \left(q + \frac{2m\omega_{LO}}{\hbar q}\right)^2\right]} \right]
$$

$$
= \frac{N_q - N_q(T_L)}{\tau_{ph}} \tag{40}
$$

We will now consider the special case where τ_{ph} is extremely short so that $N_q = N_q(T_L)$ (no hot phonon effects) and the carriers are non-degenerate so that $\eta << 0$. Under these conditions, the natural log term in (35) simplifies to $\exp(\eta) \exp(-\epsilon_{min}) (1 - \exp(-x_c))$ where x_c is given by (18) with $\hbar\omega_q = \hbar\omega_{LO}$. Also $F_j(\eta) \to \exp(\eta)$ for $\eta << 0$ for all j, and $N_q(T)$ becomes independent of q, so that Eq. (35) leads to

$$
\left< \frac{dE}{dt} \right>_{\substack{po,\ non-deg \\ no\ hot\ phonons}}
$$

$$
= -\left(\frac{2\hbar\omega_{LO}}{\pi m}\right)^{\frac{1}{2}} \left(x_c\right)^{\frac{1}{2}} \cdot eE_o \cdot \left[\frac{1}{\exp(x_c) - 1} - \frac{1}{\exp(x_L) - 1}\right]
$$

$$
\times (1 - \exp(-x_c)) \int_0^\infty \frac{dq}{q} \exp\left(-\frac{\hbar^2}{8mkT_c} \left(q - \frac{2m\omega_{LO}}{\hbar q}\right)^2\right)
$$

$$
= -\left[\frac{2\hbar\omega_{LO}}{\pi m} eE_0 \, x_c^{\frac{1}{2}} \frac{e^{-x_c} - e^{-x_L}}{1 - e^{-x_L}}\right]
$$

$$
\int_0^\infty \frac{dq}{q} \exp\left(-\frac{\hbar^2}{8mkT_c} \left(q - \frac{2m\omega_{LO}}{\hbar q}\right)^2\right) \tag{41}
$$

Here $\qquad x_L = \frac{\hbar\omega_{LO}}{kT_L}$ $\qquad\qquad$ (42)

In order to evaluate the integral, we define

$$
\frac{\hbar^2 q_0^2}{2m} = \hbar\omega_{LO} \tag{43}
$$

$$
\text{and} \quad y = \frac{1}{2}\left[\frac{q^2}{q_0^2} + \frac{q_0^2}{q^2}\right] \tag{44}
$$

With these substitutions

$$\int_0^\infty \frac{dq}{q} \exp\left[-\frac{\hbar^2}{8mkT_c}\left(q - \frac{2m\omega_{LO}}{\hbar_q}\right)^2\right]$$

$$= \int_0^{q_0} \cdots + \int_{q_0}^\infty \cdots$$

$$= -\frac{1}{2} e^{x_c/2}\left[\int_\infty^1 \frac{dy}{\sqrt{y^2-1}} e^{-\frac{yx_c}{2}} - \int_1^\infty \frac{dy}{\sqrt{y^2-1}} e^{-\frac{yx_c}{2}}\right]$$

$$= e^{\frac{x_c}{2}} \int_1^\infty \frac{dy}{\sqrt{y^2-1}} e^{-\frac{yx_c}{2}}$$

$$\equiv e^{x_c/2} K_0(x_c/2)$$

where K_0 is the modified Bessel Function of order zero. Hence Eq. (38) reduces to

$$\left<\frac{dE}{dt}\right>_{\substack{po,\, non-deg \\ no\ hot\ phonons}}$$

$$= -P_0 \frac{\exp(-x_c) - \exp(-x_L)}{1 - \exp(-x_L)} \frac{\left[(x_c/2)^{\frac{1}{2}} e^{x_c/2} K_0(x_c/2)\right]}{\sqrt{\pi/2}} \qquad (45)$$

$$\text{where} \quad P_0 = eE_0\left(\frac{2\hbar\omega_{LO}}{m}\right)^{\frac{1}{2}} \qquad (46)$$

$$\equiv \hbar\omega_{LO}\, W_0 \qquad (47)$$

W_0 in the last equation is an *effective* scattering rate whereas P_0 is an *effective* energy loss rate; both of these quantities are a measure of the carrier-phonon coupling strength. Note that $1/W_o$ is not the scattering time for a single scattering event.

We note that Eq. (45) is the standard result for energy loss rate by optical phonon scattering, for non-degenerate carriers and in the absence of hot phonon effects. The usefulness of the equation arises from the fact that it provides a simple, analytic form for the energy loss rate, especially at low temperatures ($x_c/2 > 1$) when the square bracket can be approximated by unity. If, in addition, $x_L \gg 1$ (e.g. $T_L \approx 2K$), Eq. (45) simplifies to

$$\left<\frac{dE}{dt}\right>_{\substack{po,\, non-deg \\ no\ hot\ phonons}} = -P_0 \exp(-x_c) \qquad (48)$$

$$P_0(W) = 5.67 \times 10^{-8} (\hbar\omega_{LO}(meV))^{\frac{3}{2}} \left(\frac{m}{m_0}\right)^{\frac{1}{2}} \left[\frac{1}{K_\infty} - \frac{1}{K_S}\right] \qquad (49)$$

$$W_o(sec^{-1}) = 3.54 \times 10^{14} (\hbar\omega_{LO}(meV))^{1/2} \left(\frac{m}{m_o}\right)^{1/2} \left[\frac{1}{K_\infty} - \frac{1}{K_s}\right] \qquad (50)$$

and

$$E_0(V/cm) = (1.89 \times 10^5) \hbar\omega_{LO}(meV) \left(\frac{m}{m_0}\right) \left(\frac{1}{K_\infty} - \frac{1}{K_S}\right). \qquad (51)$$

III.B Non-polar Optical Scattering

$$|M_q|^2 = \frac{1}{Y} \frac{D^2 \hbar^2}{2\rho\hbar\omega_o} \qquad (52)$$

where D is the Deformation Potential, ρ is the density and $\hbar\omega_o$ is the optical phonon energy (*TO* or *LO*). Substituting (52) into (30) gives us the energy loss rate in the general case. If we make the same assumptions, (non-degenerate, no phonon case), we obtain, after some algebra,

$$< \frac{dE}{dt} >_{\substack{npo, \, non-deg \\ no \, hot \, phonons}}$$

$$= -P_0' \left[\frac{\exp-(x_c) - \exp(-x_L)}{1 - \exp(-x_L)}\right] \left[\frac{(x_c/2)^{\frac{1}{2}} e^{x_c/2} K_1(x_c/2)}{\sqrt{\pi/2}}\right] \qquad (53)$$

where

$$P_0' = \frac{m^{\frac{3}{2}} D^2 (\hbar\omega_o)^{\frac{1}{2}}}{\sqrt{2\pi\hbar^2\rho}} \qquad (54)$$

$$P_0'(watts) = 5.72 \times 10^{-26} \left[\frac{\left(D(eV/cm)\right)^2 \left(\frac{m}{m_o}\right)^{\frac{3}{2}} \hbar\omega_o(meV)}{\rho(gm/cm^3)}\right] \qquad (55)$$

K_1 is the modified Bessel function of order 1 and

$$x_c = \frac{\hbar\omega_o}{kT_c}$$

and similar equation for x_L in terms of T_L. For $\frac{x_c}{2} > 1$ and $x_L \gg 1$, (46) simplifies to

$$\left<\frac{dE}{dt}\right>_{\substack{\text{npo non-deg} \\ \text{no hot phonon}}} = -P_0' \, e^{-x_c} \tag{56}$$

III.C Acoustic Phonon Scattering (Deformation Potential)

For acoustic phonons

$$\hbar\omega_q = \hbar s q \tag{57}$$

where s is the sound velocity: For deformation potential scattering

$$|M_q|^2 = \frac{E_1^2 \hbar^2 q^2}{2\rho V \hbar\omega_q} = \frac{E_1^2 \hbar q}{2V\rho s} \tag{58}$$

Therefore Eq. (30) reduces to

$$\left<\frac{dE}{dt}\right>_{dp} = -\left(\frac{m}{kT_c}\right)^{\frac{1}{2}} \frac{E_1^2}{(2\pi)^{\frac{5}{2}}} \cdot \frac{1}{\rho F_{\frac{1}{2}}(\eta)} \int_0^\infty dq \left[q^3 \left(N_q(T_c) - N_q \right) \right.$$

$$\left. \ln\left\{ \frac{1 + \exp\left[\eta - \frac{\hbar^2}{8mkT_c}\left(q - \frac{2ms}{\hbar}\right)^2\right]}{1 + \exp\left[\eta - \frac{\hbar^2}{8mkT_c}\left(q + \frac{2ms}{\hbar}\right)^2\right]} \right\} \right] \tag{59}$$

An expression for N_q can be written down using Eq. (32) and (58). Once again, the energy loss rate needs to be evaluated numerically in a general case. However, for the case of no hot phonons and when the temperatures are high enough to satisfy equipartition condition,

$$\hbar s q \ll kT, \tag{60}$$

so that

$$N_q(T) \simeq \frac{kT}{\hbar s q} \tag{61}$$

it has been shown that

$$\left< \frac{dE}{dt} \right>_{\substack{dp,\ \text{equipartition} \\ \text{no hot phonons}}}$$

$$= -\frac{8\sqrt{2}(m)^{5/2}E_1^2}{\pi^{3/2}\rho\hbar^4} \left[(kT_c)^{\frac{3}{2}}\right] \left(1 - \frac{T_L}{T_c}\right) \frac{F_1(\eta)}{F_{\frac{1}{2}}(\eta)} \tag{62}$$

$$= -1.68 \times 10^{-14} \frac{(m/m_o)^{\frac{5}{2}}\left[E_1(eV)\right]^2\left[T_c(k)\right]^{\frac{3}{2}}}{\rho(gm/cm^3)} \left[1 - \frac{T_L}{T_c}\right] \frac{F_1(\eta)}{F_0(\eta)}$$

$$\text{in Watts} \tag{63}$$

This is also valid for degenerate statistics so long as the equipartition condition holds.

III.D Acoustic Phonon Scattering (Piezoelectric)

While only longitudinal modes interact with electrons by deformation potential interaction, the piezoelectric interaction is allowed for longitudinal as well as transverse acoustic modes. Piezoelectric coupling has a complicated directional dependences. For simplicity we will consider only Zinc Blende Crystals in which there is only one piezoelectric constant denoted by e_{14}. After taking suitable averages for longitudinal and transverse modes, one finds

$$|M_q|^2 = \frac{e^2 e_{14}^2 \hbar\omega_q}{2q^2 V s^2 \epsilon_0 K_s \rho} \tag{64}$$

The energy loss rate for an arbitrary phonon occupation number may be obtained numerically using Eq. (30) and (32). For the case of no hot phonons and degenerate and non-degenerate distribution functions obeying equipartition condition, one obtains

$$\left\langle \frac{dE}{dt} \right\rangle_{\substack{\text{pe, equipartition} \\ \text{no hot phonons}}}$$

$$= -0.4 \left(\frac{2m}{\pi} \right)^{\frac{3}{2}} \left(\frac{ee_{14}}{\hbar e_0 K_s} \right)^2 \frac{\left(kT_c \right)^{\frac{1}{2}}}{\rho} \left(1 - \frac{T_L}{T_c} \right) \frac{F_0(\eta)}{F_{\frac{1}{2}}(\eta)}$$

$$= -1.94 \times 10^{-8} \frac{(m/m_o)^{\frac{3}{2}} \left[e_{14}(C/m^2) \right]^2 \left[T_c(k) \right]^{\frac{1}{2}}}{K_s^2 \rho (gm/cm^3)}$$

$$\times \left(1 - \frac{T_L}{T_c} \right) \frac{F_0(\eta)}{F_{\frac{1}{2}}(\eta)} \quad \text{in Watts} \tag{65}$$

For non-degenerate distributions, the ratio of the Fermi functions is unity.

IV. THEORETICAL INVESTIGATIONS IN QUASI-2D SYSTEMS

IV.A Carrier-Phonon Interactions in Quasi-2D Systems

Crucial to our understanding of hot carrier phenomena is the analysis of carrier-phonon interactions. We will restrict this review to a discussion of interaction with optical phonons because acoustic phonons do not contribute significantly to the energy loss rates of carriers above 40 K and most of the work in polar quasi-2D systems has been in the regime where optical phonons dominate. One must treat electron-phonon and hole-phonon interactions separately because the holes interact with optical phonons via polar as well as non-polar interactions whereas symmetry considerations restrict electrons in direct-gap III-V semiconductors to interact with optical phonons via only polar interactions (to the lowest order). The hole-phonon interaction is already quite complex in bulk semiconductors because of the complex character of the valence bands [9]. The valence bands in quasi-2D systems are further complicated by the effects of quantum confinement [10]. Therefore, hole-phonon interaction in quasi-2D systems has not yet been investigated from a theoretical point of view. In this section, I will briefly review some theoretical results on electron-phonon interactions which are important for hot electron phenomena in quasi-2D systems.

The problem of electron-phonon interaction in quasi-2D heterolayers has been treated by a number of authors [11-16]. There are two basic differences between quasi-2D and 3D systems that one must consider. The first is the fact that the electronic wavefunctions and density of states in quantum-confined systems are different [17] and the second is the fact that the phonons in quasi-2D systems have different character than phonons in 3D. Most of the work reported so far has assumed that the phonons can be considered 'bulk-like'. Under this assumption the principal difference in the quasi-2D case arises from the fact that momentum conservation in the direction perpendicular to the plane of the layer is not well defined. If the quasi-2D well is defined by the planes z=0 and z=L, then in the scattering by a phonon of wavevector q, the xy wavevector $k_{//}$ of electron must change by $q_{//}$. However, there is no such selection rule for k_z; instead the matrix element is multiplied by an integral which depends on the initial and final state electronic wavefunctions [13]. This is the primary reason why accurate analytical expression for the electron-optical phonon scattering rate cannot be obtained in quasi-2D systems, even under simplifying assumptions.

In order to get a qualitative appreciation of electron- optical phonon scattering rates in quasi-2D systems, we discuss here some numerical results obtained by Riddoch and Ridley [15] for the case of quantum wells with infinite potential steps and simple band structure (single, spherical, parabolic valley with effective mass m^*). Since the scattering rates depend on the thickness of the well L, these authors have introduced a dimensionless parameter $\beta = E_0/\hbar\omega_q$ where $\hbar\omega_q$ is the optical phonon energy (= $\hbar\omega_0$, assumed independent of q for small q) and E_0 is the quantum confinement energy given by

$$E_0 = h^2/(8mL^2) \tag{66}$$

$$E_0(meV) = 3.76/[(m/m_0)(L(\text{Å})/100)^2] \tag{67}$$

Fig.1 shows the calculated scattering rates for an electron in the lowest subband as a function of the electron kinetic energy E_k, for three different values of the parameter β. The scattering rates have been normalized to the bulk scattering rate W_0 given by (47) and (50). The rates in Fig.1 include both phonon absorption and emission, with the phonon occupation number

$$N_q(T) = \frac{1}{\exp(\hbar\omega_q/kT)-1} \tag{68}$$

equal to 0.3 at room temperature ($T = T_L = 300\ K$) The figure also shows the bulk rates, and the rates in the Momentum Conservation Approximation (MCA) which was introduced by Ridley [14] to obtain approximate analytical expressions for the scattering rates in the quasi-2D systems.

Several features of Fig.1 are interesting. First of all, MCA is a reasonable approximation for thick wells ($\beta < .1$) but becomes increasingly inaccurate for thinner wells, especially at larger electron energies. Secondly, the quasi-2D scattering rates can be substantially larger than the 3D rates for low energy electrons but approach the 3D rates at high electron energies. Finally, the difference at low energies diminishes with increasing thickness of the well and almost vanishes for $\beta < 0.1$. The rates shown in Fig.1 include contributions from both intra- and inter-subband scattering processes. For large β, the intra-subband processes dominate whereas for $\beta = 0.1$, the intra- and inter-subband contributions are approximately equal [15].

In order to understand the electron-phonon interaction rates in more detail, one needs to know the range of phonon wavevectors that are involved in the

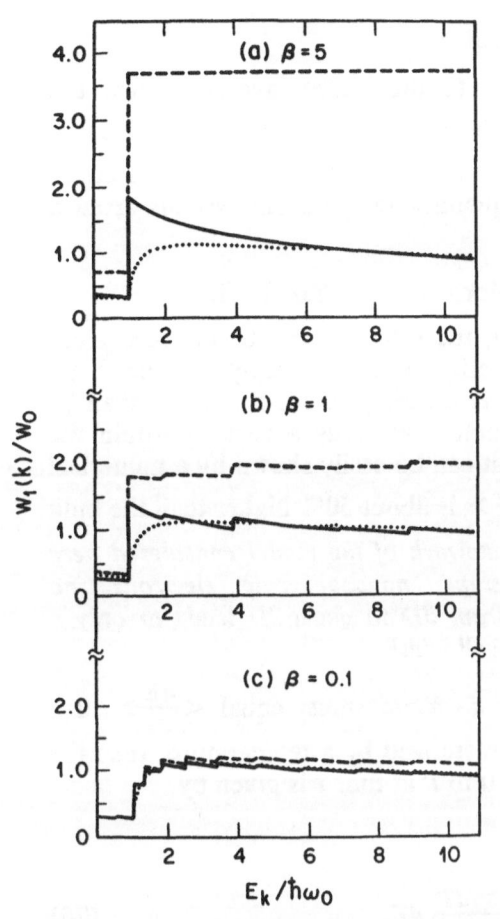

FIG. 1. Calculated normalized scattering rates as a function of kinetic energy in the plane for an electron in the lowest subband for various well widths : (a) $\beta = 5.0$ (b) $\beta = 1.0$ and (c) $\beta = 0.1$. The full curves denote Monte-Carlo results, the broken curves denote MCA results and the dotted curves denote the bulk results (from Riddoch and Ridley [15]).

interactions. For parabolic bands, the wavevector q_0 of the phonon emitted by an electron of kinetic energy $\hbar\omega_0$ is given by (43). If E is the energy of the electron before (after) emission (absorption) then the phonon wavevectors involved in the scattering will range from q_{min} to q_{max} where

$$q_{min} = q_0 \left[\left(\frac{E}{\hbar\omega_0} \right)^{1/2} - \left(\frac{E}{\hbar\omega_0} - 1 \right)^{1/2} \right] \tag{68}$$

and

$$q_{max} = q_0 \left[\left(\frac{E}{\hbar\omega_0} \right)^{1/2} + \left(\frac{E}{\hbar\omega_0} - 1 \right)^{1/2} \right] \tag{69}$$

Although the valence band structure is non-parabolic and very complicated, an order-of-magnitude estimate of phonon wavevectors involved in the scattering processes may be obtained by using the appropriate effective masses in the above formulae. For the case of GaAs, q_{min} are approximately $1 \times 10^6 \, cm^{-1}$ and $2.5 \times 10^6 cm^{-1}$ and q_{max} are approximately $6 \times 10^6 \, cm^{-1}$ and $15 \times 10^6 \, cm^{-1}$ for electrons and holes respectively.

IV.B Average Energy Loss Rates to the Lattice

The rate of loss of energy per carrier to the lattice, averaged over the distribution function of the carriers (defined as $<\frac{dE}{dt}>$) plays an important role in both steady state and transient hot carrier effects. In this subsection, we discuss what is known about this quantity in quasi-2D systems from a theoretical point of view.

For $\beta \leq 0.1$ (corresponding to a well thickness $L > 350\text{Å}$ in GaAs), $<\frac{dE}{dt}>$ for quasi-2D and 3D electrons are about the same, as can be deduced from Fig.1c. The importance of dimensionality is most simply evaluated by considering the case of large β (>5) corresponding to small L (less than 55 Å for the case of GaAs). In this extreme case, only scattering within the lowest subband need be considered and it can be easily shown by a numerical integration of the curve 1a that the $<\frac{dE}{dt}>$ is about 50% higher than the bulk value. *We conclude that, within the framework of the model considered here (parabolic bands, infinite potential steps, nondegenerate electrons, no screening), reduction in dimensionality from 3D to quasi-2D leads to only a small increase in the <ELR> for typical well widths.*

The rate of change of average energy $d<E>/dt$ must equal $<\frac{dE}{dt}>$ in a transient experiment. If DF can be characterized by a temperature, then the time to cool from a temperature T_0 at $t=0$ to T at time t is given by

$$t = \int_{T_0}^{T} \frac{d<E>/dT}{<\frac{dE}{dt}>} \, dT \tag{70}$$

Thus, if $<E>$ and $<\frac{dE}{dt}>$ are known as a function of T, one can calculate t by numerical integration of (12). Some numerical results have been given by Katayama [18].

IV.C Modifications of Average Energy Loss Rates

We have concluded above that for a ·simple non-degenerate model in the absence of screening, the average electron energy loss rates in quasi-2D systems are approximately the same as for the bulk. Measurements of this quantity in bulk and in quasi-2D systems gives values which are smaller than expected from this simple model, as we will discuss in Section IV. In this subsection, we consider the effects of degeneracy, slab modes, screening, and nonequilibrium ("hot") phonons on the $<\frac{dE}{dt}>$.

IV.C.1 Degeneracy

In order to evaluate the effect of degeneracy on $<\frac{dE}{dt}>$, we evaluate the simple case of energy independent scattering rates, given by W_0, and two-dimensional density of states $\rho(E) = m/(\pi\hbar^2)$. In this case it can be shown that

$$\langle ELR \rangle = - \frac{W_0 \times \hbar\omega_0}{1+\exp\eta} (N_q(T_c) - N_q(T_L)) \times \ell n\left[\frac{1+\exp\eta}{1+\exp(\eta - x_c)}\right] \quad (71)$$

$$\eta = E_F/kT_c = \ell n\left[\exp\left(\frac{n_s}{N_{2c}}\right) - 1\right] \quad (72)$$

$$N_{2c} = \frac{mkT_c}{\pi\hbar^2} \quad (73)$$

$$x_c = \hbar\omega_0/kT_c \quad (74)$$

Evaluation of these expressions for the case of GaAs quasi-2D electrons shows that the effect of degeneracy on the $\langle\frac{dE}{dt}\rangle$ is negligible upto a density of $1\times10^{12} cm^{-2}$. Similar conclusions can be drawn for energy dependent scattering rates by numerical calculations [19,20]. We conclude that the effect of degeneracy on the average energy loss rates is negligible for the typical densities of carriers.

IV.C.2 Slab Modes

Most of the discussion of the electron-phonon interactions in quasi-2D systems has been based on using the Frohlich coupling constant for bulk, uniform solids. The optical vibrational modes in an ionic slab are expected to be different from those in bulk [21] and evidence for such behavior has been recently obtained from Raman scattering measurements [22]. Sawaki [23] has considered the proper quasi-2D electron-phonon interaction. Riddoch and Ridley [24] have recently found that the the electron-phonon scattering rates in a single heterostructure are affected by less than a factor of 2 by slab modes. Shah et al [19] have argued that the effect of slab modes is likely to be rather small in the typical case of GaAs like materials where the product of in-plane phonon wavevector and the total thickness of one complete period is much larger than unity. While further theoretical work in this area will be useful, it appears at present that the effects of slab modes on energy loss rates are rather small for the typical multiple quantum well structures being investigated.

IV.C.3 Plasma Effects

For polar scattering, one must consider the interaction of carriers with the polarization waves of the system [19]. At low carrier density, the polarization waves are simply the polar optical phonons of the lattice. When the carrier density is high, the polarization waves of the system are the coupled phonon-plasma modes of the system. In general, one must consider the frequency, wavevector and temperature dependent dielectric constant of the coupled phonon-plasma system and calculate the matrix elements for polar interaction. Furthermore, the calculation of the average energy loss rates involves one additional complication, namely that the collective modes of the system may be Landau-damped by the single particle excitations. As a result, the energy loss rate to the lattice is determined not only by the net rate of emission of collective modes, but also by the competition between Landau damping and the decay of collective modes into acoustic phonons. While this is simple in concept, no complete calculation taking into account all these factors is available in either 3D or quasi-2D systems. In the absence of such a calculation, we will consider in this section some of the approximations that have been made to understand this problem.

Price [25] was the first to consider screening in quasi-2D systems and has given a formal treatment of the problem. Many other theoretical papers have been published on this subject recently [20, 26-29]. Some of these treatments [26-27] have considered the effects of "dynamic screening" (i.e. frequency dependent dielectric constant). However, these treatments have made a number of approximations whose appropriateness for the problem we are considering is not clear. We follow here the argument given by Das Sarma and Mason [28,29] that an *upper limit* to the plasma effects can be obtained by considering the wavevector dependent static screening (ϵ (q,0)). Das Sarma and Mason [28] have calculated the effects of such static screening and their results are shown in Fig.2. One can see from this figure that the scattering rates are reduced considerably as one goes from 2D to the quasi-2D case. However, the effect of static screening on the scattering rates is less than 40 %. Furthermore, this rather small effect is an *upper limit* to the effect of plasma on the scattering rates. These considerations lead to the conclusion that the plasma effects on the quasi-2D energy loss rates are expected to be rather small.

IV.C.4 Hot Phonon Effects

The average energy loss rate depends on the difference between the the phonon emission and absorption processes. Since hot carriers lose energy by emitting optical phonons, one must consider the possibility of what happens when the phonon lifetime is sufficiently long to lead to a build-up of a significant non- equilibrium phonon population. Such "hot phonon effects"can reduce $<\frac{dE}{dt}>$ by increasing phonon absorption and have been considered for the case of bulk semiconductors by Collet et al [30] and Kocevar and coworkers [31-32]. Nonequilibrium populations in bulk semiconductors have been experimentally observed by Shah et al [33], von der Linde et al [34], Collins and Yu [35] and Kash et al [36].

The possibility of hot phonon effects in the quasi-2D system was suggested by Shah et al [19] and considered from a theoretical point of view by Price [37]. An estimate of the hot phonon effects on the average energy loss rates can be obtained by considering a simplified 3D model [38] as discussed in Section III. The calculated $N(q)/N_q(T_c)$ is shown in Fig.3a where one sees that there is indeed a large non-equilibrium phonon population created at $T_c = 50K$ for the case of GaAs , assuming a phonon lifetime of 7 psec, as observed experimentally [34,36] and a carrier density of $2 \times 10^{14} cm^{-3}$. The energy loss rate per electron as a function of q are shown in Fig. 3b with and without including the effects of hot phonons. We see that a substantial reduction in the energy loss rate is predicted by this model. While the above simplified model calculation is made for 3D system, similar conclusions have been reached by Price for quasi-2D systems [37].

IV.D Summary of Theoretical Considerations

This brief review of theoretical investigations on carrier- phonon interactions in quasi-2D systems shows that some theoretical problems remain to be

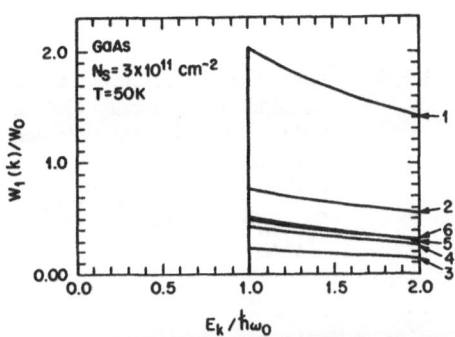

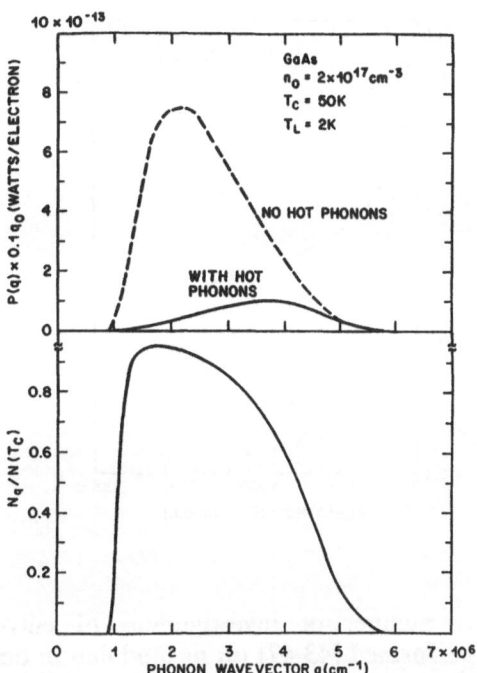

FIG. 2. Calculated normalized scattering rates as a function of kinetic energy in the plane for an electron in the lowest subband for various static screening conditions : 1. 2D result, no screening; 2-6. quasi-2D; 2. no screening; 3. Debye screening; 4. zero temperature Fermi-Thomas screening; 5. zero temperature, finite wavevector RPA screening; and 6. finite temperature, finite wavevector screening (from Das Sarma and Mason [28]).

FIG. 3. Calculated relative phonon occupation number (a) and electron energy loss rates (b) as a function of phonon wavevector for bulk GaAs; the effects of hot phonons on energy loss rates is shown in (b) (from Shah et al. [38]).

investigated in this field. These include detailed considerations of how the scattering rates in the quasi-2D systems are affected by the true phonon modes, plasma effects and hot phonon effects. However, the considerations discussed in this section provide a good basis for understanding many interesting experiments which are discussed in the next section.

V. EXPERIMENTAL INVESTIGATIONS

In this section we will consider three kinds of hot carrier experiments in quasi-2D systems : (A) Transport measurements at high fields (B) direct measurement of velocity-field characteristics and (C) measurements of the hot carrier distribution functions.

V.A Transport Measurements

One reason for the interest in high field transport properties in quasi-2D heterostructures is the possibility of obtaining ultra-high-speed devices suggested by the extremely high low-field-mobilities achieved in these systems by using the technique of modulation doping [39,40]. Low temperature mobility in excess of one million has been reported [41,42].

Since typical devices operate at high fields, the crucial question is how this high mobility varies with electric fields and how to understand this variation.

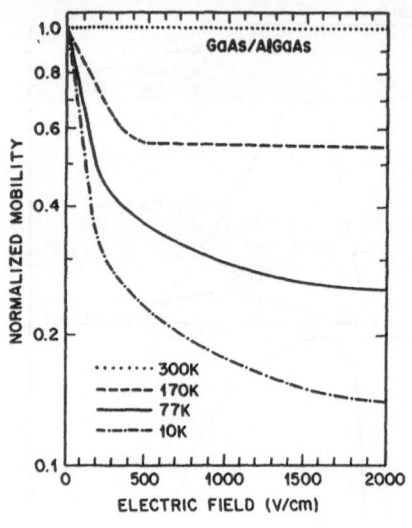

FIG. 4. Normalized mobility as a function of in-plane electric field at several lattice temperatures for a single period GaAs/Al$_{0.24}$Ga$_{0.76}$As quantum well (from Keever et al [44]).

A number of investigations of current-voltage characteristics have been performed [43-47] on n- modulation doped samples. All these measurements show a rapid decrease in the mobility of electrons with increasing electric fields. The results of Keever et al [44] are shown in Fig.4. Assuming that the electron density is not affected by high fields, the mobility decreases rather rapidly with increasing electric fields at low temperatures.

In order to obtain results that do not depend on the assumption of field-independent carrier density, two techniques have been employed. Schubert et al [48], Inoue [49] and Inoue et al [50] have performed moderate and high field Hall measurements to separate the variations of carrier density and mobility with electric fields. They find that the variation of carrier concentration with applied field is sample dependent and varies from 0 to 20 % in different samples. The reasons for the density variation is an interesting subject by itself [48]; however, it appears that the influence of this density variation on the interpretation of the I-V measurements is rather small (< 20%).

Masselink et al [51,52] have simultaneously measured the geometrical magnetoresistance and the I-V characteristics to obtain the variation of the carrier density and mobility with electric fields. The structures used in the geometrical magnetoresistance method have FET-like geometry and dimensions, and are less prone to current instabilities. The results obtained by this technique also show that the carrier density variations with electric field are rather small, in agreement with the conclusions drawn from the high field Hall measurements discussed above.

An important point of interest is the velocity of electrons at high fields. The drift velocity of electrons in III-V compounds is expected to reach a peak value and then decrease due to transfer to subsidiary valleys or real-space transfer. The I- V measurements and the high field Hall and geometrical magnetoresistance measurements discussed above have given values for peak velocities typically in the range of 1.0 to 1.8×10^{7}cm/sec[44,48,50-52]. However, values in excess of 3x10^{7}cm/sec have also been reported at helium temperatures [49]. We will discuss a direct measurement of velocity by time-of flight technique in Section IV.B.

Successful p- modulation doping has also been obtained in recent years [53,54]. Although such structures are of interest in developing complementary logic FET's [55], very little information is available about hole drift velocities at high fields. Shah et al [56] have reported I-V measurements at helium temperatures, showing a relatively rapid decline of hole mobility with increasing electric fields. In contrast, at room temperature, Hopfel et al [57] find a constant mobility up to 12 kV/cm, and a 5% decrease at 14 kV/cm. Further measurements in p- modulation doped systems would be very interesting.

V.B Direct Velocity Measurements (Time-of-Flight Technique)

The most direct method of measuring the velocity of carriers in the presence of high electric field is to excite one side of a semiconductor with a short pulse of ionizing radiation (such as an electron beam or a laser) and to measure the time it takes for one type of carriers (either electrons or holes) to arrive at the other end when a constant field is applied between the two ends. This technique [58] has been applied quite successfully to bulk semiconductors like GaAs [59]. A modification of this technique (microwave time-of-flight technique), in which the measurements are made in the frequency rather than the time domain [60,61], has also been applied successfully to bulk GaAs [62] and other III-V semiconductors [63,64].

It is very important to minimize the space-charge effect due to injected carriers in these measurements. In 3D systems, one can virtually eliminate this effect by exciting a large area with a very small carrier density. Since this is not possible in quasi-2D systems, clever design of the experiment is necessary. Cooper and Nelson [65] and Nelson et al[66] have used a resistive gate and measured the velocities of electrons and holes in Si inversion layers. Such measurements have not been performed in III-V heterostructures or quantum wells. However, Hopfel et al [57,67] have used the time-of-flight technique to measure the velocity of minority electrons in p-modulation-

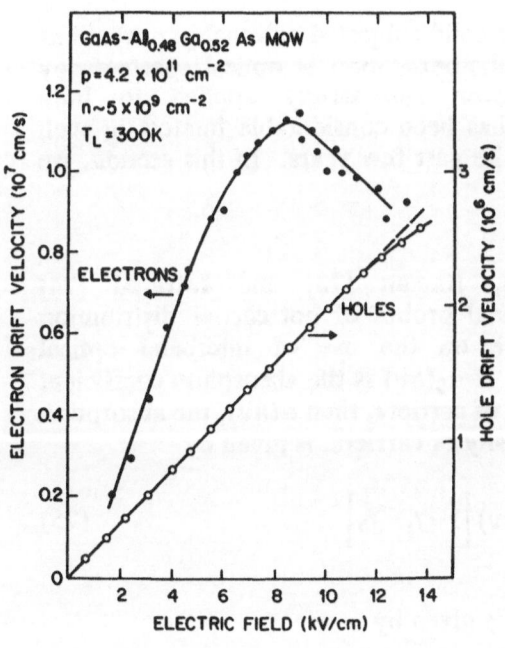

FIG. 5. Minority electron velocity in p - modulation - doped GaAs/AlGaAs as a function of applied electric field, determined by picosecond time-of-flight technique; also shown is the hole mobility deduced from I-V measurements by assuming constant hole density (from Hopfel et al [57]).

doped GaAs/AlGaAs quantum wells, since the space-charge effects are negligible for photoexcited carrier density much smaller than the background hole density.

The results of Hopfel et al [57,67] for the velocity of electrons as a function of the applied electric field are shown in Fig.5. There are several points of interest. The low-field mobility determined from these measurements ($\approx 1,500 cm^2/V-sec$) is much smaller than the typical values of about 7,000 measured for majority electrons in n-modulation-doped quantum wells and heterostructures. This was attributed to electron-hole scattering [57]. The velocity reaches its peak value at rather high fields because of this small low-field-mobility. The peak drift velocity of $\approx 1.1 \times 10^7 cm/sec$ is slightly smaller than the values deduced by other means. Finally, the velocity clearly decreases at higher fields, indicating the onset of negative differential resistance.

These direct measurements of velocity are very interesting, and provide useful information about a quantity of interest in device operation. When such measurements are combined with optical spectroscopic techniques, they provide new insights into physics of hot carriers in quasi-2D systems as we will discuss in Sec. IV.C.2.b and IV.C.2.c.

V.C Hot Carrier Distribution Functions Probed by Optical Spectroscopy

As we have discussed in the Introduction, a knowledge of the carrier distribution function (DF) is very valuable in obtaining an understanding of the microscopic scattering processes that ultimately govern the high field transport. The simplest perturbed DF is Fermi-Dirac function with a temperature T_c higher than the lattice temperature T_L. One way to estimate the carrier temperature T_c is to compare the variation of a quantity such as mobility or Shubnikov-de Haas oscillations as a function of T_L with their variation with applied electric field at very low T_L. Such measurements have indeed been made by Inoue et al [49,68] and by Sakaki et al [69]; however, they are generally restricted to a small range of temperatures and several assumptions are involved in the analysis of the data.

The best means of determining DF in a solid subjected external perturbations such as an applied electric field or photoexcitation is optical spectroscopy [4,5,70-72]. This technique has been successfully applied to bulk semiconductors [4,5,70-72] and there has been considerable interest in such measurements in quasi-2D systems in the past few years. In this section, we will review these optical measurements.

V.C.1 Optical Probing Techniques

Absorption [72], Luminescence [5], Raman [73] and Infrared [74] spectroscopies have been used as optical probes of hot carrier distribution functions. We will concentrate here on the use of interband optical absorption or luminescence transitions. If $\alpha_0(h\nu)$ is the absorption coefficient at the photon energy $h\nu$ in the absence of carriers, then $\alpha(h\nu)$, the absorption coefficient in the presence of a low density of carriers, is given by

$$\alpha(h\nu) = \alpha_0(h\nu)\left[1-f_e-f_h\right] \qquad (75)$$

$L(h\nu)$, the luminescence intensity at $h\nu$ is given by

$$L(h\nu) = A\alpha_0(h\nu)f_e f_h \qquad\qquad (76)$$

Here f_e and f_h are the electron and hole distribution functions. In general one must take into account the effects of reabsorption on the luminescence spectra. For Fermi-Dirac distributions

$$f = \frac{1}{1 + \exp[(E - E_F)/kT]} \qquad\qquad (77)$$

and for photon energies much larger than the chemical potential, the luminescence intensity decreases exponentially with increasing photon energy, with a slope characterized by the carrier temperatures,

$$L(h\nu) \propto \exp(-h\nu/kT_c) \qquad\qquad (78)$$

As we have discussed in Section II, non-equilibrium conditions can be created not only by the application of an electric field, but also by photoexcitation of the semiconductor. Analyses of interband transitions based on the above equations have been used to deduce carrier distributions functions in both these cases. We will see that optical spectroscopy provides new insight into interaction of carriers with phonons and other carriers and also into real-space transfer effects. We will first discuss the results obtained for the case of electric field heating and then for the case of optical excitation.

V.C.2 Electric Field Induced Hot Carriers

V.C.2.1 Carrier-Phonon Interactions

Southgate et al [75] and Inoue et al [76] have experimentally determined the carrier distribution function at high electric fields in bulk GaAs but did not use this technique to extract information about carrier-phonon interactions. Shah et al [19,45,46] have *directly determined the average energy loss rate* <ELR> *for both electrons and holes* in GaAs quantum wells by simultaneously measuring the luminescence spectra and I-V characteristics. As we have discussed above, <ELR> gives direct information on carrier-phonon scattering rates.

Fig. 6 shows the spectra for both n- and p- type samples at various electric fields. The spectra result from intrinsic band- to-band recombination involving photoexcited minority carriers and the majority carriers [77]. The high energy tail decreases exponentially with photon energy at each field, showing that DF is Fermi-Dirac type with a characteristic temperature T_c larger than the lattice temperature T_L. In particular, it is interesting to note that there is no kink at energy corresponding to one optical phonon energy above the Fermi energy. With increasing electric fields, the high energy tail broadens (i.e. T_c increases) and shows transitions involving additional subbands. Finally, the electrons are much hotter than holes at comparable electric fields.

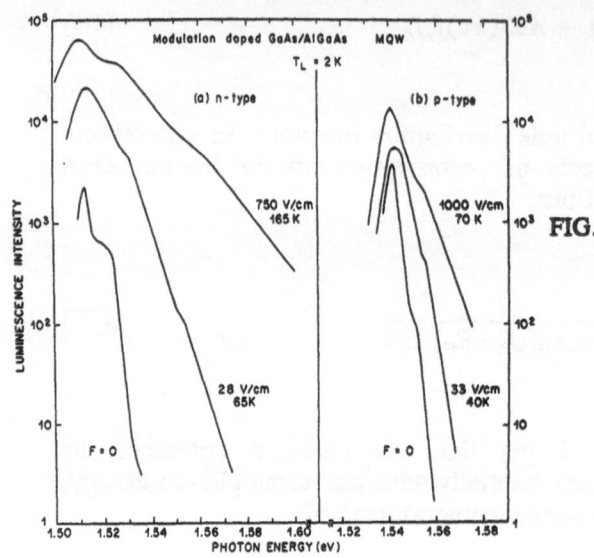

FIG. 6. Interband luminescence spectra for n- and p- modulation-doped GaAs/AlGaAs quantum wells at various electric fields; the carrier temperatures determined from the high energy tails are also indicated (from Shah et al [45,46,56]).

In this steady state experiment, the power input per carrier ($e\mu F^2$) must equal the energy loss rate per carrier to the lattice $<ELR>$. The most important result of these experiments is that, for the first time, the electron and hole $<ELR>$ were directly determined as a function of their temperatures. Fig. 7 is a plot of $1/T_c$ vs. $<ELR>$ for both electrons and holes. The most striking result is that the hole $<ELR>$ is a factor of *twenty-five* larger than the $<ELR>$ of electrons. Although some differences are expected between electron and hole $<ELR>$, the experimentally observed differences are surprisingly large.

Shah et al [19] have explored many possibilities to explain these large differences. Following the discussion given in Section III, they have concluded that the hole $<ELR>$ is close to what is expected but the electron $<ELR>$ is anomalously low. The cause for the anomalously low electron $<ELR>$ is not the reduction in the dimensionality or the plasma effects, but the presence of a large non-equilibrium population of optical phonons. A simple calculation of this "hot phonon" effect gives a reasonable agreement with the data, as shown in Fig.7. Price [37] has calculated the hot phonon effects for quasi-2D system, and finds that the effects of hot phonons on electron $<ELR>$ should be even larger than observed by Shah et al [19]. Photoexcitation experiments have shown the presence of non-equilibrium optical phonons quite clearly [33-37]. It would be very interesting to look for direct experimental evidence for the presence of hot optical phonons in electric field experiments. It should be mentioned that Yang et al [78] have recently repeated the above measurements and reported that the electron $<ELR>$ is close to its expected value. They interpreted this to mean that hot phonon effects are not important in their samples. It is not clear at this time why their results differ from all the other measurements reported in bulk and quasi-2D system in which the electron $<ELR>$ is smaller than expected.

Shah et al [45,46] also compared the dependence of the high field mobility on carrier temperature with the dependence of the low field mobility on the lattice temperature. Such comparisons are shown in Fig. 8 for the two n-type and one p-type samples used for the above study. For the first sample (Fig.8a) the high field mobility decreases much more rapidly with increasing T_c than the decrease of low field mobility with T_L. The two dependences are

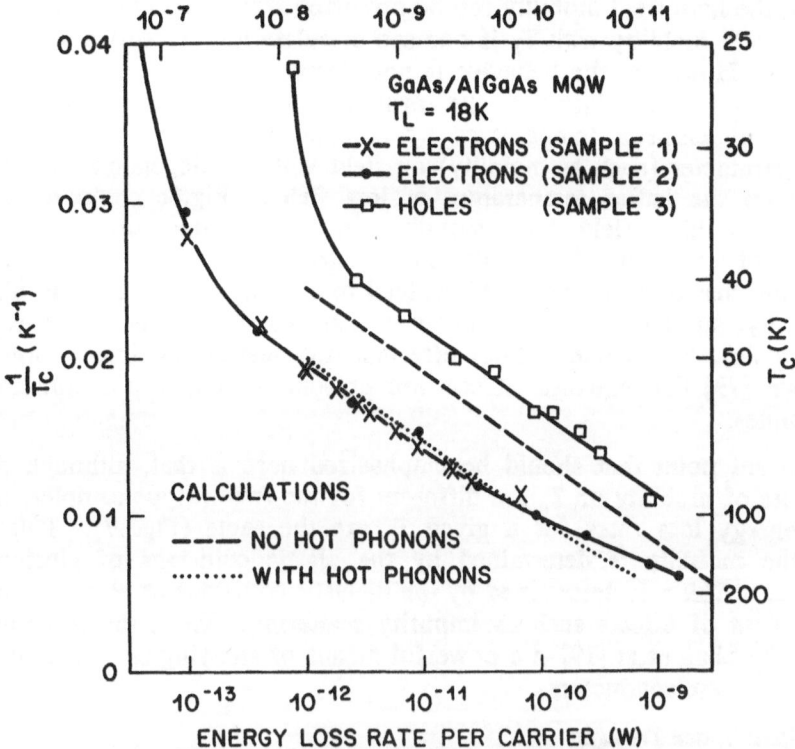

FIG. 7. The inverse of the carrier temperature as a function of the average energy loss rate <ELR>, as determined from simultaneous luminescence and transport measurements in steady state (from Shah et al [19]).

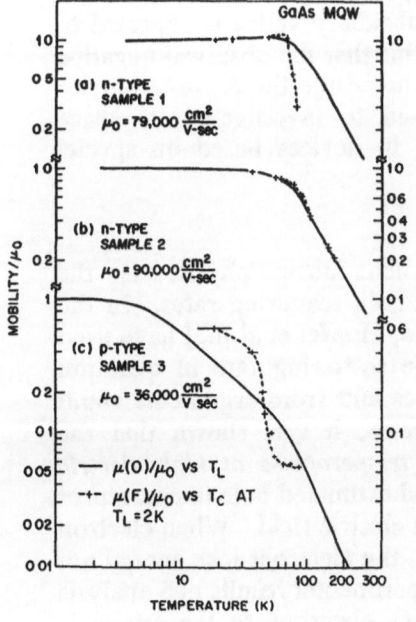

FIG. 8. Comparison of the dependence of low-field carrier mobility on lattice temperature with the dependence of high field carrier mobility on carrier temperature for the three samples in Fig.7; a. sample 1, n-type b. sample 2, n-type and c. sample 3, p-type (from Shah et al [45,46,56]).

the same for the second sample (Fig. 8b). Thus, when intrinsic behavior dominates, the high field mobility can be predicted from the known variation of the low field mobility with T_L if one can calculate how T_c varies with the applied field. However, the behavior is not always "intrinsic", as illustrated by Fig.8a; and this illustrates the problem of determining the field dependence of electron temperature by comparing the dependence of a transport parameter (such as mobility) on field with the dependence of that parameter on the lattice temperature at low fields. Fig.8c compares the dependence of high field hole mobility on hole temperature with the dependence of low field hole mobility on the lattice temperature. Once again it is apparent that some extrinsic effects lead to a rather sudden drop in high field mobility, an effect which is similar to that observed in the first n-type sample (Fig.8a). The origin of these effects is not clear at this time, although it is known [79] that extrinsic effects are present in similar p-modulation-doped samples.

The important point that should be emphasized here is that, although the dependences of mobility on T_c are different for the two n- type samples, the electron energy loss rates for a given T_c are the same (Fig. 7). This is because the mobility is determined by the elastic collisions of electrons whereas the $<ELR>$ is determined by the inelastic collisions of electrons and is independent of effects such as impurity scattering. Thus, the technique developed by Shah et al [19] is a powerful means of studying carrier-phonon interactions in semiconductors.

V.C.2.2 Real Space Transfer

A novel high field phenomenon in quasi-2D heterolayers is the transfer of carriers from the lower bandgap semiconductor into the higher bandgap semiconductor adjacent to it, i.e. real -space transfer [8]. Here we discuss the experiment reported by Hopfel et al [57,67] in which they combined luminescence spectroscopy with direct time-of-flight measurement of carrier velocity (see Section IV.B). They found that in the region of negative differential resistance (see Fig. 5), the electron temperature deduced from the luminescence spectra was about 600 K, significantly lower than the electron temperature at which k- space transfer to the subsidiary valley is expected to take place in bulk GaAs. They conclude from this that the observed negative differential resistance results from real-space transfer into the AlGaAs barrier layers. Transport studies were previously used to investigate real space transfer effects in quantum wells [80,81] and in devices based on special structures [82].

V.C.2.3 Electron-Hole Interactions

One of the outstanding problems in semiconductor physics is the determination of electron-electron and electron-hole scattering rates. In the minority carrier experiment discussed [57] above, Hopfel et al [83] have used optical spectroscopy to determine electron-hole scattering rate in quantum wells. From measurements of I-V characteristics and from arguments about hole -lattice coupling and energy exchange rates, it was shown that the *electrons and holes were at different effective temperatures at high electric fields*. Their measured electron temperatures and estimated hole temperatures are shown in Fig. 9 as a function of the applied electric field. When electron temperature is higher than the hole temperature, the electrons lose energy not only to the lattice but also to the holes. The experimental results and analysis [83] showed that the rate of energy loss from electrons to the lattice is

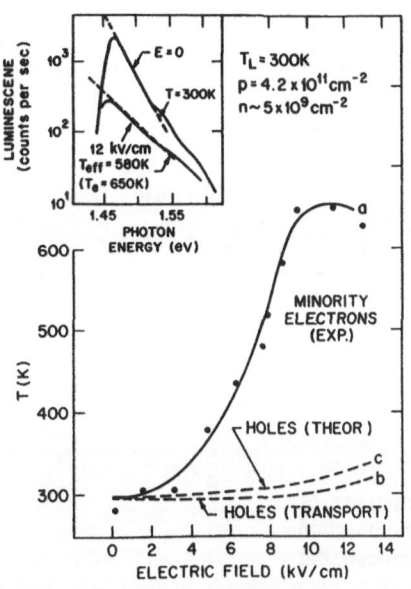

FIG. 9.   Electric field dependence of minority electron temperature (from luminescence measurements) and majority hole temperature (deduced from I-V measurements) in p-modulation-doped GaAs/AlGaAs quantum wells (from Hopfel et al [83]). The inset shows the luminescence spectra at two fields.

approximately the same as the rate from electrons to holes, under their experimental conditions. The electron-hole scattering rate was found to be approximately 10^{13} per sec..

V.C.3 Photoexcited Hot Carriers

In Section IV.C.2, we have discussed how optical spectroscopy has been used in recent years to obtain considerable information about carrier-phonon interaction, real space transfer and electron-hole scattering rates. All these experiments have investigated the steady-state properties of hot electrons under high field excitation conditions. In this section, we consider recent experiments on hot electrons in quasi-2D systems using optical excitation for generation of hot carriers and optical spectroscopy for detection of hot carriers.

V.C.3.1 General Concepts

Optical excitation and probing of hot carrier effects in semiconductors has been a very active field of research [4,5]. One reason for the appeal of the technique is its simplicity (no electrical contacts need to be made). However, one pays a price for this because quantitative interpretation of the data is hindered by the fact that the fraction of excess energy leading to the heating of the carriers is not known precisely.

Another reason for the appeal of the technique is that it allows the use of recently available ultrashort pulse lasers [7] to investigate relaxation of hot carriers on pico- and femtosecond timescales. Fig. 10 schematically illustrates what happens following femtosecond excitation of a semiconductor. The non-thermal carrier distribution initially created by the excitation pulse relaxes by carrier-carrier and carrier-phonon collisions. The carriers thermalize to a Fermi-Dirac type DF characterized by a temperature within a short time (typically less than 1 psec, determined by carrier density, excess energy, etc.). This hot plasma then relaxes to the lattice temperature by interacting with the phonons of the system. The cooling curve of this hot plasma as a function of time then provides information about $<ELR>$ and hence carrier-phonon interactions. This technique neatly bypasses the problem that the fraction of

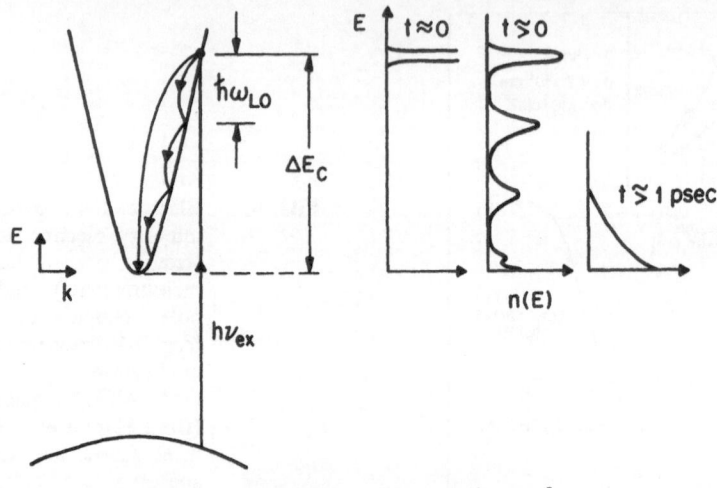

FIG. 10. Schematic representation of relaxation of photoexcited carriers after femtosecond photoexcitation.

excess energy leading to heating is not known. However, the interpretation is beset by other problems, arising from the fact that one is generally dealing with a degenerate, two-component plasma whose density is a function of time. This problem can be overcome by a careful design of experiments [84].

I will review in this section some of the many interesting experiments on quasi-2D systems performed during the last several years. Gobel et al [85] and Shah et al [86] have reviewed the subject recently.

V.C.3.2 Carrier-Carrier Interactions

Most of the studies in bulk as well as quasi-2D systems have been performed in the picosecond time range and have therefore probed the carrier-phonon interactions. However, very recently, Knox et al [87] have used the excite and probe technique [88] to measure the time evolution of the transmission spectra in an undoped GaAs/AlGaAs multiple quantum well sample following femtosecond photoexcitation. The change in the transmission spectra at a fixed delay following the photoexcitation at three different photon energies are shown in Fig.11. The spectra strikingly show increase in transmission at energies close to the excitation energies. This is the first evidence of spectral hole burning in quasi-2D systems, resulting from the initial non- thermal distribution of photoexcited carriers. Similar results in bulk GaAs have recently been reported by Oudar et al [89]. Femtosecond orientational relaxation studies in bulk GaAs have also been reported by Oudar et al [90]. The time evolution of the differential transmission spectra (Fig. 12) shows that the non-thermal distribution thermalizes in a time of the order of 200 femtosecond. The use of femtosecond spectroscopy has thus provided the first direct determination of carrier-carrier scattering time in a semiconductor. There are also many other interesting aspects of the data shown in Fig.12, related to the nature of exciton screening in quasi-2D systems [87].

Tang and co-workers [91-93] have used the technique of "equal-pulse autocorrelation" to investigate intra-band carrier scattering processes in bulk GaAs and AlGaAs as well as in GaAs/AlGaAs MQW samples. Using two orthogonally polarized, equal energy pulses of sufficient intensity to be in the saturable absorption region of the semiconductor, they measure the time-averaged transmission flux as a function of the time delay between the pulses.

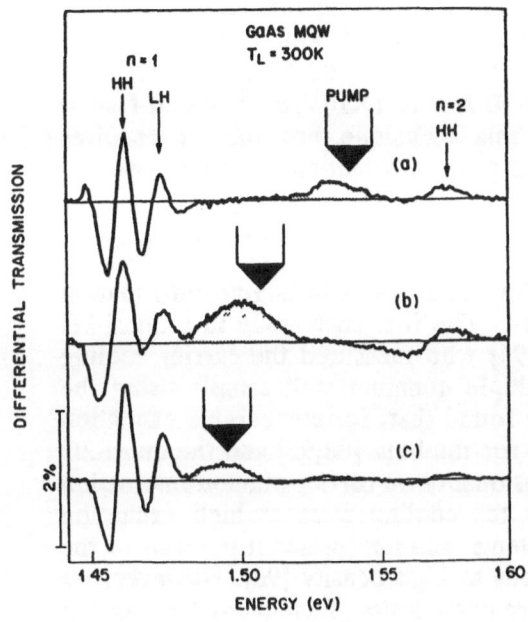

FIG. 11. Differential transmission spectra of an undoped GaAs/AlGaAs multiple quantum well sample for nominally zero delay between the excitation and the probe pulses for three different excitation photon energies (from Knox et al [87]).

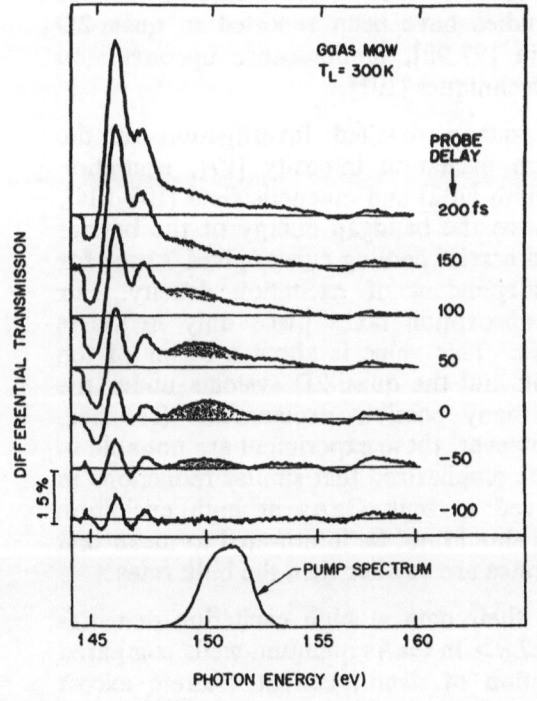

FIG. 12. Differential transmission spectra at different delays for the same sample as in Fig. 11 (from Knox et al [87]).

A "relaxation time" at the carrier energies corresponding to the laser photon energy is then obtained from the measured transmission correlation peak by first subtracting the contribution of the coherent artifact and then deconvolving the laser autocorrelation function. Using this technique, they have obtained a density independent relaxation time of 50 fsec for bulk GaAs for densities between 1 and $50 \times 10^{18} cm^{-3}$. For 150 Å thick quantum wells of GaAs, they obtain a relaxation time of 70 fsec at $1 \times 10^{18} cm^{-3}$ and 50 fsec at $5 \times 10^{19} cm^{-3}$. The disadvantages of this technique are that it requires extensive analysis of the data and that it gives information at a single energy, for rather high carrier densities.

V.C.3.3 Carrier-Phonon Interactions

We now turn to studies on picosecond timescales which give information primarily on carrier-phonon interactions. The first such study in a quasi-2D system was reported by Shank et al [94] who measured the carrier cooling curves in undoped GaAs/AlGaAs multiple quantum well sample using the excite and probe technique [88]. They found that, for comparable excitation density, the cooling rates are the same for the bulk [88,95] and the quasi-2D system; i.e. there is no effect of dimensionality on carrier-phonon interaction rates. They observed a reduction in the cooling rates at high excitation intensity in both bulk and quasi-2D systems, and interpreted it in terms of the screening of electron-phonon interactions at high density [96]. However, we know now that screening is a much more complicated phenomena (see section III.C.3) and this interpretation needs to be re-examined in light of more recent experiments discussed below.

Another technique for obtaining hot carrier cooling curves is by measuring time resolved luminescence spectra. This technique, while more difficult experimentally than the excite- and-probe techniques, offers the advantage that its larger dynamic range allows a more accurate determination of carrier temperatures. A number of such studies have been reported in quasi-2D systems recently, using streak camera [97,98], luminescence upconversion [99,100], and nonlinear luminescence techniques [101].

Ryan and coworkers have carried out a detailed investigation of the dependence of the cooling curves on excitation intensity [97], excitation photon energy [102], quantum well width [102] and magnetic field [102-104]. They reported that, for excitation above the bandgap energy of the barrier AlGaAs layers, a fit to their measured carrier cooling curves gives 7 psec for the coupling parameter τ_0 (11), independent of excitation density. For excitation photon energy such that absorption takes place only in GaAs quantum wells, τ_0 reduces to 1.5 psec. This value is about a factor of ten larger than that expected for the bulk and the quasi-2D systems under the simplest approximations. There are many possible explanations (see Sec. III.C) for this reduced cooling rate; however, these experiment are not able to distinguish between them. It should be emphasized that similar reductions in cooling rates have also been observed in bulk GaAs at high excitation densities [95], so that this observation should not be interpreted to mean that quasi-2D electron-phonon interaction rates are smaller than the bulk rates.

Xu and Tang [101] have interpreted their data at high excitation densities $(4 \times 10^{18} cm^{-3})$ in terms of reduced $<ELR>$ in GaAs quantum wells compared to bulk GaAs. However, interpretation of data taken at discrete excess energies is complicated by many factors, such as carrier lifetimes, fermi energies etc. and it is very difficult to extract information about carrier-phonon interactions from these measurements.

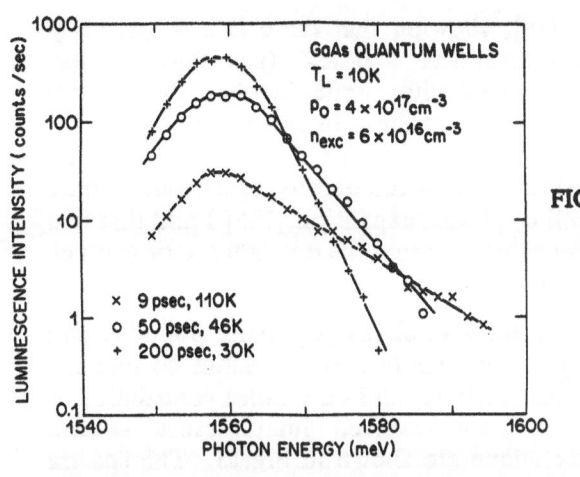

FIG. 13. Luminescence spectra from p-modulation-doped GaAs/AlGaAs multiple quantum well sample at three diffeent delays (from Kash et al [84]).

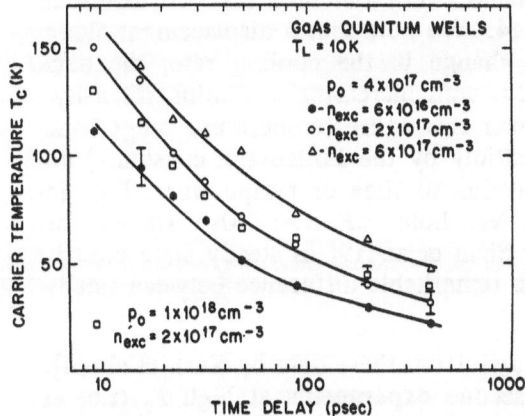

FIG. 14. Cooling curves for photoexcited holes in p-modulation-doped GaAs/AlGaAs multiple quantum well samples at different excitation intensities; from picosecond luminescence measurements (from Kash et al [84]).

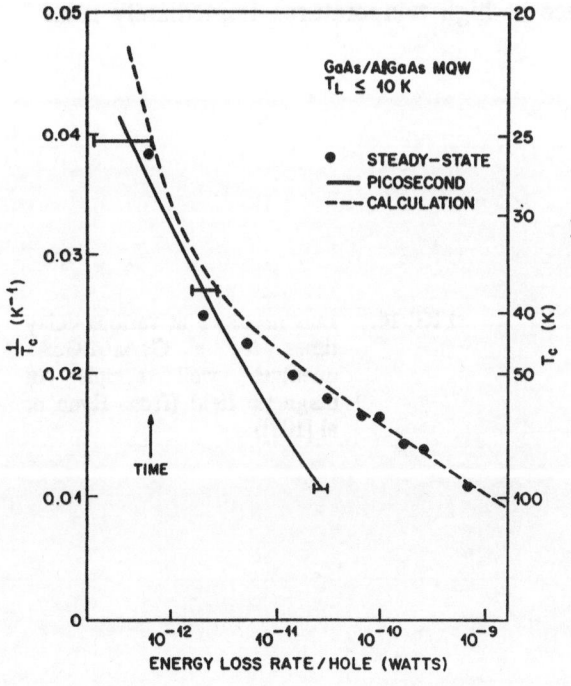

FIG. 15. Comparison of the inverse of the hole temperature on the average hole energy loss rate <ELR> for steady-state and picosecond excitation conditions (from Shah et al [19] and Kash et al [84]).

The measurements by Shah et al [19], showing that there is a surprisingly large difference between electron and hole $<ELR>$ (see Fig. 7), has profound implications for all photoexcitation experiments. Since both electrons and holes are present in equal numbers in most photoexcitation experiments, what one obtains is an effective $<ELR>$ for this system. The analysis is further complicated by the fact that the density of carriers is time dependent (because of decay as well as plasma expansion [105]) and that one or both types of carriers may be degenerate so that the dependence of average energy on carrier temperature is complicated.

These difficulties were overcome by Kash et al [84] by using p-modulation doped quantum wells and exciting at sufficiently low intensities so that the hole density is essentially independent of time, and only holes contributed to the energy loss processes. Typical time-resolved luminescence spectra obtained by using upconversion technique are shown in Fig.13. The spectra show the expected kink at the chemical potential as well as the exponential high energy tail. The hole cooling curves they obtained for two different excitation densities are shown in Fig.14. We note that a displacement along the logarithmic time axis implies a change in the cooling rate; the data therefore show a reduction in $<ELR>$ with increasing excitation intensity. Since the holes are non-degenerate over this entire temperature range, one can differentiate these curves and multiply by the Boltzmann constant k to directly obtain hole $<ELR>$ as a function of time or temperature. Fig. 15 shows their results plotted as $1/T_c$ vs. hole $<ELR>$. Also shown for comparison are the data obtained by Shah et al [19] in steady state electric field heating experiments. One sees a remarkable difference between steady state and picosecond results.

Several definitive conclusions were drawn from these data by Kash et al [84]. The hole $<ELR>$ measured in picosecond experiments at high T_c (i.e. at early times) is much smaller than the steady state $<ELR>$ at the same temperature, but the two rates become the same at lower temperatures or longer times. The large difference at high temperatures immediately rules

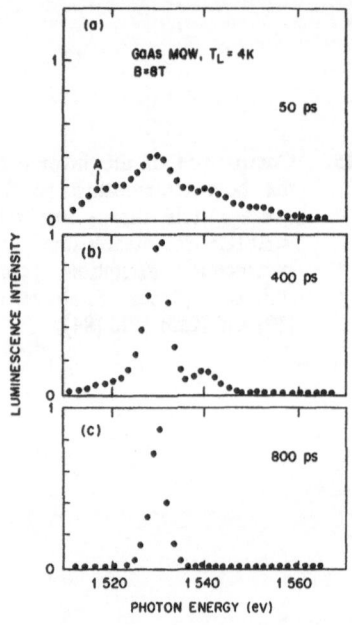

FIG. 16. Luminescence at various delay times for a GaAs/AlGaAs quantum well sample in magnetic field (from Ryan et al [103]).

out screening as the cause for the reduction in picosecond $<ELR>$ because a practical upper limit to screening, obtained by completely ignoring polar interactions, is a factor of two reduction in $<ELR>$. The fact that the steady state and picosecond $<ELR>$ become the same at low temperatures also rules out reduced dimensionality as a factor in reducing the $<ELR>$ at short times and gives further arguments against screening. Additional arguments against screening are obtained from the observed dependence of the cooling rates on excitation intensity but no dependence on doping density. All these observations point to a dynamic effect as the main cause for the reduced $<ELR>$ at short times and the authors argue that their results are qualitatively consistent with the hot phonon model. Calculation of time dependent hot phonon effects in quasi-2d system and a comparison of calculation and experiments will be very interesting.

Another interesting question is the influence of magnetic field on $<ELR>$ and whether there are any magneto-phonon resonance when energy differences between Landau levels equal the optical phonon energy. This has been experimentally investigated by Ryan et al [102-104] and Hollering et al [100]. The spectra obtained by Ryan et al [102] from a 150 Å GaAs/AlGaAs MQW sample at 8T are shown in Fig.16 for three different time delays. The carriers occupy several Landau levels at 50 psec, only two levels at 150 psec and only the lowest landau level at 800 psec. From an analysis of the data, these authors conclude that the cooling of plasma slows down with increasing magnetic field, is the slowest at about 9T and then becomes more rapid with further increase in the field. Above 14 T, Ryan et al [102] as well as Hollering et al [100] find the cooling rate to be faster than in the absence of magnetic field. The analysis of the data is quite complicated and no quantitative theory to explain these results is available at present.

VI. SUMMARY

We have reviewed hot carrier phenomena in quasi-2D systems such as quantum wells and heterostructures in polar semiconductors. Since carrier-phonon interactions play an important role in hot carrier physics, we have first reviewed electron-phonon interactions in quasi-2D systems. On the basis of current numerical calculations, we conclude that the strength of carrier-phonon interaction in quasi-2D systems is expected to be approximately the same as that in 3D semiconductors for typical well-widths. We have then considered the average energy loss rates to the lattice and how they are affected by degeneracy, slab modes, plasma effects and hot phonons. While a qualitative understanding of these phenomena is available at present, more detailed calculations and a deeper understanding of these phenomena in quasi-2D systems will be very valuable.

On the experimental side, there has been a great interest in investigating the properties of hot carriers in quasi-2D systems in the past few years. In particular, optical spectroscopic techniques have been used in conjunction with both electric field and optical excitation techniques. These experiments have directly determined carrier energy loss rates to the lattice, providing new insights into carrier-phonon interactions. In particular, unexpectedly large differences in electron and hole $<ELR>$ are observed. This and many other results indicate the importance of non-equilibrium optical phonons in carrier relaxation processes. An investigation of phonon dynamics on femtosecond timescales promises to be very interesting. First experimental results on carrier-carrier scattering rates have been obtained. Investigations of carrier-

carrier and carrier- phonon scattering rates and their dependences on electric and magnetic fields promises to be an interesting area of research.

We have concentrated exclusively on optical techniques for determining the carrier distribution function in quasi-2D polar semiconductors. Hayes et al [106] have recently developed a technique of hot electron spectroscopy which gives direct information on hot electron momentrum distribution function. Using this technique, Hayes et al [106], Levy et al [107] and Heiblum et al [108] have obtained evidence for ballistic transport of electrons through base regions whose widths approach the quantum confinements conditions. A review of this work is given by Hayes and Levy [109].

Acknowledgements

I wish to acknowledge the collaboration of A. Pinczuk, R. Hopfel, K. Kash, A. C. Gossard, W. H. Knox, C. Hirliman, D. A. B. Miller, D. S. Chemla and C. V. Shank in the work described here. I also acknowledge discussions about screening with S. Das Sarma.

REFERENCES

[1] E. M. Conwell "High Field Transport in Semiconductors" Solid State Physics, Supplement 9, edited by F. Seitz, D. Turnbull and H. Ehrenreich, Academic Press, New York 1967.

[2] "Physics of Nonlinear Transport in Semiconductors", edited by D. K. Ferry, J. R. Barker and C. Jacobani, Plenum Press, New York 1979.

[3] N. J. Shah, S. S. Pei, C. W. Tu and R. C. Tiberio, "Gate length dependence of the speed of SSI circuits using submicron selecctively doped heterostructure transistor technology", to be published in IEEE Trans. on Electron Devices, May 1986.

[4] Jagdeep Shah, "Hot electrons and phonons under high intensity photoexcitation of semiconductors" Solid-State Electronics 21, 43-50 (1978)

[5] Jagdeep Shah and R. F. Leheny, "Hot carriers in semiconductors probed by picosecond techniques" in "Semiconductors probed by ultrafast laser spectroscopy" edited by R. R. Alfano, Academic press, New York 1984; p.45- 75.

[6] Jagdeep Shah and R. C. C. Leite, "Radiative recombination from photoexcited hot carriers in GaAs" Phys. Rev. Letters 22, 1304-1307 (1969)

[7] See, for example, "Ultrafast Phenomena IV" Proceedings of the Fourth Int'l Conference, Monterey, California 1984, edited by D. H. Auston and K. B. Eisenthal, Springer-Verlag 1984

[8] K. Hess, H. Morkoc, H. Schichiijo and B. G. Streetman, "Real space transfer", Appl. Phys. Letters 35,469-471, (1979)

[9] M. Costato and L. Reggiani "Scattering probabilities for holes" Phys. Stat. Sol. (b) 58, 471-482 (1973) and Phys. Sta. Sol. (b) 58, 47-54 (1973)

[10] Y. C. Chang and J. N. Schulman, "Modification of optical properties of GaAs-AlGaAs superlattices due to band mixing", Appl. Phys. Lett. 43, 536-538 (1983)

[11] D. K. Ferry, "Scattering by polar optical phonons in a quasi-two-dimensional semiconductor", Surface Science 75, 86-91 (1978)

[12] K. Hess, "Impurity and phonon scattering in layered structures" Appl. Phys. Lett. 35, 484-486 (1979)

[13] P. J. Price, "Two-dimensional electron transport in semiconducting layers. I. Phonon scattering", Annals of Physics, 133, 217-239 (1981)

[14] B. K. Ridley,"The electron-phonon interaction in quasi-two-dimensional semiconductor quantum-well structures", J. Phys. C: Solid State Phys. 15, 5899-5917 (1982)

[15] F. A. Riddoch and B. K. Ridley, "On the scattering of electrons by polar optical phonons in quasi-2D quantum wells" J. Phys. C: Solid State Phys. 16, 6971-6982 (1983)

[16] J. P. Leburton, "Size effects on polar optical phonon scattering of 1-D and 2-D electron gas in synthetic semiconductors" J. Appl. Phys., 56, 2850-28 (1984)

[17] T. Ando, A. B. Fowler and F. Stern, "Electronic properties of two-dimensional systems", Rev. of Mod. Phys. 54, 437-672, (1982)

[18] S. Katayama, "Theory of energy relaxation of 2D hot carriers in GaAs quantum wells", to be published in Surface Science

[19] Jagdeep Shah, A. Pinczuk A. C. Gossard and W. Wiegmann, "Energy loss rates for hot electrons and holes in GaAs quantum wells" Phys. Rev. Lett. 54, 2045-2048 (1985)

[20] P. K. Basu and Sudakshina Kundu, "Energy loss of two- dimensional electron gas in GaAs-AlGaAs multiple quantum wells by screened electron-polar optic-phonon interaction", Appl. Phys. Lett. 47, 264266 (1985)

[21] R. Fuchs and K. L. Kliewer,"Optical modes of vibration in an ionic crystal slab" Phys. Rev. 140, A2076 (1965)

[22] J. E. Zucker, A. Pinczuk, D. S. Chemla, A. C. Gossard and W. Wiegmann, "Optical vibrational modes and electron-phonon interaction in GaAs quantum wells", Phys. Rev. Lett., 53, 1280-1282 (1985)

[23] N. Sawaki, "Interaction of two dimensional electrons and polar optical phonons in a superlattice", Surface Science, to be published.

[24] F. A. Riddoch and B. K. Ridley, "Electron scattering rates associated with the polar optical phonon interaction in a thin ionic slab" Physica 134B, 342 (1986)

[25] P. J. Price, "Two-dimensional electron transport in semiconductor layers II: screening", J. Vac. Sci. Technol. 19, 599-603, (1981

[26] C. H. Yang and S. A. Lyon, "Dynamical screening of the electron-optical phonon interaction in two dimensions", Physica 134B, 309-313 (1985)

[27] X. L. Lei, "Dynamical screening and carrier mobility in GaAs-AlGaAs heterostructures", J. Phys C: Solid State Phys. 18, L593-L597 (1985)

[28] S. Das Sarma and B. A. Mason "Screening of polar interaction in quasi-two-dimensional semiconductor microstructures" Phys. Rev. B, 31, 5536 (1985)

[29] S. Das Sarma and and B. A. Mason, "Degeneracy and screening effects on hot electron relaxation in quantum heterostructures", Physica 134B, 301-304 (1985)

[30] J. Collet, A. Cornet, M. Pugnet and T. Amand, "Cooling of high density electron-hole plasma", Solid State Commun. 42, 883-887 (1982)

[31] W. Potz and P. Kocevar, "Electronic power transfer in pulsed laser excitation of polar semiconductors", Phys. Rev. B 28, 7040-7047 (1983)

[32] P. Kocevar, "Hot phonon dynamics" Physica 134B, 155-163 (1985)

[33] Jagdeep Shah, R. C. C. Leite and J. F. Scott, "Photoexcited hot LO phonons in GaAs", Solid State Commun. 8, 1089-1093 (1970)

[34] D. von der Linde, J. Kuhl and H. Klingenburg, "Raman scattering of nonequilibrium LO phonons with picosecond resolution ", Phys. Rev. Lett. 44, 1505-1508 (1980)

[35] C. L. Collins and P. Y. Yu, "Generation of nonequilibrium optical phonons in GaAs and their application in studying intervalley electron-phonon scattering", Phys. Rev. B 30, 4501-4515 (1984)

[36] J. A. Kash, J. C. Tsang and J. M. Hvam, "Subpicosecond time resolved Raman spectroscopy of LO phonons in GaAs" Phys. Rev. Lett. 54, 2151-2154 (1985)

[37] P. J. Price, "Hot phonon effects in heterolayers", Physica 134B, 164-168 (1985); see also Superlattices and Microstructures 1, 255-257 (1985)

[38] Jagdeep Shah, A. Pinczuk, A. C. Gossard and W. Wiegmann, "Hot carrier energy loss rates in GaAs quantum wells: large differences between electrons and holes" Physica 134B, 174- 178 (1985)

[39] R. Dingle, H. Stormer, A. C. Gossard and W. Wiegmann, "Electron mobilities in modulation-doped semiconductor heterojunction superlattices", Appl. Phys. Lett. 33, 665-667 (1978)

[40] A. C. Gossard and A. Pinczuk "Modulation-doped semiconductors", in "Synthetic Modulated Structures", edited by L. Chang, Academic Press, New York (1985);p.215-255.

[41] J. C. M. Hwang, A. Kastalsky, H. L. Stormer and V. G. Keramidas, "Transport properties of selectively doped GaAs- AlGaAs heterostructures grown by molecular beam epitaxy", Appl. Phys. Lett. 44, 802-804 (1984)

[42] S. Hiyamizu, J. Saito,, K. Nanbu, and T. Ishikawa, "Improved electron mobility higher than $10^6 cm^2/Vs$ in selectively doped GaAs/N-AlGaAs heterostructures grown by MBE " Jpn. J. Appl. Phys. 22, L609-612 (1983)

[43] T. J. Drumond, M. Keever, W. Kopp, H. Morkoc, K. Hess, B. G. Streetman and A. Y. Cho, "Field dependence of mobility in Al0.2Ga0.8As/GaAs heterojunctions at very low fields", Electronics Lett. 17, 545-547 (1981)

[44] M. Keever, W. Kopp, T. J. Drummond, H. Morkoc and K. Hess "Current transport in modulation-doped (AlGaAs/GaAs) heterojunction structures at moderate field strengths" Jpn. J. Appl. Phys. 21, 1489-1495 (1982)

[45] Jagdeep Shah, A. Pinczuk, H. L. Stormer A. C. Gossard and W. Wiegmann, " Electric field induced heating of high mobility electrons in modulation-doped GaAs/AlGaAs heterostructures", Appl. Phys. Lett. 42, 55-57 (1983)

[46] Jagdeep Shah, A. Pinczuk, H. L. Stormer A. C. Gossard and W. Wiegmann, " Hot electrons in modulation-doped GaAs/AlGaAs heterostructures", Appl. Phys. Lett. 44, 322-325 (1984)

[47] M. Inoue, M. Inayama, S. Hiyamizu, and M. Inuishi, "Parallel electron transport and field effects of electron distribution in selectively-doped GaAs/n-AlGaAs ", Jpn. J. Appl. Phys. 22, L213-5 (1983)

[48] E. F. Schubert, K. Ploog, H. Dambkes and K. Heime, "Selectively doped $n-Al_xGa_{1-x}As/Ga/As$ heterostructures with high-mobility two-dimensional electron gas for field effect transistors", Appl. Phys. A 33, 1830193 (1984)

[49] M. Inoue, "Hot electron transport in quantum wells" Superlattices and Microstructures 1, 433 (1985)

[50] K. Inoue, H. Sakaki and J. Yoshino, "Field-dependent transport of electrons in selectively doped AlGaAs/GaAs/AlGaAs double-heterojunction systems", Appl. Phys. Letters 47, 614 (1985)

[51] W. T. Masselink, W. Kopp, T. Henderson, and H. Morkoc "Measurements of electron velocity-field characteristics of AlGaAs/GaAs modulation doped structures" IEEE Electron Device Letters,EDL-6, 539-541 (1985)

[52] W. T. Masselink, T. Henderson, J. Clem, W. Kopp and H. Morkoc "The dependence of the 77 K electron velocity-field characteristics on low field mobility in AlGaAs-GaAs modulation-doped structures", to be published

[53] H. L. Stormer, A. C. Gossard, W. Wiegmann, R. Blondel and K. Baldwin, "Temperature dependence of the mobility of two-dimensional hole systems in modulation-doped GaAs-(AlGa)As", Appl. Phys. Lett. 44, 139-141 (1984)

[54] W. I. Wang, E. E. Mendez and Frank Stern, "High mobility hole gas and valence-band offset in modulation-doped p- AlGaAs/GaAs heterojunctions" Appl. Phys. Lett. 45, 639-641 (1984)

[55] R. A. Kiehl and A. C. Gossard, "Complementary p-MODFET and n-HB MESFET (Al,Ga)As Transistors" IEEE Electron Device Lett. EDL-5, 521-523 (1984).

[56] Jagdeep Shah, A. Pinczuk, H. L. Stormer A. C. Gossard and W. Wiegmann, " Hole heating and hole-phonon interaction in modulation-doped 2D hole system ", Proceedings of the 17th Int'l Conference on the Phys. of Semiconductors, Edited by J. D. Chadi and W. A. Harrison, Springer-Verlag, New York 1985; p.345-348

[57] R. A. Hopfel, Jagdeep Shah, Dominique Block and A. C. Gossard, "Picosecond time-of-flight measurements of minority electrons in GaAs/AlGaAs quantum well structures ", Appl. Phys. Lett. 48, 148-150 (1986)

[58] L. Reggiani "Time of flight techniques" in "Physics of Nonlinear Transport in Semiconductors" edited by D. K. Ferry, J. R. Barker and C. Jacobani Plenum Press, New York 1980. p. 243

[59] J. G. Ruch and G. S. Kino, "Transport properties of GaAs" Phys. Rev. 174, 921 (1968)

[60] A. G. R. Evans, P. N. Robson and M. G. Stubbs, "New time-of-flight technique for measuring drift velocity in semiconductors" Electron. Letters 8, 195 -196 (1972)

[61] M. H. Evanno and J. L. Vaterkowski "Velocity/field characteristic measurement using a high-frequency modulated solid-state laser diode " Electron. Letters 18, 417 (1982)

[62] A. G. R. Evans and P. N. Robson, "Drift modibility measurements in thin epitaxial semiconductor layers using time-of-flight techniques" Solid State Electron. 17, 805-812 (1974)

[63] T. H. Windhorn, T. J. Roth, L. M. Zinkiewicz, O. L. Gaddy and G. E. Stillman "High field temperature dependent electron drift velocities in GaAs" Appl. Phys. Letters 40, 513 (1982)

[64] T. H. Windhorn, L. W. Cook and G. E. Stillman, " The electron velocity-field charactersitic for n-InGaAs at 300 K" IEEE Electron Device Letters, EDL-3, 18 (1982)

[65] J. A. Cooper, Jr. and D.F. Nelson, "High field drift velocity of electrons at Si-SiO2 interface as determined by a time-of-flight technique", J. Appl. Phys. 54, 1445-1456 (1983)

[66] D. F. Nelson and J. A. Cooper, Jr. and A. R. Tretola, "High-field drift velocity of holes in inversion layers on silicon" Applied Physics Lett. 41, 857-859 (1982)

[67] R. A. Hopfel, Jagdeep Shah, A. C. Gossard and W. Wiegmann, "Hot carrier drift velocities in GaAs/AlGaAs multiple quantum well structures" Physica 134B,509-513 (1985)

[68] M. Inoue, S. Hiyamizu, H. Hida, H. Hashimoto and Y. Inuishi, "Hot electron effect in a 2D electron gas at the GaAs/AlGaAs interface", J. de Physique C7, 19-24 (1981)

[69] H. Sakaki, K. Hirakawa, J. Yoshino, S. P. Svensson, Y. Sekiguchi, T. Hotta, S. Nishi and N. Miura, "Effects of electron heating on the two-dimensional magnetotransport in AlGaAs/GaAs heterostructures", Surface Science 142, 306-313 (1984)

[70] R. Ulbrich, "Low density photoexcitation phenomena in semiconductors: aspects of theory and experiments", Solid- State Electronics 21, 51-59 (1977)

[71] Jagdeep Shah, "Investigation of hot carrier relaxation with picosecond laser pulses", J. de Physique C7, 445-462 (1981)

[72] G. Bauer, "Determination of electron temperatures and of hot electron distribution functions in semiconductors" in Springer Tracts in Modern Physics 74,Edited by G. Hohler, Springer-Verlag, Berlin 1974; p.1-106

[73] Jagdeep Shah and J. C. V. Matos, "Raman scattering from photoexcited nonequilibrium excitations in semiconductors" in Proc. of the Third Int'l Conference on Light Scattering in Solids, Campinas, Brazil 1975, edited by M. Balkanski, R. C. C. Leite and S. P. S. Porto, Flammarion, Paris 1976; p.145-159.

[74] R. A. Hopfel and G. Weimann, "Electron heating and free carrier absorption in GaAs/AlGaAs single heterostructures" Appl. Phys. Lett. 46, 291-294 , (1985)

[75] P. D. Southgate, D. S. Hall and A. B. Dreeben, "Hot electron distribution in n-GaAs derived from photoluminesence measurements with applied electric field ", J. Appl. Phys. 42, 2868-2874 (1971)

[76] N. Takenaka, M. Inoue J. Shirafuji and Y. Inuishi, "Non- Maxwellian electron distribution function in n-GaAs determined from electric-field-dependent photoluminescence spectrum" J. of the Phys. Soc. of Japan 45,1630-1637 (1978)

[77] A. Pinczuk, Jagdeep Shah, R. C. Miller, A. C. Gossard and W. Wiegmann,"Optical processes of 2D electron plasma in GaAs-AlGaAs heterostructures" Solid State Commun. 50 735-738 (1984)

[78] C. H. Yang, Jean M. Carson-Swindle, S. A. Lyon and J. M. Worlock, "Hot electron relaxation in GaAs quantum wells" Phys. Rev. Lett. 55, 2359-2361 (1985)

[79] M. J. Chou, D. C. Tsui and G. Weimann, "Negative photoconductivity of two-dimensional holes in GaAs/AlGaAs heterojunctions" Appl. Phys. Lett. 47, 609-611 (1985)

[80] M. Keever, H. Shichiijo, K. Hess, S. Banerjee, L. Witkowski, H. Morkoc, and B. G. Streetman, " Measurements of hot electron conduction and real-space transfer in GaAs-AlGaAs heterojunction layers " Appl. Phys. Lett. 38, 36 (1981)

[81] M. Keever, K. Hess and M. Ludowise "Fast switching and storage in GaAs-AlGaAs heterojunction layers" IEEE Electron Device Letter EDL-3, 297-300 (1982).

[82] S. Luryi and A. Kastalsky, " Hot electron transport in heterostructure devices" Physica 134B, 453-465 (1985)

[83] Ralph A. Hopfel, Jagdeep Shah and Arthur C. Gossard, "Nonequilibrium electron-hole plasma in GaAs quantum wells" Phys. Rev. Letters 56, 765 (1986)

[84] Kathleen Kash, Jagdeep Shah, Dominique Block, A. C. Gossard and W. Wiegmann, "Picosecond luminescence measurements of hot carrier relaxation in III-V semiconductors using sum- frequency generation" Physica 134B, 189-198 (1985)

[85] Ernst Gobel, Reiner Hoger and Jurgen Kuhl, "Carrier dynamics in quantum well structures" J. de Physique, in press

[86] Jagdeep Shah, A. Pinczuk, A. C. Gossard, W. Wiegmann and K. Kash, "Steady state and picosecond investigation of hot carrier-phonon interactions in 2D systems", Proceedings of the Second Int'l;l conference on Modulated Semiconductor Structures, Kyoto, Japan 1985; to be published in Surface Science.

[87] W. H. Knox, C. Hirliman, D. A. B. Miller, Jagdeep Shah, D. S. Chemla and C. V. Shank, "Femtosecond excitation of non- thermal carrier populations in GaAs quantum wells ", to be published

[88] C. V. Shank, R. L. Fork, R. F. Leheny and Jagdeep Shah, "Dynamics of photoexcited GaAs band-edge absorption with subpicosecond resolution " Phys. Rev. Lett. 42, 112-115 (1979)

[89] J. L. Oudar, D. Hulin, A. Migus, A. Antonetti and F. Alexandre, "Subpicosecond spectral hole burning due to nonthermalized photoexcited carriers in GaAs", Phys. Rev. Lett. 55, 2074-2077 (1985)

[90] J. L. Oudar, A. Migus, D. Hulin, G. Grillon, J. Etchepare and A. Antonetti, "Femtosecond orientational relaxation of photoexcited carriers in GaAs" Phys. Rev. Lett. 53, 384-387 (1984)

[91] C. L. Tang and D. J. Erskine, "Femtosecond relaxation of photoexcited nonequilibrium carriers in AlGaAs" Phys. Rev. Lett. 51 ,840-843 (1983)

[92] D. J. Erskine, A. J. Taylor and C. L. Tang, "Femtosecond studies of intraband relaxaton in GaAs, AlGaAs and GaAs/AlGaAs multiple quantum well structures" Appl. Phys. Letters 45, 54-56 (1984)

[93] A. J. Taylor, D. J. Erskine and C. L. Tang, "Ultrafast relaxation dynamics of photoexcited carriers in GaAs and related compounds" J. Opt. Soc. Am. B, 2, 663-673, (1985)

[94] C. V. Shank, R. L. Fork, R. Yen, Jagdeep Shah, B. I. Greene, A. C. Gossard and C. Weisbuch, "Picosecond dynamics of hot carrier relaxation in highly excited multi-quantum well structures" Solid State Commun. 47, 981-983 (1983)

[95] R. F. Leheny, Jagdeep Shah, R. L. Fork and C. V. Shank, "Dynamics of hot carrier cooling in photoexcited GaAs" Solid State Commun. 31, 809-813 (1979)

[96] Ellen J. Yoffa, "Screening of hot-carrier relaxation in highly photoexcited semiconductors" Phys. Rev. B 23, 1909- 1919 (1981)

[97] J. F. Ryan, R. A. Taylor, A. J. Turberfield, A. Maciel, J. M. Worlock, A. C. Gossard and W. Wiegmann, "Time resolved photoluminescence of two-dimensional hot electrons in GaAs- AlGaAs heterostructures " Phys. Rev. Lett. 53, 1841-1844 (1984)

[98] H. Uchiki, Y. Arakawa, H. Sakaki and T. Kobayashi, "Hot photoluminescence of GaAs-AlGaAs multiple quantum well structures under high excitation by a single shot of 30 ps, 532 nm laser" Solid State Commun. 55, 311-315 (1985)

[99] Kathleen Kash, Jagdeep Shah, A. C. Gossard and W. Wiegmann, "Picosecond studies of hole-phonon interactions in GaAs quantum wells", to be published.

[100] R. W. J. Hollering, T. T. J. M. Berendschot, H. J. A. Bluyssen, P. Wyder, M. R. Leys and J. Wolter, " Effect of a strong magnetic field on the relaxation of a hot 2D electron-hole plasma studied with picosecond luminescence" Physica 14B, 422-425 (1985)

[101] Z. Y. Xu and C. L. Tang, "Picosecond relaxation of hot electrons in highly photoexcited bulk GaAs and GaAs-AlGaAs multiple quantum wells", Appl. Phys. Letters 44, 692-694 (1984)

[102] J. F. Ryan, R. A. Taylor, A. J. Turberfield and J. M. Worlock, "Time resolved luminescence from hot 2D carrier in GaAs/AlGaAs MQWs" Surface Science, to be published

[103] J. F. Ryan, R. A. Taylor, A. J. Turberfield and J. M. Worlock, "Picosecond photoluminescence measurements of Landau level lifetimes and time dependent Landau level linebroadening in modulation-doped GaAs/AlGaAs multiple quantum wells" Physica 134B, 318-322 (1985)

[104] J. F. Ryan , "Time resolved photoluminescence for quantum well semiconductor heterostructures" Physica 134B, 403-411 (1985)

[105] M. Combescot, "Hydrodynamics of an electron-hole plasma created by a pulse" Solid State Commun. 30, 81-86 (1979)

[106] J. R. Hayes, A. F. Levi, A. C. Gossard and W. Wiegmann, "Hot electron spectroscopy", presented at the APS March meeting, Baltimore, Md. 1985; Bull. Am. Phys. Soc. 30, 544 (1985)

[107] A. F. J. Levi, J. R. Hayes, P. M. Platzman and W. Wiegmann, "Injected-hot-electron transport in GaAs", Phy. Rev. Lett. 55, 2071 (1986)

[108] M. Heiblum, M. I. Nathan, D. C. Thomas and C. M. Knoedler, "Direct observation of ballistic transport in GaAs", Phys. Rev. Lett. 55, 2200 (1985)

[109] J. R. Hayes and A. F. J. Levi "Dynamics of extreme nonequilbrium transport in GaAs", IEEE J. of Quantum Electronics, Speical Issue September 1986.

ANOMALOUS CURRENT OSCILLATIONS IN SEMICONDUCTOR-INSULATOR-

SEMICONDUCTOR STRUCTURES AND RELATED DEVICES

J. P. Leburton

Coordinated Science Laboratory and
Department of Electrical and Computer Engineering
University of Illinois
Urbana, Illinois 61801

I. INTRODUCTION AND HISTORICAL BACKGROUND

During the last few years, the study of two-dimensional (2-D) systems has been intimately related to the physics of semiconductor heterojunctions.[1] One of the many different experimental techniques used to investigate the electronic properties of 2-D electron gas is tunneling across the insulator of metal- insulator-semiconductor (M.I.S.) structures.[2] This technique has been used as a means to obtain information about the carriers in accumulation or inversion at the semiconductor-insulator interface.[3] Under certain circumstances, experiments reveal intriguing secondary effects which result from fundamental transport processes in the entire M.I.S. structure. This is the case for anomalous oscillations observed in the I-V characteristics of different kinds of tunnel heterojunctions. In an early experiment, Katayama and Komatsubara[4] reported oscillatory characteristics in the tunnel conductance of a M.I.S. structure fabricated from n-InSb with a carrier concentration of 2×10^{14} cm^{-3} at liquid helium temperature. The metal electrode used was Al and the oxide layer was about 30 Å thick. Under reverse bias, electrons tunnel from the metal into the semiconductor. The voltage period of the oscillations was 25 mV which apparently corresponds to the L.O. phonon energy of 24.4 meV in InSb. The oscillatory amplitudes decrease with temperature and in the presence of magnetic fields. These authors interpreted their results as a modulation of the tunnel conductance by L.O. phonon emission: the phonons emitted by hot carriers in InSb scatter electrons in the metal and increase the tunneling probability. In a later experiment on similar structures, Cavenett[5] identified up to 30 oscillations of conductance. However, he questioned the validity of the previous interpretation because he observed conductance *minima* at multiples of the L.O. phonon energy, not *maxima* as predicted by Katayama and Komatsubara. Moreover, he showed that the amplitude of the oscillations is dependent on the barrier thickness and on the carrier concentration in the semiconductor layer. He proposed a resistance model of the structure which consists of the semiconductor in series with the tunnel barrier. Conductance oscillations arise from the resistance modulation of the semiconducting layer due to the electron-phonon interaction and are enhanced in M.I.S. capacitors which maximize the semiconductor resistance with respect to the barrier resistance. Cavenett noticed the striking similarity between conductance oscillations in InSb junctions and oscillatory photoconductivity in single InSb crystals, but did not provide any microscopic explanation of the former effect.

Very recently, Hickmott et al.[6-8] investigated the electronic properties of carrier accumulation at the GaAs-AlGaAs interface of semiconductor-insulator-semiconductor (S.I.S.) capacitors in a high magnetic field and observed remarkable periodic oscillations in the I-V characteristics of the reverse biased tunnel structures. Although the presence of the magnetic field H is essential for the onset of the oscillations, the voltage period is independent of H and corresponds to the GaAs L.O. phonon energy $\hbar\omega_{L.O.}$. Another remarkable feature is that the oscillations persist over long voltage ranges in the I-V characteristics of the tunnel barrier. Because of the unresolved issue of the origin of these oscillations, the latter experiment has stimulated a revived interest in this intriguing phenomenon. Experimental efforts have been pursued by several groups[9-13] in similar S.I.S. or M.I.S. structures and evidence the same general feature, i.e., the persistence of the oscillations over multiple voltage periods $\hbar\omega_{LO}/e$. To date, several different interpretations have been published to account for this effect, but disagreement exists among the various theoretical models.

The purpose of this chapter is to review the latest experimental developments in this field and to discuss the different interpretations which have been proposed to explain the origin of the conductance oscillations in S.I.S. and similar structures.

II. EXPERIMENTS

There is a fundamental difference between the M.I.S. and S.I.S. structures considered in these experiments and the usual tunnel structures[14] which consist of an insulating tunnel barrier sandwiched between two metallic contacts (M.I.M.) or, similarly, between two heavily doped semiconductor contacts. The presence of a lightly doped semiconductor layer of high resistivity between the insulator layer and the substrate contact makes the S.I.S. structure asymmetric and consequently divides the voltage drop unequally between the tunnel insulator and the highly resistive semiconducting buffer layer.

The Hickmott Experiment

In the Hickmott experiment,[6-8] the S.I.S. structure is a n^-–GaAs–undoped $Al_xGa_{1-x}As$–n^+ –GaAs capacitor grown by MBE on a $<100>$ oriented Si-doped GaAs substrate. The n^-–GaAs layer is about $1\mu m$ thick with a carrier concentration of $1.4 \times 10^{15}/cm^3$, the $Al_xGa_{1-x}As$ layer is about 200 Å thick with an AlAs mole fraction $x \approx 0.4$ and the n^+–GaAs gate electrode is about 0.4 μm thick with a carrier concentration of $1.5 \times 10^{18}/cm^3$. The experiment is performed at T = 1.6K. The sample area A is $4.13 \times 10^{-4} cm^2$. The band diagram of the structure under reverse bias is shown in Fig. 1. Because of the continuity of the electric displacement across the GaAs-AlGaAs interface, a depletion region of positive charge N_D^+ is formed in the n^- -GaAs buffer layer which sustains the large electric field arising from the reverse gate voltage V_G. Consequently the voltage drop across the tunnel barrier is only a fraction of V_G. For this reason, the C-V curve of the asymmetric S.I.S. capacitor is similar to those obtained from conventional MOS structures (Fig. 2). Under forward bias, i.e., when the n^+–GaAs gate has a positive external bias relative to the substrate, electrons are accumulated at the AlGaAs–n^-–GaAs interface. The resulting capacitance of the S.I.S. structure which one might expect to be the barrier capacitance (given by $C_I = \epsilon_I A/d = 205pF$ where ϵ_I is the $Al_xGa_{1-x}As$ dielectric constant and d is the barrier thickness) actually yields a lower experimental value $C_I = 123$ pF, due to the quantum mechanical extension of the accumulation layer. According to Hickmott et al.,[6] the capacitance at zero bias is reduced to $C_I \approx 25$ pF because of fixed charges in the insulator which deplete the buffer layer and add a series capacitance to C_I. Applying reverse bias further decreases the total capacitance. However, since the maximum width of the depletion layer is limited to the thickness L of the buffer layer ($\approx 1\mu m$), the capacitance reaches a minimum value $C_{min} \approx 4.7$ pF. In the presence of high magnetic fields ($H \geqslant 6.4T$) the total capacitance is reduced to its lowest possible value. Hickmott et al. point out that the condition for electrons to freeze-out onto the shallow hydrogenic donors in GaAs is achieved for $H \geqslant 6.4T$ as given by the YKA theory[15] when the dimensionless parameter $\gamma = \hbar\omega_c/2Ry$ exceeds one. Here $\omega_c = eH/m$ is the cyclotron frequency and Ry is the Rydberg energy of the impurity. Consequently the entire n^--buffer layer acts as an insulator of thickness $L \approx 1\mu m$ yielding in the minimum capacitance C_{min} of the S.I.S. structure. Similar C-V characteristics are observed with magnetic fields parallel to the barrier.

Reverse bias I-V characteristics of electrons tunneling from the gate electrode through the $Al_xGa_{1-x}As$ insulators are shown in Fig. 3.a for magnetic fields perpendicular to the tunnel barrier. The current density decreases with H, probably because of the increasing barrier height due to freeze-out of the n^-GaAs layer. The periodic structure is apparent and its amplitude increases with H. In Fig. 3.b the current seems to follow a power law relation with respect to the gate voltage i.e., $I \propto V^m$ with $m \approx 2$ as pointed out by Hellman et al.[19] Notice, however, that deviations occur at low ($V_G < 0.1V$) and high ($V_G > 0.3V$) gate voltage and that m increases with H. The oscillations appear more distinctly on the plot of the logarithmic derivative of the current with respect to V_G (Fig. 4.a). The voltage spacing is 0.036 V and is thus independent of the magnetic field. Up to sixteen periods are distinguished at 10 and 14 T. Similar periodicities are observed for H parallel to the tunnel barrier over an even greater voltage range. Current density is more strongly reduced with the magnetic field in this configuration. The C-V curves at 10 kHz also reveal the periodic structure (Fig. 2.b). However the capacitance oscillations for 6

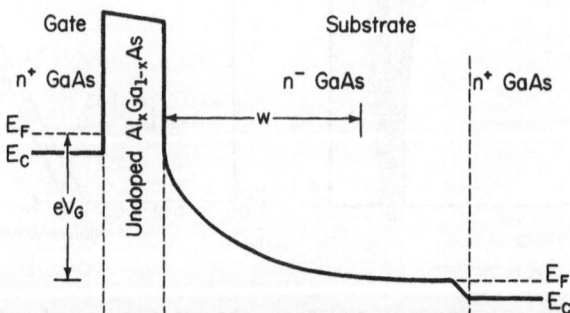

Fig. 1.   Schematic energy-band diagram of the n^+GaAs-$Al_xGa_{1-x}As$-n^-GaAs structure under reverse bias. E_c is the conduction-band edge, V_G is the gate voltage and W is the depletion width. (After Hickmott et al. ref. 6.)

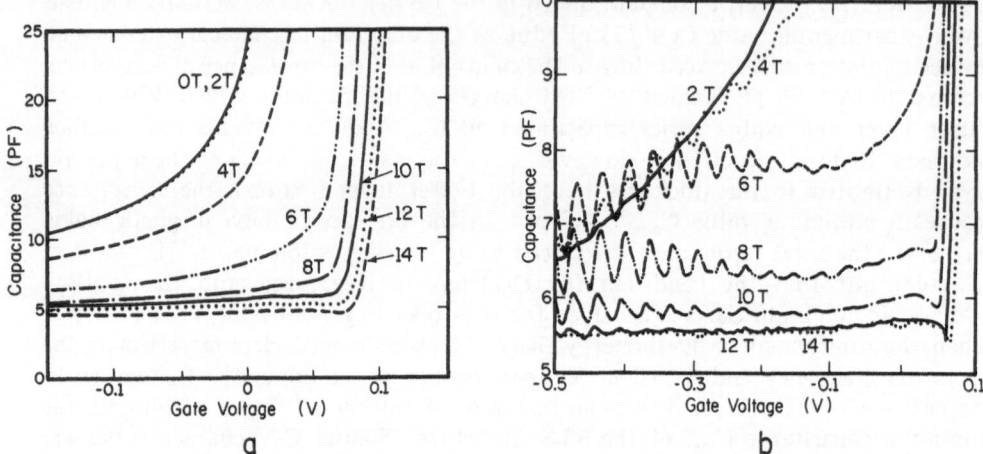

Fig. 2. Capacitance-voltage characteristics under reverse bias (depletion) for different magnetic fields. The insulator capacitance is 205pF. a) **Magnetic fields perpendicular to the barrier; the frequency is 100 KHz.** b) Magnetic fields parallel to the barrier; the frequency is 10 KHz. (After Hickmott el al. refs 6 and 7.)

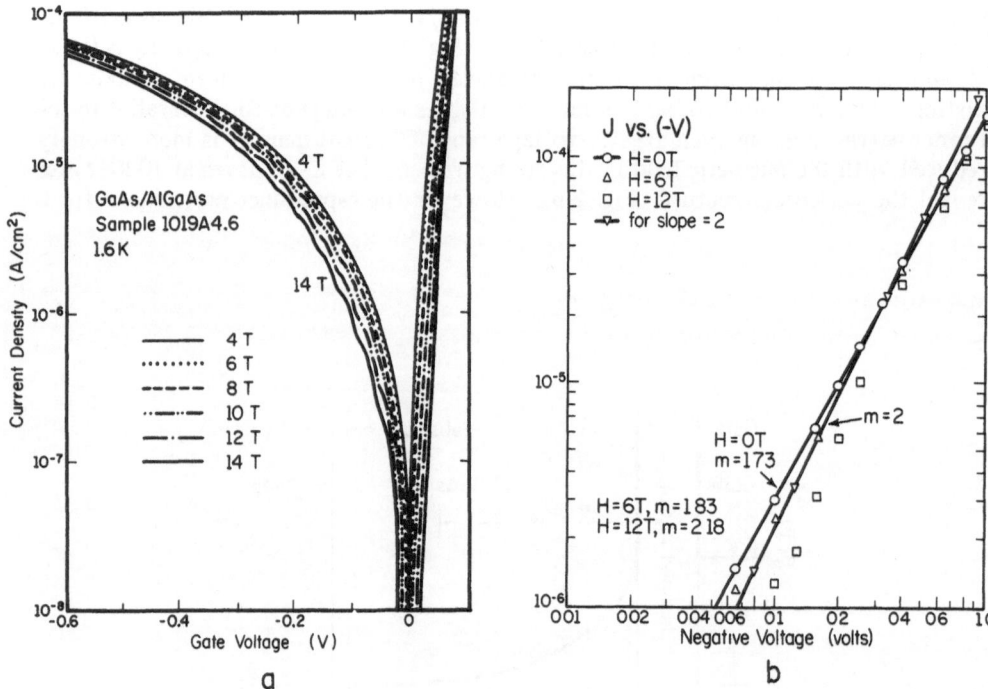

Fig. 3. a) Current density-voltage characteristics of the n^+GaAs-$Al_xGa_{1-x}As$-n^-GaAs capacitor for different magnetic fields perpendicular to the barrier (After Hickmott et al ref. 6) b) Log-Log plot of J-V curves for 3 different magnetic fields compared with the $J \propto V^2$ dependence.

and 8T begin at lower reserve bias than for higher magnetic fields. This seems to indicate that the magnitude of the current is related to the onset of the oscillations since the magnetic field affects the current density. Hickmott et al. propose that the structures in I-V and C-V curves arise because of the generation of L.O. phonons emitted by energetic electrons accelerated ballistically by the electric field in n^-—GaAs. According to these authors, the role of the magnetic field is to neutralize the positive donors in n^-—GaAs by magnetic freeze-out. Impurity scattering which is the major scattering mechanism influencing the mobility at low temperature is thereby suppressed, allowing ballistic transport. Therefore, in the absence of ionized impurity scattering, electrons are accelerated ballistically up to 36 meV, emit a L.O. phonon and are scattered back to the conduction band edge. The ballistic process is then repeated until the electrons reach the n^+—GaAs substrate. However this mechanism which attempts to relate the ballistic process in the n^-—GaAs to the current oscillations, is not plausible since the current is essentially controlled by the tunnel barrier.

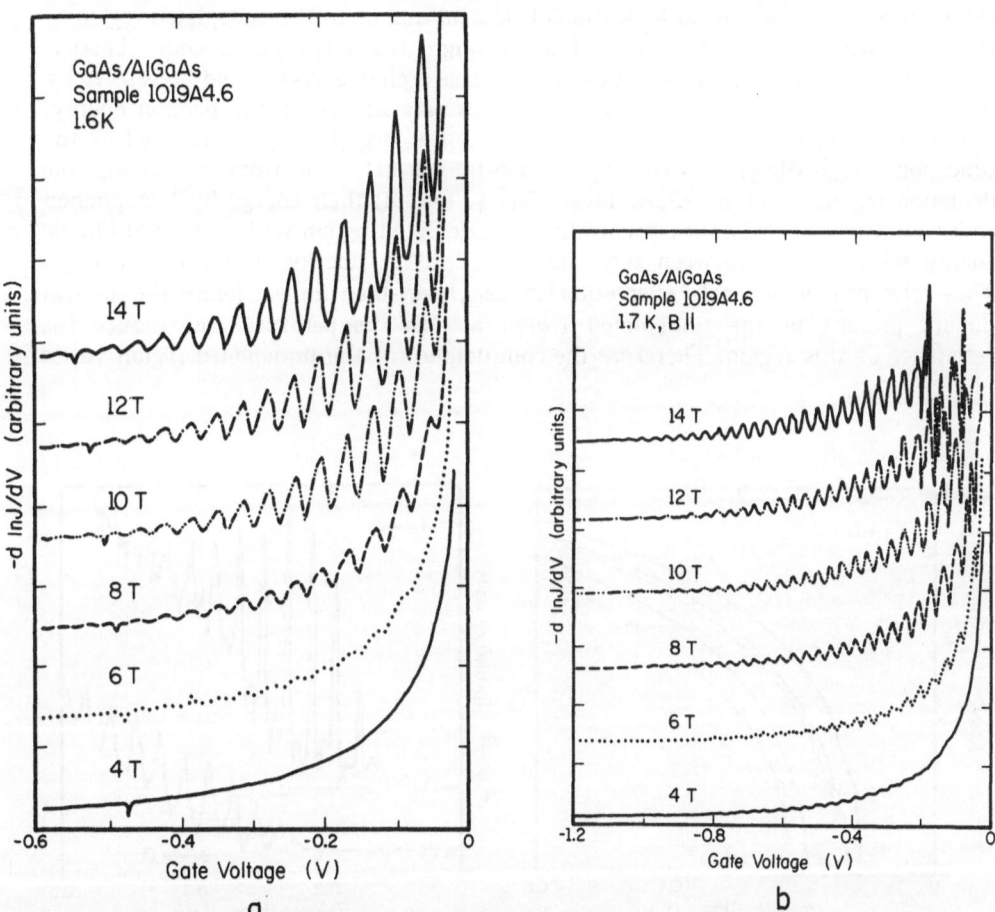

Fig. 4. Logarithm derivative of the current density versus voltage for different magnetic fields. The curves have been offset vertically for clarity. a) H perpendicular to barrier b) H parallel to the sample. (After Hickmott et al. refs. 6 and 7.)

Eaves et al.[9] have repeated the Hickmott experiment using a thinner $Al_xGa_{1-x}As$ tunnel barrier ($d \approx 170 \text{Å}$). As a consequence, the current density is several orders of magnitude larger than in the Hickmott case and depends exponentially on the barrier height and width, in agreement with the W.K.B. theory[16] (Fig. 5.a). The important finding is observation of the periodic structures at zero magnetic field. The relative amplitude of the oscillations, mostly observable in the derivatives of the current (dI/dV_G and d^2I/dV_G^2 versus V_G) is smaller than in the former case. Very sharp and narrow positive peaks in the d^2I/dV_G^2 -curves correspond to current minima (Fig. 5.b).[11] The periodic structures are visible up to 50 K. In addition, these authors observed enhancement of the amplitude and a narrower line width of some of the d^2I/dV_G^2 peaks at certain values of the magnetic fields[17] (Fig. 6). The findings lead these authors to question the conclusions reached by Hickmott et al.[6] that magnetic freeze-out is essential for the oscillatory structure to occur. Also, they refute the ballistic interpretation and argue that even with magnetic freeze-out, there is still an appreciable concentration of ionized impurities ($\geqslant 2N_A$) due to the presence of residual acceptors in the buffer layer. They extend the resistance model of Cavenett[5] and interpret the resistance variation as resulting from a mechanism of impact ionization of neutral donors in the undepleted region of the n^-—GaAs layer. After tunneling through the $Al_xGa_{1-x}As$—GaAs barrier, electrons gain energy from the field in the space charge region and release it by multiple phonon scattering. Owing to the discrete nature of the phonon energy, electrons can only emit M phonons where M is the largest integer permitted by the condition $eV_G \geqslant M\hbar\omega_{LO}$. According to Taylor et al,[10] electrons traversing the depletion region of the n^-—GaAs layer (Fig. 1) lose all their energy by L.O. phonon emission. Consequently they penetrate the undepleted region with a residual kinetic energy which varies between zero and $\hbar\omega_{L.O.}$, as a function of the gate voltage. Those electrons sufficiently energetic ($E \approx \hbar\omega_{LO}$) can then impact ionize the neutral donors present in the undepleted region at low temperature and reduce the resistance of this region. Therefore the conductance of the undepleted region varies

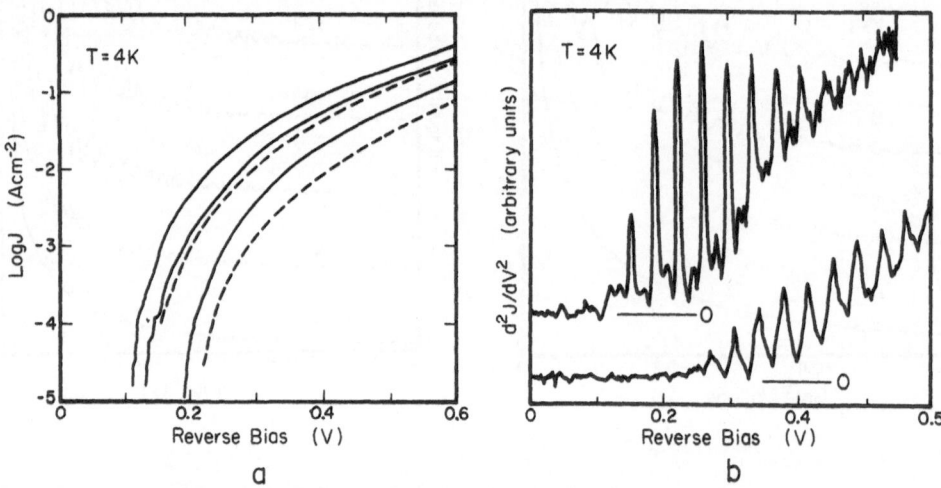

Fig. 5. a) Current density voltage characteristics for different magnetic field configurations in order of decreasing J :H=0, H=5.7T (H∥J), H=5.7T (H⊥J), H=11.4T (H∥J), H=11.4T (H⊥J). b) d^2J/dV^2-plot; upper curve H=0 lower curve H=11.4 T (H⊥J). (After Taylor et al. ref. 10.)

periodically as a function of gate voltage (with period $\hbar\omega_{LO}/e$) which determines the electron injection energy. By reference to experiments on inelastic scattering of high energy electrons by neutral hydrogen atoms,[18] Eaves et al. estimated the total cross section for impact ionization to be $\sigma \simeq 3\pi a_B^2$ where a_B is the Bohr radius of the impurity state. With a simple model, they calculated the number of carriers generated by the collision process to be $n \approx 10^{14}/cm^3$ out of a population of $N_D \approx 2 \times 10^{15}/cm^3$ neutral donors, which should be sufficient for producing the resistance variation observed experimentally. In Hickmott's experiment, owing to the low current density the resistivity of the buffer layer has to be much higher to produce an appreciable ohmic drop to modulate the tunnel current. This is achieved with the help of magnetic fields which increase the buffer layer resistance by several orders of magnitude. The disappearance of the oscillations for high gate bias and high temperature seems to confirm the donor ionization process since for $V_G \geqslant 0.8V$ the entire n^-—GaAs layer is depleted due to the space charge of the gate field, similarly for high temperature ($T \geqslant 50K$) all the donors are initially ionized. Narrowing of the peak in the presence of magnetic fields is attributed to a magneto impurity resonance (MIR)[19] which occurs when the ionization energy corresponds to electron transitions from the $p = 1$ to $p = 0$ Landau level minima where the joint density of states is maximum. Although the experimental data of Taylor et al.[10] are in good qualitative agreement with the series resistance model, the magnitude of the impact ionization mechanism has been seriously overestimated as we will see in the next section. Moreover, the argument invoking the MIR in favor of the impact ionization mechanism is invalid. First, the MIR effect, which has been previously observed in GaAs, is related to heating of free carriers accompanied by a de-excitation of the $n = 2$ to the $n = 1$ bound levels of neutral donors, i.e., the inverse process of that invoked in this case.[17] Secondly, there is no reason to limitate the

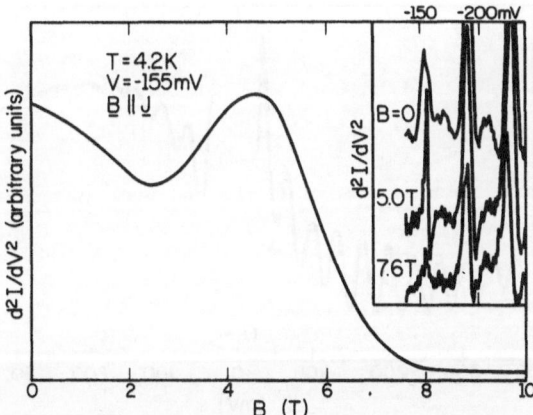

Fig. 6. Amplitude variation of the L.O. phonon peak at $V_G \simeq 155$ mV in d^2I/dV_G^2 curve at 4.2K. Inset shows the narrowing of the peak at H = 5T which is attributed to the M.I.R. effect, compared with the traces at H = 0 and 7.6T. (After Eaves et al. ref. 17.)

MIR to transitions between the p = 1 to p = 0 Landau levels since the resonance can also occur when higher Landau levels are populated, i.e., for transitions satisfying the relation $(p'-p)\hbar\omega_c = E_D(H)$ with $p'-p = 1$ and p, $p' \gg 1$. Because the occupation of the higher Landau levels is a function of the injection energy in the n^--GaAs undepleted layer, current oscillations should occur with the period $\Delta V_G = \hbar\omega_c/e$, which is not observed and thus rules out the MIR effect. Therefore the impact ionization mechanism seems unlikely and the broad maximum observed in Fig. 6 appears more like a trade-off between an enhancement of the d^2I/dV_G^2-peak caused by the magnetic field, and a disappearance of the peak due to decreasing current density with high magnetoresistance.

In a different kind of experiment, Lu et al.[12] investigated the electrical transport through metal semiconductor contacts and observed an oscillatory conductance in the I-V characteristic of In contacts to high mobility $In_{0.53}Ga_{0.47}As$ single crystals. The oscillations are observed for both bias polarities up to $\approx$ 60 K. Under negative bias, seventeen oscillation periods characterized by strong conductance variation are distinguished at 4.2 K (Fig. 7). Six weaker oscillations are observable with positive bias. The presence of a magnetic field perpendicular to the In-InGaAs interface has no effect on the oscillations, whereas for H parallel to the junctions, the oscillations vanish. The oscillations correspond to dips in the I-V

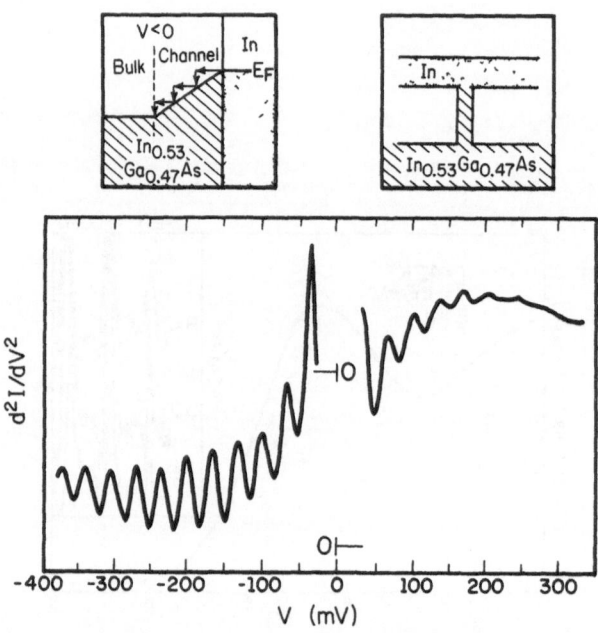

Fig. 7. d^2I/dV^2 versus V curve of an In-InGaAs contact under forward and reverse biases at 4.2K. The insets show the proposed structure of the contact with microchannels and the energy band diagrams with sequential phonon emission. (After Lu et al. ref. 12.)

characteristics in either polarity at biases equal to multiples of about 33 mV and are attributed to the emission of InGaAs L.O. phonons. Lu et al.[12] claimed that the electrical contact between the In and the InGaAs is via a large number of microchannels of InGaAs crystal protruding through an insulating oxide layer on the InGaAs surface. The channels are several hundreds of Angstroms long and few tens of angstroms wide which makes them one dimensional wires (Fig. 7). When an injected electron is at least one phonon energy above E_F, it can emit a phonon and cause a current oscillation for either bias polarity. The oscillations result from successive phonon emissions by electrons accelerated in the applied electric field through the micro-channels of InGaAs. This process occurs continuously for all biases above the phonon emission threshold $eV_G = \hbar\omega_{LO}$. However, when the emission occurs at the end of the channel, the electron suffers a sudden decrease velocity as it approaches the collecting electrode, causing a drop in the current flow through the channel. According to Lu et al. the observed asymmetry in the data reflects the asymmetry of devices where the collecting electrodes are the metal In for one polarity and bulk InGaAs for the other. The disappearance of the oscillations at T $\geqslant$ 55K is explained by increasing acoustic phonon scattering which damps out the oscillations. Moreover, the effect of a magnetic field parallel to the In-InGaAs is to bend the electron paths by the Lorentz force which smears out the oscillations, as observed experimentally.

In a more recent experiment, the same group made similar findings with rectifying (In or Sn) - InP contacts.[13] In this case three important features are observed in the I-V characteristics. First, the oscillations are seen only for bias polarity corresponding to injection of electrons into the bulk InP. Second, at low bias, a conductance gap which corresponds to carrier freeze out in InP disappears above 10K. Third, the oscillations persist above 80K although with reduced amplitude. The last two observations seem to contradict the model based on ionization of neutral donors. This lead Lu et al. to propose a point contact model similar to the theory of Vengurlekar and Inckson.[20] In this model the electrons injected through the micro-channels (or point contacts), relax to the band edge by L.O. phonons emission in the bulk semiconductor. By losing their kinetic energy, the electrons slow down, and produce a loss in the injected current contribution. In order to satisfy current continuity this current loss is compensated by the bulk electron in the semiconductor. This bulk current contribution J_b results from the fact that an electric field builds up in the bulk semiconductor due to the disturbance of the equilibrium distribution function caused by phonon emission of injected carriers. The corresponding voltage drop in the semiconductor $\delta V = \int \frac{j_b dz}{\sigma}$ where σ is the conductivity of the semiconductor feeds back to the point contact which controls the current and therefore results in a dip in the dI/dV_G vs V_G-curve. This model is consistent with high temperature observations for which the decrease of the oscillation amplitude is attributed to the onset of acoustic phonon scattering which competes with the process of L.O. phonon emission.

The fact that the oscillatory phenomenon is not observed in n^+-i-n^+ structures where the metal gate is replaced by a heavily doped semiconductor implies that the metal-semiconductor contact is essential for the effect to occur. Although the junction is different from the Hickmott tunnel structure, Lu et al. pointed out that the Cavenett requirement needs to be satisfied, i.e., that semiconductor material must be of high purity and high impedance in order to produce a voltage feedback to the metal-semiconductor contact.

Despite the similarities between observations of the two types of tunnel structures, by Hickmott and Lu et al., these are experimental differences at high temperature and in magnetic field effects.

Although the point contact model has the advantage of being consistent with the experimental data of Lu et al. it would be desirable to have a unified model which is sufficiently general to explain the oscillatory phenomenon in the two kinds of experimental structures. This is the intent of the different theories which are described in the next section.

III. THEORETICAL MODELS

In the course of investigating the origin of the conductance oscillations in S.I.S. or M.I.S. devices, a major difficulty arises because one deals with a composite structure which consists of a tunnel barrier on top of a lightly doped layer of semiconducting material. In this inhomogeneous system, it is not possible to isolate experimentally the contribution of each region of the structure to the overall current response or to the total voltage drop across the device. However the various physical variables, (current, electric field and potential) satisfy continuity conditions which make them interdependent in the different regions.

It is widely accepted that the voltage drop across the tunnel barrier is only a few percent of the total gate voltage and that the L.O. phonon emission takes place in the buffer layer of the S.I.S. structure because of the large potential variation experienced by the semi-insulating n^-–GaAs layer. The central issue is to identify the microscopic process which relates the L.O. phonon emission to the mechanism which modulates the current at the barrier injector. The general scenario leading to the current oscillations consists of three phases.

(1)   Electron tunneling and injection across the barrier.

(2)   Electron acceleration with L.O. phonon emission in the n^-–GaAs high field region.

(3)   Microscopic processes causing the current oscillations.

The tunnel barrier or the metal semiconductor contact in the Lu et al. case, operates only as a carrier injection mechanism. The two remaining phases are essential to understand the oscillatory effect. Transport in the high field buffer layer is important because it determines the energy distribution of carriers entering the undepleted region where conductance oscillations occur. Various models assume the existence of velocity or energy coherence, i.e., all the carriers travel the same distance before emitting a L.O. phonon, resulting in a significant drop of the drift velocity and of the average energy after the emission process. In order to test the validity of this concept, a Monte Carlo calculation has been carried out which simulates the transport of carriers in the high field region of the buffer layer of the S.I.S. device.[21]

Transport in the n^-–GaAs Buffer Layer (zero magnetic field case)

Individual paths of a large number (4000 in our case) of electrons are simulated using a code which incorporates the nonparabolicity of the band structure.[22] In the absence of magnetic freeze-out the n^--GaAs layer consists of a classical depleted space-charge region and an undepleted quasi neutral region extending up to the n^+ GaAs substrate. We have assumed that the potential V_b in the buffer layer has a classical profile as given by a quadratic expression as a function of the distance from the AlGaAs barrier (Fig. 1.a.), i.e., $V_b = eN_D W^2/2\epsilon_s$. Carriers are injected from the left side (AlGaAs barrier side) with zero velocity and advanced stepwise in time. After each time step, electron positions in real and momentum space are updated. By using this scheme, we keep track of the spatial dependence of scattering statistics and transport variables such as the drift velocity and average electron energy. In

Fig. 8 we have plotted the electron energy as a function of distance in the depletion region for different gate voltages. The curves have been terminated at the end of the depletion region. In the undepleted region, the electron motion is assumed to be a field-free relaxation. As can be seen, oscillatory structures which would reflect the simultaneous stop and go carrier motion caused by successive acceleration and the phonon emission processes, are almost entirely absent. This absence of coherence is mostly due to the strength of the electric field which destroys the "phase" among carriers by randomly distributing the individual scattering events. All the curves exhibit an overshoot which varies in position, magnitude and shape as a function of the gate voltage and the nonuniform electric field in the depletion region. Another important result provided by these numerical data is the value of the carrier energy at the edge of the depletion layer. Carriers penetrate the undepleted region of the GaAs-buffer layer with an energy far larger than the phonon energy $\hbar\omega_{LO}$.

Figure 9 shows the electron energy as a function of the gate voltage as the carriers penetrate into the undepleted portion of the n^--GaAs layer. Two ranges of voltage, the low bias $E_G \approx 0.1V$ and high bias $E_G \approx 0.4V$ regions have been investigated with a finer mesh size. The overall behavior of the carrier energy is a relatively smooth and monotonic function of the gate voltage. For low bias, the detailed calculation (see inset) shows a structure in the energy and velocity characteristics with a period corresponding to the LO phonon energy. The dips indicate a decrease of the velocity due to the onset of a new L.O. phonon emission sequence for low energy carriers. These carriers have already disposed of all their kinetic energy in the depletion layer, but have been reaccelerated up to an energy corresponding to $\hbar\omega_{LO}$. The dip amplitudes are less than 10 meV and 0.5×10^7 cm/s for energy and drift velocity respectively, corresponding to amplitude variations of $\approx 15\%$ and $\approx 10\%$. A close look at the carrier distribution function reveals δ-like peaks caused by the LO-phonon emission process. All the peaks in the energy distribution are separated by the well-defined phonon energy $\hbar\omega_{LO}$. The highest-energy peak is located at eV_G and corresponds to ballistic (collision free) carriers. Only a fraction of the carrier ensemble occupies the lowest level. Therefore by assuming that all the carriers penetrate the undepleted region with a residual kinetic energy on the lowest level, Eaves et al. considerably overestimate the magnitude of

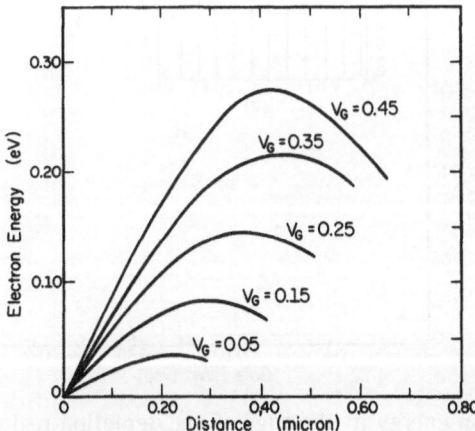

Fig. 8.   Electron energy versus distance in the depletion region of the n^--GaAs layer for different gate voltages. Curves have been terminated when carriers penetrate the undepleted layer. (After Wang et al. ref. 21.)

the impact ionization process since the impact ionization cross section strongly decrease for higher energy (Fig. 14.a).[17] At high gate voltage the electron energy decreases because carriers at the edge of the depletion region return to steady state value after the energy overshoot. The fine structure is barely noticeable. This last result clearly invalidates the concept of energy coherence.

According to the current conservation, the variation of the carrier concentration δn resulting from the velocity variation δv_d is given by

$$\delta n = -\frac{J}{e} \frac{\delta v_d}{v_d^2} \tag{1.a}$$

or

$$\delta n \approx 2J \times 10^{10}/cm^3 \tag{1.b}$$

where J is the current density in A/cm^2 and v_d is the drift velocity. If we assume that this excess charge exists all over the undepleted region of thickness D the

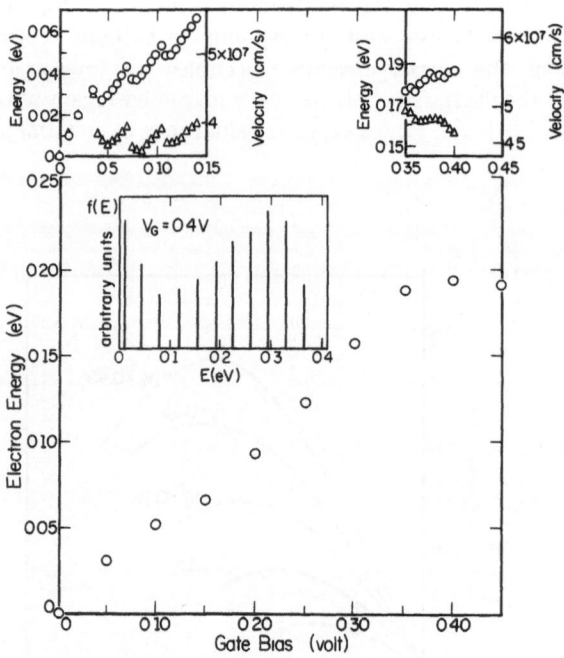

Fig. 9. Electron energy at the edge of the depletion region as a function of the gate voltage. Insets show fine analysis for two voltage ranges (0-0.15V) and (0.35-0.40V). O : Energy; Δ : Drift velocity. Also shown is the distribution function f(E) for $V_G = 0.4V$; the structure period corresponds to $\hbar\omega_{LO}$. (After Wang et al. ref. 21.)

resulting electrostatic potential variation is given by

$$\delta V = \frac{1}{2} \frac{e}{\epsilon_s} \delta n D^2 \approx 4 J \times 10^{-3} mV \qquad (2)$$

for $D \approx 0.5 \mu m$. In Hickmott's and Eaves' cases, the current density varies between 10^{-8} and $10^{-5} A/cm^2$ and between 10^{-5} and $10^{-1} A/cm^2$ respectively. In the experiment of Lu et al. the current density is larger, varying from 1 to about $30 A/cm^2$. Therefore the charge variation is really too small to produce a space charge effect resulting from electron accumulation in the buffer layer.

The Hellman–Harris–Hanna–Laughlin Model

Hellman et al. (H³L) proposed a model of electronic charge accumulation in the n^- - GaAs layer which feeds back to the tunnel barrier and modulates the current in the S.I.S. capacitor.[23] They claim that even in presence of strong magnetic fields, magnetic freeze-out cannot exist in the n^- - GaAs layer because the strength of the gate field ionizes the neutral donors and restores the classical depletion region. To demonstrate this, they assume that the high magnetic field current I_{14T}^{exp} under freeze-out of the buffer layer (i.e. under uniform electric field $E = V_{14T}/L$) should be equal to the zero magnetic field current I_{OT}^{th} at the same depletion field $E = \sqrt{2eN_D V_{OT}/\epsilon_s}$

$$I_{OT}^{th} = I_{14T}^{exp} \left[\sqrt{2eN_D V_{OT}/\epsilon_s} \right] \qquad (3)$$

if one assumes that tunneling is unaffected by the magnetic field. By noticing the disparity which exists between the theoretical value I_{OT}^{th} and the experimental data I_{OT}^{exp}, Hellman et al. conclude that magnetic freeze-out is absent in n^- - GaAs.

With current conservation, electron accumulation is not significant in the undepleted region if there is not a substantial velocity variation, as seen from Eq. 1.a. In order to make this feasible, the H³L theory considers the combined influence of two factors on the transport in the S.I.S. structure (a) inhomogeneous tunneling and (b) the velocity modulation due to magnetopolaron formation.

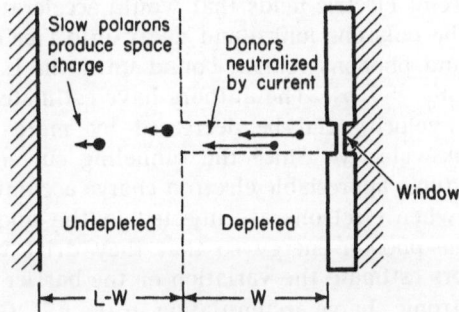

Fig. 10. "Window" model of inhomogeneous tunneling. Electrons tunnel through a high conductance region of the barrier and neutralize the ionized donors before penetrating the undepleted layer of the S.I.S. structures. (After Hellman et al. ref. 23.)

a) Inhomogeneous tunneling

Analyzing the dependence of the I-V characteristics on the magnetic field, Hellman et al. notice that the current voltage curves can be mapped onto one another by the transformation:

$$I \rightarrow \left[I^{1/2} - \beta \right]^2 \tag{4}$$

where $\beta \propto H^2$. This suggests the relation $I \propto E^2$ where E is the tunnel barrier field. Because the field E is proportional to $V_G^{1/2}$ in presence of a depletion space charge this would mean $I \propto V_G$. However, the experimental data indicate a relation $I \propto V_G^m$ where m $\approx$ 2 as seen from Fig. 3.b, with a current smaller than in classical depletion at low voltage. These authors propose that the lower electric field is obtained because of inhomogeneous tunneling across "windows" of high conductance. Current pouring through these windows neutralizes the donors and prevents depletion (Fig. 10). With a very simplified model, they estimate the area of the tunneling hole to be $\sim 1\ \mu m^2$. Therefore, with a total junction area of about $4 \times 10^{-4}\ cm^2$, they are able to obtain a local current density J which is more than 10^4 times larger than for uniform tunneling. This inhomogeneous tunnel process increases appreciably the electron accumulation in the undepleted layer, resulting from velocity variation caused by L.O. phonon emission.

b) Magnetopolaron

To explain the onset of current oscillations with high magnetic fields, the authors consider the formation of magnetopolarons.[24] Due to the electron-phonon interaction, the degenerate two-particle state consisting of 1 electron with energy E $\lesssim \hbar\omega_{LO}$ and a zero phonon state on one side, and 1 phonon and a zero energy electron on the other side, splits into 2 polaron states, with different energies. In the absence of a magnetic field, the energy splitting is negligible. However, the confinement of the quasi-particle to a one dimensional motion caused by the presence of longitudinal magnetic fields, increases the polaron binding energy which exhibits a dispersion relation shown in Fig. 11. These curves have been calculated with a variational method described by Larsen for three-dimensional phonons.[25] From these curves it can be seen that the polaron velocity given by

$$v_d(k) = \frac{1}{\hbar} \frac{\partial}{\partial k} E_{pol}(k) \tag{5}$$

reaches a maximum for $k \simeq k_{LO} = (2m\omega_{LO}/\hbar)^{1/2}$ and decreases strongly for higher wave vectors. Therefore the strong electron-phonon coupling slows the polaron motion in the presence of electric fields that would accelerate free electrons. Under high electric fields, the polarons ionize and decay into free electrons and phonons, otherwise electrons and phonons remain bound until the L.O. phonon decays into acoustic modes after $\tau_d \sim 7$ ps. The authors have estimated that during the L.O. phonon lifetime the velocity can be decreased by more than a factor of 100 compared to the peak velocity. Since the tunneling current is steady-state, this velocity reduction induces appreciable electron charge accumulation in the field free region. This occurs when electrons are injected in the undepleted layer with an energy just below the phonon energy so that they drift with very small group velocities. The authors estimate the variation of the barrier electric field E_T which results from the electronic charge accumulation in the n^- - GaAs region, to be given by

$$\delta E_T = -\frac{J}{2\epsilon_s v_d} \frac{(L-W)^2}{L} \tag{6}$$

where ϵ_s is the GaAs dielectric constant and W is the depletion width. Due to the J αE^2 dependence, this barrier field variation produces a drop in the current magnitude

$$\delta J \approx 2J \frac{\delta E_T}{E_T} \tag{7}$$

which then occurs at gate voltages a little lower than multiples of the L.O. phonon voltage $\hbar\omega_{LO}/e$. By using a 7 ps L.O. phonon decay time, the H^3L theory predicts a minimum phonon velocity $v_d \sim 4 \times 10^4$ cm/s for H = 14T. By combining the effects of the current density enhancement due to inhomogeneous tunneling and the velocity reduction caused by magnetopolaron formation, Hellman et al. obtain a relative barrier field variation $\delta E_T/E_T \approx 0.005$ at $V_G = 0.3V$ which is a little smaller than the experimental values ≈ 0.02 but still quite acceptable. In order to get a better agreement with the experimental data they assume a phonon lifetime of about 25 ps and are able to reproduce parts of the experimental I-V characteristics.

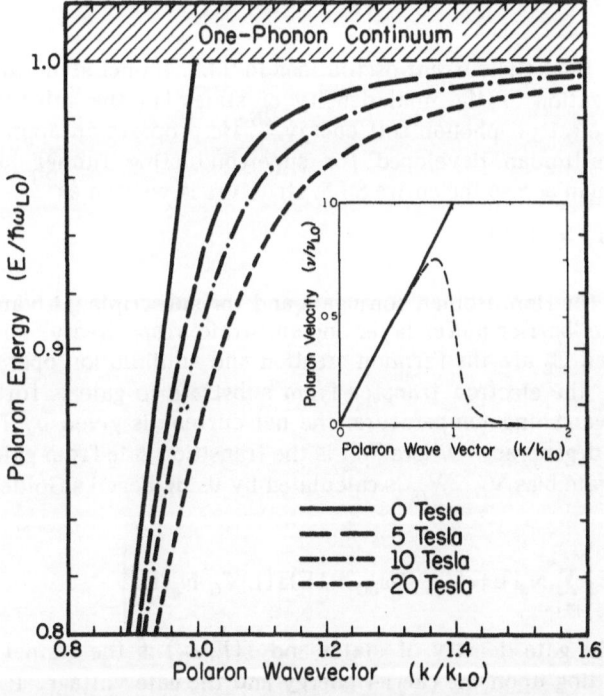

Fig. 11. Energy-momentum dispersion relation for polaron as a function of magnetic field. (low branch). The solid lines represent unperturbed zero and one phonon state. The dashed lines represent the dispersion relation of the magnetopolaron of lower energy at different magnetic fields. Inset shows the magnetopolaron drift velocity. (After Hellman et al. ref. 24.)

However, the H^3L model implicitly assumes velocity coherence since all the carriers penetrate the n^-—GaAs undepleted region with a monoenergetic distribution. We have seen in the previous section that this assumption is invalid because electrons have a velocity distribution which is spread up to the highest energies $eV_G >> \hbar\omega_{LO}$. Thus, only a fraction of the carrier population contributes to the magnetopolaron mechanism, which appreciably reduces the H^3L estimates. One can argue that the presence of the magnetic field enhances the L.O. phonon scattering, but one cannot expect a drastic change of the Monte Carlo results. Moreover, as shown by Peeters et al.,[27] collision broadening and spreading of the carrier mini-peaks in the distribution function, should also reduce the magnetopolaron effect and its influence on the charge accumulation mechanism. There is a second objection which concerns the compatibility of the H^3L model with the Eaves experiment since the oscillations occur at zero magnetic field. It can be argued that in this case the mechanism of the magnetopolaron is not necessary since the current density is three orders of magnitude larger than in Hickmott's experiment. This seems to imply that the tunneling current is similarly inhomogeneous, but this does not in fact appear to be the case because the I-V characteristics seem to follow the usual WKB dependence.[10] A more serious problem concerns the extension of the accumulated polaron charge. In the H^3L model it is assumed that accumulation occurs all over the undepleted region of thickness $D = L-W$; however, if the lifetime of the polaron is $\tau \approx 7$ ps, the classical distance traveled by the quasi- particle before decaying is $\lambda = v_d\tau \sim 30$ Å, which is two orders of magnitude smaller than the classical value $D \approx 3000$ Å of the undepleted region reducing considerably the magnitude of the space charge.

Ihm's Model

According to Ihm, the current oscillations in S.I.S. tunnel structures are caused by the renormalization of the final density of states (in the substrate) due to a correction of the electron-phonon self energy.[28] He proposes an approach based on the transfer Hamiltonian developed for superconducting tunnel junctions. The transfer Hamiltonian across the entire S.I.S. structure is written as[29]

$$H = \sum_{\vec{k},\vec{k}'} (C_{\vec{k}'}^{s+} T^b T^t C_{\vec{k}}^g + H.c) \tag{7}$$

where H.c stands for Hamiltonian conjugate and the subscripts g,t,b and s represent the gate, the tunnel barrier buffer layer and substrate, respectively, T is the transfer matrix and $C_{\vec{k}}^+$ and $C_{\vec{k}}$ are the fermion creation and annihilation operators of wave vector $\vec{k}$. Because the electron transfer from substrate to gate is forbidden under reverse bias for vanishing temperature, the net current is given by $I = 2e\ W_{sg}$ (2 accounts for spin degeneracy) where W_{sg} is the transition rate from gate to substrate evaluated under gate bias V_G. W_{sg} is calculated by using Fermi's Golden Rule which yields

$$I(V) \propto \int_{-E_F}^{0} dE \left| \sum_{j=1}^{M+1} N_S(E+eV_g - j\hbar\omega_{LO})P_j(E)T(E,V_G)N_g(E) \right| \tag{8}$$

where $N_g(E)$ is the gate density of states and $T(E,V_G)$ is the tunnel transmission probability depending upon the carrier energy and the gate voltage. P_j is the carrier density, N_S is the substrate density of states, which is a function of the carrier energy in the substrate and M is the maximum number of phonons emitted by a single carrier. According to Ihm, the renormalization of $N_S(E)$ due to the electron-phonon interaction provides the current oscillations. By calculating the real part of the self energy, the renormalized density of states is evaluated and shown in Fig. 12. Owing to the singular behavior of $N_S(E)$, the current as given by Eqn. 8 undergoes an oscillatory behavior. When $eV_G = M\hbar\omega_{LO}$ most electrons, after sequential phonon emission, have a final state energy of $\hbar\omega_{LO}$ at which the density of states of the substrate has a maximum, and the current is enhanced. By fitting the envelope of the I-V characteristics obtained by Hickmott et al. with adjustable parameters, Ihm is able to reproduce the oscillatory structures in the current with a relative amplitude $\Delta I/I \approx 0.1$. The important feature of this model however is that the oscillatory structures correspond to a mechanism which enhances the current (i.e., with maxima at $eV_G = M\hbar\omega_{LO})$ in contrast to previous interpretations. Although the role of the magnetic field is not apparent, Ihm implicitly assumes its presence by neglecting the ionized impurity scattering as in the case of magnetic freeze-out .

This model has been subjected to virulent criticism.[30-31] The main issue is the applicability of the transfer Hamiltonian formalism, which implies the coherence of the carrier wave function over the entire S.I.S. structure. According to the Monte Carlo simulation, this assumption is clearly invalid. Moreover, the alleged generality of the theory which is without device considerations is also an inherent drawback because it has been clearly evidenced that structure parameters such as the doping level, the width of the n^- -GaAs buffer layer, as well as the barrier thickness, play an important role on the occurrence of the oscillations. The model yields the unrealistic result that the current across the structure is determined by the impedance of the substrate which is several orders of magnitude smaller than that of the buffer layer. Moreover, the fact that the model predicts current enhancement rather than current reduction with respect to the background transport, is an additional drawback of the theory.

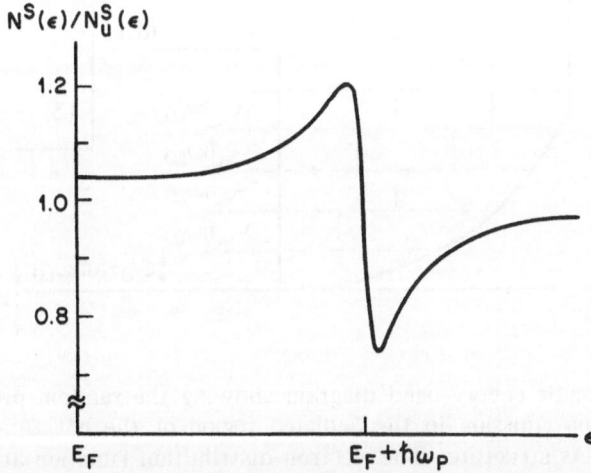

Fig. 12.   Ratio of renormalized density of states $N^S(E)$ (due to self energy correction of the electron L.O. phonon interaction) to the unrenormalized density of states $N_U^S(E)$ for n^+GaAs substrate. A Lorentzian broadening of 3.3meV corresponding to $\tau \sim 0.2$ps has been included in the calculation. (After Ihm ref. 26.)

243

In a previous version of this model, the author proposed a mechanism of space charge generation by phonon ionization of neutral donors in the n⁻ GaAs buffer layer.[32] A similar ionization process via acoustic phonons has been evidenced by Burger and Lassman in Si with superconducting tunnel junctions.[33] However the acoustic phonon mechanism is too slow to be consistent with the C-V measurements performed by Hickmott et al. at 100 kHz.[6] A generalized model where L.O. phonons participate in a mechanism of trapping of injected carriers by ionized donors and ionization of neutral donors is presented. The model shows that magnetic freeze-out is not an essential condition for phonon ionization to occur.

a) Zero magnetic field conditions

A schematic cross sectional view of the tunnel structure with hot electron mechanisms is shown in Fig. 13. At low temperatures and under reverse bias the lightly doped buffer layer consists of a depletion region and an undepleted layer containing neutral (frozen) donors, which extends up to the n⁺ GaAs substrate. Electrons are injected from the heavily doped GaAs gate through the tunnel barrier into the depletion region with a nearly monoenergetic distribution. Owing to the large electric field across the depletion region an appreciable fraction of electrons are accelerated up to the highest energy $E = eV_G$. Other electrons are scattered by emitting L.O. phonons resulting in a lower average carrier energy (Fig. 8). However, because of the dispersionless phonon relation, the energy lost $\Delta E = \hbar\omega_{L.O.}$ in each scattering event is constant. Therefore, the resulting distribution function

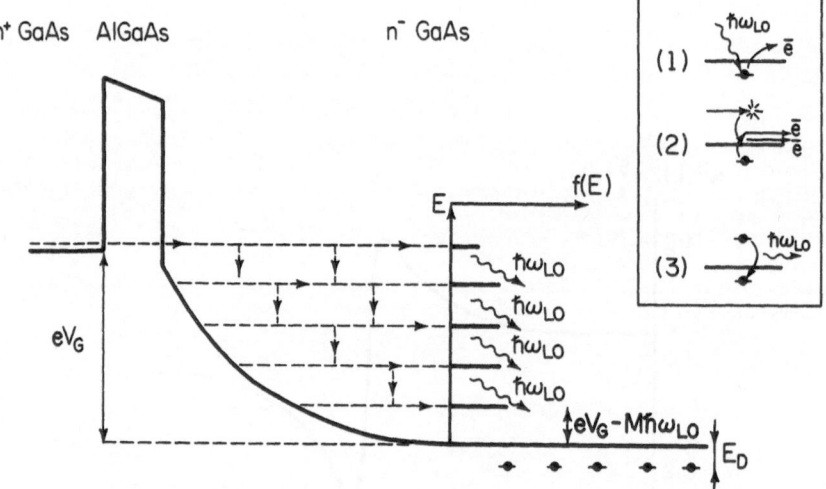

Fig. 13. Schematic energy-band diagram showing the random process of L.O. phonon emission in the depleted region of the n⁺GaAs-Al$_x$Ga$_{1-x}$As-n⁻GaAs structure. The electron distribution function at the entrance of the undepleted region is indicated on a horizontal axis as a function of energy E_D is the donor energy and M is the maximum number of L.O. phonons emitted by a single carrier. Insets show three fundamental processes (1) phonon ionization of neutral donors; (2) impact ionization of neutral donors by energetic carriers (3) resonant capture of energetic carriers by ionized donors with L.O. phonon emission.

$f(E)$ (horizontal axis in Fig. 13) of carriers leaving the high field region is characterized by discrete δ-like peaks separated by the phonon energy $\hbar\omega_{L.O.}$. Monte Carlo simulations of the transport in the depletion layer confirm this picture with a smooth distribution of carriers among the only possible energy states, (see Fig 9).[21]

a.1.) Relaxation mechanisms and carrier generation.

When carriers penetrate the undepleted region at the right of the energy axis they continue to dispose of their energy by L.O. phonon emission but impact ionization of neutral donors (Process 2 in (Fig. 13) inset) is also possible. The donor ionization rate by impact ionization $1/\tau_{ii}$ is calculated by using the Fermi Golden Rule

$$\frac{1}{\tau_{ii}} = \sum_{\vec{k}'_1,\vec{k}'_2} \frac{2\pi}{\hbar} |<1'2'|V|1,0>|^2 \delta(E_1 - E'_1 - E'_2 - E_D) \tag{9}$$

where $<1',2'|V|1,0>$ is the matrix element of the coulombic interaction between the incident electron (1-state) and the bound donor state (0-state). The two resulting states after ionization are the 1' and 2' plane wave states. E_1, E_1', and E_2' are the energies of the incident and the resulting states after ionization, respectively. This yields

$$\frac{1}{\tau_{ii}} = \frac{1}{6\pi^2} \frac{N_D^0 e^4}{(\epsilon_0 \epsilon_s)^2 \hbar^3} m a_B^3 F\left(\frac{E}{E_D}\right) \tag{10}$$

Here, a_B is the donor Bohr radius, m is the electron effective mass, ϵ_s is the dielectric constant and F is a function of the normalized incident energy E/E_D, where E_D is the donor energy (Fig. 14.a). For n$^-$ GaAs, where $N_D^0 \approx 10^{15}/cm^3$ is the neutral donor concentration and $a_B \approx 10nm$, the prefactor is about $6\times10^{10}s^{-1}$. The ionization F-function presents a maximum $F_M \approx 0.7$ at $E \approx 3.6 E_D$ which corresponds to an ionization cross section $\sigma \approx a_B^2$. Notice that the ionization rate strongly decreases at high energies; this implies that only the carriers of the first δ-like peak of the current distribution contribute to the impact ionization. The total scattering cross section is larger due to inelastic coulombic interactions causing electronic transitions between the donors hydrogenic levels. The total scattering cross section has been estimated to be as large as $\sigma \approx 3\pi a_B^2$,[17] i.e., an order of magnitude larger than the ionization process. However, the L.O. phonon emission $(1/\tau_{em} \approx 5\times10^{12}s^{-1})$ remains the only important energy relaxation mechanism in the n$^-$-GaAs buffer layer.

L.O. phonons generated by hot carriers are in excess with respect to their equilibrium distribution, and, in turn, are able to ionize neutral donors. The excess L.O. phonon population ν is a result of a detailed balance between generation by hot carriers (L.H.S. of Eqn. 11) and loss processes due to ionization and decay into acoustic phonons (R.H.S. of Eqn. 11) ,i.e.,

$$\sum_{j=2}^{M+1} \frac{n_j}{\tau_{em}} = \frac{\nu}{\tau_{pi}} + \frac{\nu}{\tau_d} \text{ or } \nu = \frac{1}{\frac{1}{\tau_{pi}} + \frac{1}{\tau_d}} \sum_{j=2}^{M+1} \frac{n_j}{\tau_{em}} \tag{11}$$

where M + 1 is the total number of δ-like peaks of the distribution function, n_j is the electron concentration of the jth-energy peak of the distribution function, $f(E)$. The j = 1 term in the sum does not contribute to the phonon generation because phonon emission is forbidden for $E < \hbar\omega_{L.O.}$. $1/\tau_d \approx 10^{11}s^{-1}$ is the L.O. phonon decay rate and $1/\tau_{pi}$ is the donor ionization rate by L.O. phonon absorption (process (1) in Fig. 13 inset). The latter is also calculated using the Fermi Golden rule

$$\frac{1}{\tau_{pi}} = \frac{2\pi}{\hbar} \sum_{\vec{k}'} |<0_{ph}|<\vec{k}|H_{e-ph}|0>|\vec{q}>|^2 \delta\left(E(\vec{k}') + E_D - \hbar\omega_{LO}\right). \tag{12}$$

where H_{e-ph} is the Frohlich electron-phonon interaction hamiltonian;[34]

$$H_{e-ph} = \sum_{\vec{q}}(C_{\vec{q}}a_{\vec{q}}e^{i\vec{q}\vec{r}} - C_{\vec{q}}^{+}a_{\vec{q}}^{+}e^{-i\vec{q}\vec{r}}) \tag{13}$$

Here $C_{\vec{q}}$ is the coupling parameter given by

$$C_{\vec{q}} = i\frac{\hbar\omega_{LO}}{q}(\frac{4\pi}{V}\alpha)^{1/2}(\frac{\hbar}{2m\omega_{LO}})^{1/4} \tag{14}$$

$\alpha = 0.068$ is the Frohlich coupling constant, $\vec{q}$ is the phonon wave vector, and V is the volume of the sample. $C_{\vec{q}}^{+}$ is the complex conjugate of $C_{\vec{q}}$ and $a_{\vec{q}}^{+}$ and $a_{\vec{q}}$ are the phonon creation and annihilation operators. In Eqn. 12 the $|0\rangle\,|\vec{q}\rangle$ -vector is the initial state consisting of the $|0\rangle$-hydrogenic donor state and the occupied $|\vec{q}\rangle$ incident phonon states, while the $|\vec{k}\rangle\,|0_{ph}\rangle$ -vector is the combination of the free electron $|\vec{k}\rangle$ states and the non-occupied $|0_{ph}\rangle$ phonon state resulting from the donor ionization process. After tedious calculations, we obtain

$$\frac{1}{\tau_{pi}} = \frac{16\pi}{3}N_D^o a_B \alpha \frac{\hbar}{m}S(qa_B) \tag{15}$$

where

$$S(x) = (\frac{\theta}{x})^3 \left\{ \frac{1}{[1+(x-\theta\sqrt{1-\frac{E_D}{\hbar\omega_{LO}}})^2]^3} - \frac{1}{[1+(x+\theta\sqrt{1-\frac{E_D}{\hbar\omega_{LO}}})^2]^3} \right\}. \tag{16}$$

Here $\theta = \sqrt{2m\omega_{L.O.}a_B^2/\hbar}$. With the same parameters as mentioned previously, the prefactor of Eqn. 15 is about $2\times10^{10}s^{-1}$. The S-function (Fig. 14b) exhibits a maximum $S_M \approx 1.5$ at $q \approx 0.80\sqrt{2m\omega_{LO}/\hbar}$ and a singularity at q = 0 as a consequence of the 1/q-dependence of the Frohlich hamiltonian. Therefore, the phonon ionization rate reaches a maximum value of about $3\times10^{10}s^{-1}$ for $q \approx 2\times10^6cm^{-1}$, which corresponds to the wave vector of phonons emitted by energetic carriers at

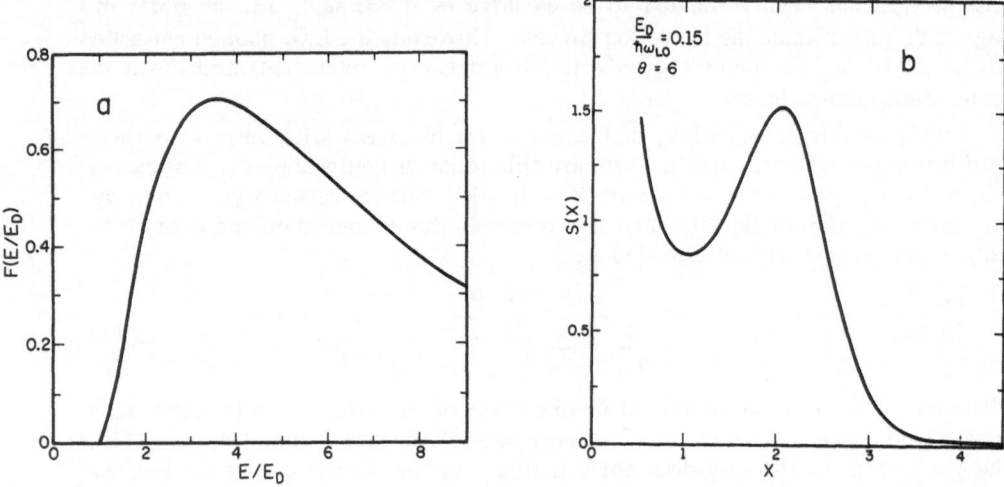

Fig. 14.   a) Plot of the impact ionization F-function versus the normalized incident energy E/E_D. b) Phonon ionization S-function versus the normalized phonon wavevector.

the emission onset. Smaller phonon wave vectors are emitted by electrons of higher energies and thus induce a lower ionization rate. It should be noted that this ionization rate is of the same order of magnitude as the impact ionization rate (Process 2 in Fig. 13 inset). Therefore under electron injection, the total nonequilibrium generation (G)-mechanism consists of three components: thermal generation G_{th}, impact ionization and donor ionization by L.O. phonon absorption, i.e.,

$$G = G_{th} + \sum_{j=1}^{M+1} \frac{n_j}{\tau_{ii}} + \frac{\nu}{\tau_{pi}} \qquad (17)$$

where ν is given by Eqn. 10.

a.2.) Recombination mechanisms.

The nonequilibrium recombination (R) mechanism is the sum of two terms:

$$R = \frac{(n_0 + \delta n)}{\tau_{rec}} + \frac{n_j}{\tau_C} \qquad (18)$$

The first term is the thermal recombination in the presence of excess carriers δn, n_0 is the equilibrium carrier concentration. The second term is resonant capture by the ionized donors with L.O. phonon emission (process (3) in Fig. 13 inset), which occurs only for the first electron level $j = 1$, i.e., when

$$E = \hbar\omega_{L.O.} - E_D \qquad (19)$$

The capture rate is given by the same expression as Eqn. 12 with the sum over $\vec{q}$ instead of $\vec{k}$

$$\frac{1}{\tau_c} = \frac{64}{3}\pi^2 N_D^+ a_B^2 \alpha \frac{(\hbar\omega_{LO})^{3/2}}{\sqrt{2m}} \frac{1}{1+k^2 a_B^2}$$

$$\times \left| \frac{3}{8} + \frac{1}{2}\frac{1}{1+k^2 a_B^2} + \frac{1}{(1+k^2 a_B^2)^2} \right| \delta(E + E_D - \hbar\omega_{LO}) \qquad (20)$$

where the δ-function accounts for the energy conservation during the resonant process and $k = \sqrt{2mE}/\hbar$. At T = 4.2K for $N_D - N_A \approx 10^{15}/cm^3$, the free carrier concentration is $n_0 \approx 2 \times 10^{13}/cm^3$, which is consistent with a resistivity value $\rho = 400\ \Omega cm$.[10] For an acceptor concentration $N_A \approx 2 \times 10^{14}/cm^3$ the charge neutrality condition yields $N_D^+ = N_A^- + n_0 \approx 2.2 \times 10^{14}/cm^3$. If we assume that the energy width of the $j = 1$ resonant peak is $\Gamma_1 \approx \hbar\omega_{LO}/50$, we can replace the δ-function by $1/\Gamma_1$ and evaluate the capture rate $1/\tau_c \approx 2 \times 10^{11} s^{-1}$ at resonance. The choice of Γ_1, is somewhat arbitrary but reasonable for low temperature since the condition for observation of the oscillations is $\hbar\omega_{LO} \gg \Gamma_1$.

a.3.) Excess carriers in the n^--GaAs layer.

At steady state, there is a balance between carrier generation and recombination processes. Therefore the resulting excess carrier concentration δn in n^- GaAs can be calculated. This excess charge concentration varies discretely because by increasing the gate bias, the lowest δ-like peak of the incident distribution function passes the threshold energy for each of the generation or recombination mechanisms. Monte Carlo simulations show that for a gate voltage $V_G \approx 0.35V$, for example, the occupation of the $j = 1$ electron state is about 1/5 of the total electron concentration up to the onset of the second $j = 2$ state and then decreases slowly for higher energy. Therefore, the carrier concentration in the lowest δ -peaks are $n_1 = n_2 \approx J/5ev_d$ in a first approximation, where $J \approx 7 \times 10^{-2} A/cm^2$ is the current density as provided by the experimental data[7] and $v_d \approx 5 \times 10^7$ cm/s is the carrier drift velocity. With a recombination lifetime $\tau_{rec} \approx 7 \times 10^{-9}$s as estimated by Melngailis et al.[35] in similar samples, the contribution of the impact ionization mechanism to the excess charge is given by

$$\delta_{n_{ii}} = \frac{1}{5} \frac{J}{ev_d} \frac{\tau_{rec}}{\tau_{ii}} \sim 5 \times 10^{11}/cm^3 \qquad (21)$$

This value is about two orders of magnitude smaller than the estimates of Eaves et al.[17] who include interatomic donor transitions resulting from carrier inelastic scattering in the total ionization rate and assume that by hopping conduction or field ionization, electrons in the donor excited states will contribute to the excess charge. However, the main difference is due to an overestimation of the number of carriers able to impact ionize the neutral donors at the threshold energy. Eaves et al. consider that the current is monoenergetic with all the carriers concentrated in the $j = 1$ lowest peak of the distribution function, while the random nature of the phonon scattering in n^- GaAs depleted layer allows only a fraction of them to be in the appropriate energy range for impact ionizing the donors.

The charge variation resulting from L.O. phonon ionization and resonant capture with L.O. phonon emission are respectively given by

$$\delta n_{pi} = \frac{1}{5} \frac{J}{ev_d} \frac{1}{1 + \dfrac{\tau_{pi}}{\tau_d}} \frac{\tau_{rec}}{\tau_{em}} \approx 10^{13}/cm^3 \qquad (22)$$

$$\delta n_c = - \frac{J}{5ev_d} \frac{\tau_{rec}}{\tau_C} \approx -2 \times 10^{12}/cm^3 \qquad (23)$$

In Eqn. 22 we have considered the excess charge resulting only from the phonons emitted by second lowest ($j=2$) peak of the distribution function. Additional ionization arises from phonons emitted by high energy carriers but they do not cause a step like charge variation. These results demonstrate that the phonon ionization process is the leading mechanism of charge generation in the undepleted region of the S.I.S. structure. The density of L.O. phonons emitted by hot carriers exceeds by at least one order of magnitude the concentration of electrons able to impact ionize the neutral donors. Notice also that the charge reduction resulting from resonant capture is comparable in magnitude to the charge generated by phonon ionization. This capture mechanism has been found responsible for oscillatory conductivity in diamond[36] and InSb.[37] However, in this experiment, the situation is more complex because both effects - resonant capture and phonon ionization - occur almost simultaneously since the resonant carrier energy is just a few meV below the threshold energy for phonon emission. The situation is complicated even more by an intermediate regime with subthreshold phonon emission for which the final electron states are the donor excited states. By varying the injection gate voltage near the resonant conditions, the current first decreases due to carrier capture, then, within 3-4 mV, increases again due to the phonon emission onset which causes donor ionization. Therefore the transition range between trapping and ionization is about the width of the peaks in the $\partial^2 I/\partial V_G^2$ -curve which corresponds to current minima observed by Guimaraes et al.[18] Moreover, the variation of electronic charge during a cycle of carrier trapping-ionization over a gate voltage period $\Delta V_G = \hbar\omega_{LO}/e$ is given by $\delta n = \delta n_{pi} - \delta n_c \approx 1.2 \times 10^{13}/cm^3$. This results in a relative resistance variation $\Delta R/R = \delta n/(n_o + \delta n) \approx 0.3$ which is in good agreement with the experimental data.[17] This resistance variation in turn corresponds to a voltage drop $\delta V_u = \Delta RI \approx 0.4mV \approx kT/e$ in the undepleted region and thus induces a variation of the AlGaAs barrier field $\Delta E_T/E_T = \Delta RI/WE_T \approx 10^{-3}$ (W is the depletion width) which modulates the tunnel current. It should be noted that this model is fully consistent with the high temperature data of Lu et al.[13] since the capture mechanism can persist above the freeze-out temperature when ionization vanishes. Although the low energy carrier distribution n_1 is broader than that shown in Fig. 9 and thus reduces the capture rate, charge trapping is still important since the current density is 10^4 times larger than in the previous experiment, resulting in an appreciable space charge in the undepleted layer, as seen from Eqn. 23.

b) Non-zero Magnetic Field Conditions.

In Hickmott's experiment the conductance oscillations are enhanced by the presence of magnetic fields. For instance, the voltage variation during an oscillation cycle $\delta V = V_G \Delta I/2I \approx 3.5$ mV, as estimated from the relation $I \propto V_G^2$,[23] (data at H = 12T with $\Delta I/I \approx 2 \times 10^{-2}$ for $V_G = 0.35$ V[6]) is more than one order of magnitude larger than the thermal voltage $kT/e \approx 0.14$mV(T $\approx$ 1.6K) and consequently cannot be attributed to an "ohmic" resistance variation of the undepleted (neutral) region, as suggested by several authors.[10,23]

We attribute this voltage variation to a space charge effects which modify the electric field profile in the GaAs⁻ undepleted region. A rapid calculation yields $\delta E = \delta V/D \simeq 100$V/cm where D = L-W $\approx$ 0.35 μm is the width of the undepleted layer and W$\approx$0.65μm is the extension of the depletion region at $V_G \approx 0.35$ V. This electric field variation is compatible with the existence of magnetic freeze-out in the undepleted region since field ionization of neutral donors occurs at higher electric fields. Therefore L.O. phonon ionization takes place and releases the frozen electrons from the neutral donors resulting in a space charge effect which reduces the potential drop in the GaAs⁻ buffer layer and consequently increases the AlGaAs barrier field causing a tunnel current enhancement.

The L.O. phonon ionization rate with electron states modified by the magnetic field is calculated in a manner similar to that of previous sections.

In longitudinal magnetic fields, the electron wave function in the conduction band is given by

$$\Psi_c = \sqrt{\frac{eH}{2\pi hc}} \frac{1}{\sqrt{L}} \exp-[\frac{eH}{4hc}(x^2+y^2)-ik_z z] \tag{24}$$

where H is the magnetic field in the z-direction, and k_z is the longitudinal wave vector. Here, for the sake of simplicity, we assume that only the lowest Landau level of the conduction band is occupied.

The ionization energies of shallow hydrogenic donors in magnetic field were first calculated by Yafet, Keyes and Adams (Y.K.A.) with a variational method.[15] Later Larsen[38] improved the model by using trial wave functions which yield more accurate ionization energies.[39] However, in the former method the donor wave functions are more practical for analytical purposes. Moreover in the range of magnetic fields achieved in the experiment, the ionization energies are not significantly lower than Larsen's results (<10%). Therefore, in the following derivation, we will use a hybrid method combining the YKA wave functions and Larsen's energies to calculate the phonon ionization rate.

The YKA donor wave function is given by

$$\Psi_i = \frac{1}{(2\pi)^{3/4}\sqrt{a_\perp^2 a_z}} \exp-[\frac{(x-x_i)^2+(y-y_i)^2}{4a_\perp^2} + \frac{(z-z_i)^2}{4a_z^2}] \tag{25}$$

where $a_\perp$ and a_z are the variational parameters and $\vec{r} = (x_i, y_i, z_i)$ is the position of the donor atom. Again, we assume that only the lowest donor state is occupied under magnetic freeze-out.

The electron-phonon interaction hamiltonian is given by Eqn. 13. After integration, the electron phonon matrix element is

$$M = 2^{7/4}\pi^{1/4}\frac{C_q}{\sqrt{L}}\frac{a_\perp^2 a_z}{[1+(\frac{a_\perp}{a_z})^2]a_c}\exp-\left|\frac{q_\perp^2 a_\perp^2}{1+(\frac{a_\perp}{a_c})^2}+(k_z-q_z)^2 a_z^2\right|$$

$$x\exp-\left|\frac{x_i^2+y_i^2}{4(a_z^2+a_\perp^2)}-\frac{iq_x x_i+iq_y y_i}{1+(\frac{a_\perp}{a_c})^2}-i(k_z-q_z)z_i\right|\qquad(26)$$

where $a_c = \sqrt{\frac{\hbar c}{eH}}$. This matrix element describes the absorption of a phonon by a bound electron at a distance $|\vec{r}-\vec{r}_i|$ from the center of the donor atom. Here $q_\perp$ is the component of the phonon wave vector perpendicular to the magnetic field. In this model it is assumed that the donor center is not necessarily on the path of the traveling phonon. The transition probability for the ionization process and the total ionization rate are given by

$$W_i(k_z,\vec{q}) = \frac{2\pi}{\hbar}M^2(\vec{q})\delta\left|\frac{\hbar^2 k_z^2}{2m}+E_D-\hbar\omega_{L.O.}\right|\qquad(27)$$

and

$$\frac{1}{\tau} = \sum_i\sum_{k_z}W_i(k_z,\vec{q})\qquad(28)$$

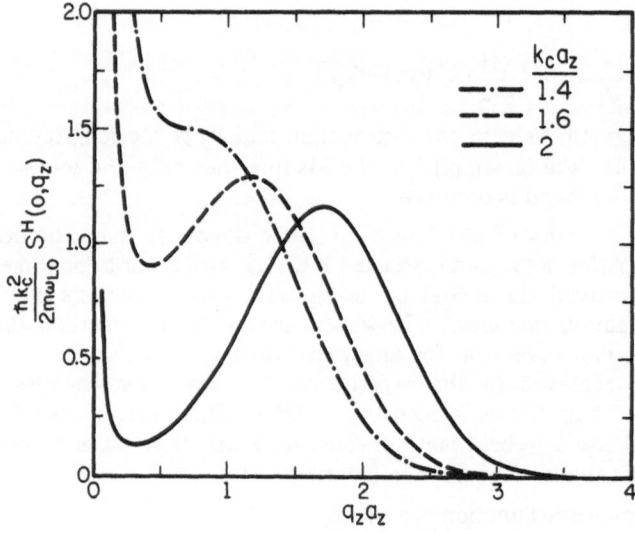

Fig. 15. Magnetic field dependent phonon ionization S^H-function versus the normalized z-component of the phonon wavevector $q_z a_z$, for different values of the parameter $k_c a_z$ decreasing with magnetic fields. From Table 1, $k_c a_z = 1.6$ for $H = 6T$. For comparison the limiting case $H = 0$ corresponds to $k_c a_z = 2.3$.

Table 1. Numerical values of the parameters used in the calculations: $\gamma = \hbar\omega_c/2Ry$ where ω_c is the cyclotron frequency and Ry is the Rydberg equal to 5.8 meV; a_z and $a_\perp$ are from Ref. 15, E_D is the ionization energy from Ref. 38, and $a_C = \sqrt{\hbar c/eH}$ is the cyclotron radius.

γ	a_z(Å)	$a_\perp$(Å)	a_C(Å)	E_D(Ry)
1	73	59	98	1.66
2	65	49	69	2.04

where the summation is carried out over all the longitudinal wave vectors k_z and all the impurity indices i. We next make the transformation from discrete to continuous variables

$$\sum_{k_z} \rightarrow \xi H \frac{L}{2\pi} \int dk_z \tag{29.a}$$

$$\sum_i \rightarrow N_D^o \int d\vec{r}_i \tag{29.b}$$

where ξH is the degeneracy of the Landau levels and N_D^o is the neutral donor concentration. We finally obtain the ionization rate for L.O. phonons.

$$\frac{1}{\tau_{LO}} = \frac{16}{\sqrt{2}} \pi^{3/2} N_D^o \frac{a_z}{1+(\frac{a_c}{a_\perp})^2} \alpha \frac{\hbar}{m} \frac{1}{\sqrt{1-\frac{E_D}{\hbar\omega_{LO}}}} S^H(q_\perp,q_z) \tag{30}$$

where

$$S^H(q_\perp,q_z) = \frac{2m\omega_{LO}}{\hbar q^2} \exp{-2\left| \frac{q_\perp^2 a_\perp^2}{1+(\frac{a_\perp}{a_z})^2)} + (k_c-q_z)^2 a_z^2 \right|} \tag{31}$$

and

$$k_c = \sqrt{\frac{2m\omega_{LO}}{\hbar}\left[1-\frac{E_D}{\hbar\omega_{LO}}\right]}$$

Table 1 gives the values of the different parameters for $\gamma = \hbar\omega_c/2Ry = 1$ and 2 corresponding to magnetic fields amplitudes H = 6.5T and 13T, respectively. Here ω_c is the cyclotron frequency and Ry is the Rydberg. The expression of the ionization rate is similar to Eqn 15. By using the parameters of Table 1, it is easily seen that the prefactor is virtually unchanged with respect to zero magnetic field conditions. The ionization G^H -function however is different from the expression in (15). A well defined peak is now identified at $q_z a_z \approx 2$ for low magnetic fields (Fig. 15). The pronounced minimum at low wave vector is a unique feature of the one dimensional (1-D) character of the electron wave function, which results from confinement caused by the magnetic field. It is a real effect which is independent of the choice of trial wave function. For high magnetic fields, the position of the peaks are shifted toward low wave vectors and are less pronounced because of the influence of the $1/q^2$ -singularity in the L.O. phonon scattering rate. For the range of magnetic fields considered in this experiment, i.e., H ⩽ 14T , the maxima are still well resolved and thus limits the ionization process to a small range of phonons

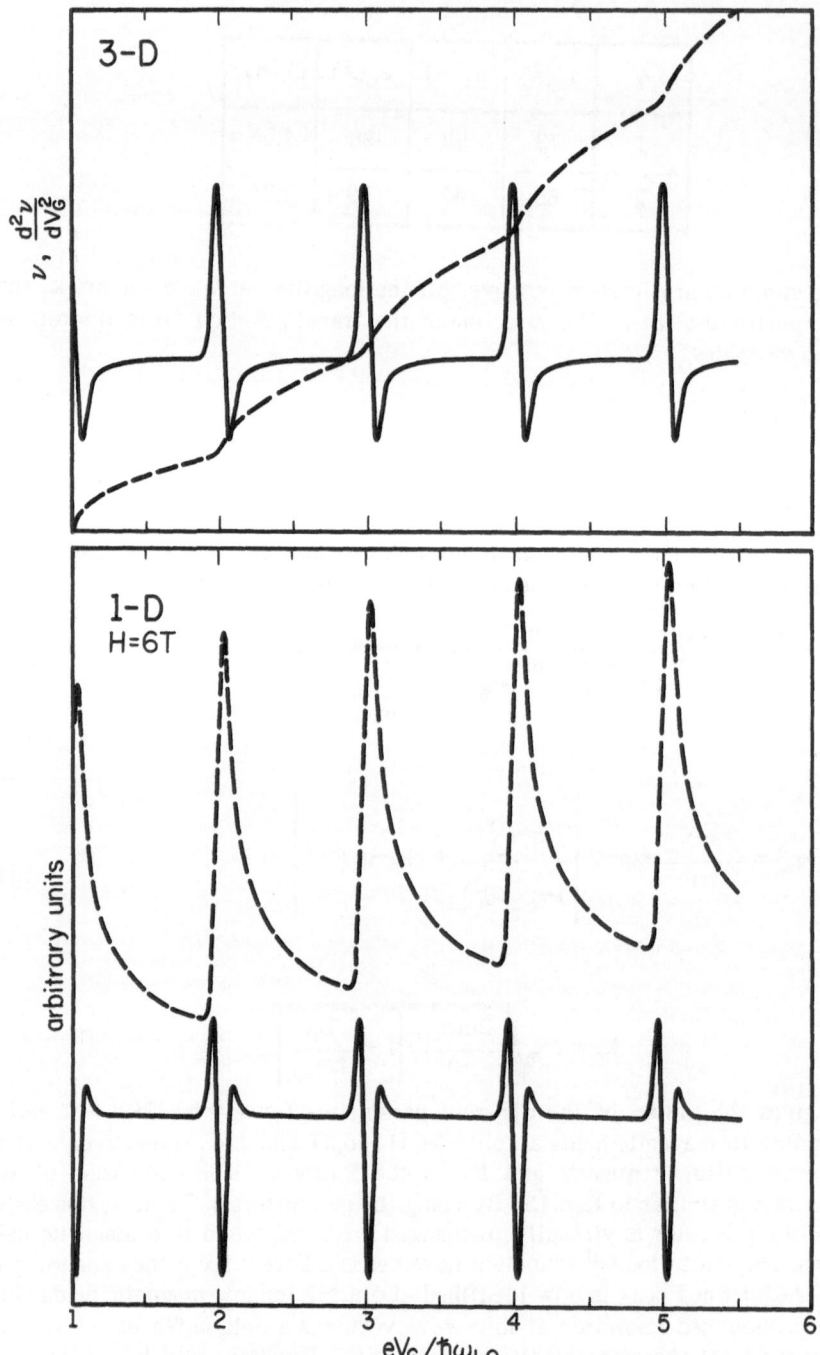

Fig. 16.   Plot of the L.O. phonon density ν and $d^2\nu/dV^2$ versus the gate voltage V_G as derived from Eqn. (32). a) Zero magnetic field (3-D), b) Non-zero magnetic field (1-D) cases. The broadening factor is $\Gamma = \hbar\omega_{LO}/20$

wave vectors around the peaks, i.e., for $q_z \approx k_c$. This is a consequence of energy conservation during the electron-phonon interaction.

In addition to this particular behavior of the ionization rate the 1-D L.O. phonon density exhibits features different from the 3-D (zero magnetic field) conditions. As a generalization of Eqn. 11, the density v of L.O. phonon emitted by hot carriers in the GaAs undepleted layer is given by the relation

$$v \alpha \sum_{j=2}^{M+1} \int_{\hbar\omega_{LO}}^{\infty} \frac{dE}{\tau_{em}(E)} N(E)f_j[eV_g-(M+1-J)\hbar\omega_{LO}-E] \qquad (32)$$

$N(E)$ is the electron density of states and f_j is the mini distribution of electrons around the jth-peak of the total $f(E)$ distribution. If for the sake of simplicity we assume a gaussian shape characterized by a broadening parameter Γ, i.e.

$$f_j \alpha n_j \exp-\frac{(eV_G-(M+1-j)\hbar\omega_{LO}-E)^2}{\Gamma^2} \qquad (33)$$

and if we neglect the weak variation of the n_j coefficients with the j-parameter, the dependence of the number of L.O. phonons v on the gate voltage has a profile shown in Fig. 16 for zero (3-D) and non-zero (1-D) magnetic fields. The second derivatives d^2v/dV_G^2 of the 3-D curve shows sharp maxima similar to those observed experimentally[17] for $V_G = Mh\omega_{LO}/e$ and which correspond to the onset of a new sequence of phonon emission when the lowest peak in the hot carrier $f(E)$ distribution function crosses the threshold energy $E = \hbar\omega_{LO}$. The weak minima on the left of the maximum peaks are also observed experimentally (peak at $V_G \approx 190$ mV in Fig. 7) when they are not washed out by the noise.

In the presence of magnetic fields, owing to the singular nature of the 1-D density of states, the trace in the second derivative changes as shown in Fig. 16.b. The maximum peak becomes narrower and is followed by a pronounced minimum. This behavior which is a characteristic of the L.O. phonon density and thus of the

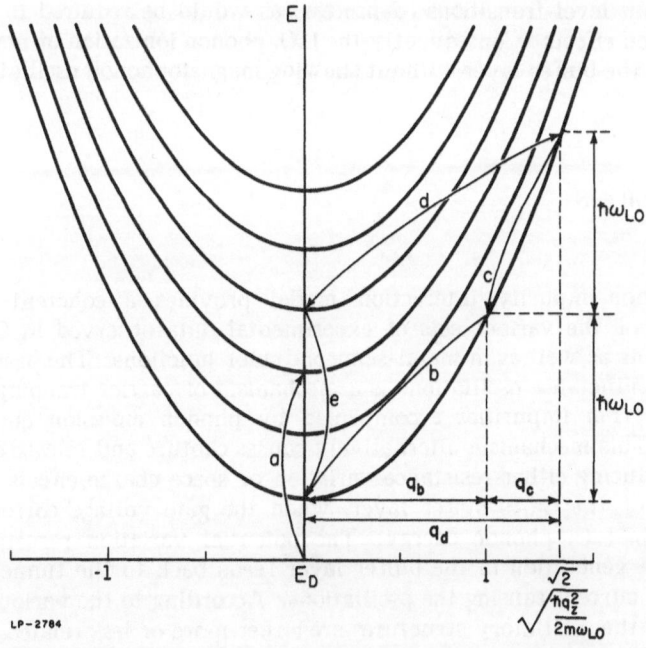

LP-2784

Fig. 17.   Schematic representation of intra- and inter-Landau level L.O. phonon emission processes. E_D is the position of the donor ground state $\hbar\omega_{LO} = 3\hbar\omega_c$.

phonon ionization process is observed in the experimental data. Hence, in Fig. 7, the experimental peak at $V_G \approx 155mV$ changes from a sharp maxima at zero magnetic field to a shape similar to that predicted by our theory at H = 5.5T. Consequently, the characteristic shape of the second derivative peak is the L.O. phonon signature in the charge ionization process. Eaves et al. have argued that L.O. phonon ionization is not possible since magneto-phonon oscillations have not been observed in the trace of the current peak.[17] Magnetophonon effects occur when L.O. phonons induce electronic transitions from donor bound states to the bottom of a Landau level when the condition

$$N\hbar\omega_c + E_D = \hbar\omega_{LO} \, ,\tag{34}$$

is satisfied with N being an integer (a-process in Fig. 17). Such <u>vertical</u> transitions occur for phonon wave vectors $q \approx o$ since no momentum is exchanged in this interaction. However, the L.O. phonons emitted by the low energy carriers which cause the current peak are characterized by rather large wave vectors $q \approx \sqrt{2m\omega_{LO}/\hbar}$ (b-process in Fig. 17). This is easily understandable since the electrons have been accelerated to the threshold emission energy on the first Landau level. Phonons of smaller wave vectors originate either from emission processes by high energy carriers (c-process) but are not related to the current peak, or from low momentum electrons on higher Landau levels (e-process). The population of such electrons is considerably smaller than that of the first Landau level because they result from emission processes of higher energy carriers (c- and d-processes). These processes favor small phonon wave vector interaction because of the $1/q^2$-dependence of the L.O. phonon emission rate.[40] As an example the c-process involves a phonon wave vector $q_c \approx (\sqrt{2}-1)\sqrt{2m\omega_{LO}/\hbar}$ while in the d-process the phonon wave vector is given by $q_d \approx \sqrt{2}\sqrt{2m\omega_{LO}/\hbar}$. Therefore the ratio

$$\left|\frac{q_c}{q_d}\right|^2 = \left|1-\frac{1}{\sqrt{2}}\right|^2 \lesssim 0.1\tag{35}$$

favors the intra-Landau level transitions (c-process) rather than transfer from low to high Landau level transitions (d-process) as would be required to observe the magnetophonon effects. Consequently the L.O. phonon ionization of neutral donors takes place in the buffer layer without showing magnetophonon oscillations.

IV. CONCLUSIONS

The phonon-impurity interaction model provides a coherent and general interpretation of the various sets of experimental data observed in GaAs-AlGaAs tunnel junctions as well as in metal-semiconductor junctions. The basic mechanism causing the conductance oscillations is a mechanism of carrier trapping and detrapping by and from impurities accompanied by phonon emission and absorption, respectively. This mechanism alternatively causes capture and release of carriers by impurities, inducing either resistance variation or space charge effects in the undepleted region of the GaAs buffer layer, when the gate voltage corresponds to multiples of the L.O. phonon energy. The potential variation resulting from the periodic charge generation in the buffer layer feeds back to the tunnel barrier and modulates the current causing the oscillations. According to the various experimental conditions, the oscillatory structures are either more or less related to one of the components of the trapping-detrapping process. Low temperature and small current conditions (Hickmott-Eaves) favor the ionization (detrapping) mechanism while for high temperature and high injection conditions (Lu et al.) the conductance oscillations are caused by the inverse process of resonant carrier capture with L. O. phonon emission since carriers are already thermally ionized. In the former case, it is not necessary to assume total magnetic freeze-out over the entire buffer layer since phonon ionization occurs in the low field region of the buffer layer with partial freeze-out. Moreover phonon ionization is perfectly consistent with the absence of magnetophonon effects since for hot electron conditions, the carriers are characterized by high momenta on the lowest Landau level rather than by low momenta on high Landau levels, which is the condition for the magneto-phonon oscillations to occur.

The phonon-impurity interaction model has the advantage of being consistent with the C-V measurements in high magnetic fields since it assumes the existence of magnetic freeze-out in the low field portion of the n^- GaAs layer, which results in the flat C-V curve as observed by Hickmott et al. Moreover it naturally provides a maximum number of oscillations since the latter disappear when the classical depletion region extends over the entire buffer layer. However some issues related to secondary experimental findings have not been considered in the model. These issues are

1) the quasi quadratic voltage dependence of the current in Hickmott's experiment, as well as the tunnel mechanism responsible for carrier injection in the n^- -GaAs buffer layer. Hellman et al.[23] have proposed an interpretation based on inhomogeneous tunneling but their demonstration is more illustrative than rigorous. Moreover, a close analysis of the I-V relation shows deviations from the quadratic dependence, which is a function of the magnetic field (Fig. 3.b).

2) The fact that more conductance structures are observed over a longer voltage range ($V_G < 1V$) in the presence of magnetic fields parallel to the barrier (Fig. 4.b). This indicates that the carrier trapping mechanism causing the current oscillations persists at higher gate voltage than in the case of perpendicular H which is consistent with the persistence of residual donor freeze-out in undepleted n^- -GaAs at higher gate voltage. Indeed the Y.K.A. theory predicts stronger magnetic confinement of donor states in the perpendicular direction (parallel to the driving electric field) then for parallel H, preventing field ionization of frozen donors.

Nevertheless, despite these unresolved questions, the microscopic mechanism of L. O. phonon assisted charge generation and recombination seems to be strongly supported by the experimental data.

Acknowledgments

We are indebted to T. W. Hickmott for providing us with his experimental data. We are also grateful to K. Hess, G. E. Stillman, M. Nayfeh and B. Mason for valuable discussions. We also wish to thank M. Wagner for critical comments and Mrs. E. Kesler, S. Briggs and D. Bailey for technical assistance. This work was supported by the Joint Services Electronics Program.

References

1. See. e.g. "Proceedings of the 6th Conference on the Electronic Properties of Two Dimensional Systems, Sept. 9-13 1985 Kyoto Japan (To be published in Surf. Sci.)

2. Ben Daniel and C.B. Duke, "Conductance anomalies due to space-charge-induced localized states," Phys. Rev. 160 679 (1967)

3. For a review, see e.g. T.W. Hickmott, "Magnetotunneling from accumulation layers in $Al_xGa_{1-x}As$ capacitors," Phys. Rev. **B32** 6531 (1985)

4. Y. Katayama and K.F. Komatsubara, "Oscillatory tunnel conductance induced by longitudinal optic phonons in InSb-Oxide-Metal structures," Phys. Rev. Lett., **19** 1421 (1967)

5. B.C. Cavenett "Electron-phonon interactions in InSb junctions," Phys. Rev. **B5** 3049 (1972)

6. T.W. Hickmott, P.M. Solomon, F.F. Fang, F. Stern, R. Fischer, and H. Morkoç, "Sequential single-phonon emission in $GaAs-Al_xGa_{1-x}As$ tunnel junctions," Phys. Rev. Lett. **52** 2053 (1984)

7. T.W. Hickmott, P.M. Solomon, F.F. Fang, R. Fischer, and H. Morkoç, "Magneto-tunneling and magnetic freeze-out in n^--GaAs-undoped $Al_xGa_{1-x}As$-n^+-GaAs capacitors," Proceedings of the 17th Int. Conf. on Physics of Semiconductors, ed J.D. Chadi and W.A. Harrison (Springer-Verlag, New York, 1985) p. 417.

8. T.W. Hickmott, P.M. Solomon, R. Fischer, and H. Morkoç, "Resonant Fowler-Nordheim tunneling in n^--GaAs-undopedAl_xGa_{1-x}-n^+GaAs Capacitors," Appl. Phys. Lett. **44** 90 (1984)

9. L. Eaves, P.S.S. Guimaraes, B.R. Snell, D.C. Taylor, and K. Singer, "Oscillatory structures in GaAs/(AlGa)As tunnel junctions," Phys. Rev. Lett. **55** 262 (1985)

10. D.C. Taylor, P.S.S. Guimaraes, B.R. Snell, L. Eaves, F.W. Sheard, G.A. Toombs, and K. Singer, "Oscillatory structures in the reverse bias J(V) plots of n^+GaAs/(AlGa)As/n^-GaAs/n^+GaAs structures," Physics **B+C 134** 12, (1985)

11. P.S.S. Guimaraes, D.C. Taylor, B.R. Snell, L. Eaves, K.E. Singer, G. Hill, M.A. Pate, G.. Toombs, and F.W. Sheard, "Tunneling and magnetotunneling effects in n^+GaAs/(AlGa)As/n^-GaAs/n^+GaAs devices," J. Phys. C. **18** L605 (1985)

12. P.F. Lu, D.C. Tsui, and H.M. Cox, "Optical-phonon emission in ballistic transport through microchannels of InGaAs," Phys. Rev. Lett. **54** 1563 (1985)

13. P.F. Lu, D.C. Tsui, and H.M. Cox, "Current oscillations in electrical transport through (In- and Sn-) InP contacts," Bull. Am. Phys. Soc. **31** 394 (1986)

14. See e.g. C.B. Duke in "Tunneling in Solides," Solid State Physics Suppl.. 10 ed. by F. Seitz, D. Turnbull, and H. Ehrenreich (Academic Press New York, 1969)

15. Y. Yafet, R.W. Keyes, and E.N. Adams, "Hydrogen atom in strong magnetic field," J. Phys. Chem. Solids, **1** 137 (1956)

16. See ref 14 p. 34

17. L. Eaves, B.R. Snell, D.K. Maud, P.S.S. Guimaraes, D.C. Taylor, F.W. Sheard, G.A. Toombs, J.C. Portal, L. Dmowski, P. Claxton, G. Hill, M.A. Pate and S.J. Bass, "Longitudinal optic-phonon emission, magnetoquantum effects and hydrostatic pressure studies in III-V tunneling heterostructures," Proceeding of the International Conference on Physics of Semiconductors, Stockholm, Sweden, 1986.

18. J.W. McGowan, J.F. Williams and E. Curley, "e-H Resonances in the 2D Excitation Channel," Phys. Rev. **180**, 132 (1969).

19. L. Eaves and J.C. Portal, " A review of the magneto-impurity effect in semiconductors," J. Phys. C. **12** 2808 (1979).

20. A.S. Vengurlekar and J.C. Inkson, "On the ballistic injection into semiconductor from point contacts and the structure in d^2I/dV^2 characteristics," Solid State Communications, **45** 17 (1983).

21. T. Wang, J.P. Leburton, K. Hess, and D. Bailey, "Absence of spatial coherence. effects of carrier energy and velocity in $GaAs^+$-AlGaAs-$GaAs^-$ tunnel structures," Phys. Rev. **B23** 2906 (1986)

22. T. Wang, K. Hess, and G.J. Iafrate, "Time dependent ensemble Monte Carlo simulation for planar-doped GaAs structures," J. Appl. Phys. 58 857(1985)

23. E.S. Hellman, J.S. Harris, C.B. Hanna and R.B. Laughlin "One dimensional polaron effects and current inhomogeneities in sequential phonon emission," Physica B+C 134 41 (1985)

24. E.S. Hellman and J.S. Harris, "Energy-momentum relation for polarons confined to one dimension," Phys. Rev. B33 8284 (1986).

25. P.W. Warmenbol, F. M. Peeters and J. T. Devreese "Effect of the polaron induced nonparabolicity of the energy-momentum relation on the dynamics of transport electrons" Phys. Rev. B 33 5590 (1986)

26. D.M. Larsen, "Polaron energy levels in magnetic and Coulomb fields," in "Polarons in ionic crystals and polar semiconductors," ed. by J.T. Devreese (North Holland Publishing Co. Amsterdam 1972) pp. 237-287.

27. F. M. Peeters, P. W. Warmenbol and J. T. Devreese, "Energy-momentum dispersion relation of a Frohlich polaron in two dimension" to be published; see also J. T. Devreese, "Electron-phonon interaction in two dimensional electron systems," these proceedings.

28. J. Ihm, "Origin of the oscillation in current voltage characteristics of GaAs-AlGaAs tunnel junctions," Phys. Rev. 55 999 (1985)

29. J.R. Schriefer in "Theory of Superconductivity," (Benjamin New York 1964) p. 78.

30. C.B. Hanna and R.B. Laughlin, "Oscillations in the current-voltage characteristics of GaAs-AlGaAs tunnel junctions," Phys. Rev. Lett. 56 2547 (1986).

31. J. Ihm, "Ihm responds," Ibid. pp. 2548.

32. J.P. Leburton, "Origin of the current oscillations in GaAs-AlGaAs tunnel junctions," Phys. Rev. B31 4080 (1985)

33. W. Burger and K. Lassman, "Energy-resolved measurements of the phonon-ionization of D^- and A^+ centers in silicon with superconducting Al tunnel junctions," Phys. Rev. Lett. 53 2035 (1984)

34. See e.g. R. Evard "The Frohlich polaron concept," in ref. 21 pp. 29-80.

35. I. Melngailis, G.E. Stillman, J.O. Dimmock, and C.M. Wolfe, "Far-infrared recombination radiation from impact-ionized shallow donors in GaAs," Phys. Rev. Lett 23 1111 (1969)

36. J.R. Hardy, S.D. Smith and W. Taylor, Proc. Inst. Conf. on the Physics of Semiconductors EXETER (1962) (London: Institute of Physics and Physical Society) pp. 521-8.

37. H.J. Stocker, H. Levinstein and C.R. Stannard, "Oscillatory photoconductivity in InSb," Phys. Rev. 150 613 (1966).

38. D.M. Larsen, "Shallow donor level of InSb in magnetic fields," J. Phys. Chem. Solids, 29 271 (1968).

39. G.E. Stillman, C.M. Wolfe, J.O. Dimmock, in Semiconductors and Semimetals, ed. by R.K. Willardson and A.C. Beer, Academic Press, NY 12 pp. 169 (1977)

40. J.P. Leburton and K. Hess, "Energy diffusion equation for an electron gas interacting with polar optical phonons," Phys. Rev. B26 5623 (1982)

k.p THEORY FOR TWO - DIMENSIONAL SYSTEMS

Rudolf Lassnig

Institut für Experimentalphysik
Universität Innsbruck
Innsbruck, Austria

Energy levels in semiconductors split into so-called energy bands with dispersion $E_1(k)$, where 1 denotes the band index and k is the wavevector. Both indices are necessary for the classification of the electronic states. k is usually restricted to the first Brillouin zone. At symmetry points, that is for k-values where the Hamiltonian exhibits some sort of rotational symmetry, the corresponding wave function $\Psi_1(k)$ transforms according to the irreducible representation of the point group. Thus the symmetry can be used for the classification of the energy bands (1).

A particularly simple description is possible at the so-called Γ-point (k=0) of cubic semiconductors. Here the electronic states are classified as s-type, p-type etc., in close analogy to atomic states. As an example, p-type states (X,Y,Z) transform as x,y, and z under rotations.

Since most electronic processes in semiconductors "occur" in the vicinity of extremal points, the band structure close to these points is of special interest. k.p-theory represents a systematic semi-empirical technique to calculate the energy levels close to the extremal points.

k.p-theory concentrates on the consistent description of the bare energy levels (i.e. rigid ions, zero temperature etc.) in the close vicinity of an extremal point. The basic principles were introduced by Luttinger and Kohn /1/, and over the years quite sophisticated models have been developed /2,3/ which are able to describe experiments in bulk (3D) semiconductors with high accuracy. Reviews on 3D work are given in ref./4/ and /5/.

In contrast, _ab initio_ calculations of the band structure, which aim at the description of the dispersion relations over the full Brillouin zone, are much more complex and require extensive numerical work. Although they are very successful in describing the overall band structure, they are usually not able to predict the detailed fine structure. However, information from these calculations is sometimes used as an input for a k.p-model.

This lecture is organized as follows: We start from a fundamental overview of k.p-theory in 3D systems, discussing the relevant principles, notions and parameters. Then k.p-theory is adapted and modified in order to account for the special features of the 2D case. The effective Hamiltonian for the envelope function is derived for the conduction band. Then first variational and then exact numerical solutions are discussed for the GaAs-GaAlAs system. After a short section about spin splitting, the lecture finishes with the discussion of tunnelling through semiconductor barriers.

Fundamentals

k.p-theory is a semi-empirical theory which requires knowledge on the properties of the symmetric wave functions at the extremal point. In Fig.1 the band structure close to the Γ-point is shown for a typical III-V semiconductor such as GaAs. The conduction band is s-type (Γ_1), and both the valence band and the higher conduction band are p-type (Γ_{15}). Including spin, double group representations are necessary: The conduction band then transforms as Γ_6, and the Γ_{15}-bands split into a Γ_7 (spin 1/2, twofold degenerate) and a Γ_8 (spin 3/2, fourfold degenerate) part.

The Hamiltonian of the system is given by:

(1) $H = \hat{P}^2/2m_0 + V_0(\vec{x}) + H_{SO} + \mu_B g_0 \vec{B} \cdot \vec{\sigma}$

where $V_0(\vec{x})$ is the crystal potential, H_{SO} is the spin-orbit interaction and $\vec{\sigma}$ are the Pauli spin matrices. m_0 and g_0 (=2) are the free electron mass and g-factor. We include a magnetic field B parallel to the z-axis from the beginning, working in Landau gauge with $\vec{A}=(-cBy,0,0)$. The generalized momentum operator $\hat{P}=\vec{p}+e\vec{A}/c$ obeys the commutation relations:

(2) $[\hat{P}_x, \hat{P}_y] = -i\hbar^2/l^2$

(3) $[\hat{P}_z, \hat{P}_x] = [\hat{P}_z, \hat{P}_y] = 0$

Ladder operators are defined as:

(4) $\hat{P}_\pm = (\hat{P}_x \pm i\hat{P}_y)/\sqrt{2}$

The relevant magnetic parameters are the Landau radius l and the cyclotron resonance energy $\hbar\omega_0$ of the free electron:

(5) $l = \sqrt{\hbar/eB}$

(6) $\hbar\omega_0 = \hbar eB/m_0 = 2\mu_B B = 2\lambda$

For the free electron system, motion in the (x,y)-plane is quantized into Landau levels $|n\rangle$, whereas it is free in z-direction (with wavevector k_z).

The calculation of the conduction band energies starts from the symmetric wave functions. At k=0 (meaning also B=0) the Bloch functions $u_1(0)$ obey the band-edge equations:

(7) $(H - \varepsilon_{1,0}) |u_1(0)\rangle = 0$

where $\varepsilon_{1,0}$ are the band edge energies. Quantizing the angular momentum in z-direction and choosing the zero of energy at the bottom of the conduction band, the following symmetric combinations are used:

(8) $\quad u_1 = iS\uparrow; \quad \varepsilon_{1,0} = 0$

(9) $\quad u_2 = iS\downarrow; \quad \varepsilon_{2,0} = 0$

(10) $\quad u_3(u_9) = R_+\uparrow; \quad \varepsilon_{3,0} = -E_0, \quad \varepsilon_{9,0} = E_1$

(11) $\quad u_4(u_{10}) = R_-\downarrow; \quad \varepsilon_{4,0} = -E_0, \quad \varepsilon_{10,0} = E_1$

(12) $\quad u_5(u_{11}) = (R_-\uparrow + \sqrt{2}Z\downarrow)/\sqrt{3}; \quad \varepsilon_{5,0} = -E_0, \quad \varepsilon_{11,0} = E_1$

(13) $\quad u_6(u_{12}) = (-R_+\downarrow + \sqrt{2}Z\uparrow)/\sqrt{3}; \quad \varepsilon_{6,0} = -E_0, \quad \varepsilon_{12,0} = E_1$

(14) $\quad u_7(u_{13}) = (-\sqrt{2}R_-\uparrow + Z\downarrow)/\sqrt{3}; \quad \varepsilon_{7,0} = -E_0 - \Delta_0, \quad \varepsilon_{13} = E_1 + \Delta_1$

(15) $\quad u_8(u_{14}) = (\sqrt{2}R_+\downarrow + Z\uparrow)/\sqrt{3}; \quad \varepsilon_{8,0} = -E_0 - \Delta_0, \quad \varepsilon_{14} = E_1 + \Delta_1$

(16) $\quad R_\pm = (X \pm iY)/\sqrt{2}$

Here the spin-orbit interaction is already included in the band edge energies. The p-type functions u_3-u_8 (u_9-u_{14}) have the same symmetry, but belong to the valence and to the higher conduction band, respectively. They are therefore orthogonal. The symbols $\uparrow$ ($\downarrow$) mean spin-up (spin-down) functions. S and (X,Y,Z) are periodic functions transforming like atomic s and p functions under the tetrahedal group at the Γ-point.

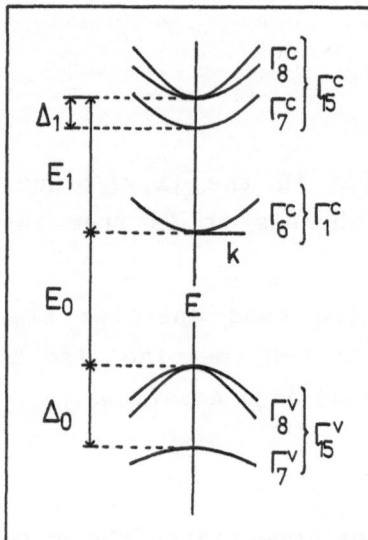

Fig.1.
Band structure at the Γ-point for GaAs. The s-type conduction band (Γ_1^c) and the p-type valence (Γ_{15}^v) and higher conduction band (Γ_{15}^c) are shown schematically. Also indicated are double group representations.

If we now "move" from the k=0 point towards finite but small k-values, the main effect will be to destroy the symmetry. In terms of the electronic states, this means that although the conduction band wave function is approximately s-type, components of different symmetry will be admixed. Interpreting this as a perturbation with respect to the "unperturbed" band edge functions $u_1(0)$, the "perturbed" wave function $\psi(k)$ will be a mixture of different $u_1(0)$-components. Basic perturbation theory tells us that the most relevant admixtures will be p-type, since the p-type bands are nearest to the conduction band.

Therefore the first step of k.p-theory is to expand the electronic wave function $\psi(k)$ in the vicinity of k=0 in terms of the symmetric Bloch functions $u_1(0)$:

(17) $\quad \psi(k) = \sum_1 f_1(k) \, u_1(0)$

The symmetric functions $u_1(0)$ oscillate rapidly, whereas the $f_1(k)$ are slowly varying envelope functions and describe the band - (or, to be more exact, the symmetry -) mixing away from the band edge. In principle, the index 1 goes over all energy bands of the crystal plus the continuum outside. For practical situations, however, only the nearest energy bands contribute. The present derivation is restricted to the calculation of the conduction band, which mainly interacts with the p-type valence and higher conduction band.

k.p-theory then takes advantage of the fact that, applying H on $\psi(k)$ from Eq.(17), the use of the band edge equations simplifies the result to:

(18) $\quad (H - \varepsilon(k))\psi(k) = 0 =$

$$= \sum_1 \left\{ u_1 \left(\frac{\hat{p}^2}{2m_0}\right) f_1 + \frac{1}{m_0} (\hat{p} u_1)(\hat{p} f_1) + (\mu_B g_0 B \hat{\sigma}_z + \varepsilon_{1,0} - \varepsilon) f_1 u_1 \right\}$$

In this expression we neglect nondiagonal spin-orbit interaction, which is a very small correction /5/.

The next step is to multiply Equ.(18) from the left with $u_m(0)$ and to integrate over the unit cells. The envelope functions $f_1(k)$ are assumed to be slowly varying and are taken to be constant over the unit cells. As a consequence, they can be taken out of the integral and one obtains a matrix equation for the $f_1(k)$:

$$(19) \quad \sum_1 \left\{ \frac{1}{m_0} <u_m|\vec{p}+\mu_B g_0 B\hat{\sigma}_z|u_1> + \delta_{1m}(\frac{\hat{p}^2}{2m_0} + \varepsilon_{1,0}-\varepsilon) \right\} f_1 = 0$$

The matrix elements dominating the symmetry mixing are momentum matrix elements between s- and p-functions:

$$(20) \quad \frac{1}{m_0} <S|\hat{p}_x|X_v> = \kappa_v \; ; \qquad \kappa_v^2 = P_v^2/2m_0$$

$$(21) \quad \frac{1}{m_0} <S|\hat{p}_x|X_c> = \kappa_c \; ; \qquad \kappa_c^2 = P_c^2/2m_0$$

The matrix element Q between the two p-type functions:

$$(22) \quad <X_v|\hat{p}_y|Z_c> = Q$$

affects the conduction band only indirectly (in third order perturbation theory); it has some influence on the spin splitting /6/ and describes the deviation from a spherical band structure (warping). We set it equal to zero at this point, but discuss it at a later point.

Further a small band mixing is introduced by the Pauli spin term.

In the present approximation, the higher conduction band and the valence band do not mix, which is the spherical approximation. In order to keep the expressions (relatively) short, we show now how to derive the band structure within a three-level model (conduction and valence band). The additional contributions from the higher conduction band have the identical form and are added only in the final formulae. For the three-level model, Equ.(19) is represented by the k.p-matrix Equ.(23) given on the next page.
Here we have written the band edge energies in the form of potentials and abbreviated the free kinetic energy operator with $\hat{T}$:

$$(24) \quad U_v = -E_0$$

$$(25) \quad U_\Delta = -E_0-\Delta_0$$

$$(26) \quad \hat{T} = \hat{p}^2/2m_0$$

$$
\begin{bmatrix}
(-\epsilon+\hat{T}+\lambda) & \kappa\hat{P}_- & \tfrac{1}{\sqrt{3}}\kappa\hat{P}_+ & -\sqrt{\tfrac{2}{3}}\kappa\hat{P}_+ & 0 & 0 & \sqrt{\tfrac{2}{3}}\kappa\hat{P}_z & \tfrac{1}{\sqrt{3}}\kappa\hat{P}_z \\[4pt]
\kappa\hat{P}_+ & (-\epsilon+U_v+\hat{T}+\lambda) & 0 & 0 & 0 & 0 & 0 & 0 \\[4pt]
\tfrac{1}{\sqrt{3}}\kappa\hat{P}_- & 0 & (-\epsilon+U_v+\hat{T}-\tfrac{\lambda}{3}) & -\tfrac{\sqrt{8}}{3}\lambda & \sqrt{\tfrac{2}{3}}\hat{P}_z\kappa & 0 & 0 & 0 \\[4pt]
-\sqrt{\tfrac{2}{3}}\kappa\hat{P}_- & 0 & -\tfrac{\sqrt{8}}{3}\lambda & (-\epsilon+U_\Delta+\hat{T}+\tfrac{\lambda}{3}) & \tfrac{1}{\sqrt{3}}\hat{P}_z\kappa & 0 & 0 & 0 \\[4pt]
0 & 0 & \sqrt{\tfrac{2}{3}}\kappa\hat{P}_z & \tfrac{1}{\sqrt{3}}\kappa\hat{P}_z & (-\epsilon+\hat{T}-\lambda) & \kappa\hat{P}_+ & -\tfrac{1}{\sqrt{3}}\kappa\hat{P}_- & \sqrt{\tfrac{2}{3}}\kappa\hat{P}_- \\[4pt]
0 & 0 & 0 & 0 & \kappa\hat{P}_- & (-\epsilon+U_v+\hat{T}-\lambda) & 0 & 0 \\[4pt]
\sqrt{\tfrac{2}{3}}\hat{P}_z\kappa & 0 & 0 & 0 & -\tfrac{1}{\sqrt{3}}\kappa\hat{P}_+ & 0 & (-\epsilon+U_v+\hat{T}+\tfrac{\lambda}{3}) & \tfrac{\sqrt{8}}{3}\lambda \\[4pt]
\tfrac{1}{\sqrt{3}}\hat{P}_z\kappa & 0 & 0 & 0 & \sqrt{\tfrac{2}{3}}\kappa\hat{P}_+ & 0 & \tfrac{\sqrt{8}}{3}\lambda & (-\epsilon+U_\Delta+\hat{T}-\tfrac{\lambda}{3})
\end{bmatrix}
\begin{bmatrix}
f_1 \\ f_3 \\ f_5 \\ f_7 \\ f_2 \\ f_4 \\ f_6 \\ f_8
\end{bmatrix}
= 0 \qquad (23)
$$

Thus we have derived an effective Schrödinger Equation for the envelope functions $f_1(k)$. The parameters of this equation are the band edges (which can be determined from spectroscopy with high accuracy) and the momentum matrix elements, which are limited in number and often zero (e.g. between two s-type bands). Usually these matrix elements are fitted from the band edge effective masses and g-factors.

The k.p matrix Equ.(23) can be diagonalised by substitution /5/ or correspondingly by the proper choice of the envelope function, which is written in the form:

$$(27) \quad \vec{f} = e^{ik_z z} \cdot \left\{ \phi_1|n>, \ \phi_3|n-1>, \ \phi_5|n+1>, \ \phi_7|n+1>, \right.$$
$$\left. \phi_2|n+1>, \ \phi_4|n+2>, \ \phi_6|n>, \ \phi_8|n> \right\}$$

with constant coefficients ($\phi_1 - \phi_8$). It is seen that band mixing occurs between different Landau levels.
For the diagonalisation, we find it useful to introduce:

$$(28) \quad T_n = (n+1/2)\hbar\omega_0, \quad T_z = \hbar^2 k_z^2 / 2m_0$$

and to write the spin-split Landau levels in the form:

$$(29) \quad E_n(\pm) = T_n + T_z \pm \mu_B B + x_n(\pm)$$

where $x_n(+,-)$ is a solution of:

$$(30) \quad x_n(\pm) = (T_n + T_z)(C_m + \frac{P^2}{3}(\frac{2}{x-U_v} + \frac{1}{x-U_\Delta})) \pm$$

$$\pm \frac{P^2 \hbar\omega_0}{6}(\frac{1}{x-U_\Delta} - \frac{1}{x-U_v}) + \gamma_\pm P_v^2 +$$

$$+ (P_v \to P_c, \ U_v \to U_c', \ U_\Delta \to U_\Delta')$$

$$(31) \quad \gamma_\pm = \frac{5\hbar^2 \omega_0^2}{18(x-U_v)^2} + \frac{\hbar^2 \omega_0^2}{9(x-U_\Delta)^2} + \frac{\hbar^2 \omega_0^2}{9(x-U_v)(x-U_\Delta)} \pm$$

$$\pm \frac{(-2\hbar\omega_0)(2T_n+T_z)}{9(x-U_v)^2} + \frac{2\hbar\omega_0(T_n-T_z)}{9(x-U_\Delta)^2} + \frac{2\hbar\omega_0(T_n+2T_z)}{9(x-U_v)(x-U_\Delta)}$$

The expresssion $\{+ \ (P_v \to P_c, U_v \to U_c', U_\Delta \to U_\Delta')\}$ means that the identical kinetic expressions must be added for the higher conduction band, with the parameters exchanged:

(32) $\quad U_C' = E_1$

(33) $\quad U_\Delta' = E_1 + \Delta_1$

In Equ.(30) a constant factor C_m has been included in order to account for possible contributions from higher bands. This rather phenomenological correction was introduced by Hermann and Weissbuch /4/ in order to optimize experimental fits. Although the relevance of C_m is not clear (and presently subject to a more detailed investigation), we include it for the sake of completeness.

The final result Equ.(29) shows that the main effect of the band mixing is to change the effective kinetic energy term, with the energies x_n appearing self-consistently in the denominators.

The simplest approximation of the band edge energies is obtained from Equ.(29) by neglecting the factors x_n in the denominators, leading to the parabolic form:

$$(34) \quad E_n^{parab.}(\pm) = (T_n + T_z)\,\frac{1}{m_0^*} \pm \frac{1}{2}\mu_B g_0^* B$$

with the band edge parameters:

$$(35) \quad \frac{1}{m_0^*} = 1 + C_m + \frac{P_v^2}{3}\left(\frac{2}{E_0} + \frac{1}{E_0+\Delta_0}\right) - \frac{P_c^2}{3}\left(\frac{2}{E_1} + \frac{1}{E_1+\Delta_1}\right)$$

$$(36) \quad g_0^* = 2 - \frac{2P_v^2}{3}\left(\frac{1}{E_0} - \frac{1}{E_0+\Delta_0}\right) + \frac{2P_c^2}{3}\left(\frac{1}{E_1} - \frac{1}{E_1+\Delta_1}\right)$$

Since the band edge parameters m_0^* and g_0^* can be measured experimentally with high accuracy, Equs.(35,36) are generally used for the determination of the momentum matrix elements P_v, P_c. Values of the factor C_m can only be resolved by very detailed investigations. Adequate values are given in ref./4/.

Thus the semi-empirical character of k.p-theory becomes evident: Within the calculation several band edge parameters appear (the gaps and the momentum matrix elements), which are determined from the band edge parameters m_0^* and g_0^*. Based on this empirical input, the great success of the theory is then to describe the dispersion relation E(k) away from the band edges with high accuracy. The relevant equations for this purpose are (29-31).

Being interested in an analytical $E(k)$-relation, one can expand the energy denominators until orders of x_n^2. For materials such as InSb and CdHgTe, where the fundamental gap E_0 is very small, a better procedure is to multiply first Equ.(30) with (x_n+E_0) in order to obtain good accuracy /5/. However, here we do not try to give analytical formulae for the energies, since Equ.(30) can be easily solved by iteration with a small computer. Instead, we give simple, linearized approximations for the effective masses and the energy levels.

Experiments, such as cyclotron resonance, usually measure energy differences, and for the description of the nonparabolicity very often the effective mass picture is used:

$$(37) \quad E_{n+1}(\pm) - E_n(\pm) = \hbar\omega_0/m^*(n,\pm)$$

This so-called cyclotron resonance mass $m*(n,\pm)$ is a handsome quantity which describes the deviations of the transition energies from their parabolic values. For many materials (especially GaAs) and experimental situations, this variation can be described with high accuracy by the form:

$$(38) \quad m^*(n,\pm) = m_0^* + 2n\hbar\omega_0/E_m^* \pm \hbar\omega_0/E_s^*$$

where two constants E_m^* and E_s^* have been introduced in order to give a feeling for the quantitative influence of the nonlinearities. The quantity E_m^* can be interpreted as an "effective gap", since it is always of the order of the fundamental energy gap E_0.

Using the effective band gap E_m^*, a good approximation of the energy levels $E_n(\pm)$ is given by:

$$(38a) \quad E_n(\pm) = \frac{E_m^*}{2}\left\{-1 + \sqrt{1 + (\frac{4(n+1/2)\hbar\omega_0}{m_0^*} \pm 2g_0^*\mu_B B)/E_m^*}\right\}$$

For GaAs, we take the band edge energies and g_0^* from Ref./4/. A precise value for the effective band edge mass was recently determined by Lindemann et al./7/ to be $m_0^*=.065$;

268

taking $C_m=-1.8$ /4/, the band edge parameters are given in Table 1. The nonparabolicity parameters are then obtained as $E_m{}^*=1.375eV$ and $E_s{}^*=-15eV$.

However, the factor $E_s{}^*$, which describes spin splitting in cyclotron resonance, does not describe experiments very accurately. This is due to the above mentioned fact that the spin splitting is rather strongly influenced by the mixing between the two p-type bands (via Q). This effect was recently calculated by Zawadzki et al./6/ and increases the spin splitting in GaAs by 20-30% with respect to the Q=0 value.

Table 1

	E_O (eV)	E_1 (eV)	Δ_O (eV)	Δ_1 (eV)	m_o^*	g_o^*
GaAs	1.519	3.140	.341	-.171	.065	-.44
Ga$_{.7}$Al$_{.3}$As	1.900	2.842	.310	-.171	.086	.45

	Cm	P_v^2 (eV)	P_c^2 (eV)	E_m^* (eV)	E_s^* (eV)	
GaAs	-1.8	29.43	6.139	1.375	-15	
Ga$_{.7}$Al$_{.3}$As	-1.8	29.48	6.574	/	/	

If we are interested in scattering matrix elements for the full wave function vector {f_1}, it is seen that the components (f_1, f_2) dominate. The other components ($f_3 - f_{14}$) are comparably small, of the order of $\mathcal{E}/E_0$. If one is interested in simple scattering matrix elements, it is sufficient to retain only the first two components. Of course, this statement holds true only in the situations where $\mathcal{E} \ll E_0$, which condition will not be fulfilled for semiconductors with a very small gap, such as InSb and CdHgTe. In these

materials, a correct theory should include matrix elements
for the full envelope function vector.

The great success of k.p theory is first of all to describe
the deviation of the dispersion relation E(k) from the
parabolic shape. Once E(k) is known, one can calculate
effective masses, optical transition energies (cyclotron
resonance), scattering matrix elements and the densitiy of
states.

k.p-theory thus represents a highly accurate method to
determine the band structure close to the extremal points.
If further fine structure effects are studied, such as polaron
effects (which also lead to nonlinearities in the energy
dispersion), the underlying structure of the bare energy
states is a fundamental requirement for any quantitative
analysis.

Finally, a short recipe is given how to construct a k.p-
model:

<u>(i)</u> Expand the exact states in terms of the symmetric band
edge Bloch functions and slowly varying envelope functions
$(u_l f_l)$.

<u>(ii)</u> Multiply the resulting Schrödinger Equation from the
left with the band edge functions u_m. Integrate over the
unit cells, approximating the slowly varying quantities
(envelope functions f_l) by their mean values within the unit
cells.

<u>(iii)</u> Diagonalize the most important parts of the resulting
k.p-matrix equation.

<u>(iv)</u> If further corrections are necessary, treat higher order
contributions by perturbation theory.

<u>(v)</u> Try to express the results for E(k) in terms of quantities
which can be easily handled for comparison with experiments.

In the remaining part of the lecture, this prescription will
be applied to 2D systems.

k.p-Theory for 2D systems

Attempting to formulate a k.p-theory in 2D systems, several new problems (compared to the 3D case) are immediately discovered:

"2D system" means that the electron motion is confined in a plane, with the other two directions free. Except for doping superlattices and delta-doping systems, the realisation of all 2D systems relies on the confinement of the carriers to (or between) surfaces and interfaces. The abrupt variation of the material at the boundary destroys the translational symmetry. The Bloch functions $u_l(0)$ are different on both sides of the interface. Thus it is necessary to describe the transition region between the two materials by constructing adequate matching conditions.

In contrast, slowly varying potentials due to space charges or a magnetic field, which also destroy the symmetry, can be treated as weak perturbations, varying only weakly over the unit cells. Evidently, an interface represents a qualitatively much more serious symmetry breaking.

A second new feature is that the electron densities are usually high and a single-particle picture is not sufficient. This means that a self-consistent electronic potential must be derived. Then, within k.p theory, the single particle equation of motion is diagonalized.

In the following we shall see that, once it is realized that the relevant quantities of k.p theory are matrix elements between band edge functions <u>within</u> unit cells, the interface problem can be directly solved.

A major simplification in the case of GaAs-GaAlAs or InP-GaInAs structures is that the lattice matching is very good and that the materials are qualitatively very similar, namely with the same symmetry and consequently similar Bloch functions.

Therefore, k.p theory can be developed for 2D systems based on the following arguments /8/: Without any detailed knowledge on the real wave functions, k.p theory generally relies on the matrix elements taken over the individual unit

cells (that is over 2-3Å), where slowly varying quantities
are approximated by their mean values. Therefore the unit
cells on both sides are described by the unperturbed bulk
functions, with the exception of a thin interface layer.
The "mixed" Bloch functions in the unit cells at the interface
are linear combinations of the "unperturbed" bulk functions.
If, as is the case in GaAs-GaAlAs systems, the band edge
functions on both sides are quite similar, the transition
between the two media can be described by a continuous
function h(z):

$$(39) \quad u_1(0) = u_1^{GaAs}(0) \cdot h(z) + u_1^{GaAlAs}(0) \cdot (1-h(z))$$

leading from the "GaAs"- to the "GaAlAs" values. The
transition is, of course, not an abrupt one; however, for
practical purposes, a step function will often be an adequate
approximation.

Here, an analogy exists with ternary semiconductors:
Although for GaAlAs a translational symmetry with identical
unit cells does not exist, the averaging procedure over many
unit cells yields a well-defined band structure (leading
rather linearly from GaAs values to AlAs values). Deviations
from the average composition are described as alloy disorder
scattering.

Using band edge functions of the form (39), and taking
the slowly varying potentials to be constant within the unit
cells, a k.p-matrix can be derived which has the identical
form as that derived for the 3D case /8/. The only
modification is the appearance of a diagonal conduction band
potential $U_c(z)$, and that also the potentials U_v, U_Δ etc.
are now z-dependent. Also the momentum matrix elements become
z-dependent, and the ordering of the momentum operators $\hat{P}_z$
must be arranged as in the matrix Equ.(23) in order to ensure
Hermiticity.

Diagonalizing the k.p-matrix for the 2D system, the
envelope function can be taken to be of the same form as
in 3D (Equ.27). The main difference is that the plane wave
$\exp(ik_z z)$ and the constants ϕ_1 are replaced by z-dependent
envelope functions $\phi_1(z)$. Solving again the k.p-matrix by
substitution, the following one-dimensional effective
Schrödinger Equation is obtained:

272

$$(40) \quad \left\{ U_c(z) + \frac{(n+1/2)\hbar\omega}{\mu(\varepsilon,z)} + \alpha(\varepsilon,z)(\hbar\omega_0)^2 + \right.$$

$$\left. + \hat{P}_z \frac{1}{2m_0\mu(\varepsilon,z)} \hat{P}_z \pm \frac{\lambda}{2} g^*(\varepsilon,z) - \varepsilon \right\} \phi(z) = 0$$

Here a small mixing term between ϕ_1 and ϕ_2 (which is irrelevant in high magnetic fields) has been neglected. The functions μ, $g*$ and α are given by:

$$(41) \quad \frac{1}{\mu(\varepsilon,z)} = 1 + C_m + \frac{P^2}{3}(\frac{2}{\varepsilon-U_v-T_n-\hat{T}_z} + \frac{1}{\varepsilon-U_\Delta-T_n-\hat{T}_z}) + (h.c.b)$$

$$(42) \quad g^*(\varepsilon,z) = 2 + \frac{2P^2}{3}\left\{ \frac{1}{\varepsilon-U_\Delta-T_n-\hat{T}_z} - \frac{1}{\varepsilon-U_v-T_n-\hat{T}_z} + \frac{4(T_n-\hat{T}_z)}{3(\varepsilon-U_\Delta)^2} + \right.$$

$$\left. + \frac{4(T_n+2\hat{T}_z)}{3(\varepsilon-U_v)(\varepsilon-U_\Delta)} - \frac{4(2T_n+\hat{T}_z)}{3(\varepsilon-U_v)^2} \right\} + (h.c.b)$$

$$(43) \quad \alpha(\varepsilon,z) = P^2_v\left\{ \frac{5}{18(\varepsilon-U_v)^2} + \frac{1}{9(\varepsilon-U_\Delta)^2} + \frac{1}{9(\varepsilon-U_v)(\varepsilon-U_\Delta)} \right\} + (h.c.b)$$

$$(44) \quad T_n = (n+1/2)\hbar\omega_0; \quad \hat{T}_z = \hat{P}^2_z/2m_0; \quad U_v=U_v(z); \quad U_\Delta=U_\Delta(z)$$

In these equations the expression (+h.c.b.) means that the identical terms for the $\underline{h}$igher $\underline{c}$onduction $\underline{b}$and must be added, with the parameters exchanged. Small corrections in the denominators have been neglected.

Equ.(40) describes the motion of two-dimensional electrons, which are quantized by the conduction band potential $U_c(z)$. In the kinetic term as well as in the spin-splitting term the z-dependent quantities $\mu(\varepsilon,z)$ and $g*(\varepsilon,z)$ appear. As can be seen from Equ.(41,42), their z-dependence comes from the local variation of the band edges ($U_v(z)$ etc.). In addition, the kinetic operator $\hat{T}_z$ appears in the denominators, which will be approximated by its expectation value.

Here, it is useful to discuss the notion "effective mass": The free kintic energy operator $\hat{T}$ contains the free electron mass m_0. If the electron moves through a crystal, its dispersion is changed due to the periodic potential. Within k.p-theory, an <u>effective</u> equation of motion is derived, which includes the influence of the crystal potential automatically, exploiting the band edge equations. If now the band edges vary due to space charges or material changes, these local variations will be also "seen" in the effective kinetic term - via $\mu(\mathcal{E},z)$. From Equ.(41) it is seen that only the potentials of the p-type bands appear in the local mass. This is due to the fact that the modification of the kinetic energy term is only due to momentum matrix elements.

$\mu(\mathcal{E},z)$ is not an experimentally observable quantity, but just that expression which appears in the effective kinetic term. In contrast, the effective masses determined from experiments are only parameters which are used for the phenomenological description of the data. They are correlated with $\mu(\mathcal{E},z)$ only in the sense that the theory solves the eigenvalue problem Equ.(40). The expectation value $\langle 1/\mu(\mathcal{E},z)\rangle$ will not directly lead to the effective masses, since the problem is self-consistent via the energies in the denominators. Only the calculated energy differences can then be described via the effective mass parameter (Equ.35).

Finally we stress the general form of the effective Schrödinger Equation (40). The potentials U(z) have not been specified until now. They can describe a heterostructure, a single quantum well or a superlattice. Superlattices have been recently reviewed by Bastard /9/. In this lecture, I will concentrate on heterostructures, applying the theory to the GaAs-GaAlAs system. Furthermore, Equ.(40) can also be used for the description of non-confined electrons in the presence of potential barriers. This corresponds to the tunneling problem, which will be treated in the last section.

Heterostructures

Heterostructures allow the realisation of very high quality electronic systems. Their band structure as well as many single- and many-particle effects can be resolved experimentally with high accuracy. Therefore the underlying bare energy levels of this system are of essential interest.

The quantisation of the electron motion results in a series of electric subbands. In very small gap materials (InSb, CdHgTe), usually many subbands are be occupied, which can then be analyzed only by detailed numerical calculations /10/. In the GaAs-GaAlAs system, mostly only the lowest electric subband is occupied, since the electron concentration seldom exceeds $5 \times 10^{11}/cm^2$. For the lowest subband, variational as well as numerical results will be derived and compared.

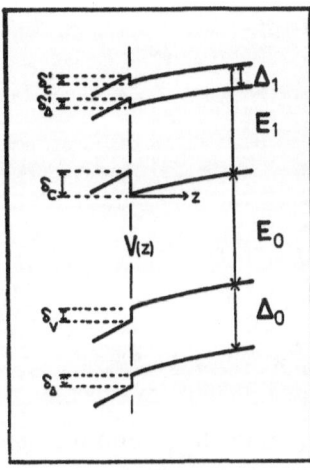

Fig.2.

Band edge potentials ($U_c(z)$, $U_v(z)$ etc.) for a GaAs-GaAlAs heterostructure, indicating the five-level approximation. The electrons are confined to the discontinuity of the conduction band δ_c.

Fig.(2) shows the local (z-dependent) band structure for a GaAs-GaAlAs heterostructure. The electrons are confined to the discontinuity of the conduction band, and the valence band and the higher conduction band are also shown. The typical height of the barrier is 300meV and the electronic energies are of the order of 20-50meV, counted from the bottom of the discontinuity. The effective mass is around 0.07, and the mean distance of the wave function from the interface is 100Å. The band discontinuities are denoted as δ_c, δ_v etc., and the resulting band edge potentials for the effective Schrödinger Equation are:

(45) $\quad U_c(z) = V(z) + \delta_c h(-z)$

(46) $U_V = V(z) - E_0 - \delta_V h(-z)$

(47) $U_\Delta = V(z) - E_0 - \Delta_0 - \delta_\Delta h(-z)$

(48) $U_C' = V(z) + E_1 + \delta_C' h(-z)$

(49) $U_\Delta' = V(z) + E_1 + \Delta_1 + \delta_\Delta' h(-z)$

V(z) denotes the slowly varying space charge potential:

(50) $V(z) = V_d(z) + V_H(z)$

including the depletion:

(51) $V_d(z) = 4\pi e^2 N_d z / \kappa_s$

and the Hartree potential:

(52) $V_H(z) = 4\pi e^2 N_{el} \left\{ z - \int_{-\infty}^{z} dz' \rho(z')(z-z') \right\} / \kappa_s$

κ_s is the static dielectric constant, N_d and N_{el} denote the depletion and the inversion charge densities (per cm²) and $\rho(z)$ is the normalized local carrier density.

Inserting Equs.(45-50) into the effective Schrödinger Equation (40), the electronic energy levels can be calculated. Through the electronic charge density in the Hartree potential, the problem requires a self-consistent solution.

Wave Function Matching at the Interface

Already Ben Daniel and Duke /11/, and more recently Morrow and Brownstein /12/ in a more general study have pointed out that the derivative of the envelope functions need not necessarily be continuous across the interface. Their results are best described as follows: Since the right hand side of:

$$(53) \quad \left\{ \frac{d}{dz} \frac{1}{\mu(\varepsilon,z)} \frac{d}{dz} \right\} \phi(z) = \frac{2m}{\hbar^2}0 \left(U_c - \varepsilon - T_n / \mu(\varepsilon,z) \right) \phi(z)$$

is finite for all z, the left hand side must be similarly well behaved (g-factor corrections have been neglected). This implies that, integrating from $-\delta$ to $+\delta$ and taking the limit $\delta \to 0$, the expression:

$$(54) \quad \frac{1}{\mu(\varepsilon,z)} \left\{ \frac{d}{dz} \phi(z) \right\}$$

must be continuous at z=0. In other words, although the absolute value of ϕ is continuous, the derivative $d\phi/dz$ can be discontinuous if the local effective mass varies discontinuously. For abrupt interfaces one obtains:

$$(55) \quad \frac{1}{\mu(\varepsilon,z)} \left\{ \frac{d}{dz} \phi(z) \right\} \Big|_{z=0_-} = \frac{1}{\mu(\varepsilon,z)} \left\{ \frac{d}{dz} \phi(z) \right\} \Big|_{z=0_+}$$

Thus the envelope function need not necessarily be differentiable at the interface, if material parameters change abruptly. This effect accounts for the mixing of the different band edge functions at the interface.

Variational Solutions

A very simple technique for approximating the eigenvalues of the Schrödinger Equation (40) is to propose a variational solution for $\phi(z)$ and to take the expectation value of /8/:

$$(56) \quad <U_c(z)> + T_n <\frac{1}{\mu(\varepsilon,z)}> + (\hbar\omega_0)^2 <\alpha(\varepsilon,z)> +$$

$$+ \frac{1}{2m_0} <\hat{P}_z \frac{1}{\mu(\varepsilon,z)} \hat{P}_z> \pm \frac{\lambda}{2} <g^*(\varepsilon,z)> - \varepsilon = 0$$

From this equation, the (implicitly given) single particle energies ε_n are calculated and the total energy is minimized variationally in a Hartree- or Hartree-Fock approximation.

A particularly simple choice for the variational wave functions is the Fang-Howard form /13/:

$$(57) \quad \phi(z) = \sqrt{b^3/2}\, z\, e^{-bz/2}$$

where the variational parameter b is given by:

$$(58) \quad b = \left(\frac{48\pi m e^2 (N_d + 11 N_{el}/32)}{\kappa_s \hbar^2}\right)^{(1/3)}$$

For this variational function, the kinetic and potential expectation values are given by /14/:

$$(59) \quad <\hat{T}_z> = \hbar^2 b^2/8m_0$$

$$(60) \quad <V(z)> = \frac{12\pi e^2}{\kappa_s b}\left(N_d + \frac{11}{16}N_{el}\right)$$

An improved version of the wave function, which accounts for the penetration into the barrier (and thus represents a more relaxed solution) is /15/:

$$(61) \quad \phi(z) = \sqrt{Bb}\,(zb+k)e^{-bz/2} \quad \ldots z>0$$

$$\phi(z) = \sqrt{Cc}\, e^{cz/2} \quad \ldots z<0$$

Using the normalisation condition and the matching conditions for the wave function at the interface, this trial function remains with two variational parameters.

Results for the two-parameter variational function Equ.(61) have been given in ref./8/. Here, we only derive a simpler expression using the Fang-Howard function. In this case the GaAlAs side does not contribute. A further simplification is introduced by replacing the slowly varying potential and the kinetic energy operator in the denominators by the expectation values $\langle V(z) \rangle$ and $\langle \hat{T}_z \rangle$. The 2D single particle energies are then obtained as:

(62) $\quad E_n(\pm) = \langle V(z) \rangle + \langle \hat{T}_z \rangle + T_n \pm \lambda + x_n(\pm)$

(63) $\quad x_n(\pm) = (T_n + \langle \hat{T}_z \rangle)(C_m + \frac{P_v^2}{3}(\frac{2}{x+E_0} + \frac{1}{x+E_0+\Delta_0})) \pm$

$$\pm \frac{P_v^2 \hbar \omega}{6} {}_0(\frac{1}{x+E_0+\Delta_0} - \frac{1}{x+E_0}) + P_v^2 \gamma_\pm +$$

$$+ (P_v \to P_c; \ E_0 \to -E_1; \ \Delta_0 \to -\Delta_1)$$

(64) $\quad \gamma_\pm = \frac{\hbar^2 \omega^2}{18} {}_0 \left\{ \frac{5}{(x+E_0)^2} + \frac{2}{(x+E_0+\Delta_0)^2} + \frac{2}{(x+E_0+\Delta_0)(x+E_0)} \right\} \pm$

$$\pm \frac{2\hbar\omega}{9} {}_0 \left\{ \frac{T_n - \langle \hat{T}_z \rangle}{(x+E_0+\Delta_0)^2} - \frac{2T_n + \langle \hat{T}_z \rangle}{(x+E_0)^2} + \frac{T_n + 2\langle \hat{T}_z \rangle}{(x+E_0)(x+E_0+\Delta_0)} \right\}$$

where Equ.(63) can be resolved by expanding the denominators or by numerical iteration.

The interesting result is that the 2D equations have the identical form as the 3D result (Equ.29-31); the only difference is that the electric potential $\langle V(z) \rangle$ appears additively and that the free kinetic energy in z-direction $\hbar^2 k_z^2/2m_0$ is replaced by the expectation value $\langle \hat{T}_z \rangle = \hbar^2 b^2/8m_0$ for the variational function. So the deviation of the band structure from the parabolic form is mainly determined by the kinetic terms. The effect of the electric potential is only to quantize the motion, thus increasing the kinetic terms and thus indirectly "creating" nonparabolicity.

These self-consistent variational results will be subsequently compared to numerical solutions.

Numerical Results

For the analysis of the energy fine structure in 2D systems, it is often quite relevant to know the contribution of nonparabolicity as exact as possible (such as for the polaron effect). The main reason is that k.p theory yields the bare band structure underlying all other physical effects. Since the other effects are often smaller or of the same order as the nonparabolicity, their analysis and interpretation depends strongly on the exact knowledge of the bare levels.

Therefore we have performed a self-consistent calculation of the electronic energy levels in GaAs-Ga$_{.7}$Al$_{.3}$As heterostructures, including the effects of exchange and correlation. The calculation is performed in analogy to that performed by Das Sarma and Stern /16/. Their approach is improved by using the correct expression in the kinetic energy according to Equ.(40) (where they have taken the band edge masses of the two materials).

The main variables of the calculation for the 2D system are the depletion charge and the inversion charge. The relevant band edge parameters can be taken from Table 1, and a band offset $\delta_c = 0.7(E_{0,GaAlAs} - E_{0,GaAs})$ is used for the calculation.

The essential result of our calculations is that, in analogy to the 3D case, the 2D band structure can be expressed in terms of an effective 2D mass $m_0^*(2D)$ and an effective 2D band gap $E_m^*(2D)$ with good accuracy. Such a description is most suitable for a comparison with experimental results. Neglecting spin splitting, we write:

(65) $\quad E_{n+1} - E_n = \hbar\omega_0 / m_2^*(n)$

(66) $\quad m_2^*(n) = m_0^*(2D) + 2n\hbar\omega_0 / E_m^*(2D)$

The "effective 2D band edge mass" as well as the "effective 2D gap" depend on N_d and N_{el}. $m_0^*(2D)$ includes mainly the influence of the kinetic energy in z-direction (and is different for each electric subband), whereas $E_m^*(2D)$ describes the influence of motion parallel to the interface.

Fig.3 shows $m_0^*(2D)$ as a function of the inversion charge N_{el} and for two depletion charge densities: $N_d = 8 \times 10^{10}/cm^2$

(full line) and $N_d = 4.7 \times 10^{10}/cm^2$ (broken line). It is seen that the band edge mass increases by approximately 1% for the condsidered density range; the higher depletion potential creates also a significantly higher nonparabolicity. For comparison, the dotted curve shows the variational result according to Equ.(62) for the Fang-Howard trial function.

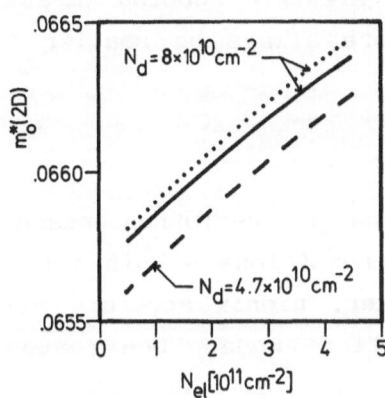

Fig.3.

Effective 2D band edge mass $m_o^*(2D)$ for a $Ga_{.7}Al_{.3}As$ heterostructure, plotted versus the electron concentration for two depletion charge values.

Full and dashed curve: numerical result;

dotted curve: variational result

It is seen that the variational result exhibits qualitatively the same behaviour, but somewhat overestimates the nonparabolicity. However, with regard to the simplicity of the Fang-Howard function, the analytical result Equ.(62-64) for the 2D band structure represents a great success. The physical background of this success is that mainly the kinetic energy terms contribute to nonparabolicity, which are well approximated by the trial function.

The effective gap $E_m^*(2D)$ is changed only weakly by the two-dimensionality, and we find a good average fit to be 1.405eV.

If these results are compared to experimental data, it should be kept in mind that here we calculate the bare masses. In bulk GaAs, the polaron effect leads to an additional increase of the 3D effective mass by 1-1.1%, plus nonlinear corrections, which are of the order of the nonparabolicity /7/. These polaron corrections are also present in the 2D system, although they are reduced by occupation number effects and screening /17/. Therefore, if one aims at describing 2D polarons in detail, the bare band structure is quite relevant.

If more than one electric subband is occupied, it is necessary to define an effective mass for each subband. In this case, the experiments usually measure the effective mass of particles at the Fermi energy. However, at the Fermi energy the electrons in the higher subband will always have a relatively larger amount of potential energy, and a smaller amount of kinetic energy. According to the above discussion, the effective mass for the higher electric subband (measured at the Fermi energy) will therefore always be smaller than that of the lower subband.

Spin Splitting

Spin splitting, which is observed in cyclotron resonance and in electron spin resonance (transitions within the same Landau level), represents a further, highly accurate method to test the validity of a theoretical model. One reason is that electron spin resonance measures directly the single particle spin splitting. This is in contrast to transport experiments, where also many-particle effects are measured.

In Fig.4 the spin splitting is plotted versus the magnetic field for a transition within the first Landau level. Experimental points are from Stein et al./18/; the dashed curve shows the spin splitting calculated with the Fang-Howard function using Equs.(62-64). The discrepancy can be resolved by performing higher order perturbation theory, including also the p-p coupling (Q=0). Such an improved calculation has been performed by Lommer et al./19/; their results are shown by the full curve. Again, the relative success of the simple variational model becomes evident.

The dotted curve shows the transition within the zeroth Landau level (variational result), which is closer to the parabolic (band edge) value. The reason for this behaviour is evident from the analytical structure of the effective g-factor Equ.(42), which decreases (in absolute units) as the energy increases.

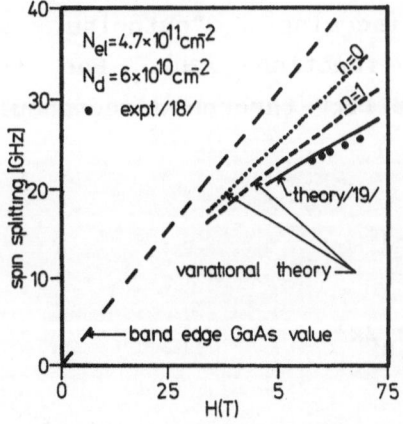

Fig.4.
Spin splitting for a GaAs-GaAlAs hetero-structure, plotted versus the magnetic field. Experimental data from Stein et al./18/ are compared to variational results (Equ.62) and to higher order perturbation theory /19/.

Finally, we point out that not only the magnetic energy levels, but also the zero-field energy levels exhibit spin splitting. In the derivation of Equ.(40) it has been stated that a small mixing term between the components ϕ_1 and ϕ_2 has been neglected. At zero magnetic field this expression becomes important and leads to a spin splitting of states with parallel wavevector $k_\parallel$ /8/:

$$(67) \quad |E(k_\parallel\uparrow) - E(k_\parallel\downarrow)| = \frac{\hbar^2 k_\parallel}{3m_0} \left| \frac{d}{dz} \left\{ P_v^2 \left(\frac{1}{\varepsilon-U_v} - \frac{1}{\varepsilon-U_\Delta} \right) + P_c^2 \left(\frac{1}{\varepsilon-U_c'} - \frac{1}{\varepsilon-U_\Delta'} \right) \right\} \right|$$

This splitting of the spin-up and spin-down level is a consequence of the asymmetry of the band edge potentials of the p-type bands ($U_v(z)$ etc.), which is felt by the electrons, even in the present "spherical" approximation.

Tunneling Through Semiconductor Barriers

Tunneling through barriers, which would be impenetrable by classical particles, is one of the most fundamental quantum processes. Therefore a basic question is, what is the equation of motion which governs the tunneling process. In the simplest approximation of stationary tunneling it is the time-independent Schrödinger Equation, describing eigenstates with a certain amount of "incoming", "outgoing" and "reflected" part of the wave function /20/. For free electrons, the transmission probability through a rectangular potential barrier is given by:

$$(68) \quad T = \left\{ 1 + \frac{(1+\beta^2)^2}{4\beta^2} \sinh^2(\kappa d) \right\}^{-1}$$

$$(69) \quad \kappa = \sqrt{2m(U-\varepsilon) + k_\parallel^2}; \qquad \beta^2 = k_z^2/\kappa^2$$

Here ($k_\parallel$, k_z) denote the momentum components parallel and normal to the barrier, and ε is the energy of the incoming particle. U and d are the height and the thickness of the barrier. The coefficient $\varkappa$ describes the decay of the wave function into the barrier and determines essentially the tunneling probability.

However, Equ.(68) does not describe a realistic problem. Solid state Hamiltonians have generally a more complex structure than that of a free electron with a simple potential barrier. Already decades ago, it was discovered that band structure effects have a significant influence on the tunneling probabilities. A discussion of this problem is found in ref./20/.

Based on the k.p-equations derived in the previous chapters, single particle tunneling is described in a fundmental but simple way. We consider the band structure for the tunneling from n-GaAs to n-GaAs through a GaAlAs barrier (Fig.5). It is evident that, in analogy to the 2D case dicussed above, an effective Schrödinger Equation can be derived:

$$(70) \qquad \left\{ U_c(z) + \hat{P} \frac{1}{2m_0 \mu(\varepsilon,z)} \hat{P} - \varepsilon \right\} \Psi(z) = 0$$

$$(71) \qquad \frac{1}{\mu(\varepsilon,z)} = 1 + C_m + \frac{P^2}{3} (\frac{2}{\varepsilon - U_V} + \frac{1}{\varepsilon - U_\Delta}) + \frac{P^2}{3} (\frac{2}{\varepsilon - U'_c} + \frac{1}{\varepsilon - U'_\Delta})$$

In these expressions spin effects and the free kinetic terms in the denominators have been neglected.

From Equ.(70) the stationary solutions must be determined, which describe the tunneling process through semiconductor barriers. With respect to the 2D problem, only the potential levels $(U_c(z), U_v(z) \ldots)$ appearing in the local effective mass $\mu(\varepsilon,z)$ (Equ.41) are changed; they describe only the barrier, with no confining potential, as can be seen from Fig.(5).

As an example, we calculate the problem of single particle tunneling from bulk GaAs to bulk GaAs through a $Ga_{.7}Al_{.3}As$ barrier (Fig.5). The E(k) relation of the particle in the GaAs regime is given by:

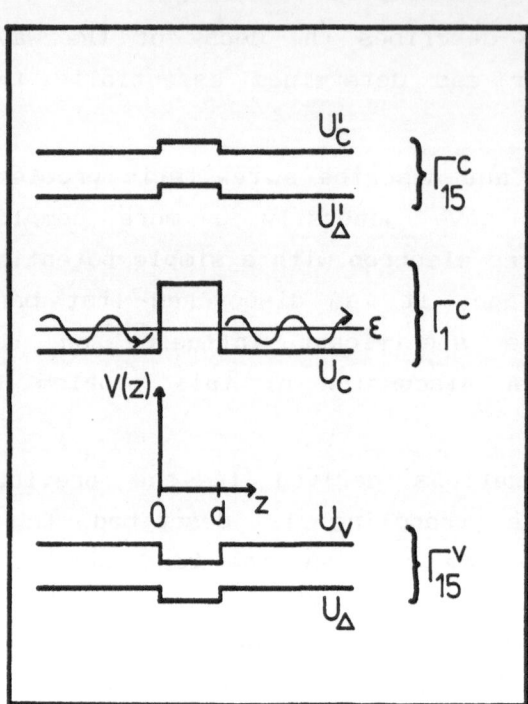

Fig.5.
Band structure "seen" by an electron tunneling from GaAs to GaAs through a GaAlAs barrier. Within the barrier, the effective mass of the electron is smaller than the band edge mass of the barrier material.

(72) $(k_z^2 + k_{/\!/}^2) = 2m_o\mu_o(\varepsilon)\cdot\varepsilon$ ... $z<0, z>d$

(73) $\frac{1}{\mu_o(\varepsilon)} = 1 + C_m + \frac{P^2}{3}(\frac{2}{\varepsilon+E_0} + \frac{1}{\varepsilon+E_0+\Delta_0}) + \frac{P^2}{3}(\frac{2}{\varepsilon-E_1} + \frac{1}{\varepsilon-E_1-\Delta_1})$

Within the barrier, characterized by a discontinuity δ_c, the particle dispersion and effective mass are:

(74) $\quad (k_z^2 + k_{\parallel}^2) = 2m_0\mu_1(\varepsilon)\cdot(\varepsilon-\delta_c) \quad \ldots \ 0<z<d$

(75) $\quad \dfrac{1}{\mu_1(\varepsilon)} = 1 + C_m + \dfrac{P_v'^2}{3}\left(\dfrac{2}{\varepsilon+E_0'-\delta_c} + \dfrac{1}{\varepsilon+E_0'+\Delta_0'-\delta_c}\right) +$

$$+ \dfrac{P_c'^2}{3}\left(\dfrac{2}{\varepsilon-E_1'-\delta_c} + \dfrac{1}{\varepsilon-E_1'-\Delta_1'-\delta_c}\right)$$

The primed quantities P_v', E_0' etc. mean the GaAlAs band edge values. Here we point out that, since the tunnelling electron has an energy much lower than the barrier band edge, $\mu_1(\varepsilon)$ is considerably smaller than the band edge value m_0^* of the Ga$_{.7}$Al$_{.3}$As. (In contrast, the effective masses increase for energies above the conduction band edge). For the GaAs-Ga$_{.7}$Al$_{.3}$ system (with a band offset of 70% of the fundamental gap difference), the band edge mass of the barrier material m_0^* is .086, whereas the "tunneling" mass $\mu_1(\varepsilon\to0)$ is 0.0708 and $\mu_1(\varepsilon=50\text{meV})=0.0735$.

Furthermore, due to the abrupt effective mass variation at the interfaces, the matching conditions for the wave function must also be satisfied:

(76) $\quad \dfrac{1}{\mu_0(\varepsilon)}\dfrac{d\Psi}{dz}(z=0^-) = \dfrac{1}{\mu_1(\varepsilon)}\dfrac{d\Psi}{dz}(z=0^+)$

(77) $\quad \dfrac{1}{\mu_1(\varepsilon)}\dfrac{d\Psi}{dz}(z=d^-) = \dfrac{1}{\mu_0(\varepsilon)}\dfrac{d\Psi}{dz}(z=d^+)$

With these ingredients, the coefficients determining the tunneling probability are evaluated straightforward:

(78) $\quad \beta^2 = k_z^2\mu_1^2(\varepsilon)/\kappa^2\mu_0^2(\varepsilon)$

(79) $\quad \kappa^2 = 2m_0\mu_1(\varepsilon)(\delta_c-\varepsilon) + k_{\parallel}^2$

In order to illustrate the relevance of a consistent k.p-theory of tunneling, Fig.(6) shows the transmission rate through a Ga$_{.7}$Al$_{.3}$As barrier with a thickness of d=50Å. The tunneling probability is plotted versus the energy of the incoming particle (with $k_{\parallel}$=0). The broken line shows the "parabolic" result using the band edge mass m_0^*=.086 in the barrier. The full line is the nonparabolic result obtained

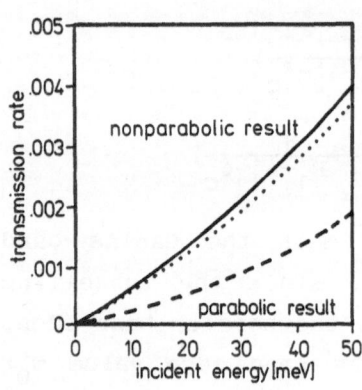

Fig.6.
Transmission rate for an electron tunneling from GaAs to GaAs through a GaAlAs barrier of 50Å, plotted against the incident particle energy. The lower effective mass in the barrier obtained with k.p-theory enhances the tunneling.

from Equs.(78,79). For comparison, the dotted curve shows the nonparabolic result, but neglecting the matching conditions for the wave function. It is seen that, using the nonparabolic mass, the transmission is significantly higher than that for the parabolic value. The matching conditions represent a smaller, multiplicative effect.

For physical situations where the barrier is either thicker or higher, it will be found that the influence of the correct nonparabolic mass it still higher than that obtained in Fig.(6). This is so because the mass appears exponentially in the tunneling probability, and therefore orders of magnitude variations are possible.

The above example is quite useful for illustration, whereas in a real system the application of an electric field (which produces the tunneling current) creates a slowly varying potential; also space charges will result in potential variations. If, in addition to the abrupt material transitions, a slowly varying contribution $V(z)$ is present in the band edge potentials $U_c(z)$ etc., similar to Equs.(45-49), then Equ.(70) is not as easily solved as with flat potentials. Within the barrier, the local effective mass is then $\mu_1(\mathcal{E}-V(z))$.

Also the matching conditions for the wave function are changed, since the local effective mass discontinuities are now different at the two interfaces.

A reasonable semiclassical solution for the tunneling is provided by WKB-theory. In this approximation, we derive the transmission for an incoming particle with energy ε:

$$(80) \quad T = \frac{16\beta_l^2}{(1+\beta_l^2)(1+\beta_r^2)} \exp\left(-2\int_0^d \kappa(\varepsilon,z)\,dz\right)$$

$$(81) \quad \kappa(\varepsilon,z) = \sqrt{2m_0\mu_1(\varepsilon-V(z))(\delta_c+V(z)-\varepsilon)+k_\shortparallel^2}$$

$$(82) \quad \beta_l = \frac{k_z}{\kappa(\varepsilon,0)}\frac{\mu_1(\varepsilon-V(0))}{\mu_0(\varepsilon-V(0))}$$

$$(83) \quad \beta_r = \frac{k_z'}{\kappa(\varepsilon,d)}\frac{\mu_1(\varepsilon-V(d))}{\mu_0(\varepsilon-V(d))}$$

k_z and k_z' denote the wavevectors of the incoming and of the outgoing particle. The simplest situation is shown in Fig.(7): A constant electric field F, which "drives" the electron, leads also to a variation of the local effective mass, because on the right side the particle energy is closer to the (barrier) band edge. In addition, the outgoing wavevector k_z' is larger than the incoming k_z.

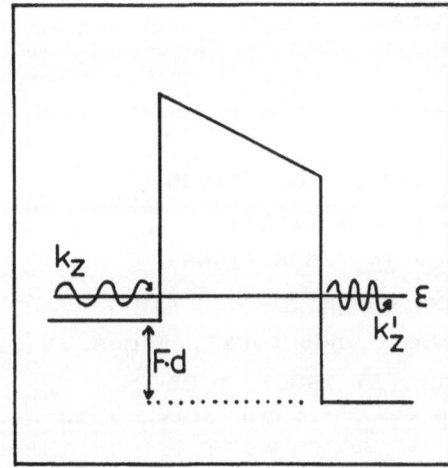

Fig.7.
Schematic description of electron tunneling, including the acceleration by an electric field F. Within the barrier, the local potential U_c as well as the local effective mass change.

Thus, the WKB-expression integrates over the local exponent $\alpha(\varepsilon,z)$ and over the local potential variation. Of course, space charges accumulating at the two interfaces will also create a slowly varying potential, and the flat potential on both sides will never occur; however, for the tunneling process, Equ.(80) is the relevant expression.

The essential result of this section is that tunneling theory must start from an adequate equation of motion, with the Hamiltonian given in Equ.(70). In this Hamiltonian the kinetic term contains a local, energy-dependent effective mass $\mu(\xi,z)$. Generally, this effective mass is smaller than the band edge value of the barrier material, because the particle energy is closer to the valence band than the barrier conduction band edge is. Since the effective mass appears exponentially in the transmission rate (via $\varkappa(\xi,z)$), the corrections can influence the result by orders of magnitude. The WKB-expression for the simple structure Fig.(7) contains the essential contributions.

Acknowledgements

I would like to thank Winfried Boxleitner for performing the numerical calculations and for support in preparing this lecture. This work was partially sponsored by the Stiftung Volkswagenwerk.

References

(1)   J.Luttinger and W.Kohn; Phys.Rev.97, 869 (1955)

(2)   E.O.Kane; J.Phys.Chem.Solids 1, 249 (1957)

(3)   J.Zak and W.Zawadzki; Phys.Rev.145, 536 (1966)

(4)   C.Hermann and C.Weisbuch; Phys.Rev.B 15, 823 (1977)

(5)   W.Zawadzki; in "Narrow Gap Semiconductors", Nimes 1979, edited by W.Zawadzki (Springer, NY 1980), p.85

(6)   W.Zawadzki, P.Pfeffer and H.Sigg; Solid State Comm.93, 777 (1985)

(7)   G.Lindemann, R.Lassnig, W.Seidenbusch and E.Gornik; Phys.Rev.B 25, 4693 (1983)

(8)   R.Lassnig; Phys.Rev.B 31, 8076 (1985)

(9)   G.Bastard; Proc. 6th Int. Conf. on Electronic Properties of 2D Systems, Kyoto 1985, p. 423

(10) Y.Takada, K.Arai, N.Uchimura and Y.Uemura;
J.Phys.Soc.Jpn. 49, 1851 (1980)

(11) D.J.Ben Daniel and C.B.Duke; Phys.Rev. 152, 683 (1966)

(12) R.A.Morrow and K.R.Brownstein; Phys.Rev.B 30, 678 (1984)

(13) F.F.Fang and W.E.Howard; Phys.Rev.Lett. 16, 797 (1966)

(14) T.Ando, A.B.Fowler and F.Stern; Rev.Mod.Phys. 54, 437
(1982)

(15) T.Ando; J.Phys.Soc.Jpn. 51, 3900 (1982)

(16) F.Stern and S.Das Sarma; Phys.Rev.B 30, 840 (1984)

(17) R.Lassnig; Proc. 6th Int. Conf. on Electronic Properties
of 2D Systems, Kyoto 1985, p. 506

(18) D.Stein and K.von Klitzing; Phys.Rev.Lett. 51, 130 (1984)

(19) G.Lommer, F.Malcher and U.Rössler; Phys.Rev.B 32, 6965
(1985)

(20) C.B.Duke; "Tunneling in Solids", Solid State Physics,
Suppl. 10, Academic Press, 1969

INVERSION ELECTRONS ON InSb IN CROSSED ELECTRIC AND MAGNETIC FIELDS

U. Merkt

Universität Hamburg
Institut für Angewandte Physik
2000 Hamburg 36, Jungiusstr.11, F.R.G.

A crossed field configuration for electrons is created by the strong electric field that is perpendicular to a space-charge layer and a magnetic field that is applied parallel to the layer. On InSb simultaneously diamagnetically shifted intersubband resonances and cyclotron resonances have been observed in inversion layers in this crossed field configuration. These experiments are reviewed and are compared with simple analytical expressions for eigenenergies and optical excitation strengths of the hybrid electric-magnetic surface band structure in the triangular-well approximation of the electrostatic potential (constant surface electric field). In addition to the one-band effective-mass approximation, we also treat a two-band model that takes into account the effects of nonparabolicity in narrow-gap semiconductors. In particular, we find that cyclotron masses of inversion electrons on InSb in crossed fields show in a rather spectacular way the analogy between electrons in narrow-gap semiconductors and relativistic electrons in vacuum and we demonstrate the destruction of the Landau quantization in strong transverse electric fields.

1. INTRODUCTION

In the bulk of metals and semiconductors the energy spectrum of electrons in a magnetic field is split into highly degenerate Landau levels. Near surfaces the degeneracy is lifted and magnetic surface levels are formed.[1] In metals and semimetals the Fermi energy by far exceeds the cyclotron energy and magnetic surface levels are occupied. In nondegenerate semiconductors the situation is different and surface levels must be populated by different means. This can be achieved by an electric field that forces the electrons towards the surface. However, a surface electric field quantizes the motion of electrons into electric subbands and coupling of the magnetic and electric quantization must be considered.[2] The resulting hybrid electric-magnetic subbands are of electric type if the electric subband energies exceed the cyclotron energy, and vice versa.[3,4]

In the hybrid subbands two different kinds of electrons have been observed, namely, surface bound electrons and bulklike electrons.[5-8] In weak magnetic fields the bound electrons behave similar to electrons in quasi two-dimensional electric subbands and their optical excitations are diamagnetically shifted intersubband resonances. In strong magnetic fields bulklike electrons exist that can be regarded as electrons in crossed electric and magnetic fields.[6,9]

In previous work on optical absorption in crossed electric and magnetic fields only interband transitions on Ge were studied.[10] Only recently corresponding experiments could be performed on intraband absorption with use of inversion layers on InSb. In particular, the destruction of the Landau quantization in strong electric fields, that has been predicted from the two-band model for narrow-gap semiconductors, can be addressed in this system on InSb.

To describe our experiments, we employ the one-band effective-mass approximation[6] and a two-band k·p theory.[9] In these models we always assume a constant surface electric field to achieve simple analytical expressions for eigenenergies and optical excitation strengths, i.e., we treat the triangular-well approximation.[2]

2. CLASSICAL TRAJECTORIES OF INVERSION ELECTRONS IN CROSSED FIELDS

It is illuminating to discuss classical trajectories of electrons in electric and magnetic fields before treating the quantum mechanical description in terms of eigenenergies, wave functions, and quantum numbers.[6] The classical motion reveals some of the very characteristic behavior of surface electrons in crossed fields and provides an intuitive and vivid illustration of the most important quantum number which is the so-called center coordinate. The motion of electrons in unlimited space in homogeneous electric and magnetic fields, respectively, is quite different in its nature: the trajectory in an electric field is a parabola and the motion is unbound, whereas in a magnetic field the motion is confined to a Landau circle. As a consequence, the quantum mechanical energy spectrum in an electric field is continuous as long as there is no barrier. In case of spatial confinement of the electrons in at least one spatial dimension, we have the well-known quantization into electric subbands. On the other hand, in magnetic fields there is discrete Landau quantization in the absence of any spatial confinement.

In Fig.1 classical trajectories of electrons near a barrier in crossed electric and magnetic fields are depicted. In Fig. 1(a) the electric limit that corresponds to electric subbands (B=0) is shown. As initial conditions we choose the coordinate $(0,0,z_i)$ and two velocities $(0,\pm v_y,0)$. In the absence of the surface at z=0 the electron accelerates in the electric field and the motion is unbound (dashed line). Quantum-mechanically this corresponds to a continuous spectrum. In the presence of a surface, that is assumed to be ideally smooth, the electron is periodically reflected from the surface and follows a so-called skipping trajectory (solid line). For the two initial velocities $\pm v_y$ the same trajectory is followed in opposite directions.

In Fig. 1(b) the magnetic limit ($F_s=0$) is depicted. We choose similar initial conditions as in the electric limit. Unlike in the electric case, the velocities $\pm v_y$ lead to completely different trajectories. Depending on the sign of the velocity, the Lorentz force either pulls the electron into the bulk of the semiconductor ($v_y>0$) or it binds it to the surface ($v_y<0$) as did the Coulomb force eF_s in the electric limit for all electrons. Thus, in a magnetic field two different trajectories are obtained, namely cyclotron circles identical with those of bulk electrons and skipping trajectories that are similar to the trajectories in a surface electric field. Most characteristic for magnetic trajectories are their center coordinates $z_0=v_y/\omega_c$. For bulk electrons they lie far inside the semiconductor, but may even lie outside the semiconductor for magnetically bound electrons ($z_0 < 0$).

Figure 1(c) shows trajectories in crossed fields. The most characteristic feature now is the constant drift velocity $v_D = F_s/B$ by which all electrons drift transverse to both fields. The trajectories for the initial velocities $v_y = \pm 2v_D$ are depicted in the figure. As in the magnetic limit in Fig. 1(b), the classical trajectory lies entirely in the interior of the semiconductor if the initial velocity is positive. Such electrons will again behave like bulk electrons. For negativ velocities the cycloide may hit the surface and the corresponding electrons cannot complete their free motion (dashed line). Again the electron is periodically reflected and its motion is represented by a relatively complicate skipping trajectory that is not included in Fig. 1(c). This is done in Fig.2 for electron trajectories with different center coordinates z_0. Quantum-mechanically it is clear that the electron energies

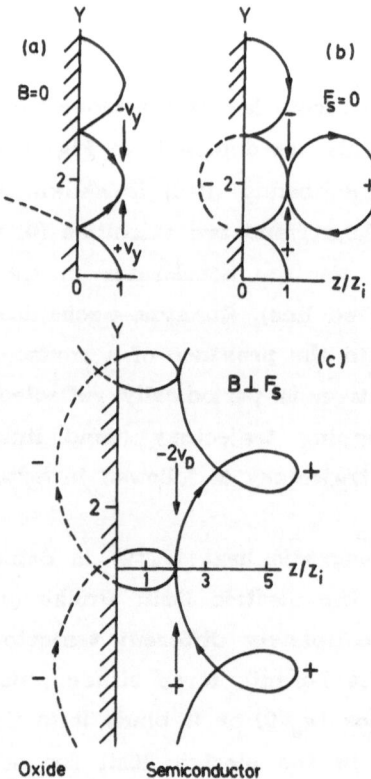

Fig.1. Classical trajectories of electrons near ideally reflecting surfaces at z=0. (a) Electric field $F_s \| z$, (b) magnetic field $B \| x$, and (c) crossed fields $F_s \perp B$. Dashed lines indicate trajectories in the absence of the surface. Arrows and signs $\pm$ indicate initial conditions and resulting trajectories. In (c) the coordinates are normalized with v_D/ω_c. From Ref.6.

will strongly increase if the coordinate z_0 of the electron is close enough to the surface. In this case the motion is restricted to a more narrow region in z-direction and the electron oscillates more rapidly. Such electrons are bound to the surface by the combined action of the electric and the magnetic field and the corresponding quantized levels are referred to as hybrid surface states.

We want to emphasize two important characteristics of surface electrons in crossed fields that are revealed by their classical trajectories. Firstly, there are two kinds of electrons. Depending on the sign of the initial velocity we have bulk electrons and bound electrons. Secondly, all electrons periodically come back to their initial z-coordinate. To achieve this kind of motion, the bulklike electrons do not hit the surface, whereas the bound electrons do. The behavior of the bulklike electrons is rather unphysical, if strong electric fields or vanishing magnetic fields are considered. In such fields an electric type of motion with continuous

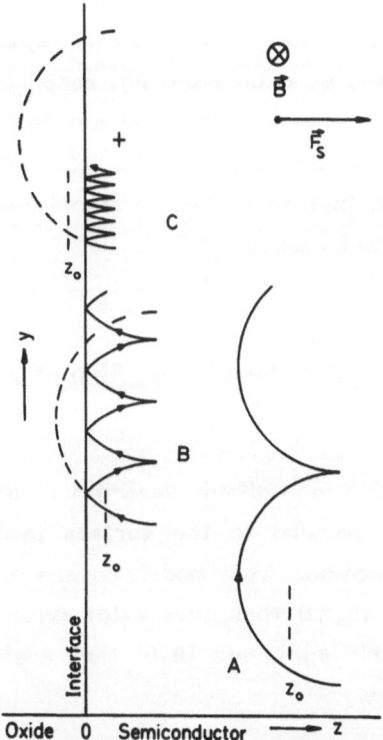

Fig.2. Classical trajectories of inversion electrons in a metal–oxide–semiconductor structure in the presence of crossed electric and magnetic fields for different center coordinates. The trajectory A lies inside the semiconductor and is not affected by the presence of the interface at z=0; trajectories B and C result from periodic reflection at the interface. From Ref.9.

acceleration along the electric field [see Fig. 1(a), dashed line] and reflection at the surface is expected for all electrons.

For a free electron in vacuum the transition to such an electric type of motion is obtained from the relativistic equation of motion. The drift velocity has a physical meaning only if it is less than the light velocity.[11] Up to the light velocity the motion remains magnetic, i.e., periodic with the cyclotron frequency ω_c. The motion becomes electric, if the drift velocity reaches the light velocity.

For semiconductor electrons an analogous transition is obtained only in a two-level approach for the valence and conduction bands. Then a limiting velocity $(E_g/2m_0^*)^{1/2}$ analogous to the light velocity in free space is obtained with the band gap E_g and the effective mass at band edge m_0^*. The limiting velocity u in the semiconductor can be deduced from the analogy between the two-band model of a semiconductor and the Dirac model for electrons in free space if the light velocity c is written as $c=(2m_0c^2/2m_0)^{1/2}$. Then the electron-positron gap which is twice the rest energy $2m_0c^2$ and the free mass m_0 have to be replaced by the gap energy E_g and the effective mass at band edge m_0^*, respectively, to obtain the expression stated above. However, it is important to note that unlike in free space an unbound trajectory like in Fig. 1(a) (dashed line) cannot be obtained near a surface. Instead of this, a transition to an electric subband state [Fig. 1(a), solid line] occurs.

3. ONE-BAND MODEL FOR CROSSED FIELDS: ENERGY LEVELS AND TRANSITION PROBABILITIES

In this chapter we give a unified description of surface electrons in magnetic fields applied parallel to the surface in terms of the one-band effective mass approximation. The model covers all cases ranging from magnetic surface levels ($F_s=0$) that have extensively been studied in metals and semimetals to electric subbands (B=0) that are important in man-made semiconductor structures.

Crossed-field results

In our model the electrostatic surface potential is approximated by a triangular potential well.[6] This simple model ignores screening and many body effects and provides only a rough approximation to a realistic space-charge potential. The merit of the triangular approximation, however, is that the hybrid surface band structure and the optical matrix elements can be calculated analytically and that many interesting aspects of the quantization in crossed electric and magnetic fields can be explained in simple terms.

In our coordinate system the $z=0$ plane separates the oxide from the semiconductor ($F_s \parallel z$, $B \parallel x$). Ignoring spin, the Schrödinger equation becomes

$$\left[\frac{(\vec{p} + e\vec{A})^2}{2m^*} + eF_s z \right] \psi = E\psi \tag{1}$$

with the boundary condition $\psi(z=0)=0$ for an infinite potential barrier. The zero of the energy scale is at the bottom of the triangular well potential. In the gauge $A=(0,-Bz,0)$ the total electric and magnetic surface potential only depends on the z-coordinate and we look for solutions in the form

$$\psi(x,y,z) = D(z) \exp(ik_x x + ik_y y) . \tag{2}$$

This ansatz reduces Schrödinger's equation to a one-dimensional differential equation. It is convenient to define dimensionless variables. Referring to the electric and the magnetic fields, the problem has two characteristic lengths and two characteristic energies. The lengths are the subband width L and the cyclotron radius l:

$$L = \left(\frac{\hbar^2}{2m^* e F_s} \right)^{\frac{1}{3}} , \qquad l = \left(\frac{\hbar}{eB} \right)^{\frac{1}{2}} . \tag{3}$$

The characteristic energies are the electric subband energy E_0 and the cyclotron energy $\hbar\omega_c$:

$$E_0 \approx \frac{\hbar^2}{m^* L^2} , \qquad \hbar\omega_c = \frac{\hbar^2}{m^* l^2} . \tag{4}$$

Dimensionless variables are introduced by the cyclotron radius l, the cyclotron energy $\hbar\omega_c$, and the wave vector $k_D=m^* F_s/\hbar B$ of the classical drift velocity $v_D=F_s/B$. The dimensionless space coordinate is defined by

$$\zeta = \sqrt{2}\,(z-z_0)/l , \qquad z_0 = l^2(k_y - k_D) , \tag{5}$$

where z_0 will turn out as center coordinate of the motion. A dimensionless energy parameter is defined by

$$\nu + \tfrac{1}{2} = (E-E_{2D})/\hbar\omega_c + (z_0/l)^2/2 \tag{6}$$

with the 2D dispersion $E_{2D}=\hbar^2(k_x^2+k_y^2)/2m^*$. The equation for $D(\zeta)$ then has the standard form of the differential equation of the parabolic cylinder (Weber) functions,[12]

$$\frac{d^2 D_\nu}{d\zeta^2} - (\tfrac{1}{4}\zeta^2 - \nu - \tfrac{1}{2})D_\nu(\zeta) = 0 \; . \tag{7}$$

The motion of surface electrons is thus described by "oscillations" $D_\nu(\zeta)$ around the center coordinate z_0.

Before we discuss the wave functions and hybrid energy levels it is illustrative to make clear the physical meaning of the two dimensionless parameters $k_0 l$ and z_0/l that enter Eq.(7) through Eqs.(5) and (6). The parameter $k_0 l$ is the ratio of the electrostatic energy $eF_s l$ and the cyclotron energy $\hbar\omega_c$ and is also related to the characteristic lengths:

$$k_0 l = \frac{eF_s l}{\hbar\omega_c} = \frac{1}{2}\left(\frac{l}{L}\right)^3 \; . \tag{8}$$

Note, that $k_0 l=1/2$ describes maximum coupling of the electric and magnetic quantization ($L=l$). The coordinate $z_0/l=k_y l-k_0 l$ of Eq.(5) is given by the momentum $\hbar k_y$ and a constant shift due to the crossed-field configuration and describes the center of the wave function or the corresponding classical trajectory.

Only parabolic cylinder functions D_ν that vanish at infinity ($z\to+\infty$) provide solutions to our problem. The allowed indices ν_i follow from the boundary condition $D_\nu(z=0)=D_\nu(-\sqrt{2}\,z_0/l)=0$. Therefore, the zeros of the parabolic cylinder functions must be known to obtain the wave functions and the energy eigenvalues. The energy eigenvalues for a particular center coordinate z_0/l are obtained from the allowed indices ν_i from Eq.(6). As is necessary for a complete solution, an infinite number $i=0,1,...$ of states exists for each coordinate. The energy spectrum in the hybrid subband i is

$$E_{i,k_x,z_0} = \frac{\hbar^2(k_x^2+k_0^2)}{2m^*} + (\nu_i+\tfrac{1}{2})\hbar\omega_c + eF_s z_0 \; . \tag{9}$$

The quantum numbers are the wave vector k_x, the hybrid subband index i, and the center coordinate z_0/l. Note that the index ν_i is a function of the center coordinate z_0/l. Unlike bulk Landau states, the levels are not highly degenerate, but the energy depends on the distance z_0/l from the surface.

The indices $\nu_i = \nu_i(z_0/l)$ have been calculated numerically and are depicted in Fig.3. Good approximate formulas exist for various ranges of the center coordinate.[13] For values $z_0/l \ll -1$

$$\nu_i = \tfrac{1}{2}(z_0/l)^2 - \tfrac{1}{2} + \tfrac{1}{2}[3\pi(i+\tfrac{3}{4})]^{\frac{2}{3}}(z_0/l)^{+\frac{2}{3}} + \tfrac{1}{15}[3\pi(i+\tfrac{3}{4})]^{\frac{4}{3}}(z_0/l)^{-\frac{2}{3}} \qquad (10)$$

and for values $z_0/l \gg 1$

$$\nu_i = i + \frac{2^i}{i!}\left(\frac{1}{\pi}\right)^{\frac{1}{2}}(z_0/l)^{2i+1}\, e^{-(z_0/l)^2} \quad . \qquad (11)$$

The first three terms of the approximation given in Eq.(10) correspond to the semiclassical Bohr–Sommerfeld result for electrons in a triangular potential that has extensively been discussed in connection with magnetic surface levels in metals and semimetals.[1] The last term is necessary for a proper description of the diamagnetic shift as will be discussed in the next section. Equation (11) is due to Kaner, Makarov, and Fuks.[13] We also note an approximation derived by Dean for center coordinates close to the surface ($z_0/l \approx 0$). It reads

$$\nu_i \approx (2i+1) - \left(\frac{4}{\pi}\right)^{\frac{1}{2}}\frac{(2i+1)!!}{(2i)!!}(z_0/l) \qquad (12)$$

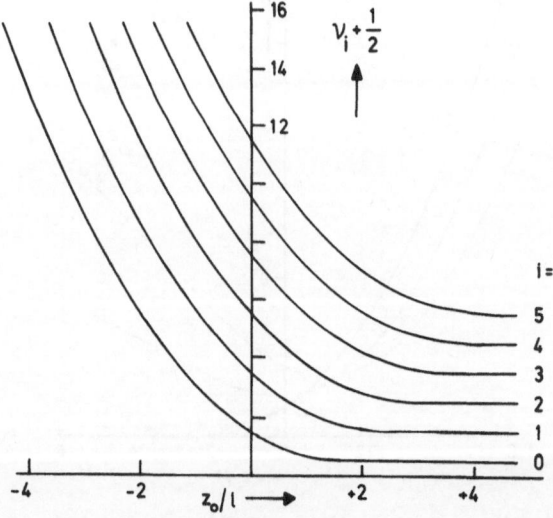

Fig.3. Zeros z_0/l of the parabolic cylinder functions $D_\nu(z)$. The most negative zero for each index ν belongs to the ground subband i=0. Correspondingly, higher subbands are obtained. From Ref.6.

in its most simple form. In Eq.(12) we make use of the abbrevations $(2i)!! \equiv (2i)\times(2i-2)\times \ldots \times 2$ and $(2i+1)!! \equiv (2i+1)\times(2i-1)\times \ldots \times 1$. In Fig.4(a) wave functions of the ground hybrid subband are depicted for various center coordinates z_0/l. For center coordinates $z_0/l \gg +1$, the wave functions are practically Gaussian and identical with bulk Landau functions. Because

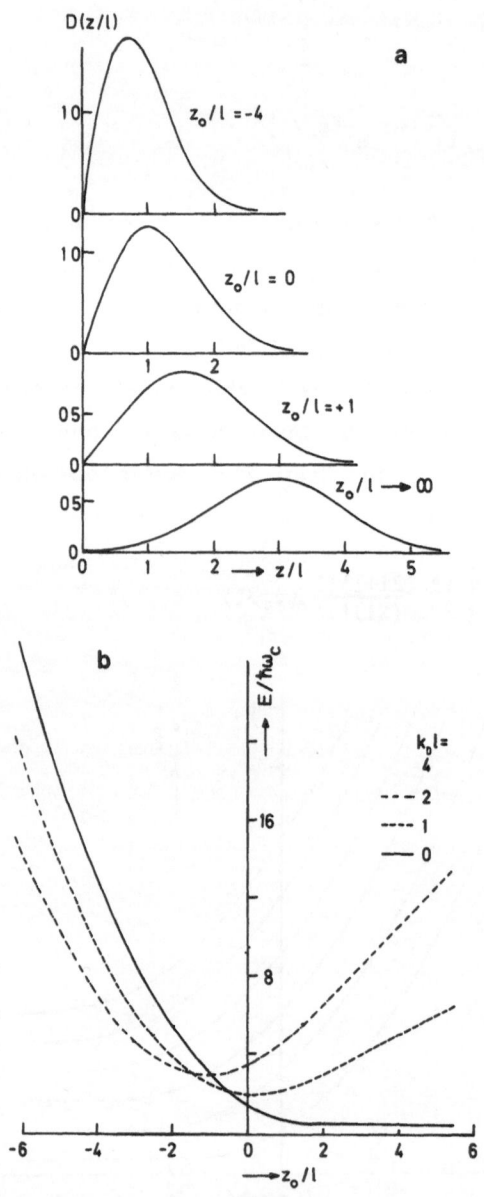

Fig.4. (a) Normalized wave functions in the ground hybrid subband for various center coordinates z_0/l. (b) Ground hybrid subbands ($i=0$, $k_x=0$) for various parameters k_0l. For $k_0l=0$ magnetic surface levels and for $k_0l \to \infty$ electric subband parabolas are obtained. From Ref.6.

the center of oscillation lies far inside the semiconductor, the electron does not feel the influence of the surface. Negative center coordinates mean that the electron is strongly bound to the surface. Then the wave function is very similar to the one of a purely magnetically or electrically bound electron. This becomes clear from a comparison with the corresponding Airy solution.[12]

The ground hybrid subband is shown in Fig.4(b) for various parameters $k_0 l$ and momentum $\hbar k_x = 0$. The energy eigenvalues are given versus the center coordinate z_0/l. The magnetic surface levels ($k_0 l = 0$) are directly given by the indices $\nu_i(z_0/l)$ of the parabolic cylinder functions. In crossed electric fields, a straight line must be added to the magnetic levels according to Eq.(9). This shifts the subband minimum to finite and even to negative center coordinates with increasing $k_0 l$ parameters, i.e., increasing electric or decreasing magnetic field strengths.

By the same geometrical method higher subbands can also be constructed.[5] This is done in Fig.5 for two hybrid subbands i=0,1. At the highest value $k_0 l = 5.0$ we approach the situation of purely electric subbands: in each subband the dispersion is nearly parabolic and the subband edge appears at the momentum $\hbar k_y \approx 0$.

We now give the transition matrix elements for electric dipole radiation in the crossed-field configuration. Two polarization vectors of the incident light are considered. The polarization can either be parallel to the surface (parallel excitation) or it can be perpendicular to it (perpendicular excitation). In both cases, the polarization is perpendicular to the magnetic field (see insert in Fig.6). Optical transitions cannot be excited with polarization parallel to the magnetic field.

From the form of the wave function in Eq.(2), we immediately obtain the selection rules

$$k_x = k_x' , \quad z_0 = z_0' , \tag{13}$$

the latter being equivalent to the conservation of the momentum $\hbar k_y$. Therefore, we have vertical transitions in the surface band structure. The matrix elements for parallel M^y and perpendicular excitation M^z are

$$M_{11'}^y = 2 \left[\frac{d\nu_i}{d\zeta} \frac{d\nu_i'}{d\zeta} \right]^{\frac{1}{2}} \frac{1}{(\nu_i - \nu_i') - 1} , \tag{14a}$$

$$M_{11'}^{z} = (\nu_1' - \nu_1) \, M_{11}^{y}, \tag{14b}$$

The derivation of these formulas can be found in the appendix of Ref.6.

The matrix elements are normalized with the matrix elements of bulk cyclotron resonance between the ground and first excited Landau level. The derivatives in Eqs.(14a) and (14b) are obtained from the zeros of the parabolic cylinder function [see Eqs.(10)–(12)] and are taken at the center coordinate. It is interesting to note, that the matrix elements only depend on the normalized center coordinate z_0/l. Therefore, optical transitions for surface electrons in crossed fields can universally be calculated. The result

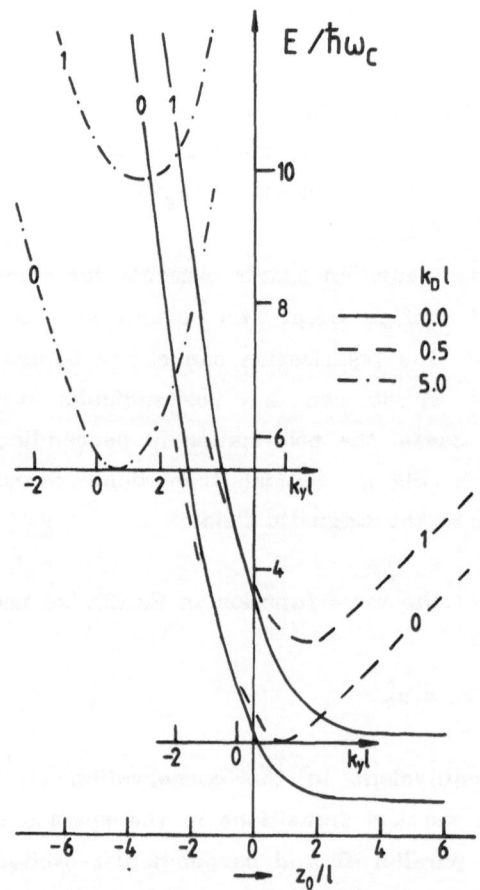

Fig.5. Energy dispersion of hybrid subbands in parallel magnetic fields. Hybrid subbands i=0,1,... are obtained by adding the straight line $k_0 l$ ($k_0 l/2 + z_0/l$) to the purely magnetic surface levels ($k_0 l = 0$). From Ref.5.

is shown in Fig.6, where the transitions strengths, i.e., squared matrix elements, are given.

For parallel excitation [Fig.6(a)], electrons with positive center coordinates ($z_0/l \rightarrow \infty$) show the highest excitation strengths, namely those of bulk cyclotron resonance: (i+1). No harmonics of CR are allowed in this limit. Through the influence of the surface, harmonics become allowed for center coordinates $z_0/l \approx 0$. For negative center coordinates all transition matrix elements vanish. This is because the corresponding states represent electrons that are strongly bound perpendicular to the surface in the z-direction.

The excitation strengths for perpendicular excitation are shown in Fig.6(b). Unlike for parallel excitation, for more negative center coordinates

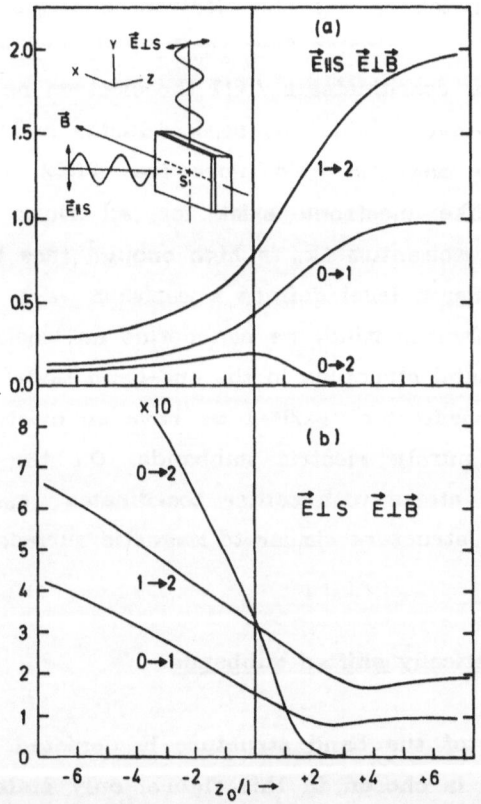

Fig.6. Optical transition strengths for (a) parallel and (b) perpendicular excitation. Parallel ($\vec{E} \| S$) and perpendicular ($\vec{E} \perp S$) excitation is defined in the inset. Transition strengths are normalized to cyclotron resonance of bulk electrons (0→1). From Ref.6.

the transition strengths increase and transitions between, e.g., the ground subband (i=0) and higher subband (i=2) become relatively strong.

It is interesting to note that, both, hybrid subband energies and transition strengths can analytically and universally be calculated as a function of the normalized center coordinate. From an inspection of Fig.5 it becomes clear that the parameter k_0l determines the center coordinates that are occupied at a given Fermi energy. This way it also determines the possible excitations in a particular surface band structure.

Therefore, there are two important parameters that determine the behavior of a surface electron in crossed electric and magnetic fields: the parameter k_0l measures the relative strength of the electric and the magnetic field [see Eq.(8)] and determines the surface band structure as a whole. The center coordinate z_0/l determines the behavior of an individual electron within this band structure.

At center coordinates $z_0/l \ll -1$ the electron is strongly bound to the surface and quantization originates from the spatial confinement (see Fig.2, trajectory C). At center coordinates $z_0/l \gg +1$ the electron no longer is bound to the interface but moves inside the semiconductor and the quantization $\Delta E = \hbar\omega_c$ is due to the magnetic field alone (see Fig.2, trajectory A). In principle, such bulklike electrons exist for all finite parameters k_0l supposed the electron momentum $\hbar k_y$ is high enough [see Eq.(5)]. However, in a real system the Fermi level defines a maximum wave vector or center coordinate. With this fact in mind, we can define an electric and magnetic limit for the surface band structure in the one-band model: if the electrons occupy only center coordinates $z_0/l \ll -1$ we have an electric type of band structure similar to purely electric subbands. On the other hand, if electrons occupy only states with center coordinates $z_0/l \gg +1$ we have a magnetic type of band structure similar to magnetic surface levels.

Electric limit: diamagnetically shifted subbands

An electric type of the band structure is depicted in Fig.7. At the Fermi energy E_F that is chosen in this figure, only states with negative center coordinates are occupied. The subband dispersion is anisotropic and only approximately parabolic. Also, there is a shift of the subband minima to wave vectors $k_y > 0$.

To a rough approximation this type of band structure can be described using the approximation that is given in Eq.(10) for the indices $\nu_i(z_0/l)$. If only the first three terms are retained, we obtain from Eq.(9)

$$E_i = E_{2D} + \left(\frac{9\pi^2\hbar^2}{8m^*} \right)^{\frac{1}{3}} (eF_s + k_y\hbar\omega_c)^{\frac{2}{3}} (i+\tfrac{3}{4})^{\frac{2}{3}} . \tag{15}$$

This equation describes purely electric subbands ($\hbar\omega_c=0$)

$$E_i^0 = E_{2D} + \left(\frac{9\pi^2}{8m^*} \right)^{\frac{1}{3}} (e\hbar F_s)^{\frac{2}{3}} (i+\tfrac{3}{4})^{\frac{2}{3}} \tag{16}$$

and purely magnetic subbands ($eF_s=0$) in the triangular well approximation.[1,2] Note, that the Coulomb force eF_s and the Lorentz force $k_y\hbar\omega_c=ev_yB$ directly appear in Eq.(15) and that Eqs.(15) and (16) are identical with the well-known semiclassical results.

Whereas Eq.(15) correctly describes purely electric or purely magnetic subbands, it does not properly account for the diamagnetic shift, i.e., for the increase of the subband minima in a parallel magnetic field. For this,

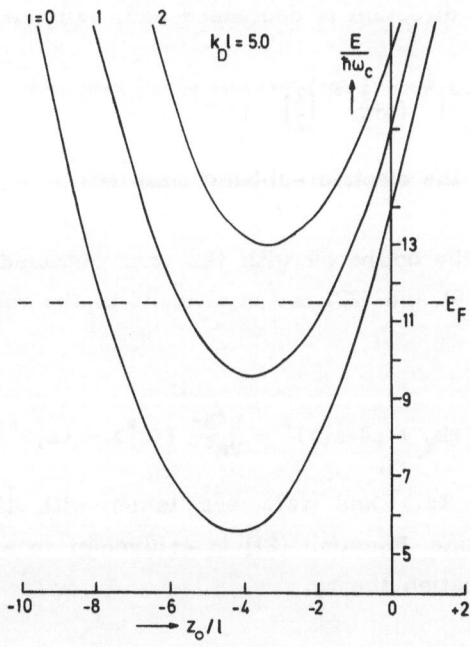

Fig.7. Surface band structure (i=0,1,2) in the electric limit ($k_Dl\gg1$). Cyclotron resonance only is possible for positive center coordinates $z_0/l\approx+1$. From Ref.6.

the last term in Eq.(10) is essential and we obtain

$$E_1 = E_{2D} + E_1^0 \left(1 - \frac{k_y}{k_0}\right)^{2/3} + \frac{3\pi^2}{10} \; (i + \tfrac{3}{4})^2 \frac{(\hbar\omega_c)^2}{E_1^0} \left(1 - \frac{k_y}{k_0}\right)^{-2/3} . \qquad (17)$$

For small wave vectors and small magnetic fields

$$k_y/k_0 \ll 1 \quad , \quad E_1^0 / \frac{\hbar^2 k_0^2}{2m^*} \ll 1 \qquad (18)$$

we obtain an even more simple description of diamagnetically shifted subbands. If we neglect terms of third and higher order, we find the following effects of a parallel magnetic field on the purely electric subband structure: a shift of the subband minima

$$\Delta k_y = \left(\frac{2\pi^2}{3}\right)^{1/3} (i + \tfrac{3}{4})^{2/3} \cdot \frac{eBL}{\hbar} \qquad (19)$$

to higher wave vectors k_y (see also Fig.7) and a shift to higher energies (diamagnetic shift)

$$\Delta E_1 = \frac{1}{10} \left(\frac{2\pi^2}{3}\right)^{\frac{2}{3}} (i + \tfrac{3}{4})^{4/3} \cdot \frac{e^2 B^2 L^2}{m^*} \qquad . \qquad (20)$$

Whereas the dispersion in the direction of the magnetic field (B∥x) remains unchanged, the curvature of the parabola of the dispersion perpendicular to the magnetic field direction is decreased and, equivalently, the mass

$$m_y = m^* \left[\; 1 + \frac{1}{3} \left(\frac{16\pi^2}{3}\right)^{\frac{1}{3}} (i + \tfrac{3}{4})^{\frac{2}{3}} \left(\frac{L}{l}\right)^4 \; \right] \qquad (21)$$

becomes higher than the electric subband mass m^*.

Our results can be compared with the ones obtained from perturbation theory.[2] Equations (19) and (20) are equivalent to the results of first order perturbation theory

$$E_1 = \frac{\hbar^2 k_x^2}{2m^*} + \frac{1}{2m^*} \; (\hbar k_y + eB\langle z_1 \rangle)^2 + \frac{e^2 B^2}{2m^*} \; (\langle z_1^2 \rangle - \langle z_1 \rangle^2) \; , \qquad (22)$$

where the averages $\langle z_1 \rangle$ and $\langle z_1^2 \rangle$ are taken with the purely electric subband wave functions. Equation (21) is equivalent to a result obtained in second order perturbation theory.

From the masses $m_x = m^*$ and m_y the density of states[2]

$$D(E) = \sqrt{m_x m_y} \; / \; \pi\hbar^2$$

for diamagnetically shifted subbands in the parabolic approximation is obtained:

$$D(E) \propto \frac{m^*}{\pi \hbar^2} \left[1 + \frac{1}{6} \left(\frac{16\pi^2}{3} \right)^{\frac{1}{3}} \left(i + \frac{3}{4} \right)^{\frac{2}{3}} \left(\frac{L}{l} \right)^4 \right] . \tag{23}$$

In this approximation, the density of states is independent of energy E as it is the case in purely electric subbands. However, the density is higher in higher subbands and increases with magnetic field. The latter effects have recently been studied by magnetotunneling in MOS-structures on Si.[14]

The optical excitations in the purely electric limit ($\hbar\omega_c$=0) are intersubband resonances. Parallel excitation is not possible since the matrix element M^y vanishes as can be seen from Fig.6(a) for center coordinates $z_0/l \rightarrow -\infty$. This means that in the absence of a parallel magnetic field intersubband excitations are only possible for polarization vectors perpendicular to the surface. The corresponding matrix element M^z [see Fig.6(b)] can be calculated in closed form:

$$M^z_{i i'} = A_0 \frac{e^2 \hbar F_s}{m^*} \cdot \frac{1}{E_{i'} - E_i} , \tag{24}$$

where the subband energies E_i are given by Eq.(16) and A_0 is the amplitude of the incident light wave.

Excitation of diamagnetically shifted intersubband resonances becomes possible in parallel excitation since the magnetic field deflects the oscillations of the electron that were directed strictly perpendicular to the surface in the purely electric limit. The matrix element for the parallel excitation of diamagnetically shifted intersubband resonance is

$$M^y_{i i'} = \sqrt{2} \frac{32}{9} \left(\frac{1}{3\pi^2} \right)^{\frac{2}{3}} \left(\frac{E_0}{E_{i'} - E_i} \right)^2 \cdot \frac{L}{l} . \tag{25}$$

This matrix element is normalized with the one for the 0→1 bulk cyclotron transition (see Eq.(28), i=0) and the energies E_i are the purely electric ones. Note that the absolute transition strength increases quadratically with magnetic field intensity. The ratio of the transition strengths of diamagnetically shifted intersubband resonance and purely electric intersubband resonances (L≪l) is to a fairly good approximation given by the ratio $(L/l)^4$.

A magnetic type of bandstructure is depicted in Fig.8 at a Fermi energy where only positive center coordinates are occupied. In such a situation almost all electrons will show bulklike behavior and their energy spectrum is given by[15]

$$E_i = \frac{\hbar^2 k_x^2}{2m^*} + (i+\tfrac{1}{2})\hbar\omega_c + eF_s z_0 + \tfrac{1}{2}m^* v_D^{\;2} \quad . \tag{26}$$

The first two terms in this relation describe Landau levels as in the bulk, the term $eF_s z_0$ represents the electrostatic energy of the cyclotron orbit in the electric field, and the last term describes an energy due to the drift in crossed fields (see Sec.2). In nearly magnetic subbands, the density of states is approximately

$$D(E) \sim \frac{m^*}{\pi\hbar^2} \left(\frac{2E}{\hbar\omega_c}\right)^{1/2} \frac{1}{k_0 l} \quad . \tag{27}$$

This relation is a rough approximation and is obtained if only the area enclosed by the Fermi line at wave vectors $k_y \geqslant 0$ is considered.

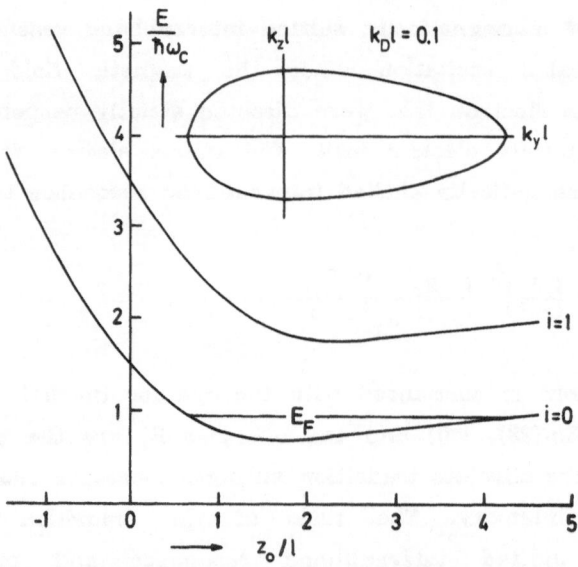

Fig.8. Surface band structure (i=0,1) in the magnetic limit ($k_0 l \ll 1$). The inset shows the Fermi line for $E_F = 1/4\hbar\omega_c$. Cyclotron resonance is possible for most of the center coordinates. From Ref.6.

The optical excitations become bulk cyclotron resonances $i \to i+1$ with

$$M_{11'}^{y} = M_{11'}^{z} = A_0 \frac{el\omega_c}{\sqrt{2}} \sqrt{i+1} \tag{28}$$

and the bulk cyclotron energy $E_{i+1}-E_i = \hbar\omega_c$. This is clear, since the drift velocity $v_0 = F_s/B$ and the center coordinate [see Eq.(13)] are conserved and the transition occurs well away from the interface (see Figs.6,8).

4. EXPERIMENTS: DIAMAGNETICALLY SHIFTED INTERSUBBAND RESONANCE VS CYCLOTRON RESONANCE

Some of the most characteristic features of magneto-optical spectra of surface electrons in crossed fields can qualitatively be understood in terms of the one-band model discussed in the preceding section.[6] At low inversion electron densities, when only the lowest hybrid subband and spin level is occupied, effects of nonparabolicity and spin are not very important and the one-band model is applicable to a good approximation.

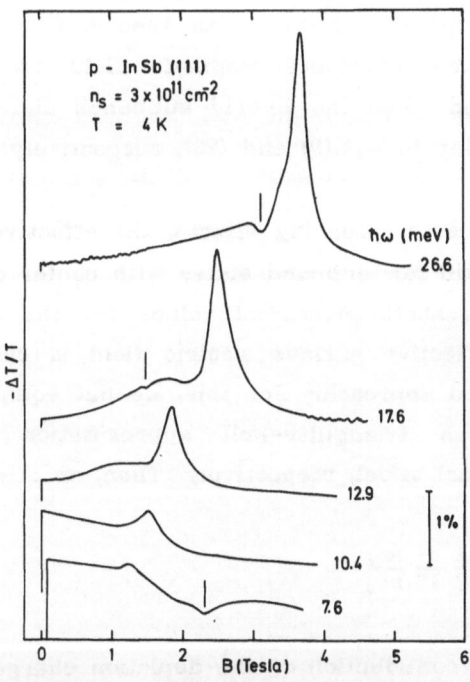

Fig.9. Experimental spectra of inversion electrons in parallel magnetic fields at various laser energies $\hbar\omega$. The traces have been successively displaced upward. The dashes mark cyclotron resonance of bound holes in the p-type InSb substrate. From Ref.6.

Figure 9 shows experimental spectra at various infrared laser energies for such a low density ($n_s = 3 \times 10^{11} cm^{-2}$). The most prominent structure is an absorption maximum that becomes stronger when the laser energy $\hbar\omega$ is increased. The maximum is caused by cyclotron resonance of bulklike electrons in the hybrid subband structure. With increasing laser energy ($\hbar\omega = 7.6$–26.6meV), cyclotron resonance is obtained at higher resonance magnetic fields ($B = 1.25$–3.68T). Correspondingly, the parameter $k_0 l$ at cyclotron resonance decreases since the surface electric field $F_s = e n_s / \varepsilon_0 \kappa_s$ is constant at a fixed inversion electron density. Simultaneously, there is a change from a more electric to a more magnetic type of surface band structure [see Fig.4(b)] and more electrons can contribute to cyclotron resonance. With the bandedge mass $m_0^* = 0.0138$ of InSb we obtain values $k_0 l = 7.3$–1.4 from Eqs.(3) and (8).

Assuming Lorentzian line shapes for cyclotron resonance, we can estimate the number of bulklike electrons from the measured change in transmission $\Delta T/T$. The number increases from about 1.2–11% of the totally induced electrons ($n_s = 3 \times 10^{11} cm^{-2}$). This means that in the trace with the lowest resonance magnetic field ($\hbar\omega = 7.6$meV) cyclotron resonance of only about $4 \times 10^9 cm^{-2}$ electrons is detected. The increase of the number of electrons is in qualitative agreement with theory. However, for the laser energy $\hbar\omega = 7.6$meV, i.e., resonance magnetic field $B = 1.25$T, no bulklike electrons are expected when the hybrid subbands and the Fermi energy are calculated according to Eqs.(9) and (23), respectively.

We think that due to screening effects, the effective parameter $k_0 l$ is lower than 7.3 and still few subband states with center coordinates $z_0/l \gg 1$ are occupied. More realistic numerical values for the $k_0 l$ parameter are obtained when an effective surface electric field is calculated along the lines of the variational approach.[2] For this, we put equal the average $\langle z_0 \rangle$ calculated within the triangular-well approximation and within the Fang-Howard variational model, respectively. Then, an effective field

$$F_{eff} = \frac{11}{32} \frac{\pi^2}{12} \frac{e n_s}{\kappa \varepsilon_0} \tag{29}$$

is obtained when the contribution of the depletion charge is ignored which is approximately valid for inversion layers on InSb. Thus, the effective surface electric field is about a factor of three less than the field just inside the semiconductor.

For the energy, e.g., $\hbar\omega=26.6$meV, the most positive center coordinates lie about two cyclotron radii inside the semiconductor and we expect that 7% of the induced electrons contribute to cyclotron resonance. The experimental value is 11%. Again, more quantitative agreement is obtained if the effective surface electric field $F_{eff}\approx 1/3 F_s$ is taken.

There is also background absorption in the spectra of Fig.9 that changes with laser energy. At low laser energies, the background slightly decreases with increasing magnetic field. This is due to intraband absorption of Drude electrons as will become clear from the Fourier spectra discussed below. As is expected, Drude absorption at zero magnetic field strongly decreases with increasing laser energy. In addition to Drude absorption, a broad background resonance is observed. It is most clearly present in the traces for the energies $\hbar\omega=12.9$ and 17.6meV. This broad absorption is caused by transitions different from cyclotron resonance between the ground and the first excited hybrid subband, i.e., it is caused by electrons with center coordinates $z_0/l<\approx 1$. The transition energies of these electrons depend on their center coordinates (see Fig.7) causing

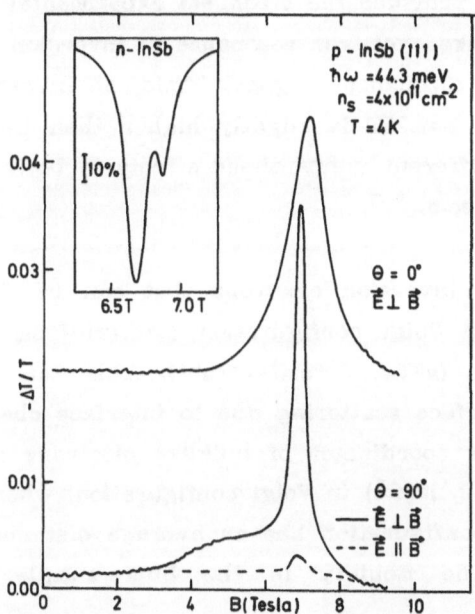

Fig.10. Comparison of bulk cyclotron resonance (conduction-band and impurity-shifted resonances) in n-type InSb (inset) with surface cyclotron resonance in inversion layers on p-type InSb. Surface cyclotron resonance is shown for the Faraday configuration ($\theta=0°$) and Voigt configurations ($\theta=90°$). The light vector is always perpendicular to the surface ($\vec{k}\perp S$). In the Voigt configuration ($\vec{k}\perp\vec{B}$) the light can be polarized parallel ($\vec{E}\parallel\vec{B}$) or perpendicular ($\vec{E}\perp\vec{B}$) to the magnetic field. From Ref.6.

broad structures. The background is much weaker than cyclotron resonance since the corresponding excitation strengths $(z_0/l < \varkappa +1)$ are weaker in parallel excitation [see Fig.6(a)] and the joint density of states is less than for the parallel hybrid subbands at $z_0/l > \varkappa +1$. The dashes in Fig.9 indicate cyclotron resonance of bound holes in the p-type substrate.

Figure 10 compares cyclotron resonance of inversion electrons in the Voigt and Faraday configurations with cyclotron resonance in a bulk sample of n-type InSb (inset of Fig.10). The figure also demonstrates the dependence on the polarization of the incident light. Cyclotron resonance in the bulk sample $(d=110\mu m)$ was found to be identical in the Voigt and Faraday configurations, as is expected for the low density $(n=6\times10^{13}cm^{-3}$ at 77K). Due to magnetic freeze-out at liquid-helium temperature, the majority of electrons shows impurity cyclotron resonance (B=6.7T);[16] only few electrons show conduction band cyclotron resonance (B=6.9T).

The resonance position of cyclotron resonance in inversion layers in Voigt configuration is nearly the same as for conduction-band electrons in the bulk sample. This provides the strongest experimental evidence that we actually observe bulklike cyclotron resonance in inversion layers in parallel magnetic fields. The resonance magnetic field of inversion electrons in Faraday configuration B=7.25T is slightly higher than in Voigt configuration. This is due to different nonparabolic effects in both configurations as will be discussed in Sec.5.

The mobility of inversion electrons that can be deduced from the linewidth is higher in Voigt configuration $(\mu \approx 6.7\times10^{+4}cm^{+2}V^{-1}s^{-1})$ than in Faraday configuration $(\mu \approx 1.5\times10^{+4}cm^{+2}V^{-1}s^{-1})$. This can be explained by reduced effects of surface scattering due to interface charges and surface roughness. The center coordinates of bulklike electrons lie far inside the semiconductor $(z_0/l \approx 2-3, l=95\text{Å})$ in Voigt configuration, whereas the envelope function in Faraday configuration has an average distance of about 150 Å from the surface. The mobility in the bulk sample is still higher: $\mu \approx 20\times10^{+4}cm^{+2}V^{-1}s^{-1}$.

The dependence on polarization of the incident light is also demonstrated in Fig.10. Since the light impinges in the direction perpendicular to the sample in the present experiments, the light is always polarized perpendicular to the magnetic field in Faraday configuration $(\theta=0°, E \perp B)$. In Voigt configuration, the polarization can either be parallel to the magnetic field $(E \| B)$ or perpendicular to it $(E \perp B)$. Cyclotron resonance is only significantly present in the perpendicular polarization as expected for bulklike cyclotron resonance. The small structure that is left in parallel polarization (dashed line) is probably due to imperfect polarization. Interband transitions different from cyclotron resonance have an onset at about B≈3T and again are only present measurably when the light is polarized perpendicular to the magnetic field. In contrast to this, the Drude background does not significantly depend on polarization.

In the spectra shown in Figs.9 and 10 the bound electrons with center coordinates close to the surface do only weakly contribute to the measured absorption since the polarization of the incident light was parallel to the surface [see Fig.6(a)]. To excite these electrons more strongly, we employed a transmission line in which there is strong polarization perpendicular to the surface. Corresponding spectra at a fixed laser energy are shown in Fig.11(a,b). In sweeps of the electron density at various magnetic fields we observe strong maxima at zero magnetic field that are due to purely electric intersubband resonances 1→2 and 2→3. These structures become less pronounced in increasing magnetic fields where the spectra approach a behavior that is dominated by Drude intraband absorption. Simultaneously, the resonances shift to lower densities n_s. This corresponds to a shift to higher energies as it is indeed expected for the diamagnetic shift. In fact, the surface band structure underlying the experiments in Fig.11(a,b) is of electric type and we observe diamagnetically shifted intersubband resonances that are inhomogeneously broadened at higher magnetic fields reflecting the distribution of center coordinates. Figure 11(b) presents a spectrum taken at a fixed density $n_s=5\times10^{11}\text{cm}^{-2}$ in a sweep of the magnetic field. This spectrum shows diamagnetically shifted intersubband resonance (B≈0.7T) and cyclotron resonance (B≈1.4T).

Except for cyclotron resonance, quantitative evaluation of the spectra taken in sweeps of the electron density n_s or the magnetic field B is difficult since in such experiments the surface band structure is not kept constant. This becomes also clear from a comparison of the spectrum shown in Fig.11(b) with the spectra shown in Fig.11(a) at density $n_s=5\times10^{11}\text{cm}^{-2}$. Therefore, we have measured diamagnetically shifted intersubband resonances in a Fourier spectrometer.[8] However, it is not possible in the spectrometer to measure in a transmission line due to intensity problems and hence, all the Fourier spectra are taken in parallel excitation (E⊥S, E⊥B). We could reliably detect diamagnetically shifted intersubband resonances only at magnetic fields B>≈2T, since the transition strength of diamagnetically shifted intersubband resonance is rather small in this configuration [see Eq.(25)]. A typical Fourier spectrum is shown in Fig.12.

In this spectrum Drude absorption $(\tilde{\nu} \lesssim 100 \text{cm}^{-1})$, cyclotron resonance $(\tilde{\nu} \sim 130 \text{cm}^{-1})$, and diamagnetically shifted intersubband resonance $(\tilde{\nu} \approx 240 \text{cm}^{-1})$ are observed. This interpretation is suggested by itself since at the low electron density $n_s = 2 \times 10^{11} \text{cm}^{-2}$ only the ground subband i=0 is

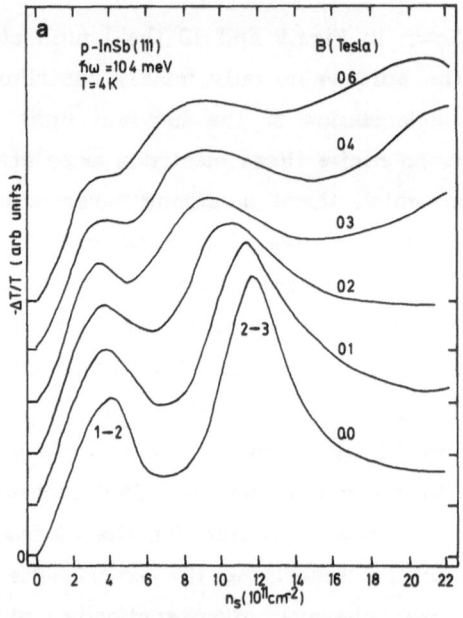

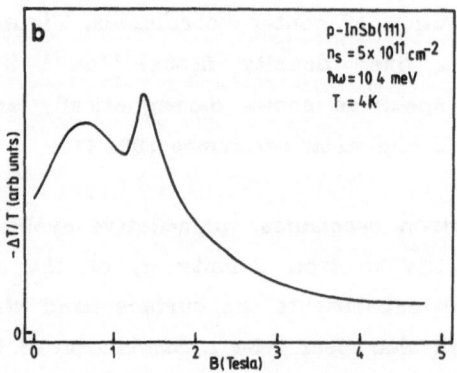

Fig.11 (a) Purely electric intersubband resonances (B=0) and diamagnetically shifted intersubband resonances (ISR) measured in a transmission-line assembly at a fixed laser energy $\hbar\omega$ versus electron density n_s. (b) Diamagnetically shifted ISR and cyclotron resonance measured versus magnetic field at a fixed electron density $n_s = 5 \times 10^{11} \text{cm}^{-2}$. Note, that the maximum due to ISR at $B \approx 0.7 T$ is much more pronounced than in parallel excitation (see Fig.9).

occupied [see Eqs.(23) and (27)]. However, in view of Fig.5 it is not obvious at all that intersubband resonance appears as a distinct resonance. This seems to be related to the joint density of states in a more realistic surface band structure.

Figure 13 depicts spectra showing cyclotron resonance and intersubband resonance for various electron densities n_s and magnetic fields B. From the spectra for $n_s=2\times10^{12}cm^{-2}$ it is seen that the excitation strength of the $1\rightarrow2$ intersubband resonance transition increases with magnetic field (B=4,5 and 6T). At magnetic fields B<$\approx$2T we could not clearly observe the relatively weak intersubband resonance. This indicates that the parallel magnetic field plays an important role in the excitation mechanism as was discussed above. We also find that the excitation strength is enhanced when the intersubband resonance is close to the reststrahlen band 183-194cm^{-1} of InSb (see Fig.12). A similar enhancement has recently been observed in space-charge layers on HgCdTe and has been explained by coupling of the intersubband resonance to the lattice polarization.[17]

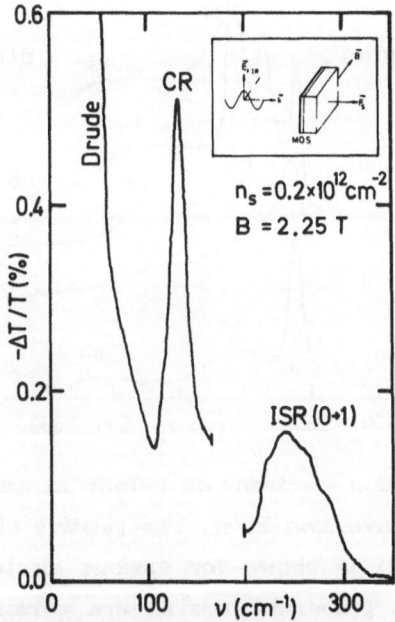

Fig.12. Typical spectrum of inversion electrons in a parallel magnetic field showing Drude absorption, cyclotron resonance (CR) and diamagnetically shifted intersubband resonance (ISR). The geometrical configuration of the experiment is indicated. From Ref.8.

Reflecting the diamagnetic shift of the electric subbands, the intersubband resonance energies increase with magnetic field as it is shown in Fig.14. From an extrapolation (B→0) the purely electric intersubband resonance energies can be obtained. They have previously been measured directly (see arrows in Fig.14) in the absence of magnetic fields as a doublet pair of peaks for the transition 0→1.[18] The doublet has convincingly been explained as resonance $\tilde{\epsilon}_{01}$ excited with light polarized perpendicular to the interface and shifted by the macroscopic depolarization field, and as resonances ϵ_{01} without depolarization shift excited with light polarized parallel to the interface via the nonparabolicity mechanism. This doublet could not be observed in the present experiments,

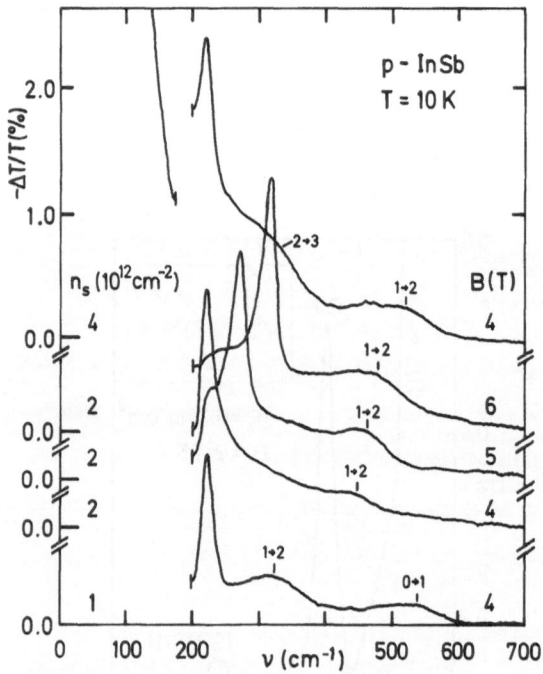

Fig.13. Spectra of inversion electrons on p–InSb in magnetic fields that are applied parallel to the inversion layer. The relative change of transmission $-\Delta T/T = 1 - T(n_s, B)/T(n_s = 0, B)$ is shown for various electron densities n_s and magnetic fields B. The prominent maxima are cyclotron resonances, the relatively weak structures are diamagnetically shifted intersubband resonances 0→1, 1→2, and 2→3, as indicated. From Ref.8.

presumably due to a combination of two effects: In our rather homogeneous gate insulators the light polarization perpendicular to the interface produced by scattering should be suppressed, and inhomogeneous broadening of the intersubband resonance in strong magnetic fields might smear out additional structures.

However, we observed a weak splitting of the intersubband resonance at high magnetic fields B>≈7T (see Fig.14) in some samples. This splitting is interpreted as spinsplitting in qualitative agreement with the k·p theory that is discussed in the following section.

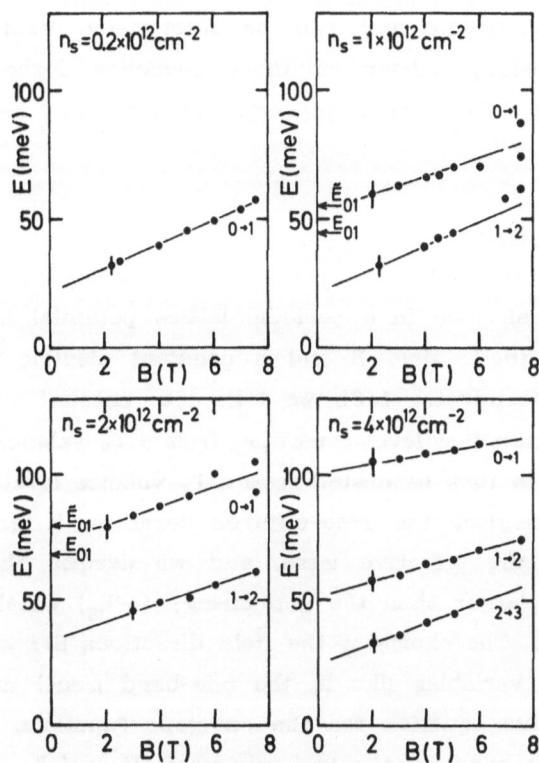

Fig.14. Diamagnetic shift of intersubband resonances in parallel magnetic fields at various electron densities n_s. The arrows indicate the values of purely electric intersubband resonances (B=0) that were found as a doublet pair of peaks E_{01} and $\tilde{E}_{01}$ separated by the depolarisation effect in Ref.18. At magnetic fields B>≈7T, spin-splitting of the diamagnetically shifted resonances is observed. The straight lines only serve as guides to the eye. From Ref.8.

5. TWO-BAND MODEL IN CROSSED FIELDS: ANALOGY TO RELATIVISTIC ELECTRONS

The experimental results presented in the previous section could qualitatively be described within the framework of the one-band effective-mass approximation and the simple model of a triangular-well potential. However, there are phenomena that even qualitatively cannot be understood within the one-band model. These phenomena are related to the destruction of the Landau quantization in crossed fields above a critical ratio of the electric and magnetic field strengths F_s/B. This destruction is experimentally observed via a strong broadening and ultimately by a disappearence of the cyclotron resonance. Simultaneously, the well-known selection rule for cyclotron resonance ($\Delta n=1$) becomes strongly violated and we observe harmonics of cyclotron resonance ($\Delta n=2$) that are comparable in intensity to the fundamental transitions. These phenomena can in fact be described by a two-band k·p theory and can be interpreted using the relativistic analogy. This analogy is based on the equivalence of the energy-momentum relation of electrons in narrow-gap semiconductors and relativistic electrons in free space.[7],[9]

Three-level k·p theory

We consider an electron in a periodic lattice potential in the presence of the external magnetic field B and a constant electric field F_s. For the narrow-gap semiconductor InSb we take into account three levels at the Γ-point: a Γ_6 conduction level separated from a Γ_8 valence level by the gap energy E_g, this in turn separated from a Γ_7 valence level by the spin-orbit energy Δ. We neglect the free-electron term as it gives only a small contribution to the effective mass, and we assume that the spin-orbit energy is much larger than the gap energy ($\Delta \gg E_g$) which is approximately the case in InSb. The choice of the field directions $B\|x$ and $F_s\|z$ allows us to separate the variables like in the one-band model discussed in Sec.3. The final effective equation for the envelope functions $\phi_\pm$ related to the S-like conduction band for the two spin projections is[9]

$$\left[-\frac{\hbar^2}{2m_0^*}\frac{d^2}{dz^2} + \frac{m_0^*}{2}\omega^2(z-z_0)^2 \right]\phi_\pm(z) = \beta_\pm\phi_\pm(z) \tag{30}$$

where

$$\omega^2 = \left(\frac{eB}{m_0^*}\right)^2 - \frac{2e^2F_s^2}{m_0^* E_g} , \tag{31}$$

320

$$z_0 = - \frac{\alpha}{m_0^{*}\omega^2} \quad , \tag{32}$$

$$\alpha = - \hbar k_y \frac{eB}{m_0^{*}} + eF_s \left(1 + 2\frac{E}{E_g} \right) \quad , \tag{33}$$

$$\beta_{\pm} = \lambda_{\pm} + \frac{\alpha^2}{2m_0^{*}\omega^2} \quad , \tag{34}$$

$$\lambda_{\pm} = E \left(1 + \frac{E}{E_g} \right) - \frac{\hbar^2 k_x^2}{2m_0^{*}} - \frac{\hbar^2 k_y^2}{2m_0^{*}} \pm \frac{1}{2}g_0^{*}\mu_B B \quad . \tag{35}$$

The effective mass m_0^{*} and the Landé g_0^{*} factor are defined at band edge. To arrive at Eq. (30) we have neglected a term resulting from the noncommutation of the P_z and eEz operators. This term is responsible for the Zener tunneling between the bands and it is of no importance for the electron energies as long as the energy gap is not very small. Note, that Eq.(30) is similar to the eigenvalue problem for the harmonic oscillator with an effective frequency ω [see Eq.(31)].It is very convenient to distinguish a magnetic $(\omega^2 > 0)$ and an electric case $(\omega^2 < 0)$ by introducing a dimensionless parameter

$$\gamma = \left(\frac{F_s}{B} \right) \bigg/ \left(\frac{E_g}{2m_0^{*}} \right)^{\frac{1}{2}} \quad , \tag{36}$$

which is the ratio of the drift velocity in crossed fields $v_D = F_s/B$ and the so-called limiting velocity $u = (E_g/2m_0^{*})^{1/2}$ of the semiconductor. The limiting velocity is the maximum velocity $v_i = \partial E/\partial p_i$ possible in the conduction band according to the two-band model and is easily deduced from the simplified Kane formula

$$E(p) = \left[\left(\frac{E_g}{2} \right)^2 + E_g \frac{p^2}{2m_0^{*}} \right]^{1/2} - \frac{E_g}{2} \tag{37}$$

when the kinetic energy by far exceeds the gap energy $(p^2/2m_0^{*} \gg E_g)$. Note, that we can also write

$$\gamma^2 = \left(\frac{2\hbar\omega_c}{E_g} \right) (k_D l)^2 = \frac{2m_0^{*}v_D^2}{E_g} \tag{38}$$

and that the square of the effective frequency

$$\omega^2 = \omega_c^2 \left(1 - \gamma^2 \right) \tag{39}$$

is positive in the magnetic case ($\gamma < 1$), but is negative in the electric one ($\gamma > 1$).

As long as we are in the magnetic case, we deal with electrons in a parabolic potential well centered at the coordinate z_0 and, similar to the situation in the one-band model, the influence of the barrier on the electron eigenenergy depends strongly on its center coordinate. Whereas the center coordinate determines the electron position with respect to the barrier, the effective frequency ω gives the width of the well, i.e., the extent of the wave function. The latter effect is illustrated in Fig.15. In the two-band model, the extent (oscillator length) of the wave function is given by ($\gamma < 1$)

$$R = \left(\frac{\hbar}{2m_0^* \omega} \right)^{\frac{1}{2}} = \frac{1}{\sqrt{2}} \left(1 - \gamma^2 \right)^{-\frac{1}{4}} . \tag{40}$$

As the parameter γ approaches unity ($\omega \to 0$), the electron wave functions are progressively extended ($R \to \infty$), and the existence of the barrier influences the motion more strongly. Also, the electrons whose initial positions are well within the bulk are now reflected from the barrier, so that their motion is basically electriclike, only weakly deflected by the magnetic field. In other words, the magnetic field is in this case not strong enough to prevent the electrons from reaching the interface. Finally for the parameter $\gamma = 1$, the effective frequency $\omega = 0$ and the oscillator length $R = \infty$, so that without the barrier, the electron motion becomes unbound and the quantization disappears. Thus, in the unlimited space a sufficiently strong transverse electric field destroys the Landau quantization. However, in the presence of an interface the electrons may not run away, since they are reflected from the barrier. For $\omega_c^2 < 2e^2 F_s^2 / E_g m_0^*$ we have, in Eq.(30), a parabolic potential barrier instead of a parabolic potential well and the electrons oscillate between this barrier and the interface. The shape of the potential and the classical electron orbits also are shown schematically in Fig.15.

Electric and magnetic case

We now present the eigenenergies of the effective Schrödinger equation given in Eq.(30).[9] We start with the electric case ($\gamma > 1$), where we have a potential barrier [see Fig.15(c)]. This potential clearly resembles the purely electrostatic triangular-well potential. The energies may be quantized using the zeros of the Weber functions of the second kind. However, it is much more simpler to use the semiclassical quantization procedure

$$\int_0^{z_t} p_z dz = \hbar\pi \left(i + \frac{3}{4} \right) \tag{41}$$

thus treating the momentum p_z as a c-number. This semiclassical quantization provides an astonishingly good approximation to the eigenenergies, up to a fraction of a percent, as has been realized some time ago for the purely electrostatic triangular-well potential. In Eq.(41) we have the right-hand turning point z_t and the phase $^3/_4$ which is the sum of the left-hand contribution $^1/_2$ of the infinite barrier and the right hand one $^1/_4$ of the nearly linear potential.

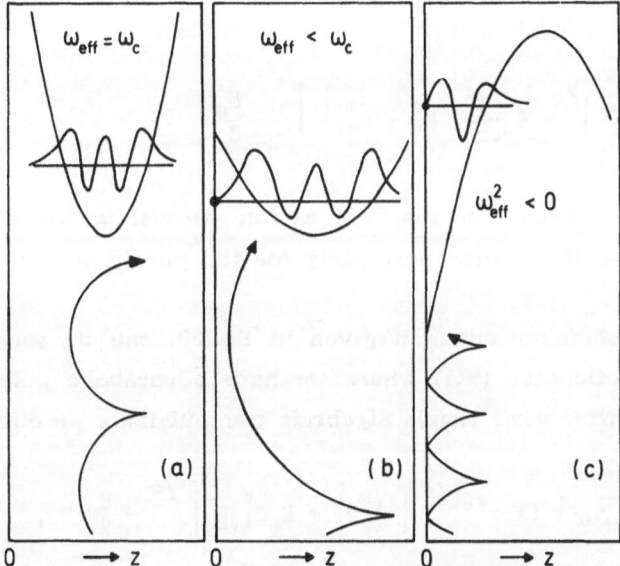

Fig.15. Potential energies, wave functions, and classical orbits for magnetic (a,b) and electric (c) type of motion for inversion electrons in crossed fields. From Ref.9.

The integration can be carried out analytically and the result is

$$(A+B)A^{1/2}B^{1/2} - \frac{1}{2}(B-A)^2 \ln\left(\frac{B^{1/2}-A^{1/2}}{B^{1/2}+A^{1/2}}\right) = 4euE'\hbar\pi(i + \frac{3}{4}) , \quad (42)$$

where

$$E' = E(1-\gamma^{-2})^{1/2} , \quad (43)$$

$$A = eE'z_0 - a_\pm , \qquad B = eE'z_0 + a_\pm , \quad (44)$$

and

$$a_\pm = E_g^{1/2}\left(\frac{\alpha^2}{4b^2} - \lambda_\pm\right)^{1/2} , \qquad b^2 = \frac{e^2E'^2}{E_g} . \quad (45)$$

Equation (42) presents a transcendental equation for the energies $E_i(k_x,k_y)$ in the electric case ($\gamma>1$). The general case of crossed fields may be reduced to the purely electric limit by putting the magnetic field B=0. This gives $\gamma^{-1}=0$ and $F'_s=F_s$ and after simple manipulation A and B become

$$A = E - E_\parallel , \qquad B = E_g + E + E_\parallel \quad (46)$$

where

$$E_\parallel = \left[\left(\frac{E_g}{2}\right)^2 + E_g\frac{\hbar^2}{2m_0^*}(k_x^2 + k_y^2)\right]^{1/2} - \frac{E_g}{2} \quad (47)$$

represents the energy of the free motion parallel to the surface. This is exactly the result obtained previously for the purely electric case.[19]

The Schrödinger equation given in Eq.(30) can be solved analytically for the magnetic case ($\gamma<1$) where we have a parabolic potential well [see Fig.15(a,b)]. After some simple algebraic manipulations we obtain

$$E = \hbar k_y v_D + (1-\gamma^2)^{1/2}\left[\left(\frac{E_g}{2}\right)^2 + E_g D_{i,k_x}^\pm\right]^{1/2} - \frac{E_g}{2} , \quad (48)$$

where

$$D_{i,k_x}^\pm = (\nu_i + \frac{1}{2})(1 - \gamma^2)^{1/2}\hbar\omega_c + \frac{\hbar^2 k_x^2}{2m_0^*} \pm \frac{1}{2}g^*\mu_B B . \quad (49)$$

Note that the indices ν_i are functions of the center coordinates (see Fig.3).

324

With Eqs. (48) and (49), one can calculate the electron energies as a function of the subband index i, the normalized center coordinate $q_0=z_0/R$, the momentum $\hbar k_x$, and the spin orientation. However, for the purpose of optical considerations, it is more useful to know the energies as a function of the momentum $\hbar k_y$ rather than of the center coordinate q_0, since $\hbar k_y$ is conserved in direct optical transitions. Once the energy is obtained from Eqs. (48) and (49) for a given coordinate q_0, one can calculate the corresponding k_y-value from

$$k_y l = \frac{1}{\sqrt{2}} (1 - \gamma^2)^{3/4} q_0 + k_0 l \left(1 + 2\frac{E}{E_g} \right) . \tag{50}$$

Purely magnetic surface levels are obtained by putting the parameter $\gamma=0$:

$$E_i = \left[\left(\frac{E_g}{2}\right)^2 + E_g \left(\hbar\omega_c(\nu_i + \frac{1}{2}) + \frac{\hbar^2 k_x^2}{2m_0^*} \pm \frac{1}{2}g_0^* \mu_B B \right) \right]^{1/2} - \frac{E_g}{2} \tag{51}$$

For magnetic surface levels the simple relation $z_0/l=k_y l$ holds. In the bulk limit $z_0 \gg l$, the energies do not depend on the electron position and we obtain the widely used formula for Landau levels in narrow-gap semiconductors ($\nu_i=0,1,...$).

In Fig.16(a-c) the surface band structure in the approximation of the triangular-well potential is depicted for (a) purely electric subbands, (b) crossed fields ($\gamma<1$), and (c) purely magnetic subbands. The electric subbands are symmetric with respect to the momentum $\hbar k_y=0$ and the subband dispersions $E_i(k_y)$ are first quadratic and then become linear. There is no spin splitting since the magnetic field B=0 and we have ignored the inherent spin splitting due to the inversion asymmetry of the potential. The Fermi lines in the upper part of the figure are circles and have been calculated numerically for the electron density $n_s=1.5\times10^{12}$cm^{-2}. In crossed fields, the subbands become asymmetric and are spin-split. Also, their density of states is higher than in the purely electric case as is seen from the position of the Fermi energy E_F. At positive momenta the subbands become parallel to each other and we have Landau levels in an electric field. These Landau levels exist since we have chosen a parameter $\gamma<1$ (magnetic case) for Fig.16(b). In the electric case $\gamma>1$ no Landau levels are formed even at very high momenta ($\hbar k_y \gg 1$). This is in clear contrast to the one-band model as was explained above. The magnetic surface levels ($F_s=0$) are only qualitatively similar to the ones of the one-band model. They are spin-split and clearly show effects of nonparabolicity: the Landau spacings become higher at higher levels and simultaneously the spin splitting is decreased. We have not calculated the Fermi lines for the purely magnetic subbands in Fig.16(c) since this would require knowledge of the thickness of the sample or periodic boundary conditions together with a bulk electron concentration.

The diamagnetic shifts of the subbands are visible in Figs.16(a–c) even if this effect is somewhat obscured by the spin splitting in Fig.16(b). For the spin-split subbands i=$0^{\pm}$, $1^{\pm}$, and $2^{\pm}$ the diamagnetic shifts are depicted in Fig. 17(a–c) versus the magnetic field for three values of the electric field intensity. The minima of the dispersion curves $E_i(k_y)$ are taken starting from the values of the purely electric subbands (B=0). It can be seen that the subband energies are continuous functions of the magnetic field although they are calculated somewhat differently for the electric ($\gamma>1$) and magnetic ($\gamma<1$) case.

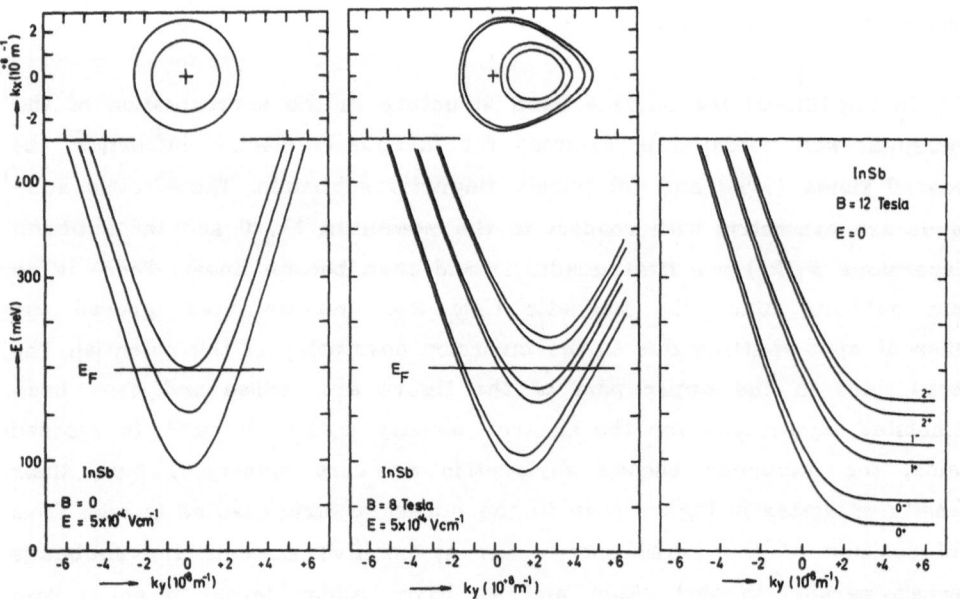

Fig.16. Hybrid subband structure in crossed electric and magnetic fields calculated for InSb band parameters. The Fermi lines are determined for an electron density n_s=1.5x10^{12}cm^{-2}. The transition from a purely electric to a purely magnetic band structure is shown according to the three-level model.

The diamagnetic shift of a particular subband is strongest at low electric fields and higher subbands show stronger shifts at all electric fields. This tendency can already be understood with the result of the perturbation theory given in Eq.(22) since the wave function is more extended in lower electric fields and in higher subbands. At the highest surface electric field $F_s=5x10^{+4}Vcm^{-1}$ the spin splitting causes a shallow minimum of the 0^+ subband at magnetic fields $B\approx4T$. The observable shift of the intersubband resonance $0^+\to1^+$, however, always increases with magnetic field. All the qualitative features of the experimental values for diamagnetically shifted intersubband resonances depicted in Fig.14 can be explained with the present model. Quantitative agreement, however, is only obtained for the lowest electron density studied in Fig.14. According to Eq.(29) we obtain a surface electric field $F_s=0.7x10^{+4}Vcm^{-1}$ for the density $n_s=2.10^{11}cm^{-2}$. With this electric field we can in fact reproduce the solid line in Fig.14 that was drawn to guide the eye.

We now consider electrons with center coordinates inside the semiconductor ($z_0/l\gg1$) that are practically not affected by the barrier at the interface. Their excitations are cyclotron resonances in crossed fields and it is convenient to define an apparent cyclotron mass

$$m^* = e\hbar B \ / \ (\ E^{\pm}_{n+1} - E^{\pm}_n \ ) \ . \tag{52}$$

This definition is independent of the center coordinate since the Landau levels run parallel to each other at center coordinates $z_0/l\gg1$ [see Fig.16(b)]. In the one-band effective-mass approximation this definition leads to the constant effective mass m^* which does neither depend on the magnetic field nor on the electric field. In the three-level description the apparent cyclotron mass does depend on the field strengths and also depends on the Landau and spin index $n^{\pm}$ of the cyclotron transition $n^{\pm} \to(n+1)^{\pm}$. Most important, the cyclotron mass only can be defined in the magnetic case ($\gamma<1$).

Cyclotron masses in electric fields $F_s=1$ and $2x10^{+4}Vcm^{-1}$ are depicted in Fig. 18 versus magnetic field B for the spin-up $0^+\to1^+$ and spin-down $0^-\to1^-$ transition. For comparison the masses in the absence of an electric field are included ($F_s=0$). The results shown in Fig.18 may be interpreted in the following way: At high magnetic fields ($\gamma\ll1$), the electric term in Eq.(31) is negligible and the effective frequency is approximately equal to the cyclotron frequency ($\omega\approx\omega_c$). At high magnetic fields the increase of the mass and the spin splitting approach the well-known bulk behavior. At lower magnetic fields the electric term in Eq.(31) becomes important. As the

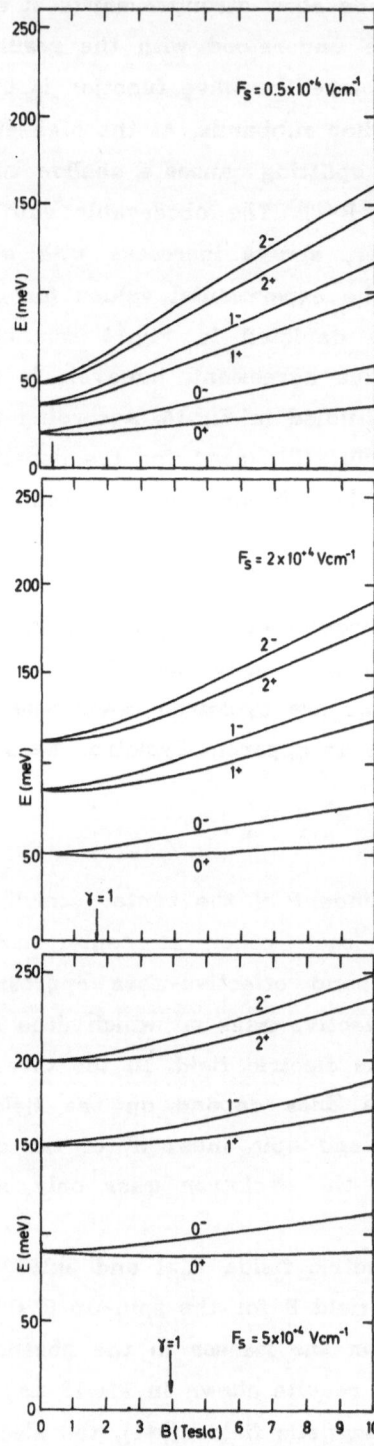

Fig.17. Eigenenergies of inversion electrons in crossed fields for three electric field strengths. The magnetic field at which the transition from the electric to the magnetic case occurs ($\gamma=1$) is indicated.

magnetic field decreases the effective frequency ω becomes smaller and smaller ($\gamma\to 1$) resulting in the strong enhancement of the cyclotron mass as defined in Eq.(52).

Relativistic analogy

It was noticed some time ago that there exists a striking and far-reaching analogy between the behavior of electrons in narrow-gap semiconductors and relativistic electrons in vacuum.[20] The crossed-field configuration can serve as a spectacular illustration of the analogy.

In the absence of external fields, the energy-momentum relation resulting from the two-band model is given by the simplified Kane formula Eq.(37). This equation has the form of the relativistic dispersion relation for electrons in vacuum with the following correspondences: $E_g \leftrightarrow 2m_0 c^2$ and

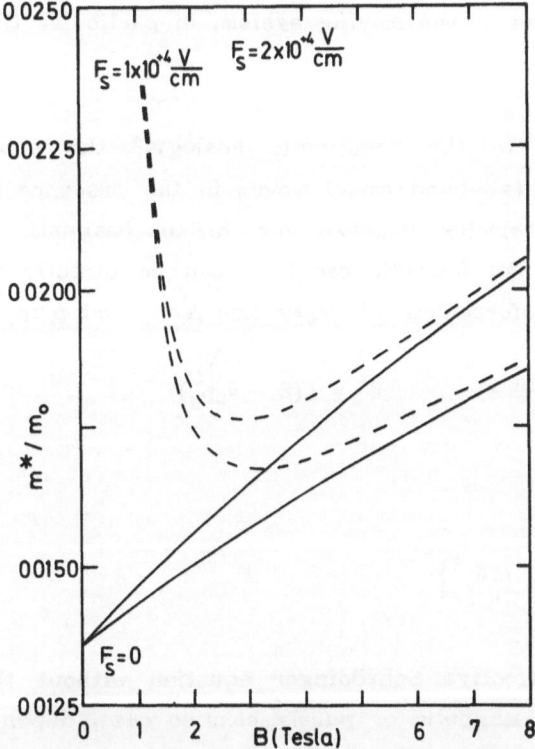

Fig.18. Electron cyclotron masses in crossed fields calculated for InSb band parameters. The masses are given for the purely magnetic case ($F_s=0$) and for two constant electric fields (magnetic case, bulk limit). From Ref.9.

329

$m_0^* \leftrightarrow m_0$. It follows immediately that the limiting velocity $u = (E_g/2m_0^*)^{1/2}$ corresponds to the light velocity c. In the presence of an external magnetic field, the expression for the electron energies in the bulk limit [see Eq.(51)] also has the form analogous to the expression for relativistic Dirac electrons. Only the magnitude of the spin term is somewhat different, as the latter in semiconductors is determined by the spin–orbit interaction which is of the atomic origin and has no correspondence in the free-electron case.

It is well known that in the presence of crossed magnetic and electric fields, the electron drifts with a constant velocity v_D transverse to both fields. For free relativistic electrons, this velocity is $v_{dm} = F_s/B$ for velocity $F_s/B < c$ (magnetic case) and $v_{de} = c^2 B/F_s$ for velocity $F_s/B > c$ (electric case) (see, e.g., Ref.11). If a Lorentz transformation is made from the laboratory system to a system moving with the drift velocity, one eliminates from the equation of motion one of the two fields. In the magnetic case the electric field disappears and the magnetic field becomes $B' = B(1-v_{dm}^2/c^2)^{1/2}$. In the electric case, the magnetic field disappears and the electric field becomes $F_s' = F_s(1-v_{de}^2/c^2)^{1/2}$. Clearly, all other quantities of interest should be transformed as well to the moving system, in particular the four vector of momentum and energy.

According to the relativistic analogy,[9] the conduction electron described by the two-band model moves in the presence of crossed fields with the drift velocity $v_{dm} = E/B$ for $E/B < u$ (magnetic case) and with $v_{de} = u^2B/E$ for $E/B > u$ (electric case). It can be directly verified that the Lorentz-type transformation (for $v_D \parallel y$ axis $p_x' = p_x$ and $p_z' = p_z$)

$$p_y' = \Gamma\left(p_y - \frac{v_D}{u^2} E\right) \quad , \qquad E' = \Gamma(E - v_D p_y) \tag{53}$$

with

$$\Gamma = \left[1 - \left(\frac{v_D}{u}\right)^2\right]^{-1/2} \tag{54}$$

transforms the effective Schrödinger equation without the spin term to either the purely magnetic or purely electric case, depending on whether the drift velocity v_{dm} or v_{de} is used.

To give an example for the relativistic analogy, the result for electrons in the bulk of the semiconductor (magnetic case) given in Eqs.(48) and (49) may be interpreted in the following way: When a transformation to the moving system is made, only the magnetic field is left. The quantization can now be carried out with the result given in Eq.(51) in which $B'=B(1-\gamma^2)^{1/2}$ in the cyclotron energy: $\hbar\omega_c=eB'/m_0^*$. However, the energy is observed in the laboratory system, so that a transfomation back to this system is necessary according to Eq.(53). This gives the final result [Eqs.(48) and (49)] and the energy difference measured in the cyclotron resonance experiment $E_{n+1}-E_n=\hbar\omega$ may be now interpreted as "the relativistic Doppler shift" $\omega=(1-v_{dm}^2/u^2)^{1/2}\omega_0$ where ω_0 is the frequency in the moving system (see, e.g., Ref.11). This corresponds, in Eq.(48), to the factor $(1-\gamma)^{1/2}$ in front of the square root. In the special theory of relativity, the relativistic Doppler shift is regarded as a direct manifestation of the time dilation. A similar reasoning may be applied to the electric case. Note, that in all cases the boundary condition at $z=0$ remains unchanged since the drift velocities are parallel to the y-direction.

The scheme of the relativistic analogy for electrons in crossed fields is summarized in Table 1. For the fields F_s and B the velocity $v_D=F_s/B$, which is the physical drift velocity only in the magnetic case, is calculated and is compared to the limiting velocity $u=(E_g/2m_0^*)^{1/2}$ of the semiconductor. If the velocity v_D exceeds the limiting velocity $(\gamma=v_D/u>1)$ we are in the electric case: the purely electric subbands $E_i'(k_y')$ are calculated in the moving system with the transformed fields $F_s'=(1-\gamma^{-2})^{1/2}F_s$ and $B'=0$ according to Eqs.(42),(46) and (47). If the velocity v_D is less than the limiting velocity $(\gamma=v_D/u<1)$ we are in the magnetic case: the purely magnetic subbands $E_i'(k_y')$ are calculated in the moving system with the transformed fields $F_s'=0$ and $B'=B(1-\gamma^{+2})^{1/2}$.

In both cases, the energies obtained in the moving system have to be Lorentz transformed back to the laboratory system to obtain the final result $E_i(k_y)$. It is interesting to note, that the description of electrons in crossed fields in the one-band effective-mass approximation is obtained as the nonrelativistic limit of the magnetic case in Fig.18. Since the dispersion in the one-band model is parabolic at all energies, we always are in the magnetic case $(u\to\infty)$. The Lorentz transformations in the nonrelativistic limit become the Galilei transformations and we obtain the result of the one-band description as given in Eq.(9).

Table 1. Schematic diagram of how the relativistic analogy can be used to
calculate the hybrid subband structure in crossed electric and magnetic
fields. Note that the spin-term cannot be included into this scheme.

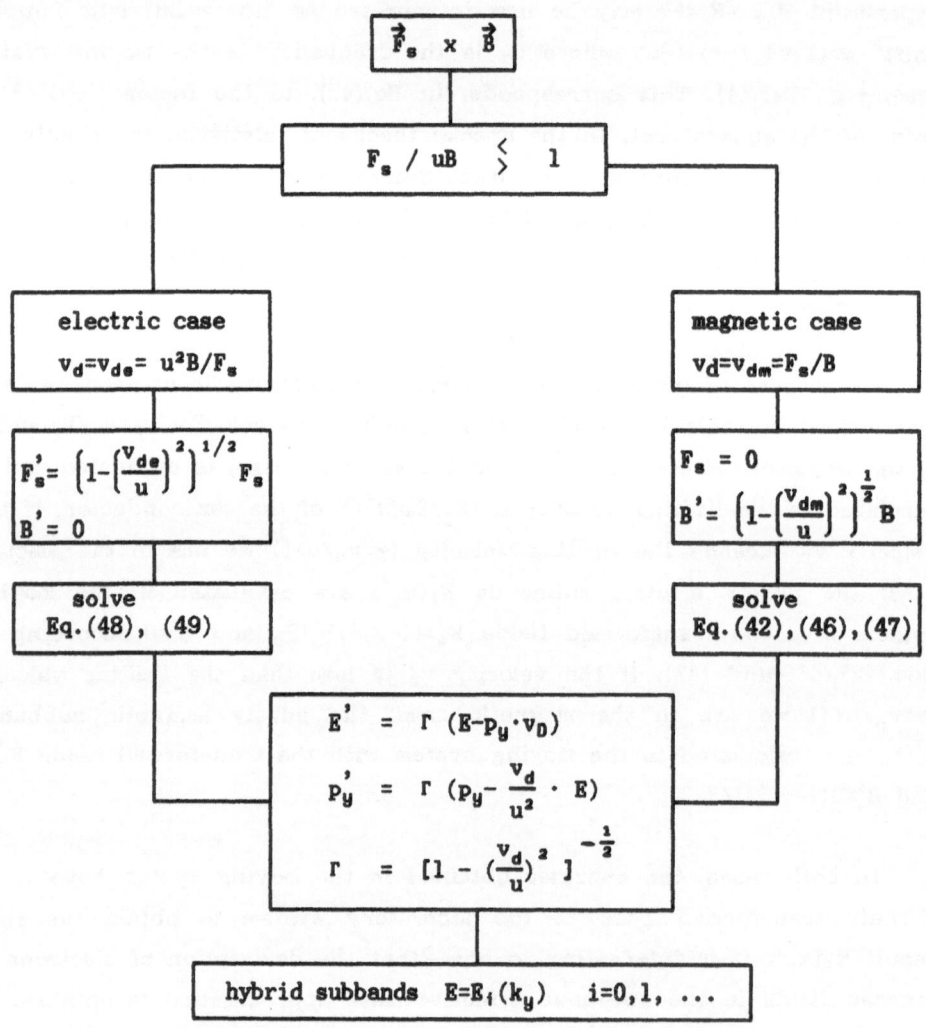

6. EXPERIMENTS: DESTRUCTION OF LANDAU QUANTIZATION IN STRONG ELECTRIC FIELDS

Perhaps the most exciting consequence of the two-band model is the prediction that cyclotron resonance in crossed electric and magnetic fields can be destroyed. This prediction has been verified experimentally by demonstrating that the cyclotron energy tends to zero ($\hbar\omega_c \rightarrow 0$) and that the well-known selection rule for cyclotron transitions ($\Delta n=1$) becomes clearly violated in strong electric fields. The first feature is detected by a divergence of the apparent cyclotron mass when the ratio of the electric and magnetic field intensities approaches its critical value $F_s/B=u$. The second feature is observed through the appearance of harmonics ($\Delta n=2$). We again like to stress that these effects cannot be described within the one-band model: here the cyclotron energy is constant and is not affected by the electric field and cyclotron transitions are only allowed when the Landau index is changed by one.

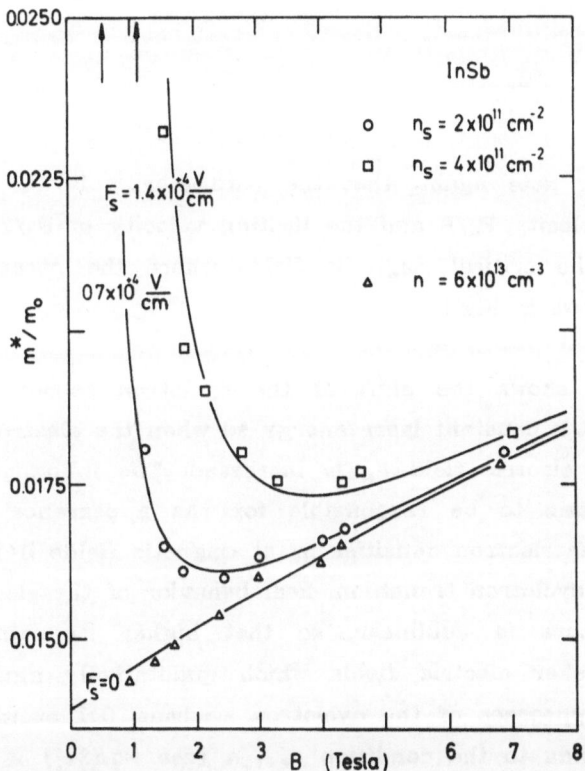

Fig.19. Cyclotron masses in InSb in crossed electric and magnetic fields for two inversion electron densities n_s, i.e., electric field strengths F_s. Masses of n-type InSb ($F_s=0$) are included for comparison. From Ref.7.

The mass divergence

Cyclotron masses in crossed fields obtained with the use of inversion layers on InSb are depicted versus magnetic field in Fig.19 for two inversion electron densities $n_s=2$ and $4 \times 10^{11} cm^{-2}$. From the densities n_s the surface electric field intensities $F_s=0.7$ and $1.4 \times 10^{+4} Vcm^{-1}$ are calculated using Eq.(29). The figure also contains masses in the absence of an electric field ($F_s=0$) measured in a n-type bulk sample of low doping ($n=6 \times 10^{13} cm^{-3}$). In both systems the masses have been extracted from the resonance positions of $0^+ \rightarrow 1^+$ transitions. The solid lines are calculated from Eq.(48) and good agreement is found between experiments and theory.

The cyclotron masses in crossed fields approach the values for purely magnetic fields at high magnetic field intensities but are clearly different at low magnetic fields: whereas the $F_s=0$ masses extrapolate to the band edge mass m_0^*, the masses in crossed fields show a strong increase. The behavior is theoretically described by Eq.(48) that predicts an apparent cyclotron mass

$$m^* \propto \frac{m_0^*}{1 - \gamma^2} \tag{55}$$

in the limit $\gamma \rightarrow 1$. Note again, that the parameter γ is the ratio of the magnetic drift velocity F_s/B and the limiting velocity $u=(E_g/2m_0^*)^{1/2}$ of the semiconductor. The critical magnetic fields where the parameter $\gamma=1$ are indicated by arrows in Fig.19.

Figure 20 shows the shift of the cyclotron resonance to higher magnetic fields at a constant laser energy $\hbar\omega$ when the electron density n_s, i.e., the surface electric field F_s is increased. The inhomogeneity of the electric field seems to be responsible for the appearance of a second maximum at higher electron densities n_s at magnetic fields $B<1.5T$ which we interpret as $1 \rightarrow 2$ cyclotron transition. Real behavior of the electric potential in a MOS-structure is sublinear, so that higher magnetic states are subjected to weaker electric fields which qualitatively agrees with the observation. Disappearace of the cyclotron maximum $0 \rightarrow 1$ at magnetic fields $B > \approx 2.5T$ corresponds to the condition $\omega_{eff}=0$ [see Eq.(31)] if scattering is neglected, or, more realistically, to the condition $\omega_{eff}\tau<1$, where τ is the electron relaxation time. The increase of the cyclotron mass with increasing electric field and the disappearance of cyclotron resonance are spectacular manifestations of the relativistic analogy. The factor $1-\gamma^2$ in the denominator of Eq.(55) is the consequence of two "relativistic" effects: the magnetic field has to be Lorentz transformed to the system moving with the magnetic drift velocity F_s/B which gives one factor $(1-\gamma^2)^{1/2}$ and the energy has to be transformed back to the laboratory which gives another factor $(1-\gamma^2)^{1/2}$ that corresponds to the relativistic transverse Doppler shift (see Table 1).

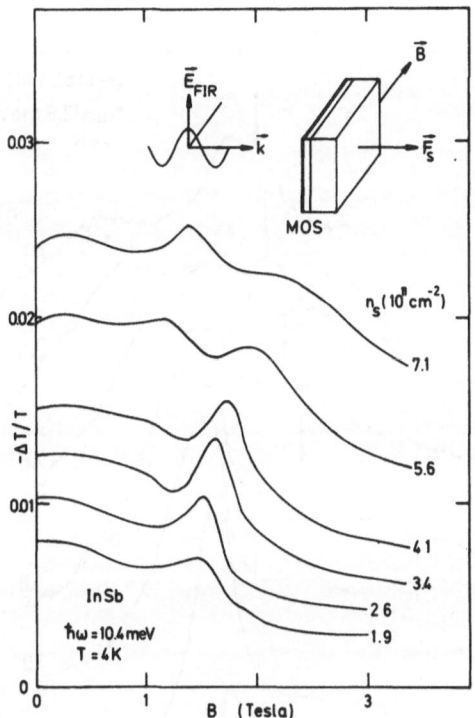

Fig.20. Cyclotron resonance spectra of inversion electrons at various electron densities n_s vs magnetic field parallel to the inversion layer at a fixed laser energy $\hbar\omega$. The geometical configuration of the crossed field configuration is shown in the inset. From Ref.7.

Breakdown of the oscillator selection rule

Some time ago it has been predicted from a perturbation theory ($\gamma \ll 1$) that harmonics of cyclotron resonance should be observable in strong transverse electric fields in a narrow-gap semiconductor.[21] This prediction could only recently been verified, again with the use of inversion layers on InSb in the crossed field configuration.[22] To demonstrate this, Fig.21 shows cyclotron spectra measured at a constant laser energy $\hbar\omega$ versus magnetic field for various electron densities n_s. At the lowest densities $n_s = 1.5 - 3.0 \times 10^{11} \text{cm}^{-2}$ the predominant maximum at $B \approx 1.8\text{T}$ is caused by $0 \to 1$ cyclotron transitions. At higher densities the $0 \to 1$ resonance strongly shifts to higher magnetic fields and vanishes as was described in the last section. Cyclotron resonances $1 \to 2$ are also present in the spectra, most clearly at the density $n_s = 5.4 \times 10^{11} \text{cm}^{-2}$. At densities $n_s \gtrsim 3.0 \times 10^{11} \text{cm}^{-2}$ an additional resonance occurs in the spectra (see arrows). If the resonance position of

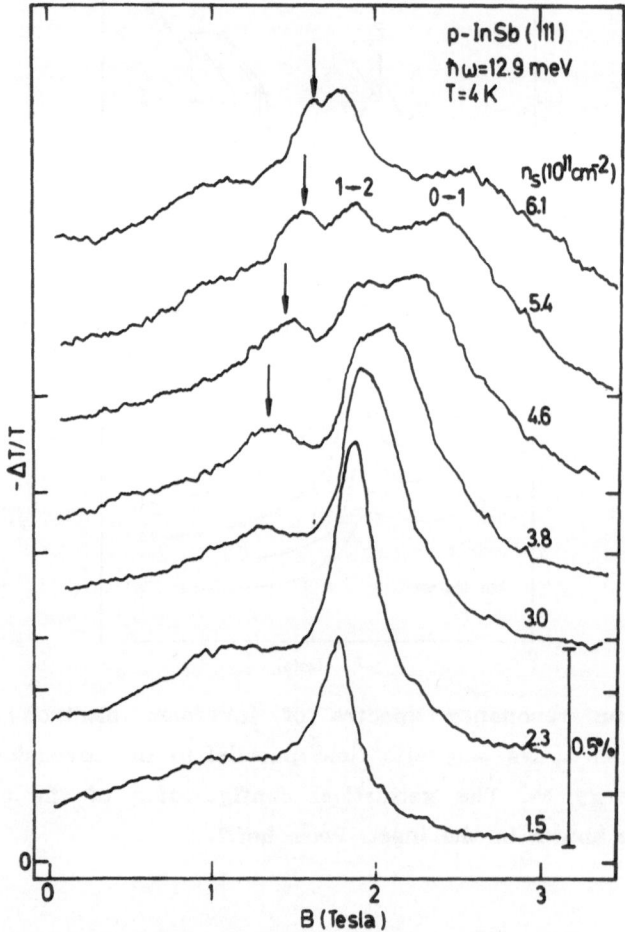

Fig.21. Spectra for various electron densities n_s, i.e., electric field strengths F_s taken in the crossed field configuration at a fixed laser energy $\hbar\omega$. Cyclotron resonance $0 \to 1$ is the most prominent structure at densities $n_s = 1.5 - 3.0 \times 10^{11} \text{cm}^{-2}$ but $1 \to 2$ transitions are also present. The maxima that are indicated by arrows are interpreted as harmonics of the fundamental cyclotron resonance. Note that their intensity is quite comparable to the one of the fundamental transition at the density $n_s = 5.4 \times 10^{11} \text{cm}^{-2}$. From Ref.22.

this maximum is extrapolated to zero density ($n_s=0$), it agrees with the position of the harmonic transition $0\to2$ in bulk crystals. In the bulk of InSb harmonics are rather weak, their oscillator strength is only about 10^{-4} of the fundamental resonance, and they are induced by effects of band warping and inversion asymmetry.[23]

We have calculated the matrix elements for the fundamental ($0\to1$) and the harmonic ($0\to2$) transitions in the full regime of Landau quantization ($0\leqslant\delta\leqslant1$). To obtain analytical results we have treated the ratio of cyclotron and gap energy $\hbar\omega_c/E_g=e\hbar B/m_0^* E_g$ as a small parameter. This is justified in view of our experiments ($\hbar\omega_c/E_g<0.1$) depicted in Fig.21. Like in the one-band model we have normalized the matrix elements with the one for the $0\to1$ transition $M_{01}(\delta=0)$ in the absence of an electric field. The normalized transition strength of the fundamental transition shown in Fig.22 is an universal function of the parameter γ, i.e., it is of zero order with respect to the parameter $\hbar\omega_c/E_g$. The strength for the harmonic $0\to2$ transition is proportional to $\hbar\omega_c/E_g$ and is given for InSb parameters in the

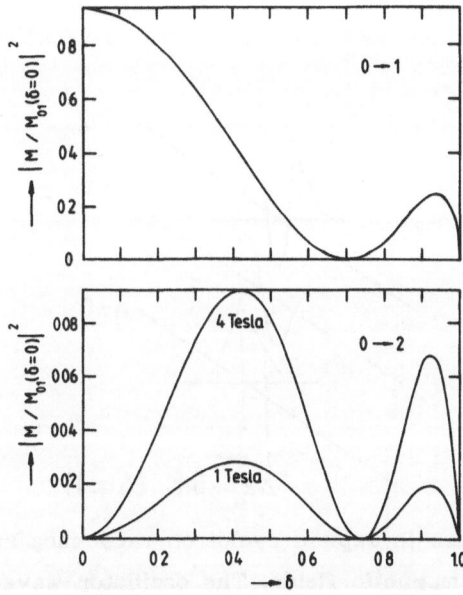

Fig.22. Transition strengths, i.e., squared matrix elements for the fundamental $0\to1$ and first harmonic $0\to2$ transition calculated in the full range of Landau quantization $0\leqslant\delta\leqslant1$ in crossed fields. The transition strengths have been normalized with the one of cyclotron resonance $0\to1$ in the absence of an electric field $M_{01}(\delta=0)$. From Ref.22.

figure. Since the transitions are calculated in different orders with respect to the   ratio $\hbar\omega_c/E_g$, the sum rule for oscillator strengths cannot directly be applied to the results shown in Fig.22.

An intuitive picture why harmonics arise in crossed fields is presented in Fig.23. Landau levels n=0,1,2 are depicted away from the interface together with oscillator wave functions. According to Eqs.(32),(33) there is a shift Δz of the center coordinate when the energy of the electron is changed. This shift can be calculated explicitly in the limit $\gamma \ll 1$ and

$$\Delta z = - \delta \cdot 1 \cdot \Delta n \tag{56}$$

for a transition with change Δn of the Landau index. Intuitively this means that the Landau electron steps in the electric field direction during a cyclotron transition. Then it is clear, that the usual selection rules for the harmonic oscillator no longer are valid. However, since the simplified picture does not take into account the proper multi-component wave function of the two-band model, it is not appropriate to describe the results in Fig.22 quantitatively.

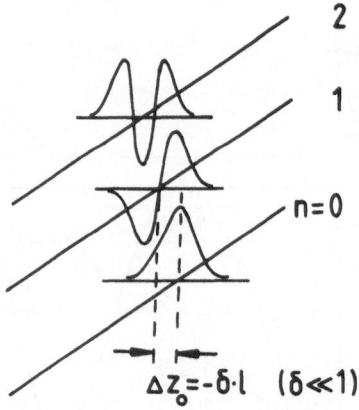

Fig.23. Intuitive picture to explain cyclotron resonance and its harmonics in crossed electric and magnetic fields. The oscillator wave functions that are related to the conduction band are schematically shown. Whereas the momentum is conserved in a transition, the center coordinate z_0 is not leading to a violation of the usual oscillator selection rule. Note that this simple picture does not take into account the full multi-component wave function of the two-band model.

7. CONCLUDING REMARKS

Inversion layers on InSb in magnetic fields applied parallel to the surface are interesting systems for magneto-optical studies. The externally applied magnetic field and the internal surface electric field of the space-charge potential provide a crossed-field configuration with strong electric fields. This way, for the first time intraband optical absorption could be studied experimentally in the crossed-field configuration on a narrow-gap semiconductor and theoretical predictions that have been put forward some time ago could be verified. In particular, the destruction of the Landau quantization in strong transverse electric fields could be demonstrated with use of inversion layers on InSb.

ACKNOWLEDGEMENTS

This review is based on work which I did together in course of time with J.H. Crasemann, M. Horst, S. Klahn, S. Oelting, and W. Zawadzki. I thank J.P. Kotthaus for his continuous encouragement and the Deutsche Forschungsgemeinschaft for financial support.

REFERENCES

1. M. S. Khaikin, Adv. Phys. 18:1 (1969).
2. T. Ando, A. B. Fowler, and F. Stern, Rev. Mod. Phys. 54:437 (1982).
3. F. Koch, in: "Physics in High Magnetic Fields – Proceedings of the Oji International Seminar, Hakone, Japan, 1980", S. Chikazumi and M. Miura, ed., Springer, Berlin, 1981.
4. H. Schaber and R. E. Doezema, Phys. Rev. B 20:5257 (1979).
5. J. H. Crasemann, U. Merkt, and J. P. Kotthaus, Phys. Rev. B 28:2271 (1983).
6. U. Merkt, Phys. Rev. B 32:6699 (1985).
7. W. Zawadzki, S. Klahn, and U. Merkt, Phys. Rev. Lett. 55:983 (1985).
8. S. Oelting, U. Merkt, and J. P. Kotthaus, Surf. Sci. 170:402 (1986).
9. W. Zawadzki, S. Klahn, and U. Merkt, Phys. Rev. B 33:6916 (1986).
10. M. Reine, Q. H. F. Vrehen, and B. Lax, Phys. Rev. 163:726 (1967).
11. J. D. Jackson, "Classical Electrodynamics", Wiley, New York (1975).

12. J. C. P. Miller, in: "Handbook of Mathematical Functions", M. Abramowitz and I. A. Stegun, ed., Dover, New York (1965), Chap. 19, pp. 685-720.

13. E. A. Kaner, N. M. Makarov, and I. M. Fuks, Zh. Eksp. Teor. Fiz. 55:931 (1968) [Sov. Phys.-JETP 28:483 (1969)]; P. Dean, Proc. Cambridge Philos. Soc. 62:277 (1966).

14. U. Kunze, Surf.Sci. 170:353 (1986).

15. W. Zawadzki, Surf. Sci. 37:218 (1973).

16. J. R. Apel, T. O. Poehler, C. R. Westgate, and R. I. Joseph, Phys. Rev. B 4:436 (1971).

17. J. Scholz, F. Koch, H. Maier, and J. Ziegler, Solid State Commun. 45:39 (1983).

18. K. Wiesinger, H. Reisinger, and F. Koch, Surf. Sci. 113:102 (1982).

19. W. Zawadzki, J. Phys. C 16:229 (1983).

20. W. Zawadzki and B. Lax, Phys. Rev. Lett. 16:1001 (1966)

21. W. Zawadzki, in: "Proceedings of the 9th International Conference on the Physics of Semiconductors" Nauka, Leningrad (1968), Vol.1, p.312.

22. S. Klahn, M. Horst, and U. Merkt, in: "Proceedings of the 18th International Conference on the Physics of Semiconductors", Stockholm (1986), in press.

23. C. R. Pidgeon, in: "Handbook on Semiconductors", T.S. Moss, ed., North-Holland, Amsterdam (1980), Vol. 2, Chap. 5, pp. 223-328.

ELECTRONIC PROPERTIES OF II-VI COMPOUND SUPERLATTICES

J.M. Berroir and M. Voos

Groupe de Physique des Solides de l'Ecole Normale Supérieure
24 rue Lhomond
75005 Paris, France

INTRODUCTION

Semiconductor superlattices (SL), which have been first proposed by Esaki and Tsu[1] in 1970, consist of thin alternate layers of two different semiconductors. In these periodic structures, the layer thickness ranges roughly from 10 Å to a few hundred Å, smaller than or comparable to the electron mean free path or to the de Broglie wavelength, but larger than the interatomic spacing. In this case, the considered system is two-dimensional, at least in first approximation, and quantum effects can be expected to occur, changing the electronic structure of the involved materials and leading to unusual transport and optical properties.

In the picture of Esaki and Tsu, the different band gaps E_g of the host semiconductors create a one-dimensional periodic potential in the z direction perpendicular to the plane of the layers. This leads to the creation of an artificial periodicity longer than the atomic spacing and superposed to the natural periodicity of the crystal, which is the origin of the name given to these systems. Such a potential causes a folding of the Brillouin zone, resulting in a series of quantized mini-bands or sub-bands in both the conduction and valence bands (Fig. 1). One can consider that the carriers are confined in quantum potential wells and that their motion is quantized in the z direction, the conduction (valence) subbands such as E_1, E_2 ... (H_1, H_2 ...) lying at higher (lower) energy than the bottom (top) of the conduction (valence) band of the involved bulk semiconductor. Tunneling interactions between successive potential wells can lead to an appreciable width of the subbands in the z direction, corresponding to a certain dispersion relation of the energy as a function of the

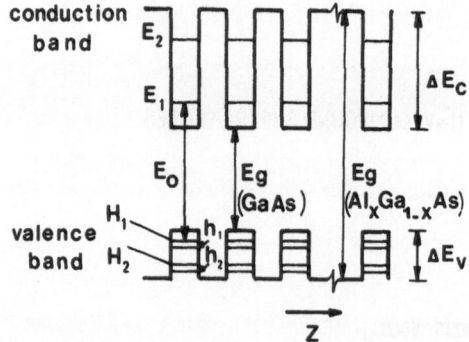

Fig. 1 . Variation of the conduction and valence bands in a GaAs–Al$_x$Ga$_{1-x}$As superlattice in the z direction perpendicular to the plane of the layers. Several conduction (E$_1$, E$_2$), heavy–hole (H$_1$, H$_2$) and light hole (h$_1$, h$_2$) subbands are represented.

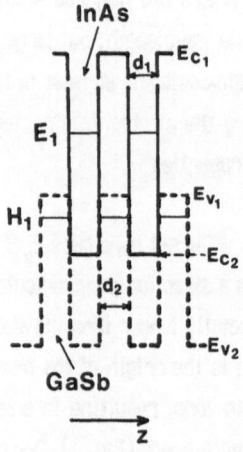

Fig. 2 . Conduction (E$_{c_1}$, E$_{c_2}$) and valence bands (E$_{v_1}$, E$_{v_2}$) in an InAs–GaSb superlattice.

electron wavevector k_z. Furthermore, the carrier motion is not quantized in the x, y plane of the layers, and each subband presents also a dispersion relation as a function of k_x and k_y which may be very different from that of the bulk semiconductor.

Until recently two types of superlattices, I and II, have been investigated. GaAs-Al$_x$Ga$_{1-x}$As structures correspond[2] to type I, and the situation is schematically represented in Fig. 1. The GaAs conduction band edge is at lower energy than that of Al$_x$Ga$_{1-x}$As, while its valence band edge is at higher energy than that of Al$_x$Ga$_{1-x}$As. As a result, the GaAs layers are potential wells for both electrons and holes which are thus confined in these layers. This situation is very frequent, and corresponds, in fact, to many of the structures which have been studied up to now. In InAs-GaSb systems[3], which belong to type II structures, the InAs conduction band edge lies at lower energy than the GaSb valence band edge, as illustrated in Fig. 2. It follows that the InAs and GaSb layers serve as potential wells for electrons and holes, respectively. Electrons and holes are thus spatially separated, but photon absorption and emission are nevertheless possible because there is a slight overlap of the electron and hole wave functions.

In addition to the spatial separation of electrons and holes, there is another important difference between InAs-GaSb and GaAs-Al$_x$Ga$_{1-x}$As superlattices. Indeed, the electron effective mass in InAs (0.023 m$_0$[4]) is much lighter than in GaAs (0.066 m$_0$[4]), leading to stronger tunneling interactions between InAs layers than between GaAs layers. As a result, in InAs-GaSb structures even for rather large values of the layer thicknesses ($\sim$ 200 Å), the E$_1$ subband has an appreciable width ΔE_1 in the z direction[5], while ΔE_1 is often negligible in GaAs-Al$_x$Ga$_{1-x}$As systems. Besides, due to the large heavy-hole effective mass in III-V compounds ($\sim$ 0.3 m$_0$[4]), the H$_1$ subband is essentially flat in the z direction.

Most of the superlattices investigated up to now involve III-V semiconductor compounds but, recently, II-VI compounds have been used to grow such heterostructures, namely HgTe-CdTe[6,7], HgTe-ZnTe[7], Hg$_{1-x}$Cd$_x$Te-CdTe[7], CdTe-Cd$_{1-x}$Mn$_x$Te[8,9], CdTe-ZnTe[7] and ZnSe-Zn$_{1-x}$Mn$_x$Se[10] SL's. It is likely that the Hg$_{1-x}$Cd$_x$Te-CdTe (for x > 0.16) CdTe-Cd$_{1-x}$Mn$_x$Te, CdTe-ZnTe and ZnSe-Zn$_{1-x}$Mn$_x$Se SL's correspond to type I structures, while the HgTe-CdTe, HgTe-ZnTe and Hg$_{1-x}$Cd$_x$Te-CdTe (for x < 0.16) ones belong to a new category, i.e. type III superlattices, whose essential characteristic will be given in the next Section.

We will focus here on the HgTe-CdTe system which presents a great technical and fundamental interest. It has been, for example, proposed as a novel material[11] for infrared detectors, especially for wavelengths around 10 μm. Besides, the band structure of HgTe-CdTe superlattices has been calculated recently in the LCAO[12] and in the envelope function[13] models, and experimental studies of the electronic properties of these heterostructures have been undertaken by different groups[14-17]. Finally, we will describe briefly some aspects of CdTe-Cd$_{1-x}$Mn$_x$Te superlattices and of Cd$_{1-x}$Mn$_x$Te-Cd$_{1-y}$Mn$_y$Te double quantum well heterostructures.

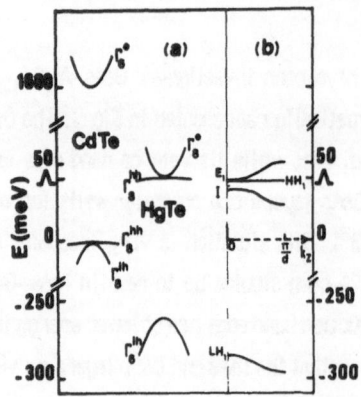

Fig. 3 (a) : Band structure of bulk HgTe and
CdTe. (b) Calculated band structure of
sample SLO along k_z with $d = d_1 + d_2$.

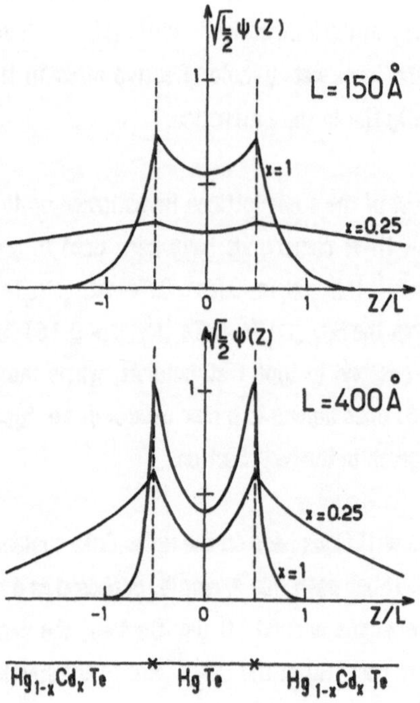

Fig. 4 : Interface wavefunction in
HgTe–$Hg_{1-x}Cd_xTe$ hetero
structures for different
values of x. L is the thickness
of the HgTe layer and z is
perpendicular to the plane of
the layers.

$\sqrt{\dfrac{L}{2}}\,\psi(z)$

L = 150 Å

x = 1

x = 0.25

Z/L

$\sqrt{\dfrac{L}{2}}\,\psi(z)$

L = 400 Å

x = 0.25

x = 1

Z/L

$Hg_{1-x}Cd_xTe$ HgTe $Hg_{1-x}Cd_xTe$

344

The HgTe–CdTe system is particularly interesting from the point of view of basic physics because of the specific band structure of bulk HgTe and CdTe and of the band line-up of the two host materials (see Fig. 3(a)). CdTe is a wide gap semiconductor ($\sim$ 1.6 eV at low temperature) with a direct gap at the Brillouin zone center (Γ point). At k = 0, the conduction band edge has a s-type symmetry (Γ_6) while the upper valence band edge (Γ_8) is fourfold degenerate (p type symmetry, J = 3/2). The spin–orbit split-off Γ_7 band (p type symmetry, J = 1/2) is located below the Γ_8 states with $\Delta = E_{\Gamma_8} - E_{\Gamma_7} \sim 1$ eV. HgTe is a zero gap semiconductor as a result of the inversion of the relative positions of the Γ_6 and Γ_8 edges. What is the Γ_8 light hole band in CdTe becomes a conduction band in HgTe, and the Γ_6 conduction band in CdTe forms a light–hole band in HgTe. The ground valence band is the Γ_8 heavy–hole one, so that the Γ_8 states correspond to the top of the valence band and to the bottom of the conduction band, yielding a zero gap configuration with a spin orbit separation Δ which is also $\sim$ 1 eV. The superlattice band structure depends on the offset between the conduction and valence band edges at the HgTe–CdTe interface. From Harrison's common anion argument[18], one can think that the offset Λ between the Γ_8 band edges of HgTe and CdTe is small. As shown later from the analysis of the experimental data, Λ is positive and smaller than 100 meV, this parameter being measured from the top of the CdTe valence band (Fig. 3(a)). This value of Λ implies that the HgTe layers are potential wells for heavy holes, but the situation for light particles (electrons or light holes) is more complicated because the bands which contribute most significantly to the light–particle SL states are the Γ_8 conduction band in HgTe and Γ_8 light valence band in CdTe. These two bands have opposite curvatures and the same Γ_8 symmetry. This mass–reversal for the light particles at each of the HgTe–CdTe interfaces is a unique property of the HgTe–CdTe system. An important consequence of these very unusual features is the existence of interface states[19] which are evanescent in both the HgTe and CdTe layers with a wavefunction peaking at the interfaces, as shown, for example, in Fig. 4 in the case[20] of a HgTe layer clad by two CdTe layers. This special situation, which corresponds to type III structures is thus very different from the more common one met, for instance, in GaAs–Al$_x$Ga$_{1-x}$As SL's (type I) where the SL states arise mainly from bands in GaAs and AlGaAs displaying the same curvature.

For bulk Hg$_{1-x}$Cd$_x$Te, the band structure is similar to that of HgTe for x < 0,16 at low temperature and corresponds therefore to a band gap equal to zero. For x > 0.16, the band structure of these alloys is similar to that of CdTe, which is also the case for bulk Cd$_{1-x}$Mn$_x$Te.

GROWTH OF II–VI COMPOUND SUPERLATTICES

There are essentially two techniques which are currently used nowadays to grow semiconductor superlattices, namely the molecular beam epitaxy (MBE) and the metalorgnic chemical vapor deposition (MOCVD). In the case of II–VI compound superlattices, good results have been obtained by MBE[6-10,21] and, in the case of HgTe–CdTe, also by a laser flash evaporation

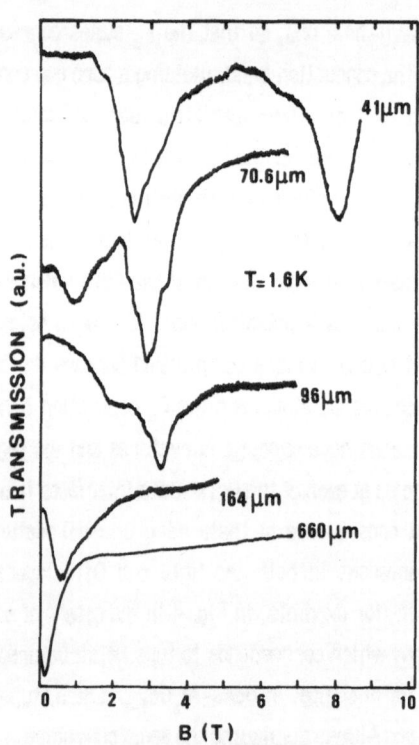

Fig. 5 . Transmission spectra observed in sample SLO as a function of B.

technique[22]. It is clear that the most difficult systems to grow are those which involve Hg, and a rather low growth temperature should be used to minimize interdiffusion between the HgTe and CdTe layers and to obtain abrupt interfaces, as shown by Faurie et al. who reported[6] in 1982 the first growth of HgTe-CdTe superlattices by MBE. The structures used in the studies described below have been generally grown along the [111] direction on (111) CdTe substrates at 185°C but also, in some cases, on (100) GaAs and (111) $Zn_{0.04}Cd_{0.96}Te$ substrates. Some characteristics of superlattices grown by[7] MBE by Faurie and his group are given in Table I.

In the case of $Hg_{1-x}Cd_xTe$-CdTe superlattices[7], there is an additional difficulty since the composition of the alloy should be very well controlled. However, a complete description of the growth of II-VI compound superlattices is clearly beyond the scope of this article and, for more details, the reader is referred ro Refs 6-10 and 21-22.

MAGNETO-OPTICAL INVESTIGATIONS OF HgTe-CdTe SUPERLATTICES

The superlattices used in these studies are essentially samples SL0 and SL7, and some of their parameters are given in Table I. The interdiffusion between HgTe and CdTe is estimated[23] to be ~ 10 Å for the first grown layers. Sample SL0 is p-type below 20 K, and SL7 is n-type in the whole temperature range investigated (2-300 K).

The infrared magneto-absorption experiments[17,24] reported here were done at liquid helium temperatures using a grating monochromator (3μm ≤ λ ≤ 5 μm), a CO_2 laser (9 μm ≤ λ ≤ 11 μm), a far-infrared laser (41 μm ≤ λ ≤ 255 μm), and carcinotrons (600 μm ≤ λ ≤ 1 μm). The transmission signals, observed at fixed photon energies in the Faraday configuration, were detected by a carbon bolometer. The magnetic field, B, which was provided by a superconducting coil, could be varied from 0 to 10 T and was applied perpendicularly to the plane of the SL layers.

Figs. 5 and 6(a) show typical transmission spectra obtained[17] in sample SL0 for several infrared wavelengths. Figs. 6(b) and 7 give the energy positions of the transmission minima (i.e., absorption maxima) as a function of B from the data presented in Figs. 5 and 6(a). Similar data[24] obtained more recently in sample SL7 are described in Figs. 8 and 9. It can be seen that they are different from those observed in sample SL0 as could be expected since the layer thicknesses are not the same in both samples. These results show also that no transmission spectra have been detected around 20 meV which corresponds to the LO phonon energy in CdTe and to the restrahlen band of the substrate.

Interpretation of the results

To interpret the data, we should first calculate the band structure of HgTe-CdTe SL's in the envelope function formalism[13,25]. The HgTe and CdTe band structures are described by the 6×6 Kane Hamiltonian[25] taking into account the Γ_6 and Γ_8 band edges. The interaction with remote bands is

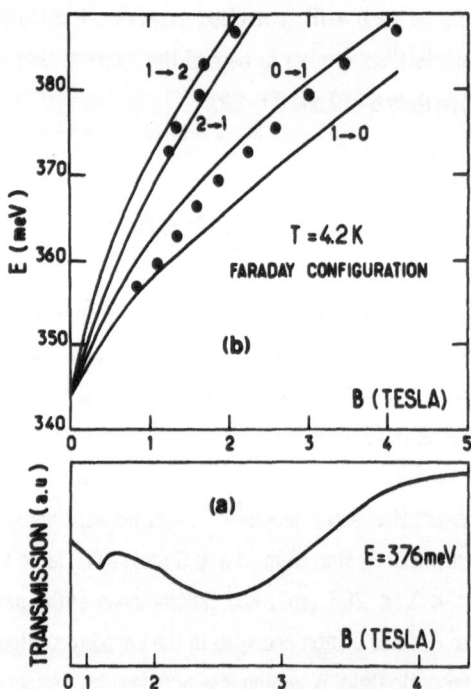

Fig. 6 . (a) Trasmission spectrum in
sample SLO for an infrared
photon energy of 376 meV.
(b) Energy position of the
observed transmission
minima (o) versus B. The
solid lines are theoretical
fits of the data.

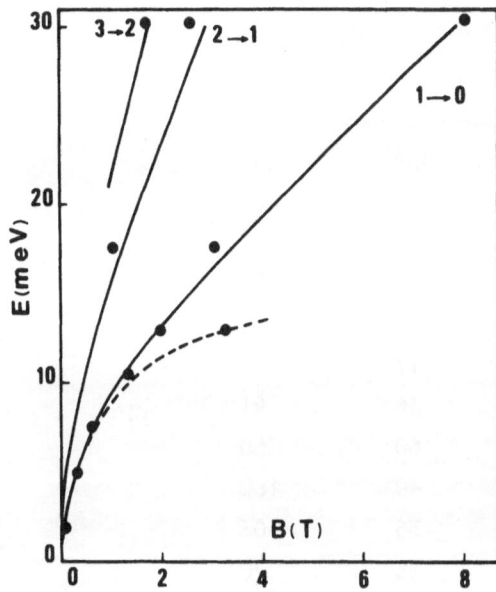

Fig. 7 . Energy position of
the transmission
minima shown in
Fig. 5 versus B (o).
The solid lines are
theoretical fits. The
dashed line is only
an eye-guide.

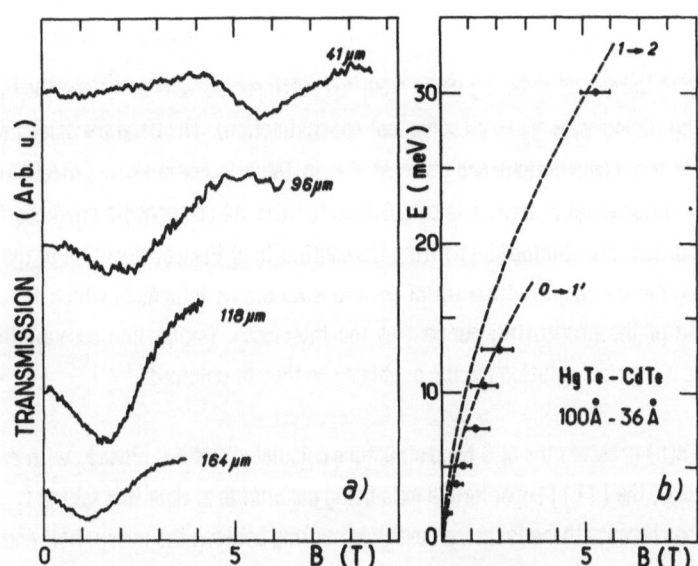

Fig. 8 . (a) Typical transmission spectra obtained at 1.6 K in sample
SL7 as a function of B for several infrared wavelengths.
(b) Energy position of the observed transmission minima
versus B (full dots); the dashed lines correspond to
theoretical fits to the data as described in the text.

Table 1 . Characteristics of some HgTe–CdTe superlattices grown by MBE. The HgTe and CdTe layer thicknesses are d_1 and d_2, respectively, and n is the number of periods.

Sample	d_1(Å)	d_2(Å)	n
SL0	180	44	100
SL1	38	20	250
SL2	40	60	100
SL3	45	17	112
SL4	74	36	91
SL5	97	60	250
SL6	110	40	180
SL7	100	36	103
SL8	70	45	70
SL9	58	35	150
SL10	47	30	110
SL11	85	45	170

considered up to the second order by using modified Luttinger parameters[26] and the Γ_8 band warping is neglected by taking $\gamma_2 = \gamma_3 = \gamma$ (spherical approximation). The band parameters of HgTe[27] and CdTe[28] used in these calculations are given at 4 K in Table 2. For a HgTe–CdTe heterostructure, a system of six differential equations is established for the six components envelope function[25]. The boundary conditions are obtained by writing the continuity of the wavefunction at the interfaces and by integrating the six coupled differential equations across an interface, which is compatible with the continuity of the probability current at the interfaces. Taking into account the superlattice periodicity, a numerical solution for the problem can thus be obtained.

Figure 3(b) presents the SL0 band structure calculated[17] at 4 K along k_z with $\Lambda = 40$ meV and $\vec{k}_\perp = (k_x, k_y) = 0$, the [111] superlattice axis being parallel to z. Note that taking $\vec{k}_\perp = 0$ simplifies the calculations because there is then an exact decoupling between the heavy–hole and light–particle states in HgTe and CdTe, so that the heavy complications due to the intricate Γ_8 bands kinematics are avoided in this case. The lowest conduction band E_1, the ground light–particle band I, and the first heavy–hole band HH_1 are shown in Fig. 3(b). Besides, one can also see another SL band, LH_1, which is the topmost SL band arising from the deeper Γ_6 HgTe states (Fig. 3(a)). The I band lies in the forbidden energy gap $[0-\Lambda]$ for the light particles, and corresponds to evanescent states in both kinds of layers. The SL wavefunction associated to I is found to peak at the interfaces instead of peaking at the center of the layers as in usual SL's, so that I corresponds to interface states[19,29]

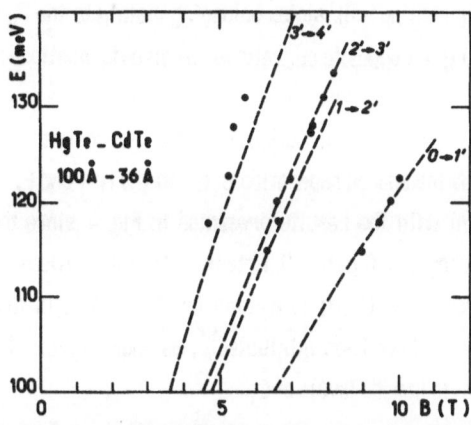

Fig. 9 . Energy position of transmission minima observed in sample SL7 at 1.6 K (solid dots). The dashed lines are theoretical fits to the data explained in the text.

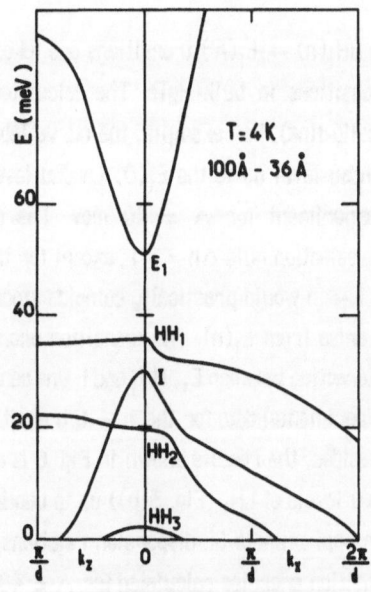

Fig. 10 . Calculated band structure of sample SL7 at low temperature along k_z and k_x. $d=d_1+d_2$ is the superlattice periodicity.

which are a consequence of matching bulk states belonging mainly to the Γ_8 bands of HgTe and CdTe displaying the same symmetry but opposite curvatures, as already mentioned.

As can be seen in Fig. 3 (b), these calculations result in a zero-energy-gap structure for SLO, the SL gap E_g being defined as the energy separation between the HH_1 and E_1. SL bands at $\vec{k} = 0$. This is qualitatively in agreement with the results presented in Fig. 7 since the observed transitions seem to extrapolate to an energy $E \sim 0$ at $B = 0$. Note that these transitions cannot be due to electron cyclotron resonance because the SL (SLO) is p-type for $T < 20$ K. In fact, the magneto-optical transitions presented in Fig. 7 have been attributed[17], as shown below, to interband transitions from Landau levels of HH_1 up to Landau levels of E_1.

To be quantitative, we should calculate now the Landau levels $E_1(n)$ and $HH_1(n)$ of E_1 and HH_1, which has been done[17] using the following simple model. For finite $\vec{k}_\perp = (k_x, k_y)$, or finite B, we have used the approximate SL dispersion relations derived[13] previously, neglecting spin effects and replacing $\vec{k}_\perp^2$ by $(2n+1)eB/\hbar$, where $n = 0, 1, 2...$ is the Landau level index. For the heavy-hole Landau levels $HH_1(n)$, the following relation has been used :

$$HH_1(n) = HH_1 - (n+1/2)(\hbar eB/m_{hh})$$

where HH_1 is here the energy of the corresponding SL band for $B = 0$, and $m_{hh} = m_0(\gamma_1 - 2\gamma)^{-1}$ (Table 2) is the effective mass of heavy-holes in bulk HgTe.

The selection rules for the $HH_1(n) \rightarrow E_1(n')$ transitions are taken to be $\Delta n = n'-n = \pm 1$, as for the interband $\Gamma_8 \rightarrow \Gamma_8$ transitions in bulk HgTe. The calculated transition energies using $\Delta n = -1$ are shown in Fig. 7 (solid line). For example, the curve labelled $1 \rightarrow 0$ corresponds to transitions from the $HH_1(1)$ Landau level up to the $E_1(0)$ Landau level. A good agreement is thus obtained between theory and experiment for $\Lambda = 40$ meV. The experimental data could be interpreted equally well with the selection rule $\Delta n = +1$, except for the transition $1 \rightarrow 0$. In fact, the transitions $n-1 \rightarrow n$ and $n+1 \rightarrow n$ would practically coincide since, with the model used here, most of the transition energies arise from $E_1(n)$. The transition energies are clearly non-linear versus B as a result of the $\vec{k}.\vec{p}$ interaction between E_1, HH_1 and I. One can also see in Fig. 7 a deviation from the theoretical fit of the experimental data for the $1 \rightarrow 0$ transition around 2.5 T, but this is not understood at the moment. Besides, the results shown in Fig. 6 are interpreted as being due to interband transitions from Landau levels of LH_1 (Fig. 3(b)) up to Landau levels of E_1. These Landau levels have been calculated from approximate SL dispersion relations[13] as described above in the case of the E_1 band, and the transition energies calculated for $\Lambda = 40$ meV are given by the solid lines in Fig. 6(b) for $\Delta n = \pm 1$. In fact, the observed broad minima (Fig. 6(a)) corespond to the two symmetric transitions $n \rightarrow n+1$ and $n+1 \rightarrow n$ which are not experimentally resolved. The agreement between the theoretical and experimental slopes is rather good and the transitions converge to 344 meV at $B = 0$ while the corresponding energy $E_1 - LH_1$ at $\vec{k} = 0$ is calculated to be 330 meV. Thus, one would deduce from these investigations that Λ, which is the only parameter

Table 2. Band parameters of HgTe and CdTe at 4 K. E_0 is the interaction energy gap between the Γ_6 and Γ_8 band edges at 4 K; E_p is related to the square of the Kane matrix element; γ, γ_1 and κ are the Luttinger parameters of the Γ_8 bands.

	γ	γ_1	κ	E_0(eV)	E_p(eV)
HgTe	− 8.9	− 15.5	− 10.85	− 0.302	18
CdTe	1.89	5.2	1.27	1.6	18

entering in the calculations described here, is positive and $\sim$ 40 meV.

To interpret the data obtained in sample SL7 more sophisticated calculations have been done[24]. The theoretical band structure, obtained for Λ = 40 meV as described briefly at the beginning of this section, is presented in Fig. 10 along k_z and k_x (in the (111) plane), showing the lowest conduction band E_1, the ground light-particle band I (interface state), and the heavy-hole bands HH_1, HH_2 and HH_3. The SL gap (E_1 − HH_1 at $\vec{k}$ = 0) is found to be 17 meV at 4 K. One can see that the valence band structure is rather complicated fo $k_x \neq 0$ as a result of the strong hybridization occuring between the heavy-hole and I states. To obtain the Landau levels, and thus the energy of the observed transitions, we have to calculate the band structure when a magnetic fiel is applied along the z direction. Such calculations[24] are formally the same as those performed at B = 0, replacing $\vec{k}$ by $\vec{k}$ − e$\vec{A}$/c in the Kane Hamiltonian and taking into account the direct coupling of the electron and hole spins by introducing the additional valence band parameter κ[26]. The motion parallel to the layers is thus given by a six-component vector[20] :

$$\varphi_n = (C_1\psi_{n-1}, C_2\psi_{n-2}, C_3\psi_n, C_4\psi_n, C_5\psi_{n-1}, C_6\psi_{n+1})$$

where ψ_n is the nth harmonic oscillator function and n = −1, 0, 1, 2... For n ⩽ 1, the C_i coefficients corresponding to the negative oscillator index vanish. The calculated Landau levels[24] of E_1, I, HH_1 and HH_2 are given, also for Λ = 40 meV, in Fig. 11. The situation is obviously very complicated, and the Landau levels are strongly mixed, again as a result of the coupling between the I state and the heavy-hole ones.

The observed transitions shown in Fig. 8 extrapolate to an energy E $\sim$ 0 at B = 0, and are attributed[24] to cyclotron resonance in the E_1 band, which is consistent with the n-type nature of

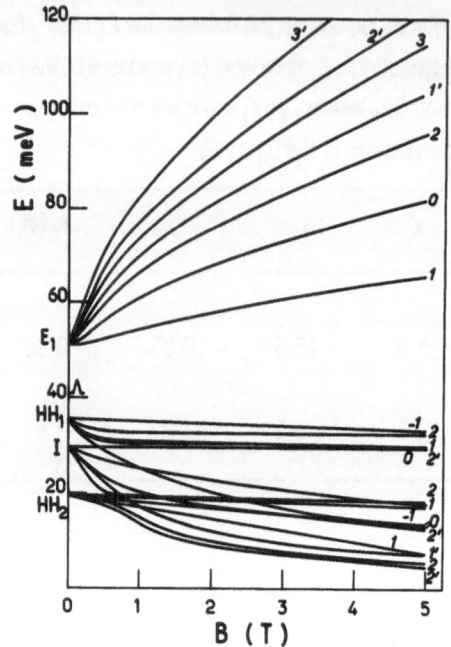

Fig. 11 . Calculated Landau levels corresponding
to the E_1, I, HH_1 and HH_2 bands (see
Fig. 10) in the case of sample SL7.

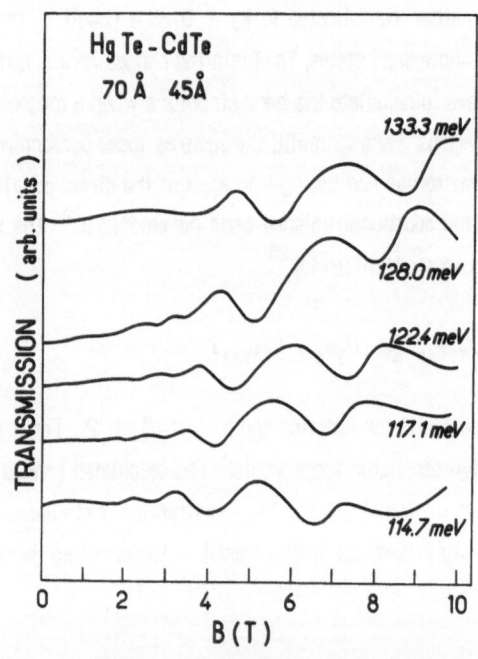

Fig. 12 . Magneto-transmission spectra observed
at 1.6 K in sample SL8.

this sample. The first intraband magneto-optical transitions corresponding to the selection rule $\Delta n = + 1$ (cyclotron resonance) are the $1 \to 2$ and $0 \to 1'$ transitions (see Fig. 11). The dashed lines in Fig. 8(b) give the energies of these transitions obtained from the previous calculations of the E_1 Landau levels shown in Fig. 11 for $\Lambda = 40$ meV. At low photon energies (< 15 meV), these calculations correspond fairly well to the observed broad absorption lines (Fig. 8(a)), showing that the $n = 1$ and $n = 0$ levels are populated. For $E = 30$ meV, the calculated magnetic field separation between the corresponding lines is wider than the observed absorption line. Only one type of transition, i.e. $1 \to 2$, is detected, which indicates that the $n = 1$ Landau level is populated for the corresponding value of the magnetic field, namely $B \sim 5$ T. We wish to emphasize that interband transitions from valence up to conduction Landau levels are not detectable in the studied infrared region (0–30 meV) because of the population of the ground conduction levels and of the value of the SL band gap E_g. Such transitions have been investigated[24] in the CO_2 laser energy region, as shown in Fig. 9. They are interpreted as being due to interband transitions from HH_1 to E_1 Landau levels with the selection rule $\Delta n = \pm 1$. The dashed lines in Fig. 9 correspond to such transitions calculated using $\Delta n = + 1$ and again $\Lambda = 40$ meV (Fig. 11), and the agreement between experiment and theory is satisfying. In fact, the experimental data could be interpreted also with the selection rule $\Delta n = - 1$ due to the width of the observed absorption lines but, for the sake of simplicity, only one type of transition has been presented in Fig. 9. Note that these investigations give essentially informations on the E_1 Landau levels because most of the transition energies arise from the conduction band. Note also that the transitions shown in Fig. 9 extrapolate to an energy $E \sim 20$ meV at $B = 0$, which is consistent with the calculated value[24] of E_g (17 meV). At this stage, it can be concluded that these investigations support the value of Λ (40 meV) obtained from the first magneto-optical experiments performed[17] on HgTe–CdTe superlattices, namely those done in sample SL0.

Note that optical and magneto-optical transitions between HH_1 and E_1 can be strong in these structures. Indeed, the HH_1 subband wavefunction, which is confined in the HgTe layers, is p-like, and the part of the E_1 subband wavefunction localized in the HgTe layers has a non-negligible s-character even at $\vec{k} = 0$.

Discussion

At first, we wish to point out that Λ is necessarily positive. Indeed, in the opposite situation ($\Lambda < 0$), charge transfer would occur between CdTe and HgTe, which is not compatible with the p-type conduction observed[31] in most HgTe–CdTe SL's at low temperature.

Considering now the analyses of the data presented above, one can argue that the interpretation proposed[17] in the case of sample SL0 is questionable because the model we used is very simple. Indeed, one should of course calculate the SL band structure along k_z, but also in the k_x, k_x directions parallel to the interface (111) plane, and then calculate the corresponding Landau levels under magnetic field. Such complicated calculations[24] have been done, as shown in the preceding Section, for sample SL7, neglecting strain effects due to the small lattice mismatch (~ 0.3 %) between HgTe and CdTe. However, it has been shown[32] by Schulman and Chang that, for such a semiconducting (111) SL, the band structure is not significantly influenced by strains. In any case[32,33] the E_1 conduction band is nearly unaffected by strains. Note that taking strains into account may be in fact an impossible task in a SL with 100 periods because strains are very likely inhomogeneous throughout the considered heterostructure.

It remains that, whatever model we use, we find that Λ is small. Therefore, we have studied the sensitivity of the fitting procedure to the value of Λ in the case of sample SL7 where the theoretical model is actually much more satisfying. At first, note[24], for instance, that E_g is less than 5 meV for $\Lambda > 100$ meV and that, in addition to cyclotron resonance, interband transitions should be observed in this case in the investigated infrared region (0–30 meV). Then, we have checked[24] that an acceptable agreement between experiment and theory could be obtained for $0 < \Lambda < 100$ meV by taking into account the possible uncertainties on the sample characteristics, the HgTe and CdTe parameters and the data themselves (i.e. the rather large width of the observed absorption lines). We are thus led to conclude from these investigations that the valence band offset Λ at the HgTe–CdTe interface is possitive and smaller than 100 meV.

Furthermore, we have done[34] recently similar experiments on other HgTe–CdTe superlattices grown on CdTe, $Cd_{0.96}Zn_{0.04}Te$ or GaAs substrates, with different values of d_1 and d_2. For instance, we have studied a superlattice (sample SL8) grown along the [111] direction on a (100) GaAs substrate and consisting of 70 periods of HgTe and CdTe with $d_1 = 70$ Å and $d_2 = 45$ Å, respectively. Note that, in this structure, a CdTe buffer layer is deposited prior to the growth of the SL because this procedure gives a high–quality superlattice despite the rather large lattice mismatch between the GaAs substrate and the HgTe–CdTe SL($\sim$ 15 %). Fig. 12 presents magneto–optical spectra obtained at 1.8 K in this superlattice. A detailed analysis of these data has not yet been done, but the observed transmission minima (or absorption maxima) are interpreted as interband magneto–optical transitions between Landau levels of HH_1 and E_1. Nevertheless, extrapolating the energy of the observed transitions to B = 0 should yield the SL band gap E_g which is found to be $\sim$ 60 meV. This result is plotted in Fig. 13 versus d_1, together with the values of E_g obtained at low temperature in samples SL0, SL1 and SL7. The solid lines in Fig. 13 give the variation of E_g as a function of d_1 calculated for $d_2 = 20$, 30 and 50 Å using $\Lambda = 40$ meV and the envelope function formalism. It can be seen that experiments and theory are actually in good agreement, evidencing again that Λ is positive and smaller than 100 meV.

OPTICAL ABSORPTION STUDIES

Optical absorption investigations have been done[35] recently at 300 K on HgTe–CdTe superlattices also grown by MBE by Faurie et al. In these experiments, infrared transmission spectra were measured between 400 and 5000 cm^{-1} and yield the absorption coefficient α as a

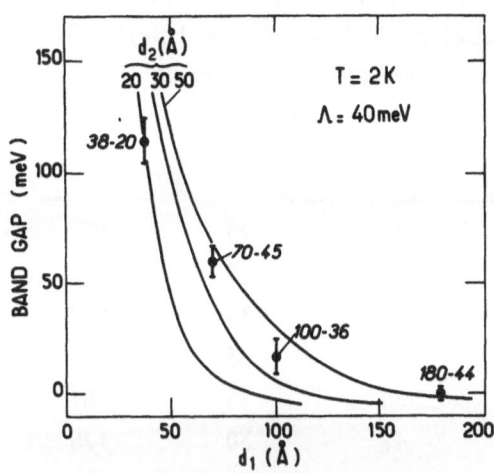

Fig. 13 . Variation of the band gap of several
HgTe–CdTe superlattices as a
function of d_1. The experimental
data are given by the solid dots.
For each sample the first number
corresponds to d_1 and the second
one to d_2. The solid lines are
theoretical fits for three values of d_2.

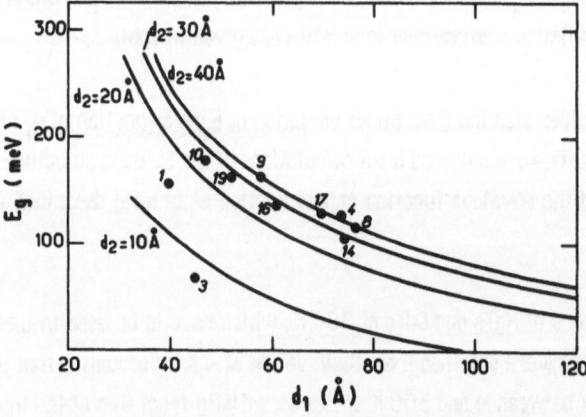

Fig. 14 . Variation of the band gap of different HgTe–CdTe superlattices
at 300 K as a function of the HgTe layer thickness d_1. The
experimental data are given by the solid dots. The numbers
1, 3... correspond to samples SL1, SL3... in Table 3. The
solid lines are theoretical fits for different values of the
CdTe layer thickness d_2.

Table 3. Some parameters of HgTe-CdTe superlattices. d_1
and d_2 are the thicknesses of the HgTe and CdTe
layers respectively; n is the number of periods.

Sample	d_1(Å)	d_2(Å)	n
SL1	40	20	155
SL3	45	17	68
SL4	74	36	152
SL8	77	38	113
SL9	58	35	161
SL10	47	30	175
SL14	74	32	104
SL16	61	25	135
SL17	70	49	128
SL19	52	34	61

function of the energy of the incident infrared photons. The SL band gap E_g was defined as being the energy where the absorption coefficient is equal to 1000 cm^{-1}. This is rather arbitrary but gives values of E_g which are in satisfying agreement with those obtained from photo-conductivity thresholds. The variation of E_g as a function of the HgTe layer thickness d_1 is shown in Fig. 14 for several HgTe-CdTe superlattices whose characteristics are given in Table 3.

Besides, Fig. 14 gives also the theoretical variation of E_g as a function of d_1 for d_2 = 10, 20, 30 and 40 Å. These results were obtained from calculations of the SL band structures done at 300 K using Λ = 40 meV and the envelope function approximation as briefly described in the previous Section.

The band parameters of HgTe and CdTe at 300 K, which have to be used in these calculations, are given in Table 4. They were obtained from their values at 4 K by assuming that the temperature variation of γ, γ_1 and κ between 4 and 300 K arises essentially from that of the interaction gap E_0 between the Γ_6 and Γ_8 band edges.

Except for samples SL1 and SL3, the overall agreement between experiments and theory is rather satisfying, especially if one takes into account the uncertainties in the experimental determination of E_g and of the layer thicknesses, and also the uncertainties in the HgTe and CdTe parameters used in the calculations. For samples SL1 and SL3, which correspond to small values of d_1 and d_2, the fact that the agreement is not satisfying may be due to the increased effect of

Table 4. Band parameters of HgTe and CdTe at 300 K

	γ	γ_1	κ	$E_0(eV)$
HgTe	– 23.55	– 44.8	– 25.50	– 0.122
CdTe	2.12	5.75	1.50	1.425

interdiffusion which, as already mentioned, is estimated to be[23] $\sim$ 10 Å for the first grown layers. It can at least be concluded from these investigations that they are consistent with the value of Λ determined in the previous Section.

HgTe–CdTe SUPERLATTICES AS INFRARED MATERIALS

The HgTe–CdTe SL band gap E_g, i.e. E_1–HH_1 at k = 0, has been calculated[36] as a function of temperature and of the HgTe and CdTe layer thicknesses (d_1 and d_2, respectively) using again Λ = 40 meV and the envelope function formalism. Fig. 15 shows E_g and the corresponding cut–off wavelength λ_g calculated at 300, 77 and 4 K for HgTe–CdTe SL's with equally thick layers ($d_1 = d_2$). It can be noted that E_g decreases as the layer thickness increases. More generally, it is found that d_1 controls essentially the SL band gap, while d_2 governs the width of the subbands and, therefore, the effective masses along the SL axis. Another important feature is that E_g increases when the temperature is increased, as observed in bulk $Hg_{1-x}Cd_xTe$ alloys with similar energy gaps, but the calculated temperature dependence is smaller for a SL than for the corresponding ternary alloy. In Fig. 15, it can also be seen that the interesting cut–off wavelength λ_g for infrared detectors around 8–12 μm should be obtained at 77 K for instance for layer thicknesses in the range 50–70 Å which are quite reasonable values from the point of view of the MBE growth.

In addition, the theoretical electron effective mass along k_z, as obtained from band structure calculations (see Fig. 10 for instance), is found to be larger than the small value occuring in the $Hg_{1-x}Cd_xTe$ bulk alloy corresponding to the same cut–off wavelength. This may be an important advantage of the SL's as infrared detector materials because, as a result, diffusion and tunneling currents should be reduced. Furthermore, the hole mobility measured in p–type SL's is large[7,31] (10^4 cm^2 V^{-1} sec^{-1} at 10 K), and this is consistent with the HH_1 subband mass obtained for small wavevectors in the (k_x, k_y) plane (see Fig. 10 for example) which is rather light as compared to the heavy–hole mass in bulk[27] HgTe ($m_{hh} \sim 0.4\ m_0$). Finally, it has been shown[35] that the SL cut–off wavelength in the far infrared should be easier to control than that of bulk $Hg_{1-x}Cd_xTe$ alloys. Thus, as predicted in Ref. 11, HgTe–CdTe SL's seem to have properties superior to those of $Hg_{1-x}Cd_xTe$ alloys as infrared detector materials for the reasons given here.

SOME ASPECTS OF CdTe–$Cd_{1-x}Mn_xTe$ SUPERLATTICES

These superlattices have been grown[8,9,37] successfully by MBE in different laboratories over a wide range of composition. In these systems, it is very likely that the CdTe layers are quantum wells for both electrons and holes, corresponding therefore to type I structures. Extensive luminescence measurements have been performed on these superlattices at low temperature, and it has been claimed[38] that exciton radiative recombination has been observed.

The energy position of the luminescence spectrum, as compared to the situation in the corresponding bulk material, is thought to result from the quantum confinement and from the strains due to the lattice mismatch between the host materials. Besides, it has been shown[39] that the observed luminescence intensity is much larger than in equivalent bulk samples which, for instance, may be due to the high quality of the samples and/or to the minimization of nonradiative surface recombination in superlattices.

Extensive studies have been done on these II–VI compound superlattices which require nevertheless more investigations. It is not our purpose to review here all the results which have been obtained on these heterostructures, but we would like to quote an interesting study. Indeed, laser action has been observed[40] in $Cd_{1-x}Mn_xTe$–$Cd_{1-y}Mn_yTe$ SL's, which provides the basis for a magnetically tuned laser due to the magnetic properties of these alloys resulting from the presence of Mn.

In conclusion, it is clear that these superlattices are promising both from the point of view of basic and applied physics.

SOME PROSPECTS ABOUT $Cd_{1-x}Mn_xTe$–$Cd_{1-y}Mn_yTe$ HETEROSTRUCTURES

Dilute magnetic semiconductors, such as $Cd_{1-x}Mn_xTe$, contain magnetic moments (Mn^{2+} ions) which interact strongly with delocalized electrons whose states at zero magnetic field B are quite similar to those found in comparable non–magnetic ternary alloys. Under magnetic field, the magnetic moments line up and, due to exchange coupling, spin–polarize the delocalized electrons[41], which can lead to interesting effects in heterostructures containing Mn.

Let us consider, for example, an asymmetric double quantum well system (see Fig. 16 (a)) consisting in a CdTe well (thickness L_1) and a $Cd_{1-y}Mn_yTe$ well (thickness $L_2 > L_1$) separated by a thin $Cd_{1-x}Mn_xTe$ barrier (thickness d) with x > y, the structure being clad by thick $Cd_{1-x}Mn_xTe$

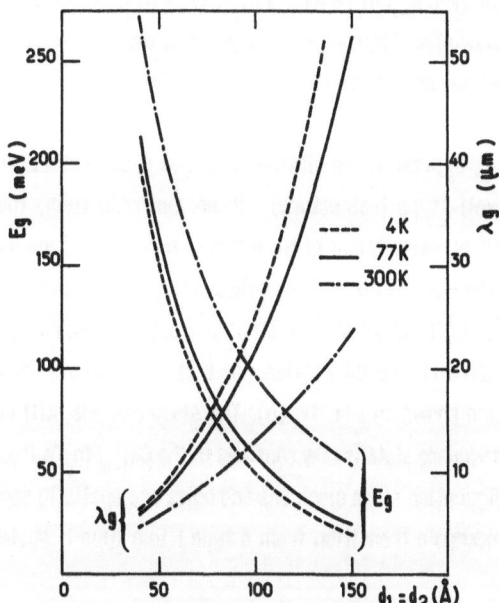

Fig. 15 . Calculated band gap and cut-off
wavelength as a function of
layer thickness for HgTe–CdTe
superlattices with $d_1 = d_2$.

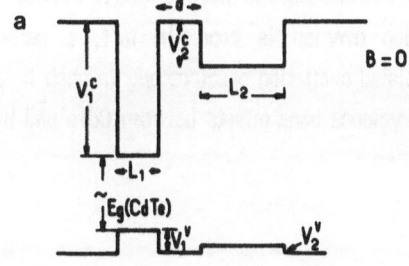

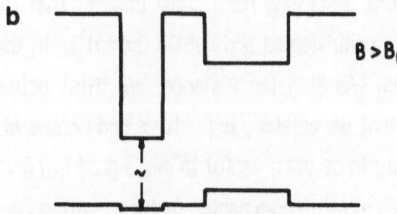

Fig. 16 . Conduction and valence band
profiles of a CdTe–
$Cd_{1-x}Mn_xTe$–$Cd_{1-y}Mn_yTe$
double quantum well at
(a) B=0; (b) $B > B_t$ (see text).

layers. Since it is likely that single $Cd_{1-x}Mn_xTe-Cd_{1-y}Mn_yTe$ (x > y) quantum wells are of type I, we assume that the alloy with the smaller Mn content is a quantum well for both electrons and holes, corresponding therefore to the heterostructure shown in Fig. 16(a).

If we choose correctly L_1, L_2, and y, the electron and hole ground states will be localized at B = 0 in the CdTe quantum well (type I structure). Under magnetic field, the situation can be completly different, as shown[42] by calculations of the effect of a magnetic field on the energy levels in such a heterostructure. If the valence band offset between $Cd_{1-x}Mn_xTe$ and $Cd_{1-y}Mn_yTe$ is not too large, a complete reorganization of the quantum well structure occurs, as a result of the effect of the external magnetic field and of the exchange interaction. For B larger than a critical value B_t, one can end up with the heterostructure shown in Fig. 16 (b). The electrons are still heavily localized in the CdTe layer, but the ground valence state is now confined in the $Cd_{1-y}Mn_yTe$ layer. In this case, we are dealing with a type II configuration since electrons and holes are spatially separated. Therefore, such a heterostructure can undergo a transition from a type I to a type II system under magnetic field.

This effect should be detectable in a luminescence experiment since it should affect strongly the magnitude of the luminescence signal. If at B = 0 the electron-hole recombination is strong, it will sensitively weaken as soon as B becomes comparable to B_t. Indeed, for $B < B_t$ one deals with the photoluminescence signal of a type I system with a good overlap of the electron and hole wavefunctions, while, for $B > B_t$, the luminescence is that of a type II structure which is faint since the electron and hole wavefunction overlap is poor. In fact, a careful analysis of the magneto-luminescence and of its related excitation spectroscopy in such heterostructures should provide useful informations on the valence band offsets between CdTe and the $Cd_{1-x}Mn_xTe$ alloys which are not known at the present time.

CONCLUSION

We believe that the investigations described here show clearly that the band structure of HgTe-CdTe superlattices is much more complicated and subtle than it is in the case of usual III-V compound systems such as $GaAs-Al_xGa_{1-x}As$ SL's for instance. We think actually that the valence band offset Λ, which is a very important parameter, is positive and ranges at most between 0 and 100 meV. These superlattices are likely to be very useful to make good infrared detectors, but it is clear these intriguing heterostructures exhibit unexpected features which require further studies to be fully understood.

We wish also to point out that we believe strongly that II-VI compound SL's widen the field of two-dimensional systems in a very interesting way, at least from the point of view of basic physics. Type I and type III configurations can be achieved, and it is likely that type II stiuations can also be otained, at least under magnetic field for heterostructures involving Mn. We think also that

$Hg_{1-x}Mn_xTe-CdTe$ SL's should prove attractive since one can expect, for example, two-dimensional spin glasses to occur in such structures.

Finally, we would like also to mention that magneto-optical experiments similar to those described here are now in progress in our laboratory on $Hg_{1-x}Cd_xTe-CdTe$ and HgTe-ZnTe superlattices. The preliminary data obtained on these structures are encouraging, showing again that II-VI compound superlattices are really interesting.

ACKNOWLEDGEMENTS

The authors would like to thank G. Bastard, and J.A. Brum from the Ecole Normale Supérieure, J.P. Faurie from the University of Illinois, A. Million from LIR-LETI in Grenoble, J.N. Schulman from Hugues Research Laboratory, and M. Altarelli from the Max-Planck Institut in Grenoble for very helpful and stimulating discussions. They wish also to point out that all the work on II-VI compound superlattices at the Ecole Normale Supérieure is done in very close collaboration with Y. Guldner and J.P. Vieren. Finally, we want to emphasize that this work was supported in part by the GRECO Expérimentation Numérique.

REFERENCES

1. L. Esaki and R. Tsu, IBM Journ. of Res. and Develop. 14 : 61 (1970).
2. See, for example, R. Dingle in : "Festkörperprobleme (Advances in Physics)", H.J. Queisser, ed., Pergamon-Vieweg, Braunschweig (1975).
3. See, for example, L. Esaki and L.L. Chang, J. Magn. Magn. Mater. 11 : 208 (1979).
4. "Handbook of Electronic Materials", M. Neuberger, ed., Plenum, New York (1971).
5. Y. Guldner, J.P. Vieren, P. Voisin, M. Voos, L.L. Chang and L. Esaki, Phys. Rev. Lett. 45 : 1719 (1980).
6. J.P. Faurie, A. Million and J. Piaguet, Appl. Phys. Lett. 41 : 713 (1982).
7. J.P. Faurie, to be published in IEEE J. Quantum Electron. (1986).
8. L. A. Kolodziejski, T.C. Bonsett, R.L. Gunshor, S. Datta, R.B. Bylma, W.M. Becker and N. Otsuka, Appl. Phys. Lett. 45 : 440 (1984).
9. R.N. Bicknell, R.W. Yanka, N.C. Giles-Taylor, D.K. Blanks, E.L. Buckland and J.F. Schetzina, Appl. Phys. Lett. 45 : 92 (1984).
10. L.A. Kolodziejski, R.L. Gunshor, T.C. Bonsett, R. Venkatasubramanian, S. Datta, R.B. Bylsma, W.M. Becker and N. Otsuka, Appl. Phys. Lett. 47 : 169 (1985).
11. D.L. Smith, T.C. McGill and J.N. Schulman, Appl. Phys. Lett. 43 : 180 (1983).
12. J.N. Schulman and T.C. McGill, Phys. Rev. B 23 : 4149 (1981).
13. G. Bastard, Phy. Rev. B 25 : 7584 (1982).
14. D. Olego, J.P. Faurie and P.M. Raccah, Phys. Rev. Lett. 55 : 328 (1985).

15. S.R. Hetzler, J.P. Baukus, A.T. Hunter, J.P. Faurie, P.P. Chow and T.C. McGill, Appl. Phys. Lett. 47 : 260 (1985).

16. N.P. Ong, G. Kote and J.T. Cheung, Phys. Rev. B 28 : 2289 (1983).

17. Y. Guldner, G. Bastard, J.P. Vieren, M. Voos, J.P. Faurie and A. Million, Phys. Rev. Lett. 51 : 907 (1983).

18. W.A. Harrison, J. Vac. Sci. Technol. 14 : 1016 (1977).

19. Y.C. Chang, J.N. Schulman, G. Bastard, Y. Guldner and M. Voos, Phys. Rev. B 31 : 2557 (1985).

20. G. Bastard, unpublished.

21. P.P. Chow and D. Johnson, J. Vac. Sci. Technol. A3 : 67 (1985).

22. T. Cheung and D.T. Cheung, J. Vac. Sci. Technol. 21 : 182 (1982).

23. D.K. Arch, P.P. Chow, M. Hibbs-Brenner, J.P. Faurie and J.L. Staudenmann, to be published in J. Vac. Sci. Technol. (1986).

24. J.M. Berroir, Y. Guldner, J.P. Vieren, M. Voos and J.P. Faurie, to be published in Phys. Rev. (1986).

25. M. Altarelli, in : "Proceedings of Les Houches Winter School on Semiconductor Superlattices and Heterojunctions", G. Allan, ed., Springer Verlag, Berlin (1986), to be published.

26. J.M. Luttinger, Phys. Rev. 102 : 1030 (1956).

27. M.H. Weiler, in "Semiconductor and Semimetals", Vol. 16, R.K. Willardson and A.C. Beer, eds., Academic Press, New York (1981).

28. P. Lawaetz, Phys. Rev. B 4 : 3460 (1971).

29. Y.R. Lin Liu and L.J. Sham, Phys. Rev. B 32 : 5561 (1985).

30. A. Fasolino and M. Altarelli, Surf. Science, 142 : 322 (1984).

31. J.P. Faurie, M. Boukerche, S. Sivananthan, J. Reno and C. Hsu, Superlattices and Microstructures 1 : 237 (1985).

32. J.N. Schulman and Y.C. Chang, Phys. Rev. B 33 : 2594 (1986).

33. G.Y. Wu and T.C. McGill, Appl. Phys. Lett. 47 : 634 (1985).

34. J.M. Berroir, Y. Guldner, J.P. Vieren, M. Voos and J.P. Faurie, unpublished.

35. J. Reno, I.K. Sou, J.P. Faurie, J.M. Berroir, Y. Guldner and J.P. Vieren, to be published.

36. Y. Guldner, G. Bastard and M. Voos, J. Appl. Phys. 57 : 1403 (1985).

37. L. Esaki and L.L. Chang, private communication.

38. A. Petrou, J. Warnock, R.N. Bicknell, N.C. Giles-Taylor and J.F. Schetzina, Appl. Phys. Lett. 46 : 692 (1985).

39. R.N. Bicknell, N.C. Giles-Taylor, D.K. Blanks, R.W. Yanka, E.L. Buckland and J.F. Schetzina, J. Vac. Sci. Technol. B 3 : 709 (1985).

40. R.N. Bicknell, N.C. Giles-Taylor, D.K. Blanks, J.F. Schetzina, N.G. Anderson and W.D. Laidig, Appl. Phys. Lett. 46 : 1122 (1985).

41. See, for example, G. Bastard, C. Rigaux, Y. Guldner, J. Mycielski and A. Mycielski, J. de Phys. 39 : 87 (1978).

42. J.A. Brum, G. Bastard and M. Voos, to be published in Sol. St. Commun.

DENSITY OF STATES OF 2 DIMENSIONAL

SYSTEMS IN HIGH MAGNETIC FIELDS

Erich Gornik

Institut für Experimentalphysik
Technikerstraße 15
Universität Innsbruck, Austria

ABSTRACT

The energy spectrum of a two dimensional electron gas in a strong magnetic field consists in the ideal case of discrete Landau levels with a degeneracy corresponding to the number of flux quanta within the area of the sample. The density of states becomes δ-function like and the Landau levels are equally spaced by the cyclotron energy.

The real form of the density of states which is changed by the presence of impurities and potential fluctuations is of great interest for the understanding of all physical phenomena observed in this system.

In this report several different experimental results which are relevant to determine the density of states in a high magnetic field are summarized. Measurements of the specific heat of GaAs/GaAlAs multilayers reveal clear evidence for a Gaussian like density of states super-imposed on a constant background. Temperature dependent measurements of the resistivity in the regime of the Hall plateaus confirms the existence of a flat, mobility dependent background between Landau levels. Magnetization data give evidence of a magnetic field dependence of the Gaussian like density of states and are consistent with significant number of states between Landau levels. Capacitance measurements in the lower magnetic field range are in agreement with the other techniques.

From cyclotron resonance transmission and emission experiments, additional information of the density of states is obtained. The origin of the broadening of the density of states is assigned to impurities in the GaAs in the range of the electron gas and to impurities in the GaAlAs close to the interface. An analysis of cyclotron resonance linewidth at a integer filling factor reveals a systematic dependence of the linewidth on the zero field mobility. The same dependence on mobility proportional to $\sqrt{1/\mu}$ is found for the dependence of the amount of flat background states.

1. INTRODUCTION

In a 2-dimensional electron system (2DES) a magnetic field perpendicular to the plane of electrical confinement leads to full quantization of the electron motion. The energy spectrum consists of Landau levels (LL) separated by the cyclotron energy $\hbar\omega_c$. In an unperturbed system the LL are discrete and highly degenerate , with $1/(\pi l^2)$ possible states in each level including both spins (with $l = \sqrt{\hbar/eB}$ the cyclotron radius). For an electron concentration n_s the filling factor is defined as $\nu = (n_s \cdot \pi l^2)$.

In a real system the LL are broadened due to scattering by impurities, phonons and other scattering mechanisms. In the simplest approximation the levels are described by a level width Γ. In the case of a high magnetic field, where $\hbar\omega_c \gg \Gamma$, real gaps appear between the LL. This leads to an oscillatory structure of practically all physical quantities as a function of the magnetic field.

The most fundamental quantity underlying all these physical properties of the system is the form of the density of states D(E). The most pronounced effects are the Quantum Hall effect[1] and the fractional Quantum Hall effect[2]. A microscopic theory of the quantum Hall effect should give a correct description not only of the quantized resistivity values $\rho_{xy} = h/ie^2$ but also of the regions between the plateaus and the values of the finite resistivity ρ_{xx}. Such transport calculations are extremely complicated since the theory itself is complicated and in addition not enough information is available about the scattering centers. The published theories are based on certain approximations and assumptions about the distribution, the strength and the range of the scattering potentials. A first test whether such assumptions are realistic should be available from a comparison between the calculated and the measured density of states D(E) since calculations of D(E) are much easier than a transport theory for ρ_{xx}(B) which includes complicated phenomena like localization and correlation. One of the first theories of the density of states D(E) assumed short range scatterers which leads within the self-consistent Born approximation (SCBA) to a broadening of the discrete energy spectrum (expected for an ideal two-dimensional electron gas without scattering) into an elliptic lineshape for the $D(E)$[3]. Higher order approximations show that an exponentially decaying $D(E)$ is expected for energies $E - E_n$ larger than the linewidth of the Landau levels E_n[4], so that a real energy gap with vanishing $D(E)$ may be not present but the $D(E)$ at midpoint between Landau levels should decrease drastically if the magnetic field (energy separation between adjacent Landau levels) is increased.

Experimental information about the $D(E)$ can be obtained from measurements of the specific heat[5], from magnetization measurements[6], from temperature dependent resistivity measurements in the regime of the Hall plateaus[7] and from cyclotron resonance spectroscopy[9].

In this paper a review on the above experimental techniques to determine the $D(E)$ is given. The different techniques are compared and their application range and accuracy are critically discussed. A brief discussion of the current theoretical understanding of the $D(E)$ will be given in the light of the experimental findings.

2. SPECIFIC HEAT

The most direct method to determine $D(E)$ is the measurement of the electronic specific heat given by

$$C_v = \frac{dU}{dT} = \frac{d}{dT} \int_0^\infty E\, D(E)\, f(E, E_F)\, dE \qquad (1)$$

where $f(E, E_F)$ is the Fermi distribution function. An externally induced temperature change leads to a reordering of the electrons at the Fermi energy. The heat capacity of the electron system is proportional to $D(E)$ at the Fermi energy, sampling localized and delocalized states.

The first calculation of the specific heat in 2D-systems was performed by Zawadzki and Lassnig[10]. They assumed a Gaussian density of states $D(E) \sim e^{-E^2/2\Gamma_G^2}$ independent of the magnetic field. Two contributions to the specific heat are found: intra- and inter-Landau level contributions. Results for a levelwidth $\Gamma_G = 0.25$ meV are shown in Fig. 1 for two temperatures. The intra LL contributions lead to an oscillatory behaviour with

367

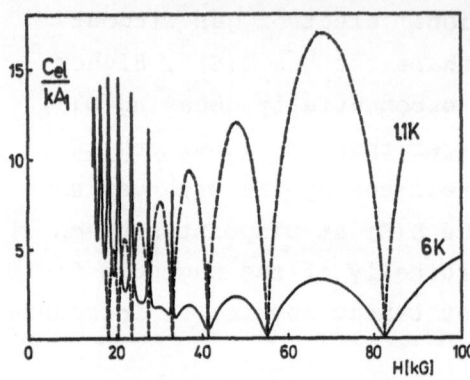

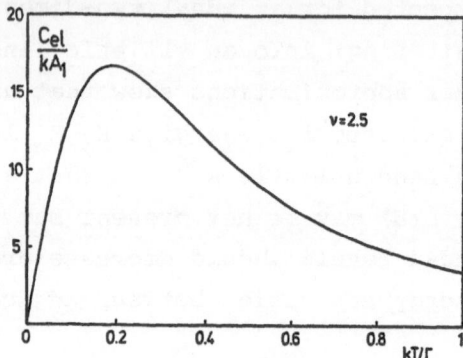

Fig. 1. Specific heat of 2D electrons versus magnetic field for two temperatures after Ref. 10. Full curve: T=6K; dashed curve: T=1.1K $n_s = 8.0 \times 10^{11} cm^{-2}$, $\Gamma_G = 0.25$ meV, $A_1 = 1/\pi \cdot (eB/\hbar)$.

Fig. 2. Intralevel contribution to specific heat calculated (for same sample as in Fig. 1) for filling factor $\nu = 2.5$ as a function of kT/Γ_G after Ref. 10.

a vanishing specific heat at integer filling factors. The inter LL contributions appear as sharp spikes at the position of integer ν-values. These spikes are only present at low magnetic fields and "high" temperatures where kT is comparable with the LL splitting. At the lower temperature (dashed curve) the inter LL peaks have disappeared. The behaviour of the intra LL contributions for a filling factor of $\nu = 2.5$ as a function of kT/Γ_G is shown in Fig. 2. A maximum is found for $kT/\Gamma_G \sim 0.2$ which means that the specific heat is not sensitive to Γ in this range.

2.1 Experimental

A heat pulse technique[11,12] was applied to determine the electronic specific heat[5]. In this technique a short heat pulse heats the sample adiabatically, which implies a controlled thermal connection between sample and heat-bath. The thermal isolation of the sample was achieved by hanging the sample on four 5 to 10 μm thick superconducting wires of 5 cm length. The whole sample was mounted in a vacuum isolated tube. The sample mount is schematically depicted in Fig. 3. A Au-Ge (8 % Au) film[13] (1000 to 2000 Å thick) served as a temperature

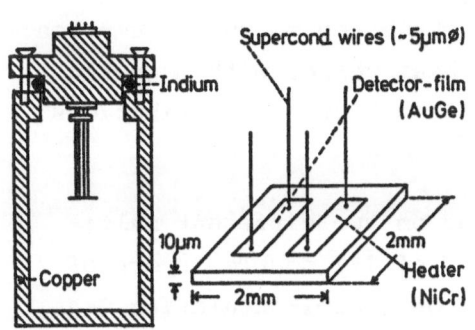
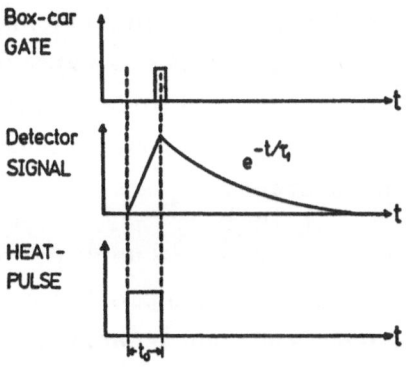

Fig. 3. Schematic of the sample geometry with the heater and detector film and the sample mount in the copper tube.

Fig. 4. Heat pulse technique: τ_1 is the thermal sample heat bath relaxation time. The voltage pulse is applied to the heater, the resistance change of the Au:Ge film measured with a boxcar.

detector, a Ni-Cr film (100 Å thick) as a heater. Both were deposited on the sample surface and contacted to the superconducting wires with silver epoxy as shown in Fig. 3.

The sample was heated with electric field pulses of 0.1 ms duration which were considerably shorter than the thermal timeconstant τ_1 of the sample-bath system which was in the range of 5 to 30 ms. The temperature change was measured with a boxcar technique at the end of the heat pulse. The schematic of the measurement is shown in Fig. 4. This technique requires the thermal relaxation time within the sample to be considerably shorter. This requirement is well fulfilled in the investigated temperature range[5].

As a temperature standard we use the vapor pressure of the surrounding liquid-helium bath. The temperature sensitivity of the detector film was about 1 MΩ/K at 4.2 K and ∿5 MΩ/K at 2 K resulting in a maximum resolution of 0.1 mK including longtime signal averaging. The absolute temperature accuracy is ∿5 to 10 mK. A sufficiently low bias current was applied to the detector film to avoid self heating. The heater film showed very little variation of resistance with temperature and magnetic field which is particularly important for measurements with the pulse method. The whole experiment is based on the assumption that only the electronic specific heat varies with magnetic field in an oscillatory manner while all other contributions remain constant.

The experiments were performed on two different multi-layer materials: Sample 1 consisted of 172 double layers of 200 Å GaAs and 200 Å GaAlAs, grown on a semi-insulating GaAs substrate. The substrate was etched away to a total sample thickness of ~ 10 μm. The mobility at 4.2 K varied between 30 000 and 40 000 cm^2/Vs for different sample pieces. The density was $n_s = (6.3 \pm 0.4) \times 10^{11} cm^{-2}$. Sample 2 consisted of 94 layers of 220 Å GaAs and 500 Å GaAlAs. The mobility at 4.2 K was $\sim 80 000$ cm^2/Vs, and the density was $n_s = (7.7 \pm 0.3) \times 10^{11} cm^{-2}$. Samples were prepared by polishing and etching the material down to a total sample thickness of 20 μm.

2.2 Experimental results

Fig. 5 shows the observed temperature change at sample 1 expressed as curves ΔR versus the magnetic field for 4.2 and 1.5 K as obtained from averaging over 10 runs. The applied heat pulse raised the sample temperature at 4.2 K by 0.5 K and at 1.5 K by 0.03 K. The dashed curves ΔR_F show the background dc resistance variation of the detector film on an extended scale. Oscillations of the sample temperature are clearly observed with a spikelike behavior for integer filling factors. Integer filling factors represent the number of fully occupied LL. They are determined from the oscillatory conductivity measured on the same sample before being thinned down, shown as dotted curve σ_{xx}.

The size of the ΔR signal is proportional to the rise in sample temperature. The data show that the sample temperature is higher for integer filling factors than for values in between, which reflects the variation of the electronic specific heat. This variation has to be considered relative to the background ΔR_F. Since the variation of the sample temperature is similar to the oscillation in σ_{xx} we can be confident that we observe the temperature change mainly due to a variation of the electronic specific heat. The oscillations which reflect

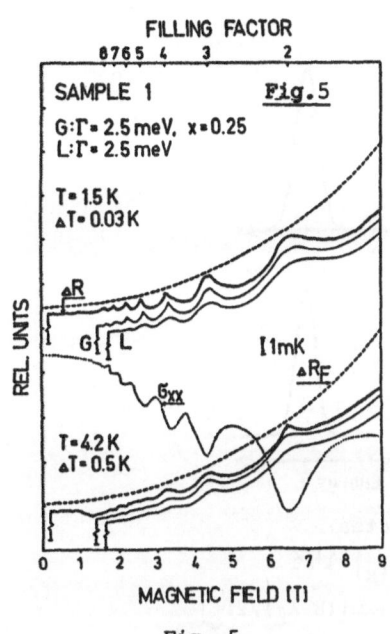

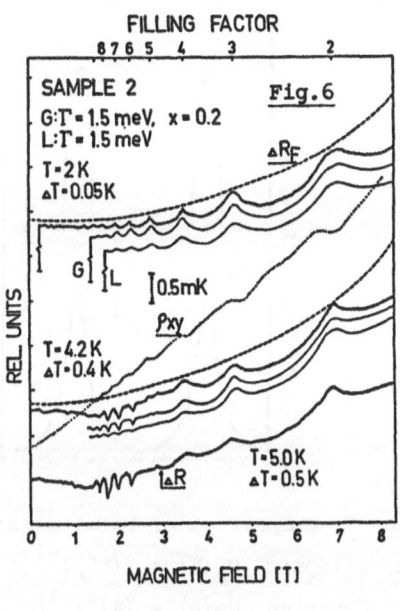

Fig. 5   Fig. 6

Temperature change of sample 1 (Fig. 5) and sample 2 (Fig. 6) measured with the Au:Ge film vs. magnetic field (curves ΔR), for a heat pulse rising the sample temperature by ΔT. ΔR_F denotes the d.c. dependence of the detector film. Theoretical calculations for a Gaussian and Lorentzian levelwidth are shown as curves G and L ($\Gamma = 2.4 \, \Gamma_G$, $\Gamma = 2\Gamma_L$)

only intra LL contributions for sample 1 are more pronounced with decreasing temperature and increasing magnetic field. The temperature changes are in the order of several mK, which amounts to less than 1 % of the total temperature change through the heat pulse at 4.2 K and to nearly 10 % at 1.5 K.

A partly similar behavior is observed for sample 2 as shown in Fig. 6. For this sample, data for 3 different lattice temperatures are given. Additional spikes are observed for 4.2 and 5.0 K as a result of inter LL contributions. The total temperature change is smaller than for sample 1 since this sample has only 92 double layers for a total thickness of 20 μm. The interpretation of the spikes at lower magnetic fields as inter LL contributions is confirmed through their temperature dependence: The inter LL contributions are only present at higher temperatures, in agreement with the theoretical prediction[10].

From the experimental data the form of the density of states can be determined by comparing the observed temperature change with calculations. The electronic specific heat is calculated using the model density of states shown in Fig. 7.

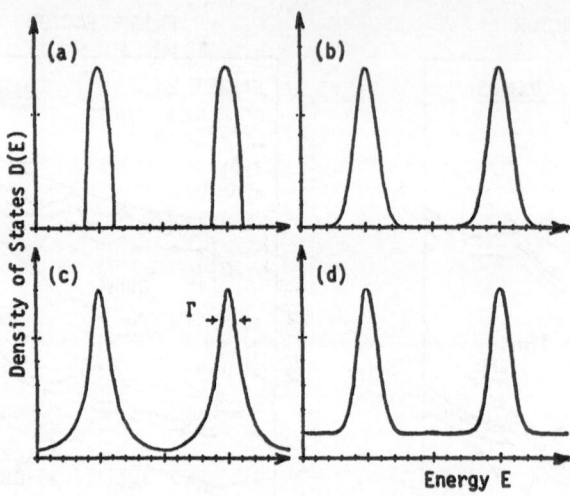

Fig. 7. Comparison of model density of states:

(a) elliptic $D_E(E) = 2(\pi\Gamma_E)^{-1}\{1-[(E-\lambda_n)/\Gamma_E]^2\}^{1/2}$

(b) Gaussian $D_G(E) = (\pi 1^2)^{-1}\sum_n (2/\pi)^{1/2} \Gamma_G^{-1}\exp[(E-\lambda_n)/2\Gamma_G^2]^2$

(c) Lorentzian $D_L(E) = (\pi 1^2)^{-1}\sum_n (\pi\Gamma_L)^{-1}\{1+[(E-\lambda_n)/\Gamma_L^2]\}^{-1}$

(d) Gaussian with background $D_{GB}(E) = D_G(E)(1-x)+(\pi 1^2)^{-1}(x/\hbar\omega_c)\,\theta(E)$

with $\lambda_n = \hbar\omega_c(n+1/2)$ and x is the percentage of background states.

In the first step the Fermi level E_F is calculated numerically for a given electron density n_s, level width Γ, and temperature. The level width Γ is defined as the total width at half maximum. The same value is used for the different level shapes which means $\Gamma = 2.4 \times \Gamma_G$ and $\Gamma = 2 \times \Gamma_L$. With the determined Fermi level the electronic specific heat is calculated for a constant temperature and level width Γ as a function of magnetic field according to equ. (1).

To compare the calculation with the experiment we have to calculate the temperature change $\Delta T(B)$ due to a change in C_{el}. A constant heat input ΔQ is applied to the sample resulting in a temperature increase ΔT, which has to be determined from the following equation:

$$\Delta Q = \frac{1}{4}\,\alpha[(T + \Delta T)^4 - T^4] + C_{el}(T,B,\Gamma)\Delta T \qquad (2)$$

with $C_{lat} = \alpha T^3 = 6.21 \times 10^{16} kT^3 (cm^{-3})$ and $\theta_D = 344$ K. The difference between $C_{el}(T + \Delta T)$ and $C_{el}(T)$ is neglected, since the temperature dependence of C_{el} is weak for the kT/Γ values considered (between 0.15 and 0.25).

The calculated $\Delta T(B)$ functions for the different densities of states are plotted in Figs. 5 and 6. The curves take into account the ΔR_F background. The best fit to the data is obtained for the curves denoted G, which are shifted for clarity. Curves G in Fig. 5 are for a Gaussian level width $\Gamma = 2.5$ meV and a constant background of x = 0.25. The curves denoted L are for a Lorentzian density and the same Γ. It is directly evident that curves G agree very well with the experimental data for both temperatures but especially with the 1.5 K data. The form and the relative size of the oscillations for the Lorentzian density never fits as well as the Gaussian density.

The question of whether a pure Gaussian density D_G is able to explain the data can be answered from Fig. 8, where C_{el} is plotted for D_G with $\Gamma = 2.5$ meV and D_{GB} with x = 0.25. It is clearly evident that the main difference is apparent at high magnetic fields where the constant background results in a flat part for integer-filling factors (solid curves). A Gaussian density D_G will result in a sharp spikelike behavior at high magnetic fields (dashed curves). The same behavior is found for an elliptic density of states D_E, the oscillations and the

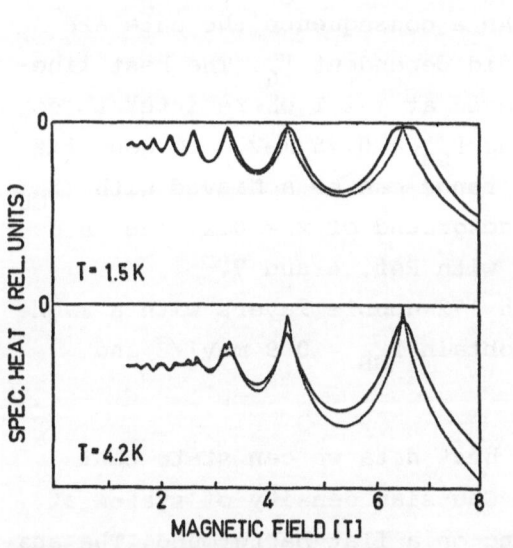

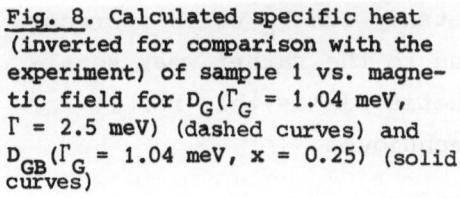

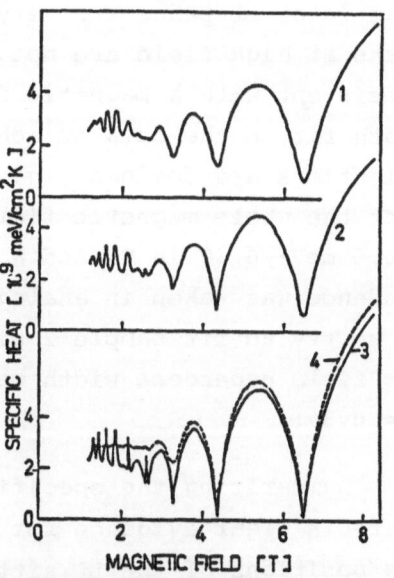

Fig. 8. Calculated specific heat (inverted for comparison with the experiment) of sample 1 vs. magnetic field for $D_G(\Gamma_G = 1.04$ meV, $\Gamma = 2.5$ meV) (dashed curves) and $D_{GB}(\Gamma_G = 1.04$ meV, x = 0.25) (solid curves)

Fig. 9. Comparison of calculated and experimental C_{el} for sample 2 vs. magnetic field at 4.2 K:
curve 1: experiment,
curve 2: $\Gamma_G = 0.6$ meV $\sqrt{B}$, x = 0.2
curve 3: $\Gamma_G = 0.75$ meV, x = 0
curve 4: $\Gamma_G = 1.5$ meV, x = 0

spikelike behavior are even more pronounced. The observed flat or rounded part of the ΔT oscillation at high magnetic fields gives a strong indication for a flat background density.

The same conclusion can be drawn from Fig. 6, where we have fitted the experimental curves for 4.2 and 2 K with a level width Γ = 1.5 meV for a Gaussian density with a background of x = 0.2 (denoted G) and a Lorentzian density (denoted L). From the fit at 2 K a certain amount of constant background density is evident again. The Lorentzian density leads to too small oscillations. At 4.2 and 5.0 K inter LL contributions at low magnetic fields are observed. Only the Gaussian density of states $D_{GB}(E)$ explains the size of the spikes and the behavior at 2 K at the same time.

The next question is whether Γ is constant with B or not. Other experimental techniques, which will be discussed in the next sections, give evidence for a magnetic field dependent level width. The influence of different level widths on C_{el} is shown in Fig. 9. The experimental result (curve 1) is compared first with a magnetic field independent Gaussian width Γ_G = = 0.75 meV (curve 3) and Γ_G = 1.5 meV (curve 4). It is evident that inter LL peaks are very sensitive to Γ_G, while intra LL peaks at high field are not. As a consequence the data are consistent with a magnetic field dependent Γ_G. The best linewidth fit to the data is achieved at $\sim$ 2 T where inter LL contributions are dominant, giving Γ_{GB} = 0.75 meV. A good fit over the whole magnetic field range can be achieved with Γ_{GB}= = 0.6 meV$\cdot\sqrt{B}$ (B in T) and a background of x = 0.2. The $\sqrt{B}$ dependence was taken in analogy with Ref. 6 and 7.
If we try to fit sample 2 with 172 double layers with a magnetic field dependent width we obtain Γ_{GB} = 0.9 meV$\cdot\sqrt{B}$ and x = 0.30.

Summarizing the specific heat data we can state that there is clear evidence for a Gaussian density of states at the positions of the LL, sitting on a flat background. The analysis of the experiments is consistent with Γ_G which increases according to a $\sqrt{B}$ law. However, due to the rather weak sensitivity of C_{el} to Γ_G at higher magnetic fields, this result relies more on other experimental techniques.

3. MAGNETIZATION

The magnetization of the 2D-electron gas is given by

$$M = - \frac{dF}{dB}$$

where F is the free energy. At absolute zero, both, the Fermi level and the magnetization exhibit a saw-tooth oscillation periodic in inverse field, with discontinuities at integer filling factors. The magnetization oscillations are of constant amplitude $M_O = n_s A \mu_B^*$, where A is the sample area and μ_B^* the effective Bohr magneton (with $m^* = 0.0665 \, m_O$ for GaAs). The first calculation of M for a Gaussian density of states was performed by Zavadzke et al.[14] A comparison of the calculated shape of the magnetization oscillation with the behavior of the Fermi level for a multilayer sample (sample 1 with 172 double layers) is shown in Fig. 10 for a pure Gaussian Γ_G and a Gaussian including background Γ_{GB} level width. It is clearly evident that a pure Gaussian level width will always show a step like behavior at high magnetic fields. The main effect of the background is to flatten the steps. With increasing level width the amplitudes of the oscillations are strongly reduced.

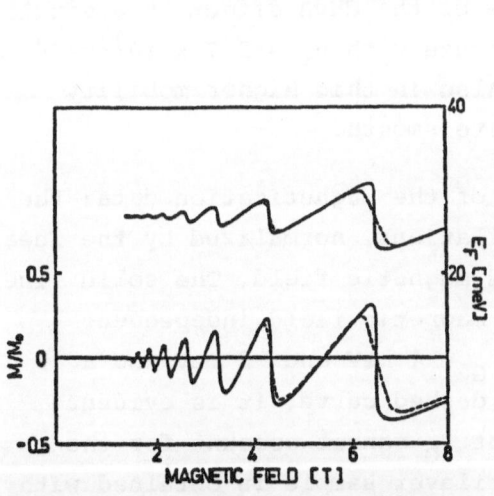

Fig. 10. Normalized calculated magnetization M/M_O and Fermi level for sample 1 vs. magnetic field at 1.5K: full curve: $\Gamma_G = 1.04$ meV, x = 0.25; dashed curve: $\Gamma_G = 1.04$ meV, x = 0.

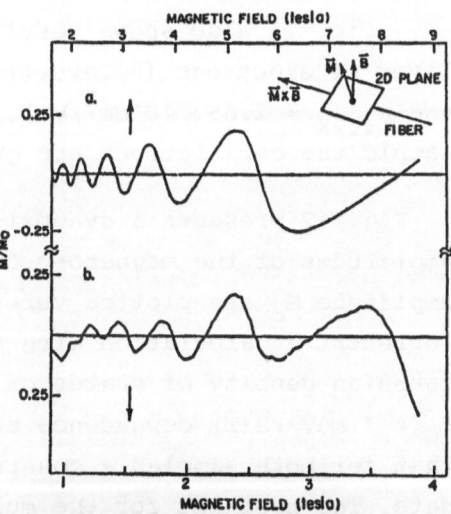

Fig. 11. Normalized measured magnetization for (a) a sample with 50 periods and (b) single heterolayer after Ref. 6. The insert shows the basic geometry of the experiment.

The first successful measurements of the magnetization were performed by Eisenstein et al.[6] with a recently developed torsional technique[15]. The samples are mounted on a thin fiber held perpendicular to the applied magnetic field and are oriented so,that the normal of the 2D plane, along which the orbital magnetic moments must lie, is tilted away from the field direction by a small angle; this geometry is seen in the insert of Fig. 11. In the experiment the magnetic field is swept very slowly; the torque is registered by a capacitive technique.

Fig. 11 shows normalized magnetization data for a GaAs/GaAlAs multilayer sample (50 periods, $\mu_{4.2K} = 8.0 \times 10^4 cm^2/Vs$; $n_s = 5.4 \times 10^{11} cm^{-2}$). A small smooth background has been subtracted from the magnetometer output. While the observed magnetization varies smoothly over the entire field range the resistivity (not shown) undergoes orders of magnitude fluctuations as the Fermi level passes between extended and localized states. The oscillations have the correct phase and periodicity to be unambiguously identified with the de Haas-van Alphen effect (dHvA) but the amplitude and general shape indicate significantly broadened LL, suggesting the absence of gaps in the density of states in agreement with specific heat experiments.

Fig. 11 also shows results of the dHvA effect in a single layer of electrons (heterostructure with $n_s = 2.7 \times 10^{11} cm^{-2}$ and $\mu_{4.2K} = 2.85 \times 10^5 cm^2/Vs$). Also in this higher mobility sample the oscillations are quite smooth.

Fig. 12 presents a synopsis of the magnetization data: the amplitudes of the magneto-oscillations, normalized by the ideal amplitude M_o are plotted versus magnetic field. The solid lines represent a calculation with a magnetic field independent Gaussian density of states of $\Gamma_G = 1$ meV and 2 meV and a $\Gamma_G = 1$ meV $\cdot \sqrt{B(T)}$ dependence as dashed curve. It is evident that for both samples a constant Γ_G cannot account for the data. The best fit for the multilayer sample is obtained with a $\Gamma_G = 1$ meV $\cdot \sqrt{B(T)}$ assuming no background. A calculation including a background of x = 20 % reduces the necessary Gaussian linewidth to $\Gamma_G = 0.8$ meV $\cdot \sqrt{B(T)}$. The single layer sample has a somewhat smaller level width. With a background of x = 10 %

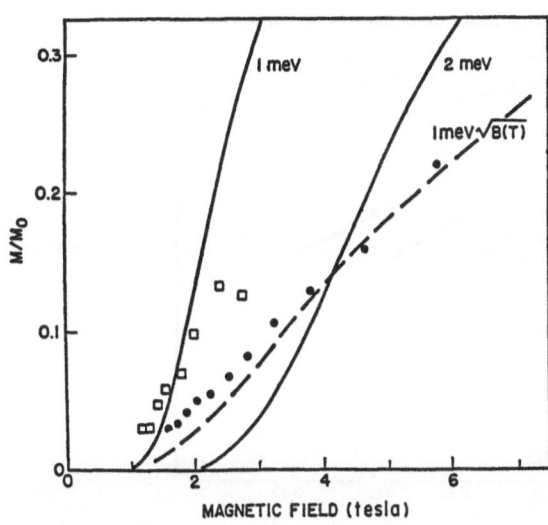

Fig. 12.
Normalized dHvA oscillation
amplitude vs. magnetic field
for the data from Fig. 11
(after Ref. 6); multilayer:
full circles, single layer:
squares.
The solid lines are calculated
with D_G: solid lines const. Γ_G,
dashed line $\Gamma_G \propto \sqrt{B}$

a good fit with $\Gamma_G = 0.65 \cdot \sqrt{B(T)}$ is achieved. However, the best fit is not as good as without background.

Summarizing the magnetization data we can state that this technique is sensitive to changes of the level width with magnetic field but no direct evidence for a flat background density is given. The experiments were performed only up to fields of 6 T where a significant overlap between LL can already be generated by purely Gaussian levels. However, it should be noted that the change in the basic level width at 1 T is rather small from 0.8 meV to 0.65 meV for the 2 samples while the mobility is increasing by a factor of 4.

4. ACTIVATED RESISTIVITY

For a 2-dimensional electron system the resistivity ρ_{xx} exhibits strong oscillation with magnetic field (Shubnikov-de Haas oscillation). Minima of ρ_{xx} occur at integer filling factors which are strongly temperature dependent. An example of the temperature dependence of ρ_{xx} for magnetic fields in the vicinity of an integer filling factor is shown in Fig. 13. As the Fermi level moves towards the center of a LL the temperature dependence of ρ_{xx} becomes weaker. The analysis of this effect was used by Weiss et al.[16] to determine the density of states between LL.

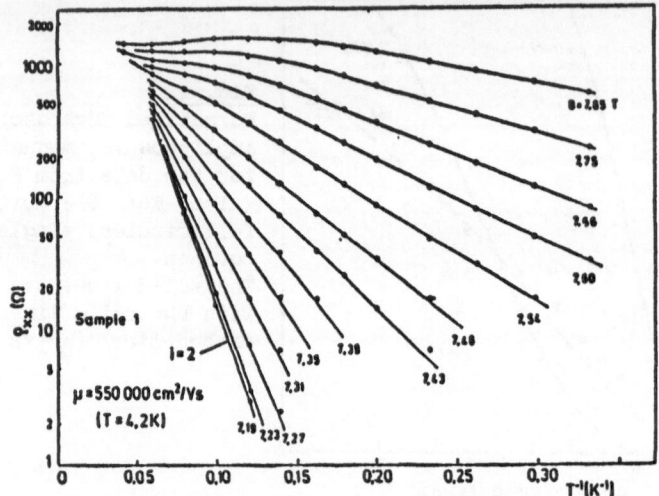

<u>Fig. 13.</u> Temperature dependence of the resistivity ρ_{xx} at different magnetic fields close to a filling factor $\nu = 2$ after Ref. 16.

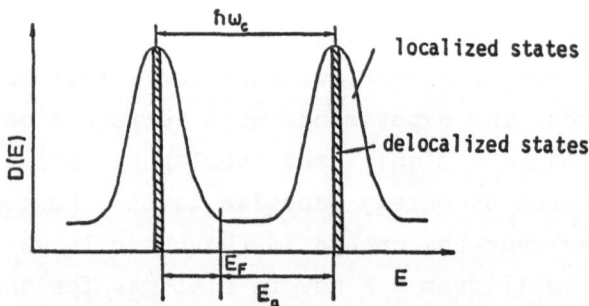

<u>Fig. 14.</u> Density of states for a given magnetic field. Schematic explanation of measured activation energy E_a due to excitation to a narrow band of extended states.

The basic idea of the analysis is shown in Fig. 14. For a given magnetic field the LL are split by $\hbar\omega_c$. It is assumed that nearly all states are localized except the states in the center of the LL.

The density of states is given by $D(E) = \Delta n_s/\Delta E$. This formalism was applied by Weiss et al.[16] to a large number

of samples with mobilities varying by a factor 30. The and-
lysis of the data from Fig. 13 is shown in Fig. 15. Curve a)
shows the determined activation energy as a function of magne-
tic field (sample 1: $n_s = 3.5 \times 10^{11} cm^{-2}$) and b) the deduced density of states in the tails of the LL. It is clearly evident
that there is a significant residual density of states at the
midpoint between LL even for a high mobility sample.

Calculations of ρ_{xx} were performed according to

$$\rho_{xx} \sim \rho_0(T) \sum_n \exp(-(E_n - E_F)/kT$$

where $\rho_0(T)$ is a temperature dependent prefactor. A Gaussian
density of states with background $D_{\infty GB}$ is included in the calculation of the Fermi-level: $n_s = \int_0 D_{GB}(E) f(E, E_F) dE$. In addition
the level width Γ_G is assumed to be $\sqrt{B}$ dependent.

The main result of this technique is a quite accurate
determination of a background density. It is interesting that
practically all samples investigated, which vary in mobility
by about a factor of 30, can be fitted with a Gaussian level
width $\Gamma_G = 0.25$ meV$\cdot\sqrt{B}$ sitting on a mobility dependent background. This indicates that the method is not sensitive to the
width of the Gaussian peaks, which is probably closely connected to the assumption of a very narrow range of extended states.

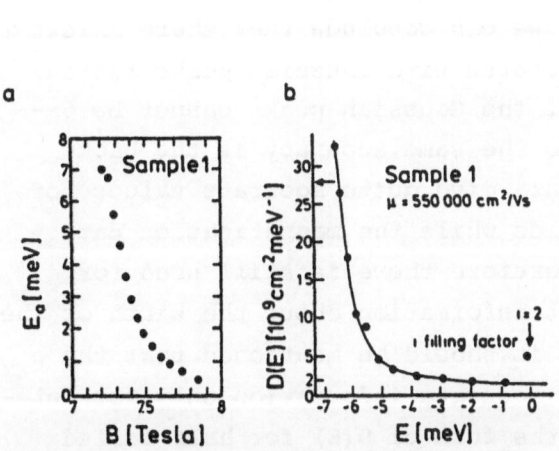

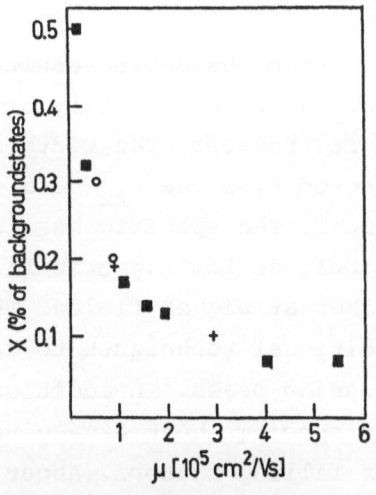

Fig. 15. Determined activation energy E_a
from the data of Fig. 14 as a function of
magnetic field (a).
Reconstructed density of states as a
function of energy (b) after Ref. 16.

Fig. 16. Summary of determined background density of
states in % of the B = 0 density. (■) from temperature
dependent ρ_{xx}; (o) from specific heat; (+) from magnetization.

A summary of the determined background values from the activated resistivity measurements together with the results from specific heat and magnetization is given in Fig. 16. The x-values are calculated from Ref. 16 by the ratio of the mid-point density to the density at B = 0 given by $D(E)_{B=0}$ = 28 x 10^9 cm^{2}meV^{-1}.

The results from all three techniques are in good agreement. It is evident that the background density of states increases very steeply for very low mobility samples while it decreases slowly for high mobilities. A hint for the origin of the background is given through the analysis of an electron irradiated sample[16] with a mobility of 14 000 cm^2/Vs showing that 50 % of all the states were contained within the background density. Annealing of the sample results in a significant reduction of the background and an increase in mobility. A direct correlation between the created point defects through electron irradiation and the background is not possible since the mobility change is even larger.

An analysis of the temperature dependence of ρ_{xx} and ρ_{xy} for integer filling factors was performed by Pudalov and Semenchinsky[17] in Si-MOSFETs. A considerable amount of density of states assigned to localized state was found in the gap between LL giving also a similar density of state as found in GaAs.

From the above results we can conclude that there exists a flat background density of states with Gaussian peaks sitting on it. However, the width of the Gaussian peaks cannot be extracted from the ρ_{xx} data to the same accuracy as the background. The specific heat data give quite accurate values for Γ_G only at low magnetic fields while the magnetization data rather at higher fields. Therefore there is still need for additional techniques to get information about the width of the Gaussian peaks. In addition it should be mentioned that the ρ_{xx} and specific heat data give accurate information only for integer filling factors. About the form of D(E) for half filled levels we have little information. The magnetization is sensitive rather for a non-integer filling factor. This gives a hint for a dependence of the background density on filling factor since the magnetization data can be fitted better without background.

5. MAGNETOCAPACITANCE

The capacitance of a metal- insulator-semiconductor structure depends on the thickness of the insulator, on the density of states in the semiconductor and on material parameters. Capacitance measurements seem to be a straight forward method to obtain direct information on D(E). A variation of the magnetic field results in oscillations of the magneto-capacitance which directly reflect changes in D(E). In a first attempt Smith et al.[18] tried to deduce D(E) from magnetocapacitance in GaAs/GaAlAs heterostructures. They analysed mainly the minima of the oscillations at 1.3 K; however the method failed for fields higher 1.6 T due to strong contributions of non-capacitive signals. Recently Mosser et al.[8] have investigated the magnetocapacitance in the same system up to fields of 8 T analysing only the maxima, thus extracting information on D(E) for half filled LL.

The experiment starts from the relation for the capacity of the GaAs/GaAlAs heterostructure after Stern[19].

$$\frac{1}{C} = \frac{1}{C_A} + \frac{\gamma z_0}{\varepsilon_s} + \frac{1}{e^2 \dfrac{dn_s}{d(E_F - E_0)}} \tag{3}$$

where C is the measured differential capacitance at a given magnetic field, C_A is the capacitance of the insulating AlGaAs layer, ε_s is the dielectric constant of GaAs, z_0 is the average position of the electrons in the channel, γ is a constant numerical factor between 0.5 and 0.7, and $dn_s/d(E_F - E_0)$ is the thermodynamic D(E) at the Fermi level, in the following denoted as dn_s/dE_F. The first two terms on the right hand side are assumed to be constant in a magnetic field, and thus changes of the capacitance are directly related to changes in the thermodynamic D(E) of the 2DEG.

The above relation is derived for the band structure shown in Fig. 17a.

The capacitance experiments were carried out using a circuit depicted in Fig. 17b. The signal proportional to the capacitance was obtained by measuring a voltage drop at the sample and at a high precision Boonton capacitance decade in the frequency range between 22 and 446 Hz.

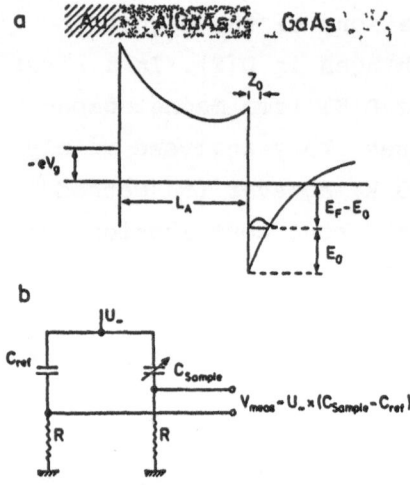

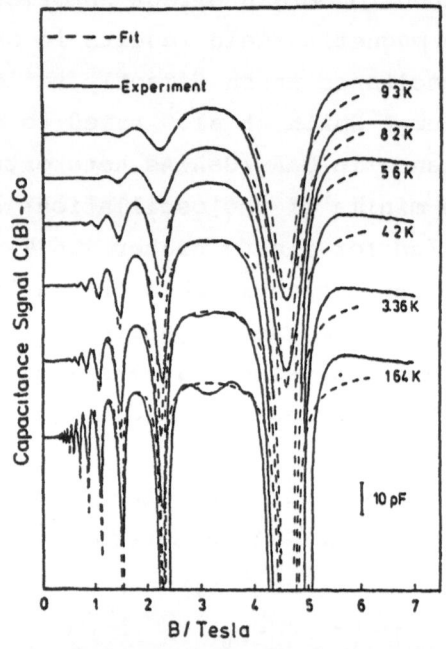

Fig. 17. Schematic diagram of the conduction band edge (a) for a gated GaAs-AlGaAs heterostructure and the experimental set up for capacitance measurements (b) after Ref. 8.

Fig. 18. Measured magnetocapacitance of a sample with n_s = 2.25x10^{11}cm^{-2} and μ = 220.000cm^2/ Vs. The fit was performed with a Gaussian width Γ_G = 0.3x$\sqrt{B(T)}$ meV and x = 0.13 (after Ref. 8)

Fig. 18 shows the capacitance data at different temperatures (full curves) for a sample with n_s = 2.25 x 10^{11}cm^{-2} and μ = 220.000 cm^{-2}Vs. Strong oscillations are clearly observed with minima at integer filling factors. From a critical analysis of the experimental situation only the maxima of the capacitance can be used to determine D(E) since only for half filled LL the above relation is defined and the signal remains mainly capacitive.

The experimental results are compared with calculations

of dn_s/dE_F using a Gaussian density of states with background $D_{GB}(E)$. The fit shown in Fig. 18 assumes $\Gamma_G = 0.3 \cdot \sqrt{B(T)}$ meV and x = 0.13 (corresponding to $D_{nG} = 3.9 \cdot 10^9 \mathrm{cm}^{-2}\mathrm{meV}^{-1}$). At all investigated temperatures the calculated maxima are in good agreement with the data for fields up to 5 T. However, the fit is not as good for a filling factor of ν = 1. The analysis of a sample with nearly the same density and a mobility of $\mu = 800.000 \mathrm{cm}^{-2}/\mathrm{Vs}$ yields a Gaussian width of $\Gamma_G = 0.25 \cdot \sqrt{B(T)}$ meV and a reduced background of x = 0.07. It is stated in the paper[8] that a constant background is even not necessary to fit the data. The obtained Gaussian width Γ_G on the other hand does not seem to depend critically on the mobility.

The result of the capacitive technique is a weakly on mobility dependent Γ_G for half filled LL, which is of similar value as obtained from the temperature dependent ρ_{xx} analysis for integer filling factors. The described thermodynamic techniques in which the main information is extracted at integer filling factor, give somewhat larger Γ_G values with a more pronounced dependence on mobility.

6. CYCLOTRON RESONANCE SPECTROSCOPY

In cyclotron resonance (CR) the observed linewidth of the transmission spectrum is determined by the broadening of both the initial and final LL. The linewidth contains information on the individual level width Γ but not in a trivial way. In the techniques described in the previous chapters the pure LL width was extracted from experiments which were not influenced by scattering processes or transitions between different LL; in the analysis the level width was always assumed to be the same for all LL.

While there is consistent information on the background, the value of the Gaussian width Γ_G and its dependence on the zero field mobility has not been derived yet in a satisfactory way. One technique which should give information on changes of Γ_G is the analysis of the CR linewidth. It is clear that this method will only give good values of Γ_G after a careful theoretical fit of the data. However, tendencies of Γ_G as a function of certain external parameters for a situation where no significant physical quantity changes can be determined.

Previous experimental[20,21,22] and theoretical[23,24] investigations have revealed a filling factor dependent CR linewidth due to screening. Maxima of the linewidth occur at integer values of the filling factor and minima in between. In the following two experimental techniques will be described which give information on the form of D(E): <u>First</u> results from cyclotron emission - the inverse effect of CR absorption - performed on several GaAs/GaAlAs heterojunction samples will be shown. These data will mainly depend on D(E) in the upper-excited level. <u>Second</u> results from CR-transmission-experiments will be analysed for a filling factor of ν = 1 and correlated to the zero magnetic field mobility.

6.1 <u>Cyclotron emission</u>

Cyclotron emission is generated by heating up the electron gas by electric field pulses. As a result a nonequilibrium carrier distribution with average carrier temperatures above the lattice temperature is obtained. Radiative transitions between neighbouring LL in the vicinity of the Fermi level occur. The basic principle of the technique is shown in Fig. 19. The emitted radiation is measured and ana-

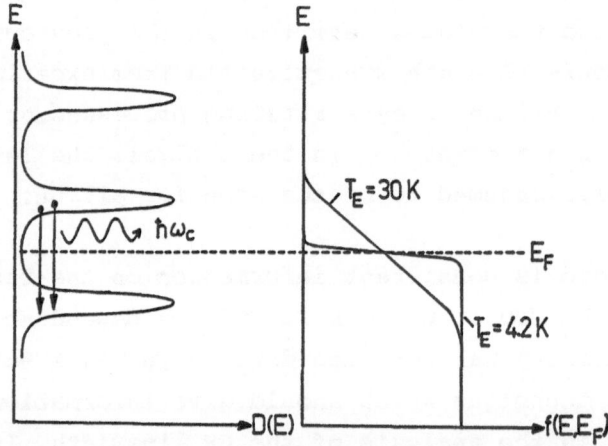

<u>Fig. 19</u>. Schematic diagram of the cyclotron emission technique. The density of states is shown with two distribution functions indicating the emission processes for the higher electron temperature.

lysed with a narrow-band and magnetic field tunable GaAs detector[9,25]. The spectrum of the detector is shown in Fig. 20 for B = 0 T and B = 3 T. The detector has a peak sensitivity at 4.4 meV (35 cm^{-1}) at B = 0. In the magnetic field this line splits into 3 lines which have a magnetic field dependence also demonstrated in Fig. 20.

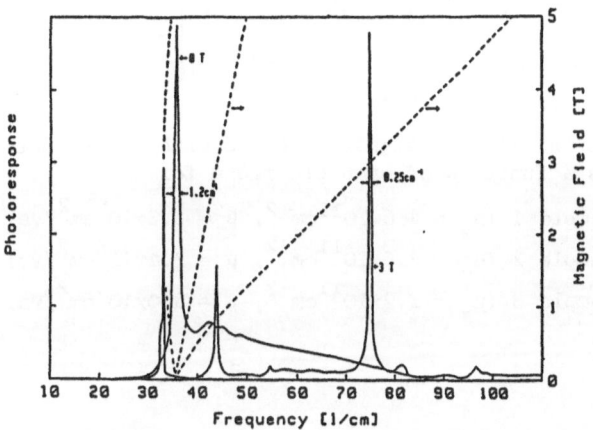

Fig. 20. Photoresponse of a n-GaAs detector at 0 T and 3.0 T obtained with a Fourier spectrometer. The tuning characteristic of the three lines with magnetic field is given by the dashed curves.

The emission technique is applied to several GaAs/GaAlAs heterostructures. Figures 21a, 21b and 21c show emission spectra as obtained with a GaAs-photoconductive detector at B = 0. In the experiment the emission signal is monitored as a function of the 2DEG magnetic field for various electric fields. The most prominent feature is that the emission spectra differ considerably from the transmission spectra, obtained on identical samples[25]. This effect is extremely pronounced for sample 1 (Fig. 21a): For very low electric fields the emission spectrum consists of a broad line, which is well below the bulk cyclotron position (corresponding to a higher energy). With increasing electric field which is equivalent to increasing the carrier temperature this peak becomes smaller and a line at the position of the transmission resonance grows until it dominates the spectrum at high fields. The dashed line separates the pure CR-line from the "other" structure. The 2D character of the structure was tested by tilting the magnetic field. For sample 2 (Fig. 21b) the peak at lower magnetic field is not that strong. Sample 2 has a higher mobility and a narrower linewidth. A shoulder on the

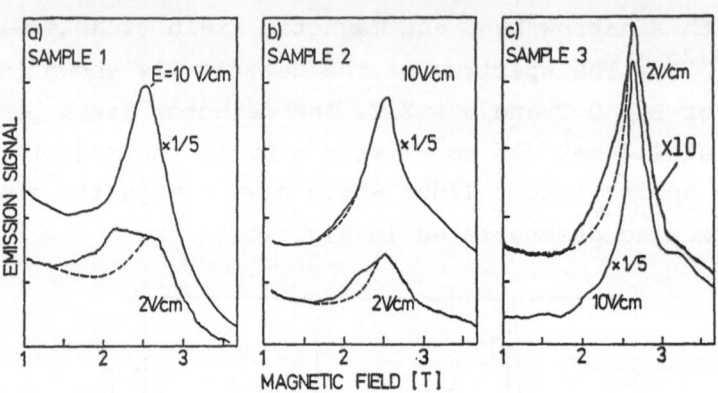

Fig.: 21. Cyclotron emission signal detected with a GaAs detector at 0 T
(see Fig. 20) as a function of magnetic field for

a) sample 1 ($n_s = 4.6 \times 10^{11} cm^{-2}$, $\mu = 1.3 \times 10^5 cm^2 / Vs$)

b) sample 2 ($n_s = 2.6 \times 10^{11} cm^{-2}$, $\mu = 2.4 \times 10^5 cm^2 / Vs$)

c) sample 3 ($n_s = 2.2 \times 10^{11} cm^{-2}$, $\mu = 5.0 \times 10^5 cm^2 / Vs$)

low magnetic field side in the emission spectrum has been
found in all investigated samples. A correlation of the emission data with the d.c. mobility is found. The higher the
mobility the less pronounced the additional peak becomes. For
the sample with the highest mobility (sample 3) a rather
narrow continuous shoulder is observed shown in Fig. 21c. This
feature is not restricted to the 3 T magnetic field range. We
have found this behaviour in the whole investigated magnetic
field ranges up to 8 T.

As the additional peak is only observed in emission it
must be correlated to the special excitation technique used:
The electric field excites carriers to the next higher LL. The
carriers first thermalize within the upper LL and occupy only
states at the lower edge. If localized states exist below the
edge, the electrons will get trapped in these states. As the
number of excited electrons is very small at low electric
fields, the recombination radiation will mainly be due to
transitions out of these localized states. The emission frequency will be shifted to higher energy due to the binding
potential similar to the bulk case, where a so called impurity

shifted CR line is observed. With increasing electric field the binding potentials are washed out and the pure CR line from extended states is observed.

If we try to assign a binding potential to the observed "impurity line" we can use a simple oscillator model were the observed frequency ω_{obs} is given by

$$\hbar\omega_{obs} = \sqrt{(\hbar\omega_c)^2 + (\hbar\omega_B)^2} \qquad \text{with}$$

$\hbar\omega_c$ the cyclotron energy and $\hbar\omega_B$ the binding energy. It is evident from Fig. 21 that the amount of "impurity line" is decreasing from sample 1 to sample 3 and that the main part of the line is shifting towards the cyclotron resonance position. From the above analysis we obtain an average binding potential of 2.8 meV for sample 1, 2.4 meV for sample 2 and about 1.5 meV for sample 3. The integral intensity of the structure normalized to the same electron density decreases proportional to the inverse mobility. The origin of the binding potentials can only be positive charges in the GaAlAs and compensating donors in GaAs. However since all the used samples are modulation doped with space layer thicknesses larger than 150 $\overset{\circ}{A}$, the high doped region in the GaAlAs cannot be responsible for the binding potentials. The impurity potentials will be smeared out in this case since the average distance between the impurities is in the same order as the distance to the 2D electron gas. A rather defined structure indicates that only impurities at a defined position can be responsible for the observed spectra. Therefore we believe that only impurities in the GaAlAs close to the interface can be responsible for some of the observed phenomena. A distinction between impurities at the interface and compensating donors in the GaAs can be made through the expected binding energy: It has been shown recently[26-29] that the impurity binding potential depends strongly on the position of the impurities in respect to the interface: At the interface the binding energy is considerably smaller than in the center of 2D channel. Since the compensating impurities in the GaAs are randomly distributed a rather broad impurity peak with higher binding energies is expected, while interface impurities will lead to smaller binding energies appearing as a shoulder close to the cyclotron energy. These two cases are clearly observed for sample 1 ($\mu = 1.3 \times 10^5 \, cm^2/Vs$) which is dominated by a large number of impurities in the GaAs and for sample 3 ($\mu = 5.0 \times 10^5 \, cm^2/Vs$) which has a considerably lower number of impurities which are positioned close to the interface.

From the emission data we can conclude that there exist binding potentials due to impurities in the GaAs and GaAlAs (close to the interface) which lead to a density of states between LL. The number of these impurities is correlated to the mobility of the sample. With increasing mobility mainly the number of impurities in the bulk GaAs is reduced while the impurities at the interface seem not to be changed significantly. From the analysis of the total intensity of the emission signal the number of impurities can be estimated in the different samples: A number of $\sim 10^9$ impurities cm^{-2} for sample 3 and about $5 \times 10^9 \, cm^{-2}$ for sample 1.

6.2. Cyclotron resonance transmission

CR transmission experiments with the use of Far infrared lasers were performed by several groups[20,21,22,30] which all found a systematic oscillation of the linewidth with filling factor. In a previous paper[21] we have shown that the cyclotron transmission spectra have Lorentzian lineshape and can well be fitted with a constant linewidth. Maxima of the linewidth are observed at integer filling factors where the influence of screening can be neglected.

Fig. 22 shows a plott of the measured linewidth as a function of filling factor for 3 different samples at a temperature of 4.2 K. All samples show a clearly defined maximum of the linewidth for $\nu = 1$. A systematic behaviour of the linewidth value at $\nu = 1$ with the sample mobility is evident, while the linewidth for a filling factor of 0.5 seems not to be correlated clearly to the mobility. There is also clear evidence for a temperature dependence of the linewidth but we only want to compare here data at one temperature.

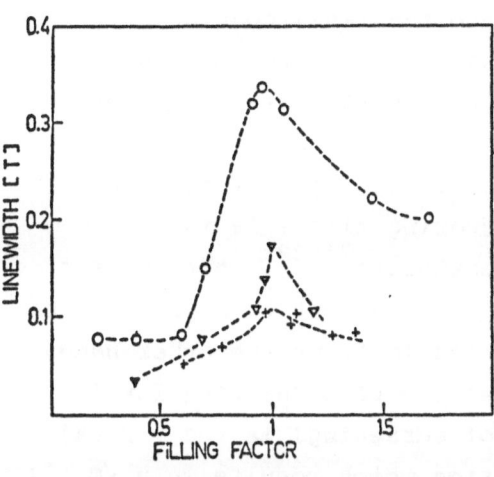

Fig. 22. Measured cyclotron resonance transmission linewidth as a function of filling factor for 3 different samples

($\triangledown$) sample 4 ($n_s = 1.2 \times 10^{11} cm^{-2}$, $\mu = 4.5 \times 10^5 cm^2/Vs$)

($\varnothing$) sample 5 ($n_s = 1.2 \times 10^{11} cm^{-2}$, $\mu = 1.2 \times 10^5 cm^2/Vs$)
data taken from Ref. 19

($+$) sample 6 ($n_s = 2.3 \times 10^{11} cm^{-2}$, $\mu = 1.0 \times 10^6 cm^2/Vs$)

To get information on D(E) from these data we have to make the assumption that the CR linewidth at integer filling factors is directly correlated to the Gaussian level width Γ_G. This correlation is demonstrated in Fig. 23 where the CR linewidth at $\nu = 1$ is plotted as a function of mobility. The linewidth decreases according to a slope of 1/2 with increasing mobility. If we plot the percentage of background states from Fig.16 on the same double logarithmic scale we find a very interesting result: The background shows the same dependence on mobility as the linewidth maxima in the CR.

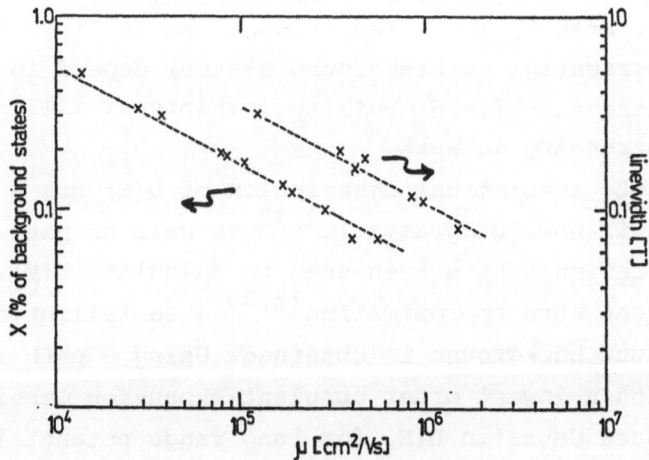

Fig. 23. Cyclotron resonance linewidth at filling factor $\nu = 1$ and percentage of background density of states as a function of sample mobility (at zero magnetic field)

389

From this plot we can come to the final conclucion:
The origin for the Gaussian peaks and the background density are the same type of impurity - scatterers since both features in D(E) show the same dependence on mobility. The zero field mobility is limited by single scattering events at the impurities and thus the scattering time τ is inversely proportional to the number of scatterers[31,32]. That means a plot as a function of mobility is equivalent to a plot versus the reciprocal number of impurities (N_i). On the other hand, the main process which determines the LL broadening for integer filling factors (absence of screening) is the virtual double scattering at the impurities which results in a $\sqrt{N_i}$ dependence of the level broadening. Thus the correct description of LL broadening has to include both at the same time the formation of Gaussian peaks at the center and a residual flat background. The broadening has to be due to a certain type of impurity potentials which are located at the interface between the GaAlAs and GaAs and in the bulk GaAs. The influence of impurity in the high doped GaAlAs region seems to be weak.

7. CONCLUSION

A consistent picture of the density of states can be drawn by summarizing the results from all the experimental techniques: The density of states consists of Gaussian peaks sitting on a flat background. The whole form of D(E) is the result of the same impurity potentials. This is evident from the fact that both Γ_G (measured through the CR linewidth) and x (the percentage of background states) depend in the same way on the zero field mobility for integer filling factors where screening is weak.

A complete theoretical description of D(E) has not been performed until now. Diagrammatic[3,22] as well as path integral[33-35] techniques have been used to calculate D(E). In the self-consistent Born approximation[22,23] a semielliptic form of D(E) without background is obtained. Using a path integral technique within lowest order cumulant expansion Gerhardts[33] obtained a pure Gaussian D(E) for long range potentials. Exact results have been obtained by Wegner and Brezin et al.[35] for some restricted types of short range scattering distributions for the lowest LL. For a white noise potential a Gaussian

D(E) is obtained. A Poisson distribution of scatterers (with nonzero higher order correlations) yields a peak density of states at the center of the LL and a weakly decaying D(E) towards the next LL.

ACKNOWLEDGEMENT

The work was partly supported by the Stiftung Volkswagenwerk (Projekt I/61840) and the European Research Office of the US Army. I would also like to thank Dr.R. Lassnig and Prof. W. Zawadzki for helpful and critical discussions.

REFERENCES

(1) K. von Klitzing, G. Dorda, and M. Pepper, Phys. Rev. Lett. 45, 494 (1980)

(2) D.C. Tsui, H.L. Störmer, and A.C. Gossard, Phys. Rev. Lett. 48, 1559 (1982)

(3) T. Ando, Y. Uemura, J. Phys. Soc. Jap. 36, 959 (1974)

(4) R.R. Gerhardts, Surf. Sci. 58, 227 (1976)

(5) E. Gornik, R. Lassnig, G. Strasser, H.L. Störmer, A.C. Gossard, W. Wiegmann, Phys. Rev. Lett. 54, 1820 (1985)

(6) J.P. Eisenstein, H.L. Störmer, V. Narayanamurti, A.Y. Cho, A.C. Gossard, Phys. Rev. Lett. 55, 875 (1985)

(7) E. Stahl, D. Weiss, G. Weimann, K. v. Klitzing, K. Ploog, J. Phys. C18, L783 (1985)

(8) V. Mosser, D. Weiss, K. v. Klitzing, K. Ploog, G. Weimann, Solid State Commun., 58, 5 (1986)

(9) E. Gornik, Physica 127B, 95 (1984)

(10) W. Zawadzki and R. Lassnig, Solid State Commun. 56, 537 (1984)

(11) R. Bachmann et al., Rev. Sci. Instrum. 43, 205 (1972)

(12) S. Alterovitz, G. Deutscher, and M. Garshenson, J. Appl. Phys. 46, 3637 (1975)

(13) B.W. Dodson, W.L. McMillan, J.M. Mochel, and R.C. Dynes, Phys. Rev. Lett. 46, 46 (1981)

(14) W. Zawadzki, R. Lassnig, Surf. Science 142, 225 (1984)

(15) J.P. Eisenstein, Appl. Phys. Lett. $\underline{46}$, 695 (1985)

(16) D. Weiss, and K. v. Klitzing, V. Mosser, Springer Series in Solid State Sciences $\underline{67}$, 204 (1986)

(17) V.M. Pudalov, S.G. Semenchinsky, Solid State Comm. $\underline{55}$, 593 (1985)

(18) T.P. Smith, B.B. Goldberg, P.J. Stiles, M. Heiblum, Phys. Rev. $\underline{B32}$, 2696 (1985)

(19) F. Stern, Phys. Rev. $\underline{B5}$, 4891 (1972)

(20) Th. Englert, J.C. Maan, Ch. Uihlein, D.C. Tsui, A.C. Gossard, Physica $\underline{117B}$ & $\underline{118B}$, 631 (1983)

(21) W. Seidenbusch, R. Lassnig, E. Gornik, W. Weinmann, Physica $\underline{134B}$, 314 (1985)

(22) W. Seidenbusch, R. Lassnig, E. Gornik, G. Weimann, Proc. 18th Int.Conf. on the Phys. of Semicond., Stockholm 1986

(23) R. Lassnig, E. Gornik, Solid State Comm. $\underline{47}$, 959 (1983)

(24) T. Ando, Y. Murayama, J. Phys. Soc. Jpn. $\underline{53}$, 693 (1985)

(25) E. Gornik, W. Seidenbusch, R. Lassnig, H.L. Störmer, A.C. Gossard, W. Wiegmann, Springer Series in Solid State Sciences $\underline{53}$, 60 (1984)

(26) G. Bastard, Phys. Rev. $\underline{B24}$, 4714 (1981)

(27) R.L. Greene, K.K. Bajaj, Solid State Comm. $\underline{45}$, 825 (1983)

(28) N.C. Jarosik, B.D. McCombe, B.V. Shanabrook, I. Comas, I. Ralston, and G. Wicks, Phys. Rev. Lett. $\underline{54}$, 1283 (1985)

(29) J.L. Robert, A. Raymond, L. Konczewicz, C. Bousquet, W. Zawadzki, F. Alexandre, I.M. Masson, J.P. André, and P.M. Frijlink, Phys. Rev. $\underline{B33}$, 5935 (1986)

(30) G.L.I.A. Rikken, H.W. Myron, P. Wyder, G. Weimann, W. Schlapp, R.E. Horstman, and J. Wolter, Surface Science $\underline{170}$, 160 (1986)

(31) F. Stern and W. Howard, Phys. Rev. $\underline{163}$, 816 (1967)

(32) W. Walukiewicz, H.E. Ruda, J. Lagowski and H.C. Gatos, Phys. Rev. $\underline{B30}$, 4571 (1984)

(33) R.R. Gerhardts, Z. Phys. $\underline{B21}$, 275 (1975)

(34) F. Wegner, Z. Phys. $\underline{B51}$, 279 (1983)

(35) E. Brezin, D.I.Gross, C. Itzykson, Nuclear Phys. $\underline{B235}$, 24 (1984)

ELECTRONS ON A LIQUID HELIUM FILM

François M. Peeters

University of Antwerp (U.I.A.)
Department of Physics
Universiteitsplein 1
B-2610 Antwerpen(Wilrijk)

I. INTRODUCTION

Electrons on a liquid helium surface are more and more recognized as a unique model system of a two-dimensional(2D) electron system[1-3]. They form the cleanest example of a two-dimensional electron gas(2DEG). The interest in this system stems from the interest in 2D phase transitions[4] which coincides with the recent observation of solidification of electrons into a *Wigner lattice* on the vapour side of liquid helium[5]. Furthermore the predicted[6,7] and recently observed[8] *self-trapping* of electrons in the plane parallel to the helium shows this system as a nice testing ground for field-theoretical calculations on the interaction of an electron with a scalar field, i.e. the surface modes of the liquid helium.

This fascinating system is rather new and has been recently the subject of a systematic study. The first theoretical[9,10] and experimental results[11,12] in this area were obtained less than 20 years ago. Nevertheless this comparitively new subject has been very quickly matured to an important area in physics.

Progress in this area has been paralleled by an explosion of interest in the field of 2DEG's in semiconductor structures[13] like Si-MOS, $GaAs/Al_xGa_{1-x}As$ heterostructures and superlattices[14],... . The latter one became possible thanks to the technological progress in material engineering and has already led to important applications[15] in e.g. optoelectronics, new fast transistors(HEMT),... . In both-type of 2D-systems the electrons are bound in their motion perpendicular to the surface and they are more or less free to move parallel to the surface.

In spite of several similarities there are some fundamental differences between a 2DEG on a liquid helium surface and a 2DEG in a semiconductor structure (see also the lecture of Dr. P. Platzman). A comparison of some typical physical parameters of these systems are shown in Table I. In this table E_B is the binding energy due to the confinement potential, $< z >$ is the average distance of the electrons from the interface which is also a measure for the width of the 2D electron layer, n_e is the electron density, E_F is the Fermi energy which in 2D is given by $E_F = \pi \hbar^2 n_e / m^*$, m^* is the electron effective mass and m_e is the mass of an electron in vacuum, τ is the relaxation time and $\mu = e\tau/m^*$ the mobility of the electrons in the 2D layer.

Table I. Some physical parameters for 2DEG in different systems

	electrons on He	Si-MOS	GaAs/Al$_x$Ga$_{1-x}$As
E_B(meV)	0.7	5 – 50	20 – 40
$< z >$(Å)	114	30	50 – 100
$n_e(cm^{-2})$	$10^5 - 10^9$	$10^{11} - 10^{13}$	$10^{11} - 10^{12}$
E_F(meV)	$10^{-7} - 10^{-3}$	1 – 50	20 – 100
m^*/m_e	1.0	0.19	0.066
τ(sec)	10^{-7}	10^{-12}	$10^{-12} - 10^{-11}$
$\mu(cm^2/Vs)$	10^7	10^3	$10^5 - 10^6$
new effects discovered	Wigner lattice[5] Self-trapping[8]	QHE[16] (1980)	FQHE[17] (1983)

From Table I it is apparent that the 2DEG on liquid helium is in a different region in the physical parameter space than the 2DEG in semiconductor structures; e.g. electrons on He are a *classical* 2DEG while electrons in semiconductor systems behave typically as a *quantum* 2DEG.

Electrons on *thin liquid He-films* may bridge the gap between the two different regions. This will be discussed in more detail in the next chapters. Two essential different aspects of the system of electrons on a liquid He-film will be discussed: 1) *Wigner crystallization* which is a many-particle characteristic of the system, and 2) *Self-trapping* which, in essence, is a one-particle property.

II. SURFACE STATE ELECTRONS ON LIQUID HELIUM

II.1 Electrons on bulk helium

An electron a distance z above the helium surface polarizes the helium atoms of the liquid which results in an attractive force towards the helium surface[18,19]. Mathematically this polarization can be described by an image charge $\Omega_0 e$ at a distance z inside the helium. $\Omega_0 = (\epsilon_{He} - \epsilon_v)/2(\epsilon_{He} + \epsilon_v) \approx 0.014$ with $\epsilon_v \simeq 1$ the dielectric constant of the medium above the liquid helium surface, which will contain some helium gas atoms, and $\epsilon_{He} = 1.0572$ is the dielectric constant of liquid helium.

A repulsive barrier of strength $V_0 \sim 1.eV$ prevents the electron from penetrating into the liquid helium. This potential barrier is a consequence of the Pauli exclusion principle which states that the electron wavefunction should be orthogonal to the core electrons of the liquid helium atoms. Because the last shell of a helium atom is completely filled the addition of an extra electron will lead to a strong energy enhancement which is responsible for this strong repulsive barrier.

Energy spectrum of the electron surface states. The repulsive potential barrier V_0 is very large in comparison to the weak attractive image potential and therefore we may take $V_0 = \infty$. In a first approximation we may assume that the helium surface is perfectly sharp and consequently the image potential becomes $V(z) = -Qe^2/z$ for $z > 0$ with $Q = \Omega_0/2$. Thus in the z-direction the electron behaves like a *one-dimensional hydrogen atom* with a strongly reduced nuclear charge of $Qe \simeq 1/144$. The energy levels of this system are readily found to be

$$E_n = -\frac{Q^2}{2n^2}\text{Ry}, \quad n = 1, 2, \dots \ , \tag{1}$$

with Ry=Rydberg=$13.6eV$. The ground-state energy is $E_1 = -0.65 meV = -7.5K$ and has the wavefunction

$$\Phi_1(z) = \frac{2}{a^{3/2}} z e^{-z/a} \ , \tag{2}$$

with $a = a_0/Q = 76\text{Å}$ the effective Bohr radius and $a_0 = 0.529\text{Å}$ the Bohr radius. The confinement potential together with the electron wavefunction are shown in Fig.1. The average distance of the electron above the helium surface is $< z >_1 = \frac{3}{2}a = 114\text{Å}$. For the excited states we find $< z >_2 = 456\text{Å}$, $< z >_3 = 1026\text{Å}$, ... and the energy difference between the first excited state and the ground-state energy is $E_2 - E_1 = 0.5meV = 5.6K$. The experiments have to be performed at temperatures lower than liquid helium (i.e. $T = 4.2K$) and are typically of the order $T = 0.1 - 1.K$. Thus we may assume that the electron is in the lowest energy level $n = 1$.

The existence of these surface state were predicted theoretically by Cole and Cohen[9] and independently by Shikin[10]. Experimentally they were observed by Grimes and Brown[12] in a beautiful spectroscopic experiment. Microwave radiation was applied in order to induce transitions between the different electron states. An electric field (called *pressing field*) is applied perpendicular to the helium surface which adds an additional component to the image potential: $V(z) = -Qe^2/z + eE_\perp z$. In this way it was possible to change the energy spectrum or equivalently the energy separation between the electron states. Grimes and Brown were able to observe transitions up to the 7th excited state. In the

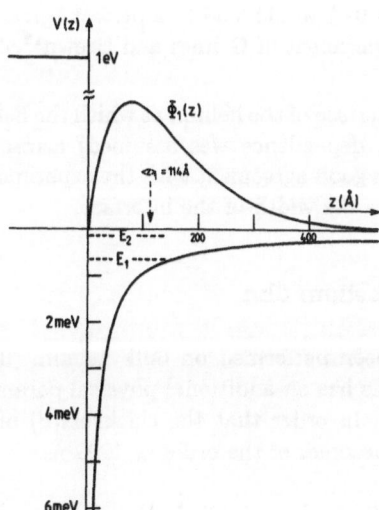

Fig. 1. Image potential and wave function for the electron in the ground-state above liquid helium.

limit of zero pressing field they found the transition frequencies $\nu_{1\to 2} = 125.9 \pm 0.2 GHz$ and $\nu_{1\to 3} = 148.6 \pm 0.3 GHz$. This compares with the results of the above one-dimensional hydrogen atom $\nu_{1\to 2} = 119.3 GHz$ and $\nu_{1\to 3} = 141.3 GHz$ (the relative difference is 5.2 and 4.9 % respectively). The agreement becomes better for higher excited states.

The small deviations between experiment and theory can be traced back to the fact that: 1) $V_0 \sim 1.04 eV$ instead of $V_0 = \infty$ and 2) the Coulomb potential $1/z$ will be disturbed in the neighborhood of the helium surface which is a consequence of: i) the fact that the helium liquid consists of atoms with a non-zero dimension, ii) the liquid surface can be deformed by the electron and consequently will not be exactly at $z = 0$, and iii) the liquid helium surface does not have an abrupt density profile at $z = 0$ but rather gradually decreases to the gas phase density.

In order to correct for these effects different potentials[20,21] have been introduced. The most important ones are:

(1) A potential with a *finite cutoff at a distance b* from the helium surface

$$V(z) = V_0 \qquad , \quad z < 0$$
$$\qquad - \frac{Qe^2}{b}, \quad 0 < z < b \qquad\qquad (3a)$$
$$\qquad - \frac{Qe^2}{z}, \quad z > b.$$

Cole[18] choose $b = 3.6\text{Å}$ which equals the interatomic distance between the helium atoms and $V_0 = 1.58 eV$ was determined by Grimes and Brown[12] by fitting the energy spectrum to the experimental results.

(2) A potential with an *effective helium surface* a distance z_0 in the helium

$$V(z) = V_0 \qquad , \quad z < 0$$
$$\qquad - \frac{Qe^2}{z + z_0}, \quad z > 0. \qquad\qquad (3b)$$

Grimes and Brown[12] found $z_0 = 1.04\text{Å}$ for $V_0 = 1.0 eV$ by fitting the experimental results. Hipólito et al[22,23] showed that the energy spectrum of this potential can be obtained exactly. They took $V_0 = 1.0 eV$ and found that $z_0 = 1.01\text{Å}$ would lead to a perfect agreement with the transition energies as measured in the experiment of Grimes and Brown[12]. This potential is shown in Fig. 1.

(3) Stern[24] introduced a model for the liquid-gas interface of the helium at which the helium dielectric constant changes continuously (a linear dependence was assumed) across the interface layer. If $V_0 = 1.0 eV$ was chosen he found a good agreement with the experimental results of Grimes and Brown[12] if 5.7Å was taken as the width of the interface.

II.2 Motivation for the use of a thin liquid helium film

Up to now most of the experiments have been performed on bulk helium. If the thickness of the liquid helium layer is decreased one has an additional physical parameter which may influence the properties of the 2DEG. In order that the thickness(d) of the liquid He-film has an effect it is necessary that d becomes of the order of 1000 Å.

For thin liquid He-films an additional problem arises which is the *flatness of the substrate*. The substrate surface has to be flat on a scale which is much smaller than

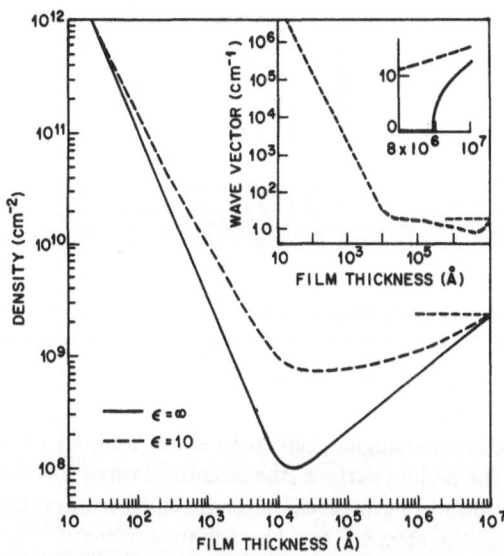

Fig. 2. Electron density at which the helium film becomes unstable for a metallic($\epsilon = \infty$) and a non-metallic($\epsilon = 10$) substrate. The inset shows the wave vector of the instability.

the helium film thickness. Otherwise the bumps in the substrate will deform the helium surface appreciably. In the following we will assume that the substrate surface is completely flat. The problem of irregular substrate surfaces is an interesting subject[25] in itself in which up to now little research has been done. The irregularities in the substrate will lead to a potential for the motion of the electrons parallel to the surface. By changing the thickness of the He-film it will thus be possible to change the strength of this potential.

The following essential *new features* are introduced by thinning down the liquid helium layer:

(1) The helium surface of a thin helium layer can support much **higher electron densities** (see Fig.2). The surface of bulk helium ($d > 2mm$) is stable for electron densities up to $n_e = 2.2 \times 10^9 cm^{-2}$. For higher electron densities an electro-dynamical instability of the charged surface occurs which leads to a macroscopic deformation of the helium surface. But for a thin He-film of thickness $d \sim 100$Å this instability occurs only when $n_e \sim 10^{11}cm^{-2}$. The stabilization of the helium surface with decreasing film thickness (at least for $d < 10^4$Å) was predicted theoretically by Gor'kov and Chernikova[26] and was later further discussed by Ikezi and Platzman[27], and is a consequence of the van der Waals forces between the helium layer and the substrate which supports the He-film. Recently Etz al[28] was able to charge a He-film of thickness 100Å with an electron density of the order of $10^{11}cm^{-2}$.

(2) **Quantum-mechanical** effects can be investigated. Because higher densities are attainable the Fermi energy $E_F = \pi\hbar^2 n_e/m$ can reach values (i.e. up to 2.8 K for $d = 100$Å) which are larger (or at least of the same order of magnitude) than the temperature, which is typically of the order of $0.1 - 1.K$. Note that for electrons on bulk helium the Fermi energy is smaller then 0.06 K and consequently they form a classical 2DEG. Thus it should be possible to investigate quantum-mechanical effects on the phase diagram for Wigner crystallization. In this way the transition from a classical Wigner lattice to a quantum Wigner lattice can be studied.

(3) The **interaction between the electrons** can be modified by changing the thickness of the He-film. For elecons on bulk helium the electron-electron interaction is described by a Coulomb interaction e^2/r, where r is the distance between two electrons. When the electrons are on a thin liquid helium film the direct Coulomb interaction between the electrons is screened by the substrate (see Fig.3). Mathematically this may be described by an image charge δe which is located a distance $d+ < z >$ from the helium surface inside the substrate ($\delta = (\epsilon_s - 1)/(\epsilon_s + 1)$ with ϵ_s the dielectric constant of the substrate). As a

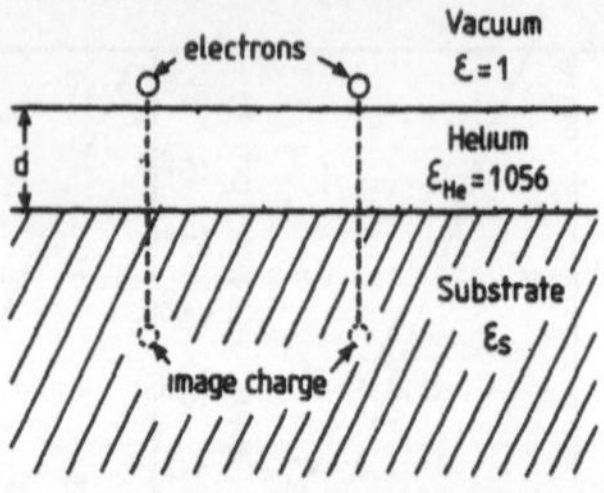

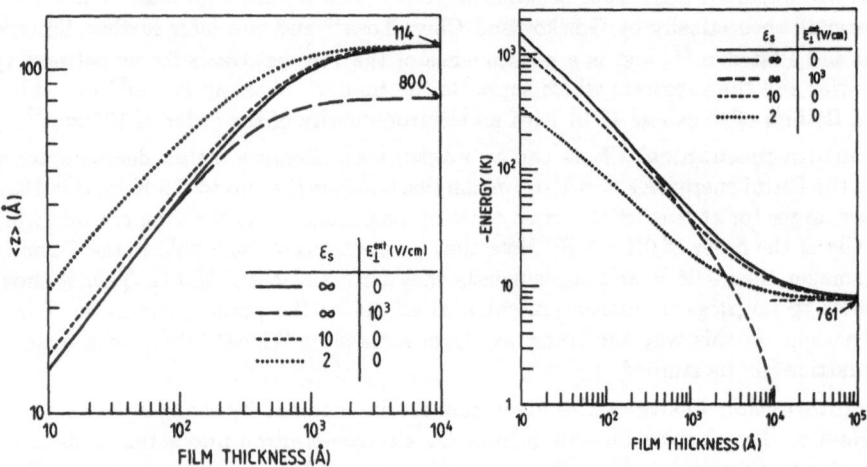

Fig. 3. Electrons above a thin liquid helium film. The image charges in the substrate are also shown.

consequence for a helium layer with a thickness smaller than the average distance between the electrons ($d \ll r$) the electron-electron interaction becomes[29]: 1) a *dipole interaction*: $2e^2 d^2/r^3$ in the case of a metallic substrate, or 2) a *renormalized Coulomb interaction*: $(1 - \delta)e^2/r$ when the substrate is non-metallic.

(4) The **electron-ripplon interaction** can be changed. Electrons above a liquid helium surface interact with the surface modes of the helium surface (the quantized version of these modes are called *ripplons*). The interaction strength of the electron-ripplon interaction increases with decreasing distance between the electron and the helium surface[3,23,30]. In Fig.4a I have plotted the average distance between the electron and the helium surface versus the thickness of the He-film for different values of the substrate dielectric constant ($\epsilon = \epsilon_s$) and in the absence and in the presence of a pressing electric field ($E_\perp^{ext}$). Note that a pressing field also decreases the average distance between the helium surface and the electron. Thus decreasing the helium film thickness or increasing the pressing field will both result in an increase of the electron-ripplon interaction. This can be shown by writing down the confinement potential of the electron in the z-direction

$$V(z) = -\frac{\Omega_0}{z} - \frac{\Omega_1}{z + d} + eE_\perp^{ext} z \quad , \tag{4}$$

where $E_\perp^{ext}$ is the pressing field. For $z \ll d$ this equation reduces to

$$V(z) = -\frac{\Omega_1}{d} - \frac{\Omega_0}{z} + eE_\perp^{eff} z \quad , \tag{5}$$

Fig. 4. Average distance of the electrons above the helium surface (Fig. 4a) and its binding energy(Fig. 4b) as function of the He-film thickness for different substrates and in the presence and in the absence of a pressing field($E_\perp^{ext}$).

with the effective electric field

$$E_\perp^{eff} = E_\perp^{ext} + E_\perp^{film} \quad , \tag{6}$$

where the *substrate induced electric field* is given by (d is expressed in Angstrom)

$$E_\perp^{film} = \frac{3.71 \times 10^8}{d^2} \frac{(\epsilon_s - 1)}{(\epsilon_s + 1)} \ V/cm$$

with

$$\frac{\Omega_1}{d} = \frac{4.18 \times 10^4}{d} \frac{(\epsilon_s - 1)}{(\epsilon_s + 1)} \ Kelvin$$

For completeness we show in Fig.4b the binding energy of the electron on the helium surface versus the helium film thickness.

III. WIGNER CRYSTAL

III.1 Wigner crystal on bulk helium

In 1934 Wigner[33] suggested that electrons in a three-dimensional Fermi system (e.g. free conduction electrons in a metal like Na) crystallize if the electron system can be expanded sufficiently such that the Coulomb potential energy dominates the Fermi energy (i.e. the average electon kinetic energy). Such an electron crystal is called a *Wigner crystal*. No clear observations have been made of such a crystal in a *three dimensional system* because no suitable experimental system has been found up to now.

In 1971 Crandall and Williams[34] realized that such a Wigner lattice could probably be realized in the *2D system of electrons on a helium surface*. The essential difference between Wigner crystallization in two dimensions with respect to Wigner crystallization in three dimensions is that in 2D the crystal is formed when the electron density is increased while in 3D the electron density has to decrease in order to reach the Wigner crystal phase. This shows clearly the importance of the dimensionality of the electron gas.

The phase diagram for Wigner crystallization in two dimensions was first calculated by Platzman and Fukuyama[35]. Later more accurate calculations were performed by applying the Kosterlitz-Thouless theory[4] for melting of a classical 2D system. The 2D crystal was for the first time observed by Grimes and Adams[5] in 1979.

The thermodynamical state of the *classical* two dimensional electon gas is determined by the dimensionless parameter $\Gamma = \sqrt{\pi n}e^2/k_B T$ which is a measure for the ratio of the average potential energy $<V> = e^2/r = e^2\sqrt{\pi n}$ versus the kinetic energy $<K> = k_B T$. For $\Gamma < 1$ the kinetic energy dominates and the system behaves like an electrongas. For intermediate densities such that $1 < \Gamma < 100$ the motion of the electrons are strongly correlated and are fluidlike. For still higher densities such that $\Gamma > 100$ the Coulomb potential energy dominates and the 2D electron system will make a phase transition to a periodic lattice structure at a critical value Γ_m. Experimentally the value of Γ_m will determine the phase diagram, and it will be this number which has to be compared with a theoretical calculation of the melting of a 2D system.

In Table II a shematic overview is given of the different theoretical determinations of the value of Γ_m for the melting of a 2D classical electron system.

Table II. Theoretical estimates for Γ_m

Theory	Γ_m	Method
Platzman and Fukuyama[35](1974)	2.7	Self-consistent phonon method
Hockney and Brown[36](1975)	95 ± 2	Molecular dynamics
Thouless[37](1979)	78	Kosterlitz-Thouless theory with T=0 results for Lamé coeff.
Gann et al[38](1979)	125 ± 15	Monte Carlo simulation
Morf[39](1979)	128.2	KTHNY-theory with $T \neq 0$ results for Lamé coeff.
Imada et al[40](1984)	120	Quantum Monte Carlo

After the theoretical prediction of the possibility of the existence of a Wigner lattice on a Helium surface in 1971 by Crandall and Williams[34] the fundamental problem arises: *how to observe such a Wigner lattice?* The electron lattice consists of electrons with a mass which is 10,000 times smaller than the mass of the constituents of a normal solid. Furthermore the lattice constant in a Wigner lattice is typical of the order of 3000Å which is several orders of magnitude larger than this for a normal solid (e.g. Al has a lattice constant of 4.05Å). The fact that the electron solid is lighter, much less dense and also much less strongly bound than a normal solid implies that different experimental techniques have to be used in order to detect whether or not the 2D electron system is in the ordered phase. Let us review some of the possibilities for detecting a Wigner crystal:

(1) Traditionally the existence of a *space ordered structure* is determined by a *diffraction experiment*. For a normal solid with a lattice constant of 3 − 4Å this will be done by a X-ray diffraction experiment. For an electron lattice with a lattice constant of 3000Å one would have to use laserlight in order to perform such a diffraction experiment. But after a calculation one finds that the cross section for scattering of light on a 2D electron lattice is too small in order to obtain an observable diffraction spectrum.

(2) The difference between a solid on the one hand and a fluid on the other hand is that a solid exhibits *phonon modes*. Thus a possiblity is to observe these transverse and/or longitudinal phonon modes by a resonant absorption of electro-magnetic radiation.

(3) A less direct method is by observing the effect of crystallization on the electron *mobility* or the *scattering time* for transport parallel to the helium surface.

Table III. Experimental results for Γ_m

Experiment	Γ_m	$T(K)$	Method
Grimes and Adams[5](1979)	131 ± 7	$0.4 - 0.7$	Longitudinal coupled electron-ripplon mode
Rybalko et al[41](1979)	137 ± 15	$0.08 - 0.3$	Mobility
Marty et al[42](1980)	$125 - 132(\pm 5)$	$0.3 - 0.9$	Longitudinal dielectric response $\chi_{xx}(k,\omega)$
Gallet et al[43](1980)	139 ± 8	$0.3 - 0.7$	Transverse sound mode
Mehrota et al[44](1982)	124 ± 4	$0.2 - 0.7$	Mobility
Kajita[45](1985)	138 ± 25	$1.4 - 2.5$	Conductivity (electrons on Ne)

Different experiments have been done over the last few years which have determined the value of Γ_m. An overview is given in Table III.

The good agreement between the experimental results and the theoretical result of Morf[39] for Γ_m gives strong support for the validity of the Kosterlitz-Thouless[4] model for the melting of a classical 2D system. According to the theory of Kosterlitz-Thouless[4], Halperin and Nelson[46], and Young[47] (KTHNY) the melting of a classical 2D solid is induced by the *formation of dislocations*. This melting process proceeds in two steps from a 2D solid with long-range orientational order and algebraic decay of positional order, through a *hexatic* phase with algebraic decay of orientational order but no positional order, to an isotropic fluid.

III.2 Wigner crystal on a thin liquid helium film

The electrostatic interaction energy[29,48] between two electrons above a helium film which is supported by a substrate with dielectric constant ϵ_s is

$$V(r) = e^2 \left(\frac{1}{r} - \frac{\delta}{\sqrt{r^2 + (2d)^2}} \right) \quad , \tag{7}$$

where $\delta = (\epsilon_s - 1)/(\epsilon_s + 1)$ and d is the distance between the 2D electron layer and the substrate. The geometrical configuration of the different layers are shown in Fig.3. The first term in Eq.(7) describes the direct electron-electron Coulomb interaction while the second term is a consequence of the screening of this interaction by the substrate.

In order to obtain the phase diagram for this 2D electron system we will apply the *Lindeman melting criterium*[49]. This criterium states that a solid will melt if the average kinetic energy $< K >$ of the electrons become a fraction $1/\Gamma_m$ larger than the average potential energy $< V >$. Thus the equation

$$\Gamma_m = \frac{< V >}{< K >} \quad , \tag{8}$$

determines the phase diagram. Within a qualitative picture the average potential energy is given by

$$< V >= e^2 \sqrt{\pi n} \left(1 - \frac{\delta}{\sqrt{1 + n/n_d}} \right) \quad , \tag{9}$$

with $n = 1/\pi r_0^2$ the electron density and $n_d = 1/4\pi d^2$. The kinetic energy of the two dimensional electron system is

$$< K >= \frac{2}{n} \int \frac{d^2 p}{(2\pi)^2} \frac{E_p}{\exp\left[\beta(E_p - \mu)\right] + 1} \quad , \tag{10}$$

with $E_p = p^2/2m$, $\beta = 1/k_B T$ where k_B is the Boltzmann constant and T the temperature.

The system of Eqs.(8-10) defines the phase diagram which is shown in Fig.5 for a metallic and a non-metallic substrate. In Fig.5 the following units where introduced: for the density $n_c = 4/\pi a_B^2 \Gamma_m^2$ and the temperature $T_c = 2e^2/\Gamma_m^2 k_B a_B$ where $a_B = \hbar^2/me^2 = 0.529$ Å is the Bohr radius. For $\Gamma_m = 137$ we find $T_c = 33.6 K$ and $n_c = 2.4 \times 10^{12} cm^{-2}$. From the above numerical results we notice that the phase diagram for Wigner crystallization on a thin helium film is strongly determined by the type of substrate which supports the helium film. This is more clearly shown in Fig.6 where we plot a T=0 cut of the phase diagram of Fig.5.

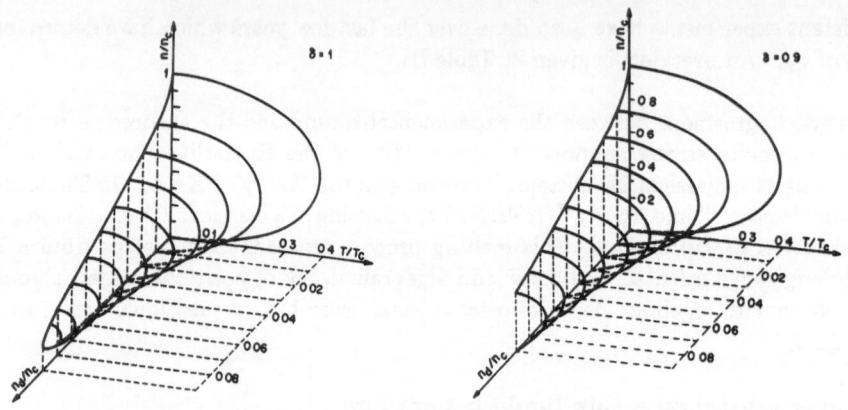

Fig. 5. Phase diagram for Wigner crystallization of electrons on a He-film which is supported by a metallic($\delta = 1$) and a non-metallic($\delta = 0.9$) substrate. Inside the curves the 2D electron system is in the solid phase.

The small-density—small-temperature region is of particular interest. Let us first consider a non-metallic substrate: the only effect of thinning down the helium film is the renormalization of the Coulomb interaction. In the limit of $d \to 0$ this implies $e \to e\sqrt{1-\delta}$ which leads to a renormalization of $n_c \to n_c^* = (1-\delta)^2 n_c$ and $T_c \to T_c^* = (1-\delta)^2 T_c$. This renormalization results in a shrinking of the phase diagram, e.g. for $\delta = 0.9$ the phase diagram for electrons on bulk helium will be reduced by a factor of 100 in the limit of $d \to 0$.

For a metallic substrate ($\delta = 1$) an important difference is found when $d \to 0$. For convenience I will consider the $T = 0$ case in order to clarify my point. For small densities ($n \ll n_d$) the potential is given by $< V > \sim d^2 n^{\frac{3}{2}}$ while the kinetic energy is $< K > \sim n$ purely quantum-mechanical. This implies that for sufficiently small densities we have $< V > \ll < K >$ and consequently the electron system is in the fluid phase. This is essentially different from the non-metallic case (i.e. $\delta \neq 1$) where $< V > \sim \sqrt{n}$ in the small density limit and thus $< V > \gg < K >$, and as a consequence the electron system will be in the crystal phase.

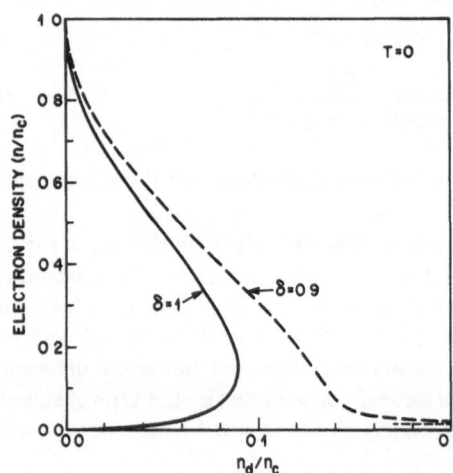

Fig. 6. T=0 cut of the phase diagrams of Fig. 5.

In the limit of high electron densities such that $n \gg n_d$ one has $<V> \sim \sqrt{n}$ and consequently $<V> \ll <K>$ which implies that the system is in the liquid phase. Thus for a fixed temperature the electron system is in the liquid state when $n < n_1$ or when $n > n_2$ and consequently the Wigner crystal can only exist for densities such that $n_1 < n < n_2$ where n_1 and n_2 are electron densities which are a function of the He-film thickness. With decreasing He-film thickness the difference $\Delta = n_2 - n_1$ decreases and reduces to zero at a critical value $d = d^*$ beyond which no Wigner crystal can exist. Our numerical results give for $\Gamma = 137$ a value $d^* = 60$ Å when $T = 0$. Note also that for a non-metallic substrate one has $n_1 = 0$ at $T = 0$ because a classical limit always exists. This is different for $\delta = 1$ where the system is always quantum mechanical at T=0!

III.3 Dynamics of the Wigner lattice

The properties of the 2DEG will be studied for the case the electron system is crystallized into a Wigner lattice. The first question to be answered is: *what is the crystal structure of the 2D electron lattice?* For electrons on bulk helium it has been shown theoretically[50] that a *hexagonal lattice*: 1) gives the lowest energy, and 2) is stable. We have shown explicitly[48] that for electrons on a thin liquid helium film this is still valid.

The energy spectrum of the Wigner lattice is obtained from the phonon spectrum. In order to calculate the phonon spectrum we define the 2×2-matrix

$$S_{ij}(\mathbf{q}) = \frac{1}{e^2} \lim_{|\mathbf{R}| \to 0} \frac{\partial^2}{\partial R_i \partial R_j} \left(\sum_l V(\mathbf{R} - \mathbf{R}(l)) e^{-i\mathbf{q} \cdot \mathbf{R}(l)} - V(\mathbf{R}) \right) \quad , \tag{11}$$

from which we obtain the dynamical matrix

$$C_{ij}(\mathbf{q}) = -\frac{e^2}{m}(S_{ij}(\mathbf{q}) - S_{ij}(\mathbf{0})) \quad , \tag{12}$$

In this harmonic analysis the normal mode frequencies are given by

$$\omega_{pm}^2(\mathbf{q}) = \frac{1}{2} \left([C_{xx}(q) + C_{yy}(q)] \pm \sqrt{[C_{xx}(q) - C_{yy}(q)]^2 + 4C_{xy}(q)C_{yx}(q)} \right) \quad , \tag{13}$$

The interaction potential $V(\mathbf{r})$ in Eq.(11) is given by Eq.(7), $\mathbf{R}(l)$ is the position of the electron at the lattice position l. The sum in Eq. (11) is over all possible lattice positions in the hexagonal lattice. This sum is slowly convergent. Therefore it is advisable to apply the *Ewald summation technique*[51] in order to convert this sum into: 1) a sum over the short-range interactions and 2) a sum over the long-range interactions, which, after a Fourier transform turns into a fast converging sum because only short wavelengths are relevant. For technical details I refer to Refs.48 and 50.

The phonon spectrum is plotted in Fig.7 along the boundary of the irreducible Brillouin zone (see Fig.8) for different values of the He-film thickness and for a metallic and a non-metallic substrate. The longitudinal (C_l) and the transverse (C_t) sound velocity $(C = \partial \omega / \partial q|_{q=0})$ are shown in Fig.9 as a function of d. The unit of frequency is $\omega_0 = \sqrt{8e^2/mb^3}$ with b the lattice vector, $r_0 = 1/\sqrt{\pi n}$ is the unit of distance and $C_0 = \sqrt{e^2/mb}$ is the unit of sound velocity. For an electron density of $n = 10^9 cm^{-2}$ one obtains $\omega_0 = 230 GHz$, $r_0 = 1800$Å and $C_0 = 2.7 \times 10^6 cm/s$.

From Figs.7 and 9 it is apparent that: 1) *screening effects* are starting to play a role when $d/r_0 \leq 1$, i.e. when the distance between the electron layer and the substrate is comparable to or smaller than the average distance between the electrons in the crystal.

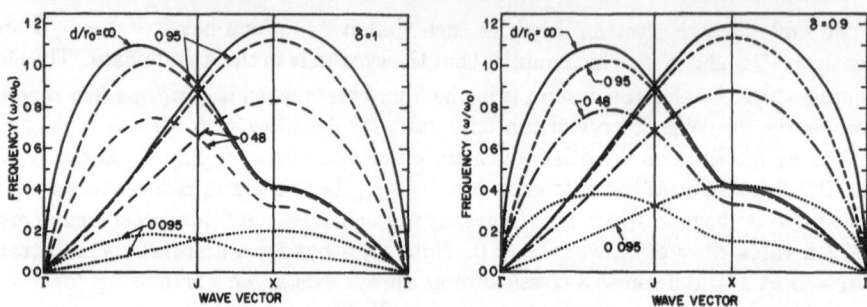

Fig. 7. Phonon spectrum along the irreducible element of the first Brillouin zone in the case of a metallic($\delta = 1$) and a non-metallic($\delta = 0.9$) substrate for different values of the He-film thickness.

Decreasing the film thickness and/or increasing the dielectric constant of the substrate softens the phonon spectrum. 2) For a non-metallic substrate the longitudinal sound velocity is always infinity. The reason is that for $\delta \neq 1$ the 2D electron system behaves, in the long-wavelength limit, as a Coulomb system and thus, as is well-known, $\omega_l \sim \sqrt{q}$ for $q \to 0$. For the case of a metallic substrate and $d/r_0 < \infty$ the two branches of the spectrum are acoustical in the long-wavelength limit, i.e. $\omega_{l,t} \sim q$ for $q \to 0$. For $\delta = 1$ the system behaves, in the $q \to 0$ limit, as a collection of *dipoles* which is the underlying reason for the acoustical nature of the longitudinal phonon mode. 3) Notice the *superlinear behavior* of the transverse mode if $d/r_0 \geq 0.5$. This was already observed in Ref.35 in the case of a 2D Wigner lattice on bulk helium.

III.4 The phase diagram for Wigner crystallization

The phase diagram as calculated in III.2 and represented in Fig.5 was presented in terms of the physical variables T, n, and $n_d = 1/\pi d^2$ which were scaled by the renormalization constants T_c and n_c. These constants depend on the parameter Γ_m which contains information on the interactions in the system. As discussed above the electron-electron interaction depends on the thickness of the helium film(d) and consequently we expect that Γ_m will depend on d.

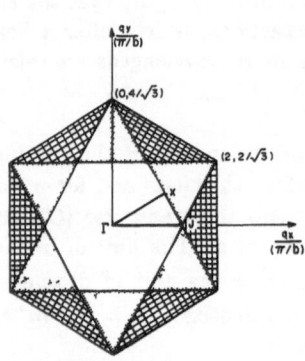

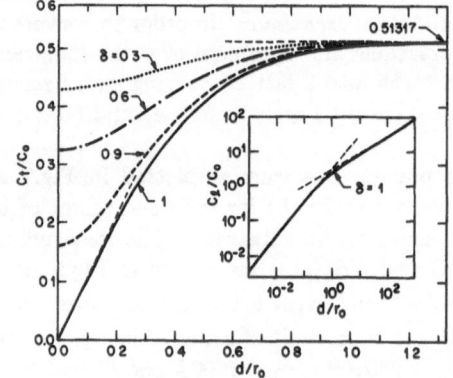

Fig. 8. First(inner blank area), second(speckled area) and third (hatchet area) Brillouin zone of the hexagonal lattice.

Fig. 9. Longitudinal(C_l) and transverse(C_t) sound velocity as function of the He-film thickness for different substrates.

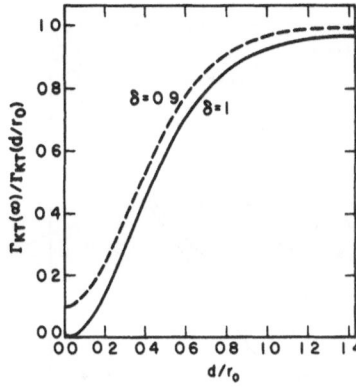

For 2D electrons on bulk He and for small temperature ($T < 1K$, i.e. the classical regime) Thouless[37] applied the Kosterlitz-Thouless[4] theory of *dislocation-mediated melting* as elaborated in greater detail by Halperin-Nelson[46] and Young[47] (KTHNY) to estimate Γ_m. He obtained

$$\Gamma_m = \frac{20.7}{(C_t/C_0)^2(1 - C_t^2/C_l^2)} \quad , \tag{14}$$

where $C_0^2 = e^2/mb$ with b the lattice constant, e the elementary charge and m the electron mass.

It should be noted that: 1) the KTHNY-theory is in essence a classical theory and is in principle not valid for a quantum-mechanical system. But, at this moment no theory exists for the melting of a quantum-mechanical 2D system. Furthermore one has found experimentally[44] that the yet found small quantum corrections to the phase diagram could well be explained by quantum corrections to the kinetic energy and did not result from a deviation of Γ_m from its classical value. 2) If one takes the zero temperature value for the sound velocities (i.e. $C_t/C_0 = 0.513, C_l = \infty$) Thouless[37] found for bulk He $\Gamma_m = 79$ which is smaller than the experimental value $\Gamma_m \simeq 137$. Morf[39] showed that the experimental value for Γ_m could be explained: i) if one corrects for the temperature dependence of the sound velocities(which is a consequence of phonon-phonon interactions), and ii) if one takes into account the polarizability of the dislocation pairs. In order to correct(roughly) for these effects for thin He-films we will define

$$\Gamma_m(d) = \Gamma_m(\infty)\frac{[C_t(\infty)/C_t(d)]^2}{1 - [C_t(d)/C_l(d)]^2} \quad , \tag{15}$$

where we will take for $\Gamma_m(\infty)$ the experimental result 137 and for C_t and C_l the zero temperature results of forgoing section. In this way we expect that the correct d–dependence is found.

The behaviour of $1/\Gamma_m$ is shown in Fig. 10 as function of the thickness of the He-film and for a metallic($\delta = 1$) and a non-metallic($\delta \neq 1$) substrate.

Once we know the numerical value for Γ_m it is possible to express the phase diagram for Wigner crystallization in real units. The result is shown in Fig.11 for a metallic and a non-metallic substrate and for different values of the thickness of the helium film. The 2D electron system is in the crystal phase above the phase boundary line for the respective He-film thicknesses. The point on each line indicates the density at which the helium surface becomes instable; in particular higher densities lead to a macroscopic deformation of the

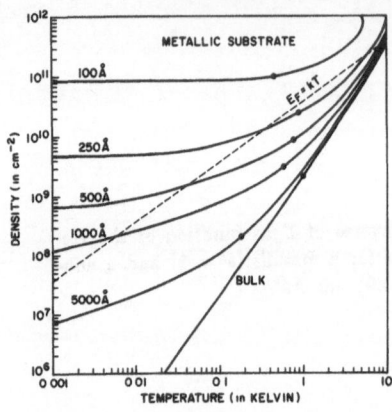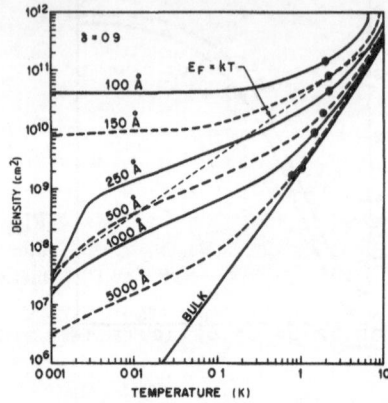

Fig. 11. Phase diagram for Wigner crystallization for a metallic and a non-metallic substrate and for different values of the He-film thickness. The thin dashed line indicates the condition $E_F = k_B T$.

He-surface. From these figures it is apparent that screening effects (as due to the substrate) are starting to become important when the He-layer is thinner than $10,000\text{Å}$. Furthermore for very thin He-layers the phase diagram is pushed into the quantum mechanical region. The division between the quantum mechanical and the classical region is indicated by the dashed line $E_F = k_B T$. For densities above this division line the 2D electron system behaves as a quantum mechanical system. From Fig.11 we note that for a temperature of $0.4K$ the phase transition will exhibit quantum mechanical effects when the He-layer is of the order of 200Å.

It is remarkable that for a metallic substrate there exists a fluid phase in the low density region even for $T \to 0$. This new interesting fact is a consequence of the dipole nature of the electron-electron interaction for $n \to 0$. From the study of Ref.32 we know that typical helium surface excitation energies are of the order of $10^{-2}K$ and thus we expect a *BCS superfluid phase transition* of the 2D electron liquid.

IV. SELF-TRAPPING

For Wigner crystallization the many-particle aspects of the 2DEG was crucial. The importance of the helium surface is only secondary in order: 1) to provide a 2D surface (i.e. to confine the electrons to two dimensions) and 2) the properties of the He-surface were used in order to observe experimentally the Wigner crystal state. In the *self-trapping* problem it is sufficient to limit ourselves to *one electron* and the deformability of the He-surface is crucial. Many-particle effects are only of secondary importance but are not expected to have any qualitative effect on the phenomenon studied.

In the present chapter we will investigate the one particle properties of one electron moving in 2D and interacting with the surface excitations of a He-film (called *ripplons*). By changing the He-film thickness or the pressing field it is possible to change the electron-ripplon interaction. This system is a nice example of a field theoretical problem in which the interaction between a particle and a (scalar) field can be varied experimentally in a continuous way and over a large range of coupling strength, i.e. from *weak to strong coupling*. Experimentally this is observed as a dramatic change of the mobility(see Figs.12 and 13) with changing He-film thickness. We will try to understand the behaviour of the mobility as shown in Fig.12.

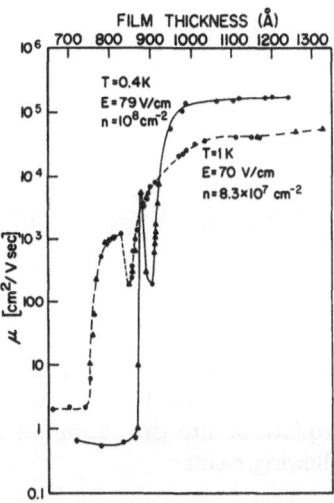

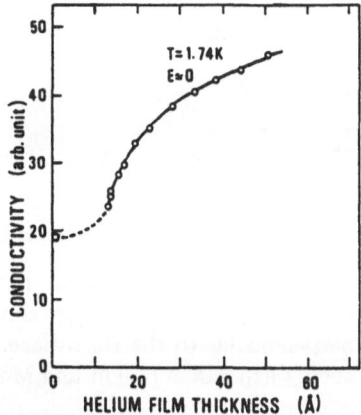

Fig. 12. Mobility as function of the He-film thickness as measured by E. Andrei[8] for electrons on a sapphire substrate.

Fig. 13. Mobility as function of the He-film thickness measured by K. Kajita[52] for electrons on a neon substrate.

IV.1 The Hamiltonian for the electron-ripplon interaction

A liquid surface in rest will be perfectly flat (if we neglect boundary effects) due to gravitational forces. A slight perturbation of such a surface will be visible as ripples on this surface. It is possible to quantize these surface excitations which then are called *ripplons*. Sometimes one also calls them *quantized capillary gravitational-waves*.

If the He-surface is not charged these ripplons have the dispersion law

$$\omega_k^2 = k \left(g' + \frac{\sigma}{\rho} k^2 \right) \tanh kd \quad , \tag{16}$$

with k the wave vector of the ripplons, d the thickness of the He-film, $\sigma = 0.378 erg/cm^2$ is the helium surface tension, $\rho = 0.145 g/cm^3$ is the helium mass density and $g' = g + 3c/\rho d^4$ is the *effective* gravitation constant with g the gravitation constant and c the *van der Waals* coupling constant of the helium to the substrate (a typical value is $c = 9.5 \times 10^{-15} erg$ in the case of a glass substrate). This dispersion law has been verified experimentally by e.g. Wanner and Leiderer[53]. In the case of a thin helium film the above dispersion law reduces to

$$\omega_k = s \cdot k \quad , \tag{17}$$

with $s = \sqrt{g'd}$.

An electron above a He-surface is pressed against this surface because of the interaction between the electron and the image charge and by a possible external applied pressing field. As a consequence the helium surface underneath the electron will be deformed.

Applying the concept of ripplons it is possible to express an arbitrary small deviation of the He-surface $\xi(\mathbf{r})$ in terms of ripplon creation $(a_{\mathbf{k}}^+)$ and annihilation $(a_{\mathbf{k}})$ operators

$$\xi(\mathbf{r}) = \frac{1}{\sqrt{A}} \sum_{\mathbf{k}} \phi(k) \left(a_{\mathbf{k}} e^{i\mathbf{k}\cdot\mathbf{r}} + a_{\mathbf{k}}^{+} e^{-i\mathbf{k}\cdot\mathbf{r}} \right) \quad , \tag{18}$$

with A the surface area and

$$\phi(k) = \sqrt{\frac{\hbar k \tanh kd}{2\rho\omega_k}} \quad . \tag{19}$$

The potential energy due to this deformation is given by

$$U = \delta U + eE_{\perp} \cdot \xi(\mathbf{r})$$

which consists of two parts: i) δU is a consequence of the polarization interaction and ii) $eE_{\perp} \cdot \xi(\mathbf{r})$ is the energy change as a consequence of the electric field $E_{\perp}$ which is perpendicular to the He-surface. Inserting the above equations into the above potential energy expression Shikin and Monarkha[54] found the following result

$$U = \sum_{\mathbf{k}} \left(V_{\mathbf{k}} a_{\mathbf{k}} e^{i\mathbf{k}\cdot\mathbf{r}} + V_{\mathbf{k}}^{*} a_{\mathbf{k}}^{+} e^{-i\mathbf{k}\cdot\mathbf{r}} \right) \quad , \tag{20}$$

with the interaction coefficient $V_{\mathbf{k}} = \phi(k)E_k(z)/\sqrt{A}$ where the function $E_k(z)$ is given by

$$E_k(z) = \Lambda_0 \frac{k}{z} \left(\frac{1}{kz} - K_1(kz) \right) + eE_{\perp} \tag{24}$$

and $K_1(x)$ is the modified Bessel function.

Putting all the above terms together the electron-ripplon system is described by the Hamiltonian

$$H = \frac{\mathbf{p}^2}{2m} + \sum_{\mathbf{k}} \hbar\omega_k a_{\mathbf{k}}^{+} a_{\mathbf{k}} + U \quad , \tag{21}$$

with U the electron-ripplon interaction as given by Eq. (20). The electron is characterized by its mass m and the position($\mathbf{r}$) and the momentum($\mathbf{p}$) operator. Because we are interested in thin liquid helium films we will approximate the ripplon frequency by its linear expression (17). The polarization interaction δU is small in comparison to the applied electric field $E_{\perp}^{ext}$ and the field due to the image charge of the substrate. The latter one is given by $e\Lambda_1/4d^2$ with $\Lambda_1 = (\epsilon_s - 1)/(\epsilon_s + 1)$ which is of the order of 1 and ϵ_s is the dielectric constant of the substrate. For a thin liquid helium film $E_k(z)$ is well approximated by $eE_{\perp}$.

For a thin liquid helium film the Hamiltonian is characterized by the ripplon frequency $\omega_k = \omega_0 \eta k/k_c$ and the interaction coefficient

$$|V_k|^2 = \frac{2\pi\alpha\eta}{A} k \frac{\hbar^3\omega_0}{mk_c} \theta(k_c - k) \quad , \tag{22}$$

where only ripplons with wave vector smaller than the capillary constant $k_c = \sqrt{\rho g'/c}$ are relevant[6](in the following k_c is taken as the unit of wave vector). The unit of energy is taken as $\hbar\omega_0 = \hbar^2 k_c^2/2m$. The parameter $\eta = sk_c/\omega_0$ is a measure of the adiabatical character of the system. In the above expression (22) we introduced the coupling constant[6]

$$\alpha = \frac{(\rho E_{\perp})^2}{8\pi\sigma} \frac{1}{\hbar\omega_0} \quad . \tag{23}$$

For a typical He-film thickness of $d = 100\text{Å}$ one obtains $k_c = 2.7 \times 10^5 cm^{-1}, \hbar\omega_0 \simeq 0.3K$ and $\eta \simeq 2.5 \times 10^{-3}$.

It is interesting to note that the Hamiltonian which describes the electron-ripplon interaction has the same structure as the *Fröhlich Hamiltonian* (see the lectures of Prof. J. Devreese) which describes polarons in polar semiconductors and ionic crystals. Furthermore in the limit of small He-film thickness the electron-ripplon Hamiltonian reduces to the Hamiltonian of a *two-dimensional acoustical polaron*.

IV.2 The ground-state properties

Jackson and Platzman[6] applied the *Feynman path-integral technique* in order to calculate the ground state energy and the effective mass of such a polaronic electron. The Feynman path-integral technique has been proven to be a powerfull technique in the investigation of statical and dynamical properties of a system in which one electron interacts with a scalar field. The power of this technique lies in the ability to eliminate the ripplon variables exactly. In this way the problem reduces to a *one particle problem* but now the electron exhibits a non-local interaction in time. In order to calculate the ground-state energy it is necessary to make an approximation. Therefore Feynman[55] introduced a *two-particle model* which approximates the non-local electron interaction by a non-local quadratic interaction. This *Feynman model* consists of an electron which interacts quadratically with a second particle (called *fictitous particle*) which, in the present problem, simulates the deformation of the He-surface. When the coordinates of this second particle is eliminated one finds a model action which is used in order to calculate the ground-state energy. Furthermore Feynman showed that the ground-state energy obtained in this way is an upper bound to the exact ground-state energy. The parameters which characterizes the Feynman model (v,w) can then be obtained by a variational calculation of the ground-state energy.

The above discussion is also valid at non-zero temperature. In that case the quantity which one has to calculate is the *Helmholtz free energy*. Applying the Feynman approximation one obtains the free energy[56] $F = F_0 - F_1 - F_2$ with

$$F_0 = \frac{2}{\beta} ln \left(\frac{\sinh \frac{\beta v}{2}}{\sinh \frac{\beta w}{2}} \right) - \frac{2}{\beta} ln \frac{v}{w} \quad , \tag{24}$$

$$F_1 = \frac{v^2 - w^2}{2v} \left(\frac{-2}{\beta v} + \coth \frac{\beta v}{2} \right) \quad , \tag{25}$$

$$F_2 = \frac{1}{2\pi} \int_0^{\beta/2} d\tau \int_0^\infty dk \, k|Q(k)|^2 \frac{\cosh \omega_k (\frac{\beta}{2} - \tau)}{\sinh \omega_k \frac{\beta}{2}} e^{-k^2 D(\tau)} \quad , \tag{26}$$

where $|Q(k)|^2 = A|V_k|^2$ and

$$D(\tau) = \frac{w^2}{2v^2} \tau (1 - \frac{\tau}{\beta}) + \frac{v^2 - w^2}{2v^3} \left(1 - e^{-v\tau} - 4n(v) \sinh^2 \frac{v\tau}{2} \right) \quad . \tag{27}$$

In the limit of zero temperature the free energy equals the ground-state energy.

A numerical variational calculation leads to the results as shown in Fig.14 where we have plotted the free energy $\Delta F = F(\alpha) - F(\alpha = 0)$ [here $F(\alpha = 0)$ is the free energy of a non-interacting electron] and the first$(\partial \Delta F/\partial \alpha)$ and second$(\partial^2 \Delta F/\partial \alpha^2)$ derivative of the free energy to α, as a function of the electron-ripplon coupling constant. Numerical results are displayed for different values of the temperature. The temperature is expressed in units

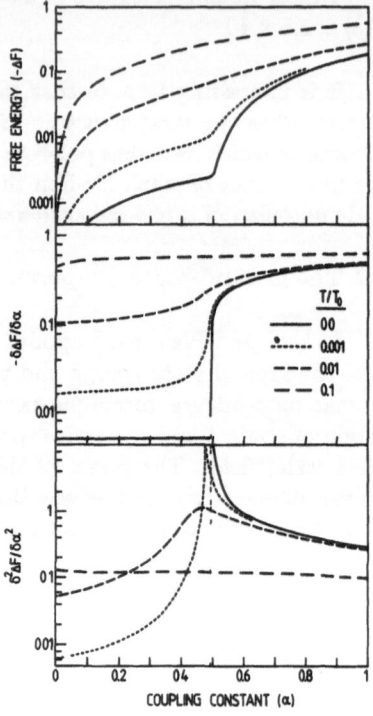

Fig. 14. The free energy and the first and second derivative of the free energy with respect to the electron-ripplon coupling constant (α) as function of α for different relative temperatures.

of $T_0 = \hbar\omega_0/k_B \simeq 0.3K$. From Fig.14 we note that for $T = 0$ the ground-state energy exhibits a discontinuous first derivative at $\alpha = 0.49 \pm 0.01$. With increasing temperature this discontinuity disappears and ΔF is continuous with continuous derivatives. For low temperatures (i.e. $T/T_0 \leq 0.01$) the second derivative $\partial^2\Delta F/\partial\alpha^2$ exhibits a peak around $\alpha \sim 0.48$. For $T = 0$ the second derivative is $\simeq 0.001$ when $\alpha < 0.49$.

This discontinuity is a consequence of a *self-trapping* transition. For $\alpha < 0.5$ the electron is weakly coupled to the ripplons and the electron is a *quasi − free* electron. The mass renormalization is only a few procent(see Fig.1.8 of Ref.3). For $\alpha > 0.5$ the electron is strongly coupled to the ripplons and is *self-trapped*. The electron effective mass is several orders of magnitude larger than the free electron mass. This self-trapping transition is found, in the present approximation, to be a discontinuous transition when $T = 0$ and becomes a continuous one for non-zero temperatures. The effect of a magnetic field on this self-trapping transition was discussed in Ref.57. For completeness we also give in Fig.15 the free energy as a function of the temperature for two values of α. The temperature dependence of the electron-ripplon contribution to the specific heat ($C = -T\partial^2 F/\partial T^2$) is depicted in Fig.16.

One refers to the above transition as a self-trapping transition because the electron is localized in a *potential well* which has been created by the electron itself. This potential well can be calculated by calculating the expectation value of the potential operator of an electron in the field of ripplons

$$V(\mathbf{r}) =< \Phi(\mathbf{x} - \mathbf{r}) > \quad , \tag{28}$$

where the potential operator is given by the electron-ripplon interaction term $\Phi(\mathbf{x}) = U(\mathbf{x})/e$ with the operator $U(\mathbf{x})$ given by Eq.(20). In order to calculate the expectation value in Eq. (23) we made use of the Feynman polaron model. The result is given by(for an explicit calculation I refer to Ref.58)

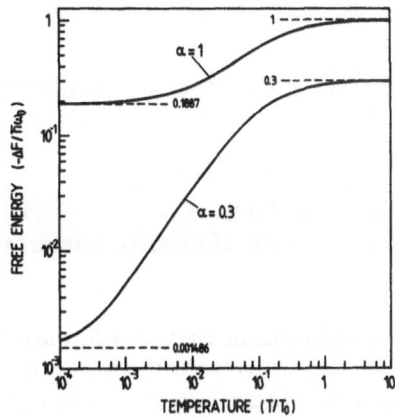

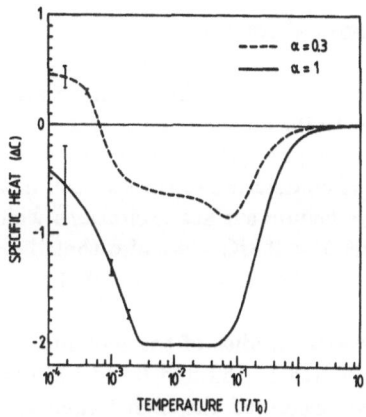

Fig. 15. The free energy as function of the temperature for an electron in the quasi-free state and in the self-trapped state.

Fig. 16. Same as Fig. 15 but now for the specific heat. In the low temperature region our numerical results are less acurate which is indicated by the error bars.

$$V(\mathbf{r}) = -\frac{2}{e} \sum_{\mathbf{k}} |V_k|^2 e^{-\mathbf{k} \cdot \mathbf{r}} \int_0^\infty d\tau e^{-\omega_k \tau} e^{-k^2 D(-i\tau)} \quad . \tag{29}$$

A computer calculation has been performed and leads to the result given in Fig.17. The potential(in units of $\hbar\omega_0/e$) is plotted as function of the distance with respect to the average position of the electron for the electron in the quasi-free state(i.e. $\alpha = 0.1$) and for the electron in the self-trapped state(i.e. $\alpha = 1.$). We note that: i) the electron potential for a self-trapped electron is much deeper (e.g. there is a factor 1000 difference between the situation with $\alpha = 0.1$ and $\alpha = 1.$), and ii) the behaviour at large distances is essentially different. Namely the potential in the quasi-free state is always negative for any r while, for the self-trapped state $V(r)$ oscillates around zero for $r/r_0 > 5$. These oscillations have a periodicity $\Delta r \sim 2\pi r_0 = 2\pi/k_c$ and are thus a consequence of the finite *cut-off*(k_c) in the ripplon spectrum. From Fig.17 we conclude that the potential for a self-trapped electron is deeper and of shorter range than that for a quasi-free electron.

Fig. 17. The electron self-induced potential versus the distance from the electron average position for an electron in the quasi-free and in the self-trapped state.

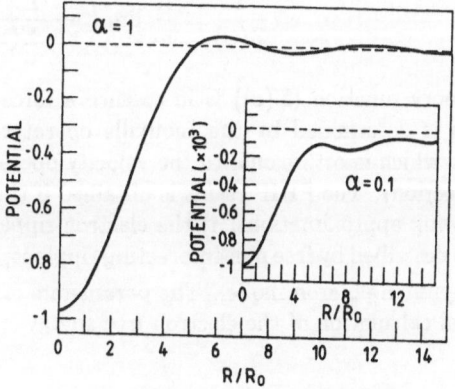

IV.3 Dynamical response

Electrons above a helium surface are free to move in the 2D plane parallel to the helium. Due to the almost ideal character of the 2DEG there are only two relevant scattering mechanisms present: i) scattering of electrons on helium gas atoms above the surface. This mechanism determines the electron mobility for $T > 0.8K$. ii) Scattering of the electrons on the helium surface excitations which limits the mobility to $\mu \sim 10^7 cm^2/Vs$ for temperatures $T < 0.8K$. Note also that the mobility is an order of magnitude larger than the highest mobility observed in 2DEG semiconductor structures.

Theoretical studies of the mobility for a time independent electric field have been performed in Refs.54, 59 and 60 in the case of electrons above bulk helium. Platzman et al[61] have calculated the high frequency conductivity. Measurements of the electron mobility have been performed by different groups[62,63] and good agreement with theory[60] was found for the temperature dependence of the mobility.

Here the response of an electron above a thin liquid helium film will be studied when an oscillating electric field is applied which is taken parallel to the He-surface. Thus a frequency dependent mobility will be calculated. Our aim is to study the effect of the self-trapping transition on the mobility and therefore we will only consider scattering with ripplons. Another motivation is that only low temperatures will be considered where scattering with the helium gas atoms is less important.

An oscillating electric field with frequency ω induces a current

$$\mathbf{j}_\omega(t) = \frac{1}{Z(\omega)} \mathbf{E}_0 e^{i\omega t}$$

in the plane parallel to the He-surface. Experimentally one measures the mobility $(\mu(\omega))$ which is defined as the real part of the inverse impedance function $(Z(\omega))$ $\mu(\omega) = \mathrm{Re}[1/Z(\omega)]$.

Applying Feynman path-integral techniques Feynman, Hellwarth, Iddings and Platzman[6] calculated the impedance function of an electron interacting with the longitudinal-optical phonons of a lattice. They determined the impedance function by calculating the current to first order in the electric field. A different but equivalent result was found recently by starting directly from linear response theory. Kubo[65] showed that the conductivity could be expressed as a velocity-velocity correlation function which has to be determined at thermodynamic equilibruim. In Ref.66 it was shown that applying Kubo's linear response theory in combination with the Mori-Zwanzig[67] projection operator technique it is possible to write the impedance function as

$$\frac{1}{Z(\omega)} = \frac{ie}{m} \lim_{\epsilon \to 0} \left(\frac{1}{z - \Sigma(z)} \right)_{z=\omega+i\epsilon} \quad , \tag{30}$$

The memory function $(\Sigma(z))$ is in essence a force-force correlation function whose time evolution is determined by the Liouville operator which is projected onto the space of operators which is orthogonal to the velocity operator $\dot{x}$ (the electric field is chosen along the x-direction). The FHIP result is obtained if the memory function is calculated within the following approximations: i) the electron-ripplon system is decoupled, ii) the ripplon system is described by free non-interacting ripplons, and iii) the electron system is described by the Feynman polaron model. The parameters of this model(v,w) were determined by a variational calculation of the electron free energy.

For the electron-ripplon system we obtained[68] the following result for the memory function

$$\Sigma(z) = \frac{1}{z} \int_0^\infty dt \left(1 - e^{izt}\right) \mathrm{Im} S(t) \quad , \tag{31}$$

where the imaginary part of the function

$$S(t) = \frac{1}{2\pi m\hbar} \int_0^\infty dk \; k^3 |\phi(k)|^2 \left((1 + n(\omega_k))e^{i\omega_k t} + n(\omega_k)e^{-i\omega_k t}\right) e^{-k^2 H(t)} \quad , \tag{32}$$

has to be calculated and where we defined the function

$$H(t) = D(-it) = \frac{w^2}{2v^2}\left(-it + \frac{t^2}{\beta\hbar}\right) + \frac{v^2 - w^2}{2v^3}\left(1 - e^{ivt} + 4n(v)\sin^2\frac{vt}{2}\right) \tag{33}$$

with $n(\omega)) = 1/(\exp\beta\hbar\omega - 1)$ the occupation number and $\beta = 1/k_B T$ the inverse temperature. In the limit of zero temperature (i.e. $\beta \to \infty$) the above results reduce to the results of Ref.58 (see also Ref.31).

It is impossible to calculate the integrals in Eq.(31) analytically. Even the numerical calculation of these integrals is quite involved because the integrand of Eq.(31) is a strongly oscillating function of the time (t). For details on the numerical evaluation of these integrals I refer to Ref.68.

IV.3.1 The Quasi-free electron state

For an electron-ripplon coupling constant $\alpha < 0.49$ the electron is quasi-free and as a consequence the mobility is large and the frequency dependence of the mobility is approximately described by a Drude-like behaviour. In Fig.18 we give the mobility as a function of the frequency of the applied electric field for different values of the temperature. The electron-ripplon coupling constant is taken to be $\alpha = 0.1$. The unit for the mobility is $\mu_0 = e/m\omega_0 \simeq 4.5 \times 10^5 cm^2/Vs$, for frequency $\omega_0 \simeq 6.3 GHz$ and for temperature $T_0 = 0.3 K$.

From Fig.18 we see that the mobility is independent of the frequency for $\omega/\omega_0 < T/T_0$. For higher frequencies the mobility decreases with increasing frequency. Further-

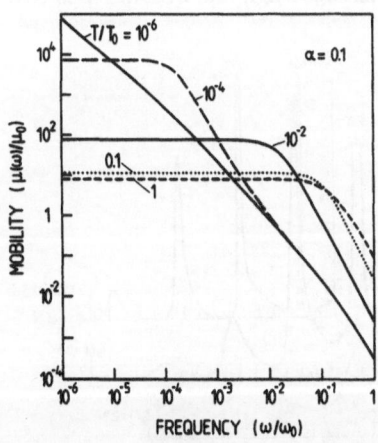

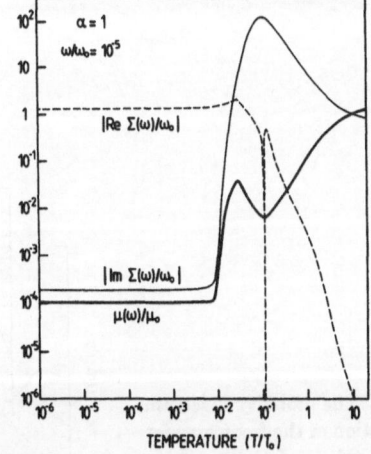

Fig. 18. The electron mobility as function of the frequency of the applied electric field in the weak coupling case for different temperatures.

Fig. 19. The electron mobility and the real and imaginary part of the memory function as function of the temperature in the strong coupling case.

more the mobility is a decreasing function of the temperature when the applied frequency of the electric field is small. This behaviour is typical for a diffuse particle.

IV.3.2 The self-trapped state

1. Mobility for small frequencies

When the electron-ripplon coupling constant is larger than $\alpha > 0.5$ the electron is localized in a self-induced potential well. In this region the electron effective mass is very large and the electron mobility is extremely small. The temperature dependence of the mobility for $\alpha = 1$ and for a small frequency of the applied field(i.e. $\omega/\omega_0 = 10^{-5}$) is shown in Fig.19. Note that the linear mobility of the self-trapped electron is temperature independent when $T/T_0 < 0.01$. If for the moment we forget about the anomalous behaviour around $T/T_0 \sim 0.01$ we find that for $T/T_0 > 0.01$ the mobility is an increasing function of the temperature. This temperature dependence of the mobility is *opposite* to the temperature dependence of the mobility when the electron is in the quasi-free state!

In order to understand the origin of the anomaly around $T/T_0 \sim 0.01$ we have plotted in Fig.19 also the imaginary and real part of the memory function. For low temperatures, i.e. $T/T_0 < 0.025$, $|\mathrm{Re}\Sigma(\omega)| \gg |\mathrm{Im}\Sigma(\omega)|$ and as a consequence the mobility is given by

$$\mu(\omega) \simeq \frac{|\mathrm{Im}\Sigma(\omega)|}{[\mathrm{Re}\Sigma(\omega)]^2}$$

Thus the mobility increase in the temperature range $0.01 - 0.025$ is a consequence of the increase in the scattering frequency $|\mathrm{Im}\Sigma(\omega)|$. The electron mobility for $T/T_0 < 0.025$ is predominantly limited by the small reactive response of the electron which is due to its large effective mass. For $T/T_0 > 0.025$ one has $|\mathrm{Re}\Sigma(\omega)| < |\mathrm{Im}\Sigma(\omega)|$ and the mobility becomes,

$$\mu(\omega) \simeq \frac{1}{|\mathrm{Im}\Sigma(\omega)|},$$

roughly the mobility of a Brownian particle with a mass equal to the mass of a free electron and a scattering time $\tau \sim 1/|\mathrm{Im}\Sigma(\omega)|$. When $0.025 < T/T_0 < 0.11$ the scattering frequency increases with increasing temperature and as a consequence the mobility decreases. At $T/T_0 = 0.11$ the real part of the memory function changes sign and becomes positive when

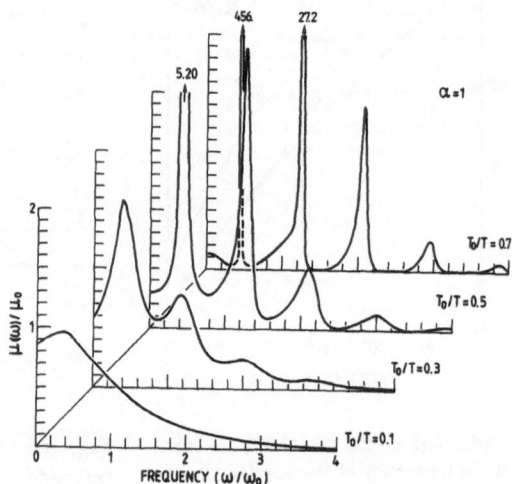

Fig. 20. The mobility spectrum as a function of the frequency for different values of the temperature for $\alpha = 1$.

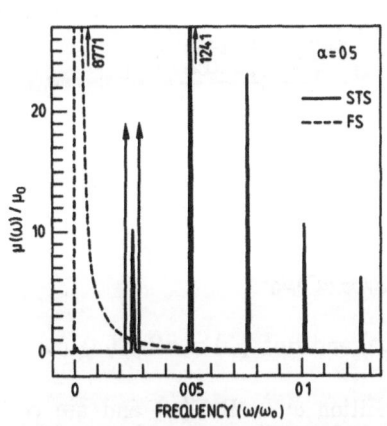

Fig. 21. Mobility at $T = 0$ for the electron in the quasi-free(FS) and in the self-trapped state(STS).

Fig. 22. Imaginary and real part of the memory function for the electron in the self-trapped state at $T = 0$.

$T/T_0 > 0.11$. The mobility increases with increasing temperature for $T/T_0 > 0.11$ as a consequence of the decreasing scattering frequency.

2. The collective electron-ripplon states (CERS)

For low frequencies $\mu(\omega)$ gives the mobility of the electron. For larger frequencies the applied electric field is due to microwave radiation and $\mu(\omega)$ should be interpreted as a measure for the absorption of this radiation.

The absorption spectrum for electrons interacting with ripplons is given in Fig.20 for $\alpha = 1$ and for different values of the inverse temperature. The spectrum has a series of peaks which corresponds to transitions of the electron within the self-induced potential well. Thus the absorption spectrum gives information on the excited state of the electron-ripplon compound. The position of the different peaks and the line widths are temperature dependent.

To get a more clear view of the origin of the different peaks in the absorption spectrum it is instructive to consider the zero temperature limit. In Fig.21 I show the absorption spectrum for $\alpha = 0.5$ at $T = 0$. The spectrum consists of two different peaks: i) peaks at a frequency $\omega/\omega_0 \simeq n\nu(n = 0, 1, 2, ...)$ with a width of the order of $4.5 \times 10^{-4}\omega_0$. These peaks correspond to transitions between Franck-Condon(FC) states. They are a result of excitations of the electron in a rigid solid potential. Here they are the states of our harmonic oscillator model potential(i.e. Feynman polaron model) which were used in the path-integral calculation of the mobility. Mathematically they correspond to peaks in the imaginary part of the memory function(see Fig.22). ii) Peaks which have the form of a Dirac-delta function(which are indicated by arrows in the figure) and which for $\alpha = 0.5$ are located at $\omega/\omega_0 = 0.02275$ and 0.02844 with a width $< 10^{-37}\omega_0$ and $1.6 \times 10^{-21}\omega_0$ respectively. These peaks occur when $\omega = \text{Re}\Sigma(\omega)$ in a frequency range where $|\text{Im}\Sigma(\omega)|$ is small. In the limit of $|\text{Im}\Sigma(\omega)| \to 0$ those peaks are represented by δ-functions

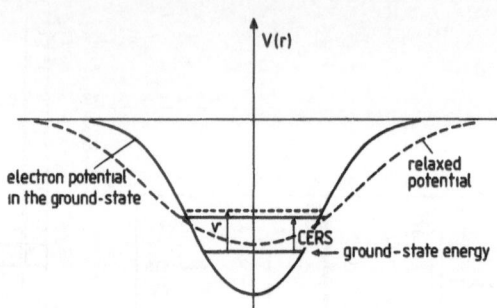

Fig. 23. The electron self-induced potential(qualitative picture) for the electron in the ground-state (solid curve) and in the RES (dashed curve).

$$\simeq \frac{\pi}{|1 - \mathrm{Re}\Sigma'(\omega^*)|}\delta(\omega - \omega^*)\mu_0$$

with ω^* the solution of the equation $\omega^* = \mathrm{Re}\Sigma(\omega^*)$ and $\mathrm{Re}\Sigma'(\omega) = \partial\mathrm{Re}\Sigma(\omega)/\partial\omega$.

The δ-function peaks result from the condition $\omega = \mathrm{Re}\Sigma(\omega)$ and are *collective electron-ripplon states (CERS)* which are the analogs of plasmon excitations in an electron gas, which occur when $\mathrm{Re}\epsilon(\omega) = 0$. These states are not unlike the *relaxed excited states*[69] *(RES)* occurring in the LO-phonon polaron problem, except that there are two CERS's on either side of the FC peak, in contrast to one RES. The CERS's result from a self-consistent cooperative interaction of the electron and ripplons. The form of the electron self-induced potential is determined by the electron wavefunction and the interaction of the electron with the ripplons. If the potential was rigid the transition frequency would be v, but the electron in the excited state will have a different wave function which in its turn leads to a different self-induced potential(see Fig.23). When the electron is excited from the ground-state to an excited state the potential will also change. The potential will relax and this is the reason why the excitation frequency $\omega \neq nv$ with $n = 1, 2, ...$ In fact one expects that the relaxed excited state has an energy which is smaller than the FC energy. This CERS peak has for $\alpha < 1.6$ also the largest oscillator strength.

IV.3.3 Influence of the self-trapping transition on the mobility

In all measurements of the mobility of electrons on a He-surface ac-electric fields are used with a frequency typically of the order of $f = 22 - 100kHz$. In our units this corresponds to $\omega/\omega_0 = (0.4 - 1.6)10^{-5}$ for $\hbar\omega_0 = 0.3K$. In Fig.24 we show the mobility as a function of the electron-ripplon coupling constant for four different values of the temperature.

When we make a qualitative comparison between our theoretical results and the experimental results of E. Andrei[8](see Fig. 12) we notice the following similarities: i) with increasing α(i.e. decreasing He-film thickness) the mobility drops by several orders of magnitude at the self-trapping transition(this is at $\alpha \sim 0.5$) when $T/T_0 \ll 1$. The drop in the mobility decreases with increasing temperature. ii) In the quasi-free state (i.e. when $\alpha < 0.5$) the electron mobility decreases with increasing temperature, while in the self-trapped state (i.e. $\alpha > 0.5$) the mobility increases with increasing temperature. iii) For very small temperature (i.e. when $T/T_0 \sim 10^{-3}$) the mobility exhibits an anomalous behaviour at the self-trapping transition point. The actual transition occurs in two steps: the mobility decreases with increasing α, but before the largest drop in μ occurs the mobility increases first with almost an order of magnitude. Because no hysteresis was found

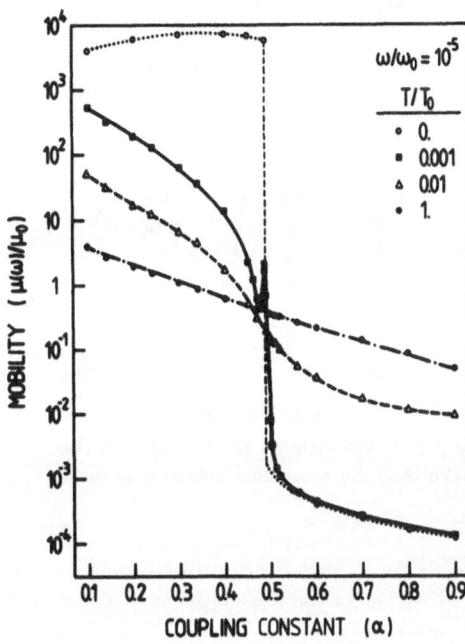

Fig. 24. The mobility as function of the electron-ripplon coupling strength(α) for different values of the temperature and at a fixed value of the frequency of the applied electric field.

experimentally it is believed that the self-trapping transition is not a first-order transition. This agrees with the Feynman path-integral approach for $T \neq 0$ which is in contrast to other theories[23,70,71] who always predict a first-order transition and consequently also a hysteresis.

At a temperature $T/T_0 \sim 10^{-3}$ we found theoretical that the mobility exhibits a local maximum at the self-trapping transition as was also found experimentally. In order to understand this surprising result we have plotted in Fig.25 the mobility and the real and imaginary part of the memory function for temperatures around $T/T_0 \sim 10^{-3}$ and for an α-range around the self-trapping transition. Notice also that the local maximum in μ shifts towards smaller α values (i.e. thicker He-films) with decreasing temperature as is in agreement with the experiment of Andrei[8]. The local maximum in the mobility can be understood in terms of the α-dependence of the real and imaginary part of the memory function. For $\alpha < 0.49$ (thus when the electron is in the quasi-free state) one has $|\mathrm{Im}\Sigma(\omega)| > |\omega - \mathrm{Re}\Sigma(\omega)|$ and the mobility is roughly given by $\mu \simeq 1/|\mathrm{Im}\Sigma(\omega)|$. It is the mobility of a Brownian particle with mass equal to the bare electron mass and with a scattering time $\tau = 1/|\mathrm{Im}\Sigma(\omega)|$. τ decreases with increasing α when $\alpha \leq 0.48$ and consequently the mobility increases. For α larger than 0.49 the scattering time increases with increasing α and consequently the mobility decreases with increasing α if $0.48 \leq \alpha \leq 0.49$. Also note that at $\alpha \simeq 0.48$ the real part of the memory function changes sign. When $\alpha > 0.49$ one has $|\omega - \mathrm{Re}\Sigma(\omega)| > |\mathrm{Im}\Sigma(\omega)|$ and thus the mobility is given by $\mu \simeq |\mathrm{Im}\Sigma(\omega)|/[\mathrm{Re}\Sigma(\omega)]^2$. This corresponds with a reactive response. For $\alpha \leq 0.49$, $|\mathrm{Re}\Sigma(\omega)|$ increases slowly with increasing coupling strength which leads to a mobility which decreases with increasing α.

The peak in the mobility occurs for $0.48 < \alpha < 0.5$. At $\alpha \simeq 0.48$ the real part of the mobility changes sign and at $\alpha \simeq 0.49$ the electron is in its self-trapped state. The change of sign in $\mathrm{Re}\Sigma(\omega)$ indicates [note that $\omega - \mathrm{Re}\Sigma(\omega) \simeq -\mathrm{Re}\Sigma(\omega)$] that the velocity response of the electron has a phase shift of $180°$. For $\alpha \leq 0.48$, $\mathrm{Re}\Sigma(\omega)$ is positive and the electron current follows the electric field with a phase shift which is smaller than $\pi/2$. This is typically the behaviour for a particle moving in a dissipative medium. When $\alpha \geq 0.48$,

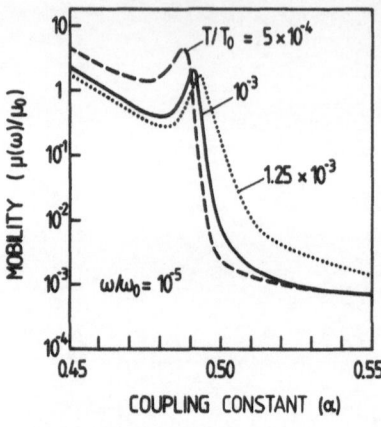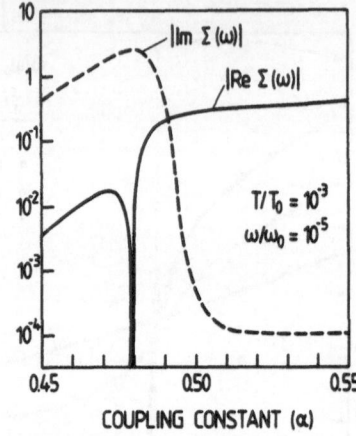

Fig. 25. Mobility and real and imaginary part of the memory function around the self-trapping transition for temperatures such that the anomalous structure is most pronounced.

$\text{Re}\Sigma(\omega)$ becomes negative and the applied electric field follows the electron motion which is characteristic for a system with a strong reactive response. We can understand this as follows: for $\alpha \geq 0.48$ the electron strongly deforms the He-surface and creates a dimple which, as far as the reactive response of the system is concerned, may be thought of as a second particle with a mass $m_f \gg m$ which is much larger than the bare electron mass. Because this second particle is coupled to the electron it will drive the electron response out of phase with the applied electric field if $v > \omega\sqrt{m^*/m}$, where v is the characteristic frequency of the coupled electron-second particle system and $m^* = m + m_f$ is the total mass of the compound system. For $\alpha = 1$ we have $v \sim 0.6$ (v is the frequency of the CERS) and $m^*/m \sim 5 \times 10^4$ which results in $\omega\sqrt{m^*/m} \sim 0.002 \ll v \sim 0.6$.

The excellent *qualitative* agreement between theory and experiment convinces us that there is some essential correctness in our theory. Unfornutaly *quantitatively* there are some serious discrepancies. The experiments of Andrei[8] were performed at $T = 0.4K$ and $T = 1K$ while our theoretical results show the large mobility drop and anomaly at the transition point for $T/T_0 \sim 10^{-3}$ and thus for temperatures in the millikelvin region. Theoretically we found that for temperatures of the order of $1K$ the low-frequency mobility does not shown any spectacular transition from the quasi-free electron state to the self-trapped state. This is different from the high frequency results where the effects of the self-trapped state on the mobility spectrum are still clearly observable in terms of collective electron-ripplon states for temperatures as high as $T \sim T_0 \sim 0.3K$.

The experimental results of Andrei[8] seems to indicate that the self-trapped state is more stable than is inferred from our theoretical analysis. In Ref.68 different mechanisms were proposed which may enhance the stability of the self-trapped state. For example it is possible that the Coulomb interaction between the electrons may enhance the stability of the electron self-trapped state. Recently[72] it was suggested that in Ref.8 a many-electron dimple was observed. In Ref.72 it was found that the *nature* of the self-trapping transition is unchanged by the presence of several electrons(results were given for 40-90 electrons) in one dimple. Only the temperature scale and the critical He-film thickness at which the self-trapping occurs are inlfluenced by the many-particle character. Although the results are promising, the temperature scale is still far too low as compared with the experimental situation.

V. ACKNOWLEDGMENTS

I wish to acknowledge the collaboration with Dr. P.M. Platzman and Dr. S.A. Jackson in the work described here. During the course of this work I have benefited from stimulating discusions with J. Devreese, C. Grimes, E. Andrei, Y. Iye, and M. Paalanen. This work was supported by the National Science Foundation of Belgium.

REFERENCES

1. C.C. Grimes, Surf. Sci. **73**, 379 (1978).
2. F.I.B. Williams, Surf. Sci. **113**, 371 (1982).
3. P.M. Platzman (in these proceedings).
4. J.M. Kosterlitz and D.J. Thouless, J. Phys. **C6**, 1181 (1973).
5. C.C. Grimes and G. Adams, Phys. Rev. Lett. **42**, 795 (1979); Surf. Sci. **98**, 1 (1980).
6. S.A. Jackson and P.M. Platzman, Phys. Rev. **B24**, 499 (1981).
7. O. Hipólito, G.A. Farias and N. Stuart, Surf. Sci. **113**, 394 (1982).
8. E.Y. Andrei, Phys. Rev. Lett. **52**, 1449 (1984).
9. M.W. Cole and M.H. Cohen, Phys. Rev. Lett. **23**, 1238 (1969).
10. V.B. Shikin, Zh. Eksp. Teor. Fiz. **58**, 1748 (1970) [Sov. Phys. JETP **31**, 936 (1971)].
11. T.R. Brown and C.C. Grimes, Phys. Rev. Lett. **29**, 1233 (1972).
12. C.C. Grimes and T.R. Brown, Phys. Rev. Lett. **32**, 280 (1974); C.C. Grimes, T.R. Brown, M.L. Burns and C.L. Zipfel, Phys. Rev. **B13**, 140 (1976).
13. T. Ando, A.B. Fowler and F. Stern, Rev. Mod. Phys. **54**, 437 (1982).
14. H.L. Störmer, Surf. Sci. **132**, 519 (1983).
15. V. Milutinović, Computer **19**, 10 (1986).
16. K. von Klitzing, G. Dorda and M. Pepper, Phys. Rev. Lett. **45**, 494 (1980).
17. D.C. Tsui, H.L. Störmer and A.C. Gossard, Phys. Rev. Lett. **48**, 1559 (1982).
18. M.W. Cole, Rev. Mod. Phys. **46**, 451 (1974).
19. V.B. Shikin and Yu. P. Monarkha, Fiz. Nizk. Temp. **1**, 957 (1975) [Sov. J. Low Temp. Phys. **1**, 459 (1975)].
20. T.M. Sanders, Jr. and G. Weinreich, Phys. Rev. **B13**, 4810 (1976).
21. M.W. Cole, Phys. Rev. **B2**, 4239 (1970).
22. O. Hipólito, J.D. De Felicio and G.A. Farias, Solid State Commun. **28**, 365 (1978).
23. N. Stuart and O. Hipólito, Revista Brasileira de Física **16**, 194 (1986).
24. F. Stern, Phys. Rev. **B17**, 5009 (1978).
25. Yu. Z. Kovdrya, F.F. Mende and V.A. Nikolaenko, Fiz. Nizk. Temp. **10**, 1129 (1984) [Sov. J. Low Temp. Phys. **10**, 589 (1984)]; F.F. Mende, Yu. Z. Kovdrya and V.A. Nikolaenko, Fiz. Nizk. Temp. **11**, 355 (1985) [Sov. J. Low Temp. Phys. **11**, 355 (1985)].
26. L.P. Gor'kov and D.M. Chernikova, Pis'ma Zh. Eksp. Teor. Fiz. **18**, 119 (1973) [JETP Lett. **18**, 68 (1973)].
27. H. Ikezi and P.M. Platzman, Phys. Rev. **B23**, 1145 (1981).
28. H. Etz, W. Gombert, W. Idstein and P. Leiderer, Phys. Rev. Lett. **53**, 2567 (1984).
29. F.M. Peeters and P.M. Platzman, Phys. Rev. Lett. **50**, 2021 (1983).
30. P.M. Platzman, Surf. Sci. **170**, 55 (1986).
31. S.A. Jackson, in *Polarons and Excitons in Polar Semiconductors and Ionic Crystals*, Eds. J.T. Devreese and F.M. Peeters (Plenum, New York, 1984), p.419.
32. Yu. P. Monarkha and V.B. Shikin, Fiz. Nizk. Temp. **8**, 563 (1982) [Sov. J. Low. Temp. Phys. **8**, 279 (1982)].

33. E.P. Wigner, Phys. Rev. **46**, 1002 (1934).

34. R.S. Crandall and R. Williams, Phys. Lett. **A34**, 404 (1971).

35. P.M. Platzman and H. Fukuyama, Phys. Rev. **B10**, 3150 (1974).

36. R.W. Hockney and T.R. Brown, J. Phys. **C8**, 1813 (1975).

37. D.J. Thouless, J. Phys. **C11**, L189 (1978).

38. R.C. Gann, S. Chakravarty and G.V. Chester, Phys. Rev. **B20**, 326 (1979).

39. R.M. Morf, Phys. Rev. Lett. **43**, 931 (1979).

40. M. Imada and M. Takahashi, J. Phys. Soc. Jpn. **53**, 3770 (1984); M. Imada, Surf. Sci. **170**, 112 (1986).

41. A.S. Rybalko, B.N. Esel'son and Yu. Z. Kovdrya, Fiz. Nizk. Temp. **5**, 947 (1979) [Sov. J. Low. Temp. **5**, 450 (1979)].

42. D. Marty, J. Poitrenaud and F.I.B. Williams, J. Physique Lett. **41**, L311 (1980); G. Deville, F. Gallet, D. Marty, J. Poitrenaud, A. Valdes and F.I.B. Williams, in *Ordering in two dimensions*, Ed. S.K. Sinha (North-Holland, Amsterdam, 1980), p.309.

43. F. Gallet, G. Deville, A. Valdes and F.I.B. Williams, Phys. Rev. Lett. **49**, 212 (1982).

44. R. Mehrota, B.M. Guenin and A.J. Dahm, Phys. Rev. Lett. **48**, 641 (1982).

45. K. Kajita, J. Phys. Soc. Jpn. **54**, 4092 (1985).

46. B. Halperin and D. Nelson, Phys. Rev. **B19**, 2457 (1979); **41**, 519(E) (1978).

47. P. Young, Phys. Rev. **B19**, 1855 (1979).

48. F.M. Peeters, Phys. Rev. **B30**, 159 (1984).

49. F. Lindeman, Z. Phys. **11**, 609 (1910).

50. L. Bonsall and A.A. Maradudin, Phys. Rev. **B15**, 1959 (1977).

51. J.M. Ziman, *Principles of the Theory of Solids*, (Cambridge University Press, Cambridge, 1964), p. 37.

52. K. Kajita, J. Phys. Soc. Jpn. **51**, 3747 (1982); J. Phys. Soc. Jpn. **52**, 372 (1983); Surf. Sci. **142**, 86 (1984).

53. M. Wanner and P. Leiderer, Phys. Rev. Lett. **42**, 315 (1979).

54. V.B. Shikin and Yu. B. Monarkha, J. Low. Temp. Phys. **16**, 193 (1974).

55. R.P. Feynman, Phys. Rev. **97**, 660 (1955).

56. S.A. Jackson and P.M. Platzman, Phys. Rev. **B25**, 4886 (1982).

57. S.A. Jackson and F.M. Peeters, Phys. Rev. **B30**, 4196 (1984).

58. F.M. Peeters and S.A. Jackson, Phys. Rev. **B31**, 7098 (1985).

59. V.B. Shikin, Zh. Eksp. Teor. Fiz. **60**, 713 (1971) [Sov. Phys. JETP **33**, 387 (1971)]; R.S. Crandall, Phys. Rev. **B12**, 119 (1978); T. Ando, J. Phys. Soc. Jpn. **44**, 765 (1978).

60. M. Saitoh, J. Phys. Soc. Jpn. **42**, 201 (1977).

61. P. M. Platzman, A.L. Simons and N. Tzoar, Phys. Rev. **B16**, 2023 (1977).

62. W.T. Sommer and D.J. Tanner, Phys. Rev. Lett. **27**, 1345 (1971); T. R. Brown and C.C. Grimes, Phys. Rev. Lett. **29**, 1233 (1972); C.C. Grimes and G. Adams, Phys. Rev. Lett. **36**, 145 (1976); A.P. Volodin and V.S. Edel'man, Sov. Phys. JETP **54**, 198 (1982).

63. Y. Iye, J. Low Temp. Phys. **40**, 441 (1980).

64. R.P. Feynman, R.W. Hellwarth, C.K. Iddings and P.M. Platzman, Phys. Rev. **127**, 1004 (1962).

65. R. Kubo, J. Phys. Soc. Jpn. **12**, 570 (1957).

66. F.M. Peeters and J.T. Devreese, Phys. Rev. **B28**, 6051 (1983).

67. H. Mori, Progr. Teor. Phys. **33**, 423 (1965); **34**, 339 (1965).

68. F.M. Peeters and S.A. Jackson, Phys. Rev. **B34**, 1539 (1986).

69. J.T. Devreese, J. De Sitter and M. Goovaerts, Phys. Rev. **B5**, 2367 (1972).

70. M. Degani and O. Hipólito, Surf. Sci. **142**, 107 (1984).

71. N. Tokuda and H. Kato, J. Phys. **C16**, 1567 (1983).

72. M.H. Degani and O. Hipólito, Phys. Rev. **B32**, 3300 (1985).

Effective energy loss rate, 193
Effective helium surface, 396
Effective mass, 133, 135, 138,
         145, 198, 199, 245, 270,
         273-275, 277, 280, 282,
         285-288, 290, 291, 293,
         294, 298, 321, 327, 331,
         343, 359
Effective scattering rate, 193
Electric intersubband resonance,
         309, 315
Electric subband, 298, 302, 306-
         309, 318, 325
Electron accumulation, 239, 240
Electron charge accumulation, 240
Electron correlation, 71
  energy, 59
Electron distribution, 221
Electron-electron interaction,
         122
Electron-electron scattering
         rates, 210
Electron heating, 222, 223
Electron-hole interactions, 210
Electron-hole pair excitations,
         122
Electron-hole plasma, 225
Electron-hole recombination, 362
Electron-hole scattering, 206
  rates, 210, 211
Electron-hole symmetry, 57, 71
Electron injection, 247
Electron-LO-phonon interaction,
         243
Electron-momentum distribution
         function, 218
Electron-optical phonon scatter-
         ing rate, 198
Electron-phonon coupling, 76
Electron-phonon interaction, 98,
         185, 197, 201, 217, 219,
         227, 240, 242, 246, 250,
         253, 256
  rates, 198, 214
Electron-phonon scattering, 220
  rates, 201
Electron-phonon self-energy, 242
Electron-ripplon coupling, 106
Electron-ripplon interaction,
         398, 407, 408, 410
Electron spin resonance, 282
Electron temperature, 210
Electron-velocity field charac-
         teristics, 221
Elliptic density of states, 372
Emission spectra, 385
Energy gap, 51, 52, 57-59, 70-
         72, 79, 82-84, 93, 350,
         352, 353, 359
Energy loss, 183, 216, 219
  function, 145

Energy loss (continued)
  rates, 185, 187, 190, 193-196,
         200-203, 207, 208, 210,
         215, 217, 219, 220
Energy momentum relation, 136,
         257, 329
Energy spectrum, 33, 295, 395,
         403
Equipartition condition, 195, 196
Exchange coupling, 360
Excitation energy, 83, 96, 111
Excitation gap, 96, 108, 124
Excitation spectrum, 84, 108,
         113, 114
Exciton, 59, 95
  radiative recombination, 360
  screening, 212
Extended states, 22, 376, 378,
         387

Fang-Howard form, 278
Fang-Howard function, 279, 281,
         283
Fang-Howard model, 312
Fang-Howard variational form,
         119
Faraday configuration, 313, 314,
         347, 348
Fast switching, 223
Femtosecond, 183, 185, 212, 224
  excitation, 211, 224
  photoexcitation, 212
  spectroscopy, 212
  timescales, 211, 217
Fermi distribution, 33, 43
Fermi-Dirac distribution function,
         185, 187, 206, 207, 211
Fermi-Thomas screening, 203
Feynman-Bijl formula, 111, 112,
         116, 119
Feynman model, 135, 151
Feynman polaron, 145
Feynman polaron model, 410, 415
Field effect transistor (FET), 5, 205
Filling factor, 30, 32, 44, 66,
         77, 78, 80, 94, 103,
         114, 166, 169, 172, 173,
         179, 181, 365, 368, 370,
         375, 377, 378, 380, 382-
         384, 388-390
Fine structure argument, 27
Fine structure constant, 15, 17,
         53, 96
Five-level approximation, 275
Fractional filling, 54, 92, 102
Fractional fitting, 87
Fractional Landau-level filling,
         51, 52
Fractional quantum Hall effect,
         11, 50-52, 54-60, 66,
         67, 69, 70, 72-76, 79,